Glycobiology Protocols

METHODS IN MOLECULAR BIOLOGY™

John M. Walker, SERIES EDITOR

384. Capillary Electrophoresis: *Methods and Protocols,* edited by *Philippe Schmitt-Kopplin, 2007*

383. Cancer Genomics and Proteomics: *Methods and Protocols,* edited by *Paul B. Fisher, 2007*

382. Microarrays, Second Edition: *Volume 2, Applications and Data Analysis,* edited by *Jang B. Rampal, 2007*

381. Microarrays, Second Edition: *Volume 1, Synthesis Methods,* edited by *Jang B. Rampal, 2007*

380. Immunological Tolerance: *Methods and Protocols,* edited by *Paul J. Fairchild, 2007*

379. Glycovirology Protocols, edited by *Richard J. Sugrue, 2007*

378. Monoclonal Antibodies: *Methods and Protocols,* edited by *Maher Albitar, 2007*

377. Microarray Data Analysis: *Methods and Applications,* edited by *Michael J. Korenberg, 2007*

376. Linkage Disequilibrium and Association Mapping: *Analysis and Application,* edited by *Andrew R. Collins, 2007*

375. In Vitro Transcription and Translation Protocols: *Second Edition,* edited by *Guido Grandi, 2007*

374. Quantum Dots: *Methods and Protocols,* edited by *Charles Z. Hotz and Marcel Bruchez, 2007*

373. Pyrosequencing® Protocols, edited by *Sharon Marsh, 2007*

372. Mitochondrial Genomics and Proteomics Protocols, edited by *Dario Leister and Johannes Herrmann, 2007*

371. Biological Aging: *Methods and Protocols,* edited by *Trygve O. Tollefsbol, 2007*

370. Adhesion Protein Protocols, *Second Edition,* edited by *Amanda S. Coutts, 2007*

369. Electron Microscopy: *Methods and Protocols, Second Edition,* edited by *John Kuo, 2007*

368. Cryopreservation and Freeze-Drying Protocols, *Second Edition,* edited by *John G. Day and Glyn Stacey, 2007*

367. Mass Spectrometry Data Analysis in Proteomics, edited by *Rune Matthiesen, 2007*

366. Cardiac Gene Expression: *Methods and Protocols,* edited by *Jun Zhang and Gregg Rokosh, 2007*

365. Protein Phosphatase Protocols: edited by *Greg Moorhead, 2007*

364. Macromolecular Crystallography Protocols: *Volume 2, Structure Determination,* edited by *Sylvie Doublié, 2007*

363. Macromolecular Crystallography Protocols: *Volume 1, Preparation and Crystallization of Macromolecules,* edited by *Sylvie Doublié, 2007*

362. Circadian Rhythms: *Methods and Protocols,* edited by *Ezio Rosato, 2007*

361. Target Discovery and Validation Reviews and Protocols: *Emerging Molecular Targets and Treatment Options, Volume 2,* edited by *Mouldy Sioud, 2007*

360. Target Discovery and Validation Reviews and Protocols: *Emerging Strategies for Targets and Biomarker Discovery, Volume 1,* edited by *Mouldy Sioud, 2007*

359. Quantitative Proteomics by Mass Spectrometry, edited by *Salvatore Sechi, 2007*

358. Metabolomics: *Methods and Protocols,* edited by *Wolfram Weckwerth, 2007*

357. Cardiovascular Proteomics: *Methods and Protocols,* edited by *Fernando Vivanco, 2006*

356. High-Content Screening: *A Powerful Approach to Systems Cell Biology and Drug Discovery,* edited by *D. Lansing Taylor, Jeffrey Haskins, and Ken Guiliano, 2007*

355. Plant Proteomics: *Methods and Protocols,* edited by *Hervé Thiellement, Michel Zivy, Catherine Damerval, and Valerie Mechin, 2006*

354. Plant–Pathogen Interactions: *Methods and Protocols,* edited by *Pamela C. Ronald, 2006*

353. DNA Analysis by Nonradioactive Probes: *Methods and Protocols,* edited by *Elena Hilario and John. F. MacKay, 2006*

352. Protein Engineering Protocols, edited by *Kristian Müller and Katja Arndt, 2006*

351. *C. elegans:* *Methods and Applications,* edited by *Kevin Strange, 2006*

350. Protein Folding Protocols, edited by *Yawen Bai and Ruth Nussinov 2007*

349. YAC Protocols, *Second Edition,* edited by *Alasdair MacKenzie, 2006*

348. Nuclear Transfer Protocols: *Cell Reprogramming and Transgenesis,* edited by *Paul J. Verma and Alan Trounson, 2006*

347. Glycobiology Protocols, edited by *Inka Brockhausen, 2006*

346. *Dictyostelium discoideum* Protocols, edited by *Ludwig Eichinger and Francisco Rivero, 2006*

345. Diagnostic Bacteriology Protocols, *Second Edition,* edited by *Louise O'Connor, 2006*

344. Agrobacterium Protocols, *Second Edition: Volume 2,* edited by *Kan Wang, 2006*

343. Agrobacterium Protocols, *Second Edition: Volume 1,* edited by *Kan Wang, 2006*

342. MicroRNA Protocols, edited by *Shao-Yao Ying, 2006*

341. Cell–Cell Interactions: *Methods and Protocols,* edited by *Sean P. Colgan, 2006*

340. Protein Design: *Methods and Applications,* edited by *Raphael Guerois and Manuela López de la Paz, 2006*

339. Microchip Capillary Electrophoresis: *Methods and Protocols,* edited by *Charles S. Henry, 2006*

338. Gene Mapping, Discovery, and Expression: *Methods and Protocols,* edited by *M. Bina, 2006*

337. Ion Channels: *Methods and Protocols,* edited by *James D. Stockand and Mark S. Shapiro, 2006*

336. Clinical Applications of PCR, *Second Edition,* edited by *Y. M. Dennis Lo, Rossa W. K. Chiu, and K. C. Allen Chan, 2006*

335. Fluorescent Energy Transfer Nucleic Acid Probes: *Designs and Protocols,* edited by *Vladimir V. Didenko, 2006*

334. PRINS and *In Situ* PCR Protocols, *Second Edition,* edited by *Franck Pellestor, 2006*

333. Transplantation Immunology: *Methods and Protocols,* edited by *Philip Hornick and Marlene Rose, 2006*

332. Transmembrane Signaling Protocols, *Second Edition,* edited by *Hydar Ali and Bodduluri Haribabu, 2006*

331. Human Embryonic Stem Cell Protocols, edited by *Kursad Turksen, 2006*

METHODS IN MOLECULAR BIOLOGY™

Glycobiology Protocols

Edited by

Inka Brockhausen

Departments of Medicine and Biochemistry
Queen's University
Kingston, Ontario, Canada

LIS - LIBRARY

Date	Fund
09/06/17	xzc-shr

Order No.
02818231

University of Chester

HUMANA PRESS ✱ TOTOWA, NEW JERSEY

© 2006 Humana Press Inc.
999 Riverview Drive, Suite 208
Totowa, New Jersey 07512

humanapress.com

All rights reserved. No part of this book may be reproduced, stored in a retrieval system, or transmitted in any form or by any means, electronic, mechanical, photocopying, microfilming, recording, or otherwise without written permission from the Publisher. Methods in Molecular Biology™ is a trademark of The Humana Press Inc.

All papers, comments, opinions, conclusions, or recommendations are those of the author(s), and do not necessarily reflect the views of the publisher.

This publication is printed on acid-free paper. ∞
ANSI Z39.48-1984 (American Standards Institute)

Permanence of Paper for Printed Library Materials.

Production Editor: Amy Thau
Cover design by Patricia F. Cleary

For additional copies, pricing for bulk purchases, and/or information about other Humana titles, contact Humana at the above address or at any of the following numbers: Tel.: 973-256-1699; Fax: 973-256-8341; E-mail: orders@humanapr.com; or visit our Website: www.humanapress.com

Photocopy Authorization Policy:
Authorization to photocopy items for internal or personal use, or the internal or personal use of specific clients, is granted by Humana Press Inc., provided that the base fee of US $30.00 per copy is paid directly to the Copyright Clearance Center at 222 Rosewood Drive, Danvers, MA 01923. For those organizations that have been granted a photocopy license from the CCC, a separate system of payment has been arranged and is acceptable to Humana Press Inc. The fee code for users of the Transactional Reporting Service is: [1-58829-553-2/06 $30.00].

Printed in the United States of America. 10 9 8 7 6 5 4 3 2 1

eISBN: 1-59745-167-3

ISSN: 1064-3745

Library of Congress Cataloging-in-Publication Data

Glycobiology protocols / edited by Inka Brockhausen.
p. ; cm. — (Methods in molecular biology ; v. 347)
Includes bibliographical references and index.
ISBN 1-58829-553-2 (alk. paper)
1. Glycoconjugates. 2. Glycoconjugates—Physiological effect.
I. Brockhausen, Inka, 1944- . II. Series: Methods in molecular biology (Clifton, N.J.) ; v. 347.
[DNLM: 1. Glycoproteins—physiology—Laboratory Manuals.
2. Molecular Biology—methods—Laboratory Manuals.
GW1 ME9616F v.347 2006 / QU 25 G5677 2006]

QP702.G577G559 2006
572'.567—dc22 2006000942

Preface

Glycobiology involves studies of complex carbohydrates and posttranslational modifications of proteins, and has become an important interdisciplinary field encompassing chemistry, biochemistry, biology, physiology, and pathology. Although initial research was directed toward elucidation of the different carbohydrate structures and the enzymes synthesizing them, the field has now moved toward identifying the functions of carbohydrates. The protocols described in *Glycobiology Protocols* form a solid basis for investigations of glycan functions in health and disease. The cloning of many of the genes participating in glycosylation processes has helped to enhance our knowledge of how glycosylation is controlled, but has also added another dimension of complexity to the great heterogeneous variety of the structures of the oligosaccharides of glycoproteins, proteoglycans, and glycolipids. A family of similar enzyme proteins exists for each glycosylation step. Glycosyltransferases are extremely specific for both the nucleotide sugar donor and the acceptor substrate, but many other factors control sugar transfer, including the localization and topology of enzymes, cofactors, possible chaperone proteins, and the availability of sugar acceptor substrates. The analysis of the intracellular organization of glycosylation and of the factors controlling the activities of the participating enzymes in the cell are important areas that need more research efforts. Another challenge for future research is to understand the glycodynamics of a cell, that is, how the cell responds to stimuli leading to biological and pathological changes in terms of alterations in glycosylation, and how this affects the biology of the cell. Because complex carbohydrates have many demonstrated and postulated tissue-specific functions, they can have an impact on the pathology of diseases such as cancer, inflammatory bowel diseases, and cystic fibrosis. Specific carbohydrate structures have been shown to be closely linked to tumor metastasis and invasiveness, inflammation, and immune functions.

The 24 chapters of *Glycobiology Protocols* highlight important methodological progress in the field of glycobiology, and will help scientists to answer specific questions on glycoprotein structures, on the biosynthesis of glycoconjugates, and on the functions of lipid- or protein-bound carbohydrates. This volume is meant to help students, postdoctoral fellows, and senior scientists, whether new or established in the field of glycobiology.

The biosynthesis of GlcNAc-Asn-linked and Man-*O*-Ser/Thr-linked oligosaccharides involves dolichol-phospho-sugars as both sugar donors and acceptor substrates. Because both the enzymes involved and their substrates are in a lipid phase they are very difficult to study, and this area has therefore been neglected in the past. Useful reagents and methods that overcome the problem inherent in assays of water-insoluble biomolecules have been suggested. Chapter 1 (by N. Gao) demonstrates a new fluorescence-based method to study dolichol pyrophosphate oligosaccharides. Chapter 2 (by J. S. Rush and C. J. Waechter) and Chapter 3 (by J. S. Schutzbach) describe how the enzymes utilizing dolichyl substrates can be purified and characterized. Chapter 3 emphasizes the importance of assaying membrane-bound enzymes within their membrane environment. In Chapter 4, T. Endo and H. Manya address the unusual mammalian type of *O*-glycosylation where the first enzyme utilizes dolichol-phospho-mannose as the donor substrate, as well as the following reaction that utilizes nucleotide sugar as the donor substrate. This chapter highlights the fact that assay systems established in yeast cannot necessarily be used in mammals, owing to differences in enzyme topology and/or specific substrates. The other unusual types of *O*-glycosylation (*O*-Fuc and *O*-Glc) also require specific acceptor substrates, at least for the first sugar added to the protein, with the donor substrates being nucleotide sugars. A. Nita-Lazar and R. Haltiwanger (Chapter 5) describe how enzymes involved in these unusual *O*-glycosylations can be assayed.

Sialic acids are a family of terminal acidic sugars of glycoproteins and glycolipids that have multiple functions in determining the overall structure and properties of carbohydrates and the exposure of recognition determinants on the cell surface. These acidic sugars exhibit great structural diversity and play important roles in the immune system, apoptosis, and in providing receptors for microbes. The nature and amount of relatively labile sialic acid linkages are regulated by biosynthesis and degradation. In Chapter 6, J. P. Kamerling and G. J. Gerwig summarize the state-of-the-art methods by which sialic acids can be isolated and analyzed, utilizing chemical methods combined with gas chromatography and mass spectrometry. S. Schrader and R. Schauer report, in Chapter 7, an assay for a *trans*-sialidase from trypanosomes that facilitates analyses of a large number of samples, and can be applied to study the pathology of trypanosomal parasitic diseases.

The biggest hurdle in the field of glycobiology has been to show unequivocally that carbohydrates have specific functions. Many biochemists still consider protein-bound glycans as unwanted modifications that interfere with protein functions rather than having a direct and recognized role. Some of the problems in studying functions are the marked heterogeneity of carbohydrate structures of glycoproteins (especially in mucus glycoproteins) and the fact that the glycosylation inhibitors in use (e.g., tunicamycin) do not prevent the synthesis of any one specific structure, but rather have a nonspecific effect and inhibit glycosylation in general. The discovery of mammalian lectins has greatly influenced our vision of the role of carbohydrates. Thus, carbohydrates not only regulate protein functions, but also have defined roles by themselves. The functions of sialic acid-containing carbohydrate determinants in cell adhesion are addressed in Chapter 23 (by M. E. Beauharnois et al.), which deals with selectin-carbohydrate interactions, and in Chapter 24 (by N. Bock and S. Kelm), which describes how the specificity of sialic acid-binding lectins in the immune system can be determined. B. S. Ireland et al. report in Chapter 22 on an assay for the lectin-like function of chaperones in the endoplasmic reticulum, which is important for our understanding of protein folding.

A sensitive, efficient, and accurate analysis of protein-bound carbohydrates of glycoproteins or proteoglycans is essential for continued progress in glycobiology. In most cases, only small amounts of materials for these analyses can be obtained, and new tools for the separation of oligosaccharides and their analysis by enzymatic methods combined with chromatography and mass spectrometry have been developed. In Chapter 8, C. Robbe et al. describe the analysis of mucin-type *O*-glycans by mass spectrometry methods. This group of researchers has accomplished the detailed structural analysis of relatively small amounts of underivatized, highly heterogeneous mucin-type *O*-glycans from complex mixtures. The exquisite approach by Royle et al. (Chapter 9) to study glycan structures involves the release of *N*-glycans from the protein, treatments with specific glycosidases, and separations by high-performance liquid chromatography using a battery of glycan standards.

The field of glycosyltransferases has been advanced by intense gene-cloning efforts during the past 20 yr, by crystallography of an increasing number of enzymes, and by methods of protein modeling. Therefore, we now have a better understanding of the characteristics and functions of enzyme-active sites, and the relationships among enzymes from different species that have similar activities. As a consequence, the dogma of "one linkage, one enzyme" has been modified to "one linkage, one enzyme family." The molecular modeling described by A. Imberty et al. in Chapter 10, illustrates that enzyme-modeling efforts have to be based on specific expertise and require specialized computer hardware and an ability to utilize specially designed programs. Because relatively few glycosyltransferases have been crystallized to date, the modeling methods, in combination with biochemical methods, are indispensable to obtain knowledge of enzyme mechanisms. The design of specific inhibitors will benefit greatly from computer modeling.

Biochemical methods for assaying specific mammalian glycosyltransferases involved in the biosynthesis of *N*- and *O*-glycans of glycoproteins have been described in Chapter 11 by F. Dall'Olio et al., in Chapter 12 (by M. Prorok-

Hamon et al.), and in Chapter 14 (by I. Brockhausen et al.). The methods presented in Chapter 12 can also be applied to measure the distribution and activity of a glycosyltransferase in cultured cells. Chapter 14 addresses the dynamic state of glycan structures and biosynthesis that can be altered by cytokines, in inflammation, or in disease. The protocol presented for bone cell studies can be used to relate biological phenomena to glycosylation in many other biological systems. Complementing these sometimes tedious enzyme assays are protocols on RT-PCR (Chapters 11 and 13). J. J. García-Vallejo et al. (Chapter 13) have designed and compiled a large library of mammalian DNA sequences that are useful for measuring the expression levels of glycosyltransferases by RT-PCR in a sensitive, efficient, and reliable fashion. Thus, many of the enzymes involved in glycan synthesis of glycolipids and glycoproteins, as well as many sulfotransferases and mannosidases, can be investigated by this method.

The lipopolysaccharides of Gram-negative bacteria are essential for bacteria and important for our encounters with bacteria. Chapter 15 (by C. L. Marolda et al.) describes a protocol to rapidly characterize lipopolysaccharides, as well as the outer carbohydrate O-chains. The method can be applied to assess the effects of gene modifications in bacteria. In the past, it has been difficult to study the biosynthesis of these lipopolysaccharides because polyprenol-phosphate intermediates are involved, which are similar to the dolichol-linked intermediates in *O*-Man- or *N*-glycan biosynthesis. In Chapter 16 (by I. Brockhausen et al.), the synthesis of a novel substrate analog that can serve as a substrate in facile assays of an O-chain glycosyltransferase is described.

Extracellular glycoconjugates, such as mucins or proteoglycans, have essential functions for tissue homeostasis but are difficult to investigate because of their large sizes, charges, and abundance of a heterogeneous mixture of large glycans. K. J. Rees-Milton and T. P. Anastassiades (Chapter 17) describe a method to quantify anionic glycoconjugates with a dye-binding method. This protocol can be useful for the analysis of the alterations of proteoglycans in arthritis or other conditions. Chapter 18 (by P. Argueso and I. Gipson) describes the analysis of small amounts of large mucins, utilizing antibodies against specific mucin peptides. These mucins can be isolated from mucus secretions or from cellular material. Antibodies as well as lectins, can also be used to detect mucins in tissues. F. Kan reports on sensitive ultrastructural analyses (Chapter 19) using colloidal-gold labeling that has successfully demonstrated the subcellular and extracellular distribution of zona pellucida glycoproteins.

Glycolipids are difficult to study because of their hydrophobic character. C. Lingwood et al. (Chapter 20) have established methods by which glycolipid mimics can be synthesized and utilized to study glycolipid function. Finally, glycosidases are essential for the metabolic handling of glycocolipids and abnormalities in these enzymes can lead to severe pathological conditions. J. Callahan and A. Skomorowski (Chapter 21) describe how a lysosomal storage disease (Krabbe disease), characterized by deficiency in galactocerebrosidase, can be diagnosed. This simple protocol is suitable for a routine laboratory test.

The protocols in *Glycobiology Protocols* contain specific methods for the analysis of the structures or functions of glycoconjugates, as well as of glycosyltransferases and glycosidases involved in the biosynthesis of glycans. The methods described for a specific system can usually be modified for investigations of similar biomolecules and tissues or cell types.

I am grateful to all contributors for taking time and effort to share their valuable expertise.

Inka Brockhausen

Contents

Contributors

MAAN ABUL-MILH, PhD • *Section of Infection, Immunity, Injury, and Repair, Research Institute, The Hospital for Sick Children, Toronto, Ontario, Canada*

TASSOS ANASTASSIADES, MD, PhD • *Department of Medicine, Department of Biochemistry, Queen's University, Kingston, Ontario, Canada*

PABLO ARGÜESO, PhD • *The Schepens Eye Research Institute and Harvard Medical School, Boston, MA*

MARK E. BEAUHARNOIS, PhD • *Chemical and Biological Engineering, State University of New York at Buffalo, Buffalo, NY*

CHRISTELLE BRETON, PhD • *Centre de Recherches sur les Macromolécules Végétales, CNRS, Grenoble, France*

NADINE BOCK, PhD • *Centre for Biomolecular Interactions Bremen, University of Bremen, Bremen, Germany*

INKA BROCKHAUSEN, PhD • *Departments of Medicine and Biochemistry, Queen's University, Kingston, Ontario, Canada*

JOHN W. CALLAHAN, PhD • *Genetic-Metabolic Laboratory, Department of Pediatric Laboratory Medicine, The Hospital for Sick Children, and Department of Biochemistry, University of Toronto, Toronto, Ontario, Canada*

CALLIOPE CAPON, PhD • *Unité de Glycobiologie Structurale et Functionelle, Université des Sciences et Technologies de Lille, Villeneuve d'Ascq, France*

MARIELLA CHIRICOLO, FRS • *Department of Experimental Pathology, University of Bologna, Bologna, Italy*

FABIO DALL'OLIO, PhD • *Department of Experimental Pathology, University of Bologna, Bologna, Italy*

RAYMOND A. DWEK, FRS • *Department of Biochemistry, Glycobiology Institute, University of Oxford, Oxford, England*

ASSOU EL-BATTARI, PhD • *Faculté de Medécine, INSERM U-559, and Université de Provence, Marseille, France*

TAMAO ENDO, PhD • *Glycobiology Research Group, Tokyo Metropolitan Institute of Gerontology, Tokyo, Japan*

NINGGUO GAO, PhD • *Department of Pharmacology, University of Texas Southwestern Medical Center, Dallas, TX*

JUAN J. GARCÍA-VALLEJO, MD, PhD • *Department of Molecular Cell Biology and Immunology, Free University Medical Center, Amsterdam, The Netherlands*

GERRIT J. GERWIG, PhD • *Bijvoet Center, Department of Bio-Organic Chemistry, Utrecht University, Utrecht, The Netherlands*
ILENE K. GIPSON, PhD • *The Schepens Eye Research Institute and Harvard Medical School, Boston, MA*
SONJA I. GRINGHUIS, PhD • *Department of Molecular Cell Biology and Immunology, Free University Medical Center, Amsterdam, The Netherlands*
ROBERT S. HALTIWANGER, PhD • *Department of Biochemistry and Cell Biology, State University of New York at Stony Brook, Stony Brook, NY*
MARK HARRISON, MD • *Department of Surgery, Human Mobility Research Centre, Queen's University, Kingston, Ontario, Canada*
ANNE IMBERTY, PhD • *Centre de Recherches sur les Macromolécules Végétales, CNRS, Grenoble, France*
BREANNA S. IRELAND, BSc • *Department of Immunology, University of Toronto, Toronto, Ontario, Canada*
JOHANNIS P. KAMERLING, PhD • *Bijvoet Center, Department of Bio-Organic Chemistry, Utrecht University, Utrecht, The Netherlands*
FREDERICK W. K. KAN, PhD • *Department of Anatomy and Cell Biology, Queen's University, Kingston, Ontario, Canada*
SØRGE KELM, PhD • *Centre for Biomolecular Interactions Bremen, University of Bremen, Bremen, Germany*
JAROSLAV KOČA, PhD • *National Centre for Biomolecular Research, and Department of Biochemistry, Masaryk University, Brno, Czech Republic*
PIYA LAHIRY, BSc • *Department of Microbiology and Immunology, University of Western Ontario, London, Ontario, Canada*
CLIFFORD LINGWOOD, PhD • *Section of Infection, Immunity, Injury, and Repair, Research Institute, The Hospital for Sick Children, and Department of Biochemistry, University of Toronto, Toronto, Ontario, Canada*
NADIA MALAGOLINI, PhD • *Patologia Spermentale, University of Bologna, Bologna, Italy*
HIROSHI MANYA, PhD • *Glycobiology Research Group, Tokyo Metropolitan Institute of Gerontology, Tokyo, Japan*
CRISTINA L. MAROLDA • *Department of Microbiology and Immunology, University of Western Ontario, London, Ontario, Canada*
SYLVIE MATHIEU, PhD • *Faculté de Médecine, INSERM U-559, Marseille, France*
KHUSHI L. MATTA, PhD • *Department of Cancer Biology, Roswell Park Cancer Institute, Buffalo, NY*
JEAN-CLAUDE MICHALSKI, PhD • *Unité de Glycobiologie Structurale et Functionelle, Université des Sciences et Technologies de Lille, Villeneuve d'Ascq, France*

MURUGESPILLAI MYLVAGANUM, PhD • *Section of Infection, Immunity, Injury, and Repair, Research Institute, The Hospital for Sick Children, Toronto, Ontario, Canada*
SRIRAM NEELAMEGHAM, PhD • *Department of Chemical and Biological Engineering, State University of New York at Buffalo, Buffalo, NY*
MONIKA NIGGEMANN, PhD • *Department of Biochemistry, University of Toronto, Toronto, Ontario, Canada*
ALEKSANDRA NITA-LAZAR, PhD • *Department of Biochemistry and Cell Biology, State University of New York at Stony Brook, Stony Brook, NY*
MARCUS PETER, PhD • *Section of Infection, Immunity, Injury, and Repair, Research Institute, The Hospital for Sick Children, Toronto, Ontario, Canada*
MAËLLE PROROK-HAMON, PhD • *Faculte de Medecine, INSERM U-559, Marseille, France*
CATHERINE M. RADCLIFFE, BSc • *Department of Biochemistry, Glycobiology Institute, University of Oxford, Oxford, England*
KAREN J. REES-MILTON, PhD • *Departments of Medicine and Biochemistry, Queen's University, Kingston, Ontario, Canada*
JOHN G. RILEY, PhD • *Department of Chemistry, Queen's University, Kingston, Ontario, Canada*
CATHERINE ROBBE, PhD • *Unité de Glycobiologie Structurale et Functionelle, Université des Sciences et Technologies de Lille, Villeneuve d'Ascq, France*
LOUISE ROYLE, PhD • *Department of Biochemistry, Glycobiology Institute, University of Oxford, Oxford, England*
PAULINE M. RUDD, PhD • *Department of Biochemistry, Glycobiology Institute, University of Oxford, Oxford, England*
JEFFREY S. RUSH, PhD • *Department of Molecular and Cellular Biochemistry, University of Kentucky College of Medicine, Lexington, KY*
SKANDA SADACHARAN, PhD • *Section of Infection, Immunity, Injury, and Repair, Research Institute, The Hospital for Sick Children, Toronto, Ontario, Canada*
SOLEDAD SALDÍAS, PhD • *Department of Microbiology and Immunology, University of Western Ontario, London, Ontario, Canada*
ROLAND SCHAUER, MD, Dipl Biochem • *Biochemisches Institut, Christian Albrechts-Universität, Kiel, Germany*
SILKE SCHRADER, PhD • *Biochemisches Institut, University of Köln, Köln, Germany*
JOHN S. SCHUTZBACH, PhD • *Human Mobility Research Centre, Queen's University, Kingston, Ontario, Canada; and Department of Microbiology, University of Alabama, Birmingham, AL*

MARIE-ANNE SKOMOROWSKI • *Genetic–Metabolic Laboratory, Department of Pediatric Laboratory Medicine, The Hospital for Sick Children, Toronto, Ontario, Canada*
WALTER A. SZAREK, PhD • *Department of Chemistry, Queen's University, Kingston, Ontario, Canada*
MIGUEL VALVANO, MD • *Department of Microbiology and Immunology, University of Western Ontario, London, Ontario, Canada*
IRMA VAN DIE, PhD • *Department of Molecular Cell Biology and Immunology, Free University Medical Center, Amsterdam, The Netherlands*
WILLEM VAN DIJK, PhD • *Department of Molecular Cell Biology and Immunology, Free University Medical Center, Amsterdam, The Netherlands*
ENRIQUE VINÉS, PhD • *Department of. Microbiology and Immunology, University of Western Ontario, London, Ontario, Canada*
JASON Z. VLAHAKIS, PhD • *Department of Chemistry, Queen's University, Kingston, Ontario, Canada*
CHARLES J. WAECHTER, PhD • *Department of Molecular and Cellular Biochemistry, University of Kentucky College of Medicine, Lexington, KY*
DAVID B. WILLIAMS, PhD • *Departments of Immunology and Biochemistry, University of Toronto, Toronto, Ontario, Canada*
MICHAELA WIMMEROVÁ, PhD • *National Centre for Biomolecular Research and Department of Biochemistry, Masaryk University, Brno, Czech Republic*
XIAOJING YANG, MD, PhD • *Departments of Medicine and Biochemistry, Queen's University, Kingston, Ontario, Canada*

1

Application of Fluorophore-Assisted Carbohydrate Electrophoresis for the Study of the Dolichol Pyrophosphate-Linked Oligosaccharides Pathway in Cell Cultures and Animal Tissues

Ningguo Gao

Summary

Defects in the synthesis of dolichol-linked oligosaccharide (or lipid-linked oligosaccharide [LLO]) cause severe, multisystem human diseases called type 1 congenital disorders of glycosylation (CDG type 1). LLOs are also involved in another disease, neuronal ceroid lipofuscinosis. Because of the low abundance of LLOs, almost all studies of LLO synthesis have relied upon metabolic labeling of the oligosaccharides with radioactive sugar precursors such as [^{3}H]mannose or [^{14}C]glucosamine, and therefore have been limited almost entirely to cell cultures and tissue slices. A procedure is presented for a facile, accurate, and sensitive nonradioactive method for LLO pathway analysis based on fluorophore-assisted carbohydrate electrophoresis (FACE). It is feasible to analyze almost any component in the LLO pathway with the application FACE, from sugar precursors to mature LLO ($Glc_3Man_9GlcNAc_2$-P-P-dolichol).

Key Words: Glycosylation; lipid-linked oligosaccharide; dolichol; fluorophore-assisted carbohydrate electrophoresis.

1. Introduction

In the lumenal space of the endoplasmic reticulum (ER), asparagine (*N*)-linked glycoproteins are formed by *en bloc* transfer of preformed oligosaccharide (OS) units from lipid-linked oligosaccharide (LLO) donors to nascent polypeptides with asparaginyl residues in the context Asn-X-Ser/Thr (*see* **refs.** ***1*** and ***2***; *see also* Chapters 2 and 3). Completed LLOs have the structure $Glc_3Man_9GlcNAc_2$-P-P-dolichol, the preferred substrate for oligosaccharyl

From: *Methods in Molecular Biology, vol. 347: Glycobiology Protocols*
Edited by: I. Brockhausen © Humana Press Inc., Totowa, NJ

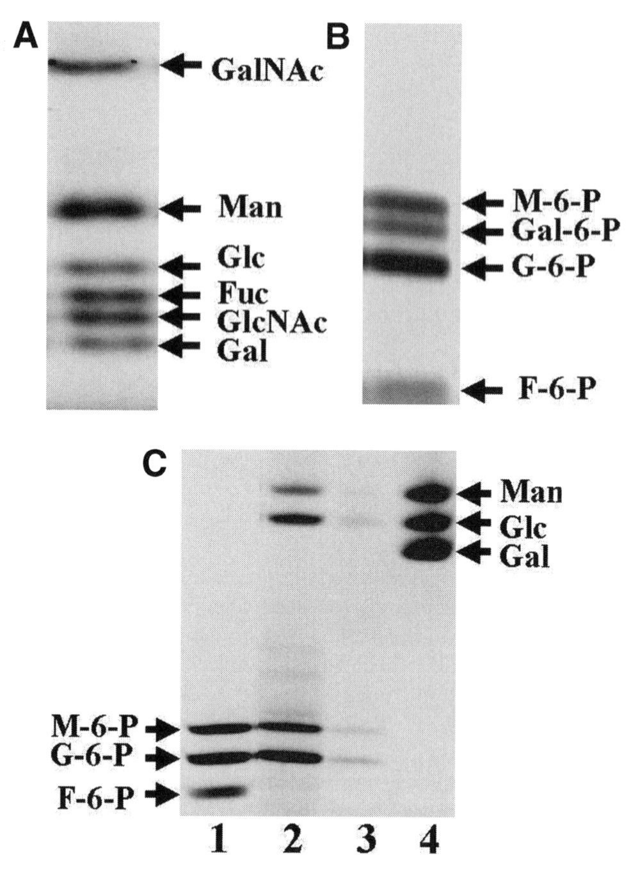

Fig. 1. Fluorophore-assisted carbohydrate electrophoresis (FACE) analysis of free monosaccharides and monophosphate sugars standard. **(A)** The separation of 2-aminoacridone (AMAC)-labeled free monosaccharides on a FACE mono gel. **(B)** The separation of AMAC-labeled sugar-6-phosphates on a FACE mono gel. **(C)** The separation of sugar-1-phosphates and sugar-6-phosphates on the same FACE mono gel. Sugar-1-phosphates form free sugar while sugar-6-phosphates remain intact with the weak acid treatment. Free sugars correspond to sugar-1-phosphates on this mono gel. **(Lane 1)** Sugar-6-phosphate standard. **(Lane 2)** Four sugar phosphates standard, Man-1-P, Glc-1-P, Man-6-P, and Glc-6-P recovered from an AG1-2× column with 4 *M* of formic acid elution. **(Lane 3)** 2 *M* of PAC elution of the same column after 4 *M* of formic acid. There are very few sugar phosphates detected after formic acid elution. **(Lane 4)** Monosaccharides standard.

transferase compared with premature LLO intermediates *(3)*. LLO synthesis is highly conserved and requires at least 34 known genes *(4)*.

A thorough knowledge of LLO assembly is essential to understand both the ER function and the pathophysiology of diseases such as type 1 congenital disorders of glycosylation (CDG type 1) and neuronal ceroid lipofuscinosis (NCL; *see* **ref.** *5*). Because abundance of LLOs is low, typically on the order of 1 nmol/g of tissue *(6,7)*, it is difficult to directly measure the OSs by physical and/or chemical means. It is common to study LLOs that have been made radioactive by incubations of cells, organelle preparations, or tissue slices with appropriate [^{3}H]- or [^{14}C]-labeled precursors. The incorporated isotopes then permit facile detection of the OSs.

Although isotopic approaches have been extremely useful for LLO analysis, they have a number of limitations.

1. Isotopic labeling in normal culture medium is often inefficient unless low-glucose medium is used (5- to 10-fold below the physiological range), subjecting the cultures to potential glucose deprivation effects and causing artifacts *(8)*.
2. Because metabolic labeling is typically done for a brief incubation period (20–60 min), the results obtained only reflect the LLOs made during that short time period and the true steady-state LLO compositions might be quite different.
3. It is difficult to quantify each LLO species from the amount of radioactivity incorporated because of isotope dilution. Intermediates with few sugars may be difficult to detect.
4. Metabolic labeling is impractical for living animals. Pool dilution and catabolism would require the use of very large quantities of radioactive compounds.

The use of fluorophore-assisted carbohydrate electrophoresis (FACE; *see* **refs.** *9–12*) to circumvent these problems is described in this chapter. With the application of FACE, it is simple to quantitatively analyze the LLO pathway from sugar phosphate precursors to mature $Glc_3Man_9GlcNAc_2$-P-P-dolichol in cultured cells or animal tissues (*see* **Figs. 1–3**). First, sugar-6-phosphate, free monosaccharide released from sugar-1-phosphate and nucleotide sugars, or glycans released from LLO by mild acid, are labeled with fluorophores 2-aminoacridone (AMAC), 8-aminonophthalene-1,3,6-trisulfonate (ANTS), or 7-amino-1,3-naphthalenedisulfonic acid (ANDS); the labeled sugars are then resolved by a high percentage of a special kind of acrylamide gel. Multiple samples are easily processed with a sensitivity of 1–2 pmol with the use of a commercial fluorescence scanner.

2. Materials

2.1. Phosphate Sugars and Nucleotide Sugars Assay

2.1.1. Partial Purification and Separation of Phosphate Sugars and Nucleotide Sugars

1. 70% Ethanol.
2. AG1-2X anion exchange resin, 200–400 mesh, formate form (Bio-Rad, Hercules, CA). Store at 4°C (do not freeze).

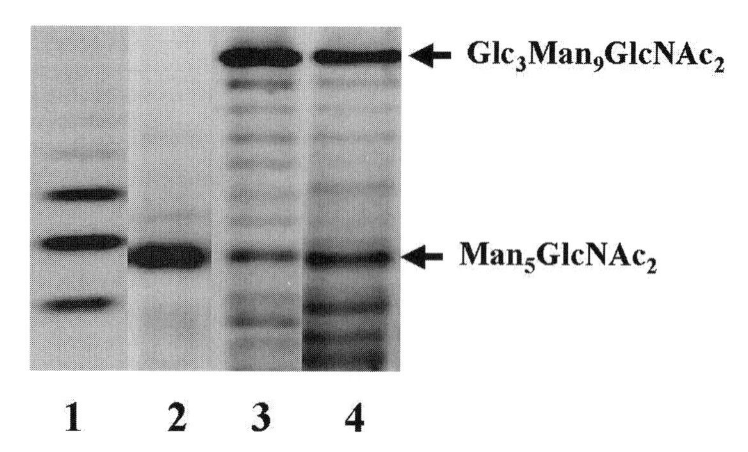

Fig. 2. Lipid-linked oligosaccharide (LLO) pattern in cultured cells determined with fluorophore-assisted carbohydrate electrophoresis. **(Lane 1)** Glucose oligomers standard. From bottom to top: Glc5, Glc6, and Glc7. **(Lane 2)** Lec35. **(Lane 3)** Chinese hamster ovary-K1. **(Lane 4)** *Saccharomyces cerevisiae*. In Chinese hamster ovary and yeast cells the major LLO is $Glc_3Man_9GlcNAc$, while in Lec35 cells, the major LLO is $Man_5GlcNAc_2$. (LLOs were labeled with 7-amino-1,3-naphthalenedisulfonic acid).

3. 4 *M* of formic acid. Store at room temperature.
4. 2 *M* of pyridine acetate (PAC), pH 5.0. Store at room temperature in the dark.
5. Poly-Prep Columns (Bio-Rad).

2.1.2. AMAC Labeling

1. 0.1 *M* of AMAC (Molecular Probes) in dimethyl sulfoxide (DMSO) containing 15% acetic acid, stable at –80°C for at least 3 mo (*see* **Note 1**).
2. 1 *M* of sodium cyanoborohydride (SCB; Aldrich) in DMSO (stable at –80°C for at least 3 mo).

2.1.3. Monosaccharide FACE

1. 4X Resolving buffer: 0.75 *M* of Tris-HCl and 0.5 *M* of boric acid, adjusted to pH 7.0 with concentrated HCl. Store at 4°C.
2. 4X Stacking buffer: 0.5 *M* of Tris-HCl and 0.5 *M* of boric acid, adjusted to pH 6.8 with concentrated HCl. Store at 4°C.
3. 5X Running buffer: 0.5 *M* of glycine, 0.6 *M* of Tris-HCl base, and 0.5 *M* of boric acid, final pH 8.3. Store at 4°C.
4. Resolving gel stock solution: 38% (w/v) acrylamide and 2% (w/v) *N,N′*-methylene-*bis*-acrylamide. Store at 4°C in the dark.
5. Stacking gel stock solution: 10% (w/v) acrylamide and 2.5% (w/v) *N,N′*-methylene-*bis*-acrylamide. Store at 4°C in the dark.
6. 10% Ammonium persulfate (APS; prepared daily).

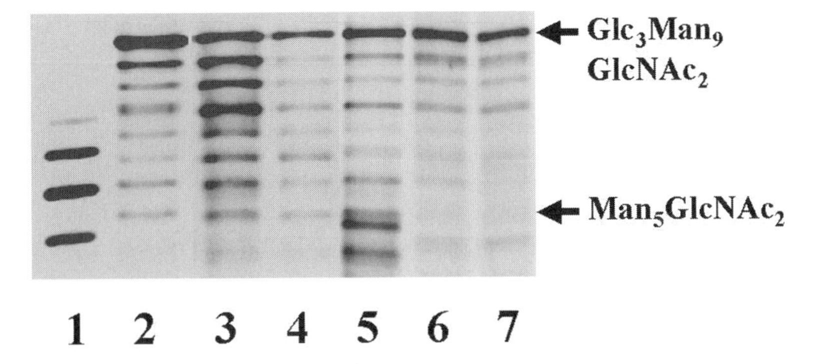

Fig. 3. Lipid-linked oligosaccharide (LLO) pattern in mouse tissues detected by fluorophore-assisted carbohydrate electrophoresis. **(Lane 1)** Glucose oligomers standard. From bottom to top: Glc5, Glc6, and Glc7. **(Lane 2)** Kidney. **(Lane 3)** Kidney. **(Lane 4)** Liver. **(Lane 5)** Brain. **(Lane 6)** Testis. **(Lane 7)** Muscle. The major LLO in mouse tissues is $Glc_3Man_9GlcNAc_2$-P-P-dolichol with the exception of the kidney, in which two kinds of LLO pattern have been detected.

7. 100% *N,N,N′N′*-tetramethylenediamine (TEMED). Store at 4°C.
8. 2X Loading buffer: 0.01% thorin I (Sigma) in 30% DMSO.
9. Gel box with cooling system (*see* **Note 2**).
10. 1-mm-Thick glass plates with very low ultraviolet absorption (*see* **Note 3**).
11. Imager: Bio-Rad Fluor-S Multi-imager with a 530DF60 filter and Quantity-One software supplied with the imager (*see* **Note 4**).

2.2. LLO FACE Analysis

2.2.1. LLO Extraction and Partial Purification

1. Chloroform/methanol (CM) mixed at a ratio of 2/1(v/v). Store at room temperature in a dark bottle.
2. Chloroform/methanol/water (CMW) mixed at a ratio of 10/10/3 (v/v). Store at room temperature in a dark bottle.
3. Diethylamino ethanol (DEAE)-cellulose (Sigma), converted to acetate form by washing with 10 bed volume (BV) of 1 *M* acetic acid (in CMW). Store at room temperature in methanol.
4. 3 *M* of Ammonium acetate (NH_4OAc) stock solution, prepared by dissolving 3 mol of NH_4OAc into a total volume of 1000 mL of methanol with 3% acetic acid. Store at room temperature.
5. 0.1 *N* of HCl in 50% isopropanol. Store at room temperature.
6. Water-saturated butanol: Shake equal volumes of water and butanol in a glass bottle and allow to separate. Store at room temperature.
7. AG50-8X Cation exchanger resin (Bio-Rad), hydrogen form. Store at 4°C.
8. AG1-8X Anion exchanger resin (Bio-Rad), formate form. Store at 4°C.

9. 15-mL Conical disposable centrifuge glass tubes with caps (Kimble Glass).
10. Water bath sonicator (Fisher).
11. Nitrogen gas evaporator or centrifuge vacuum evaporator (CVE).

2.2.2. Derivatization of Released OSs With ANTS and ANDS

1. 0.15 *M* of ANTS (Molecular Probes) in 15% acetic acid, stable at –80°C for at least 3 mo.
2. 0.15 *M* of ANDS (Aldrich) in 15% acetic acid, stable at –80°C for at least 3 mo.
3. 1.0 *M* of SCB (Aldrich) in DMSO. Stable at –80°C for at least 3 mo.
4. Fluorescent OS standards: A known amount of maltoOS mixture, formerly offered by Phanstiehl (cat. no. M-138), is used as the OS standard. It will remain stable at –80°C for at least 12 mo. Fragments of glycogen or other glucose polymers can also be used.

2.2.3. OS Profiling FACE

All the reagents and equipments are the same as in **Subheading 2.1.3.** except the following:

1. 8X Resolving gel buffer: 1.5 *M* of Tris-HCl, pH 8.9.
2. 8X Stacking gel buffer: 1.0 *M* of Tris-HCl, pH 6.8.
3. 10X Running buffer: 1.92 *M* of glycine and 0.25 *M* of Tris-HCl base, pH 8.3.

3. Methods

3.1. Phosphate and Nucleotide Sugar Assay

3.1.1. Partial Purification and Separation of Phosphate Sugars and Nucleotide Sugars

1. Grow cells on 15-cm dishes until 80–90% confluence.
2. Wash twice with ice-cold PBS.
3. Harvest the cells in 10 mL of 70% ethanol with a cell lifter. Transfer the suspension to a glass tube.
4. Sonicate in a water-bath sonicater for 10–15 min.
5. Put the tube on ice for 15 min.
6. Centrifuge at 3000*g* at 4°C for 15 min.
7. Collect the supernatant and dry with CVE.
8. Reconstitute the dried samples in 1 mL of water and load onto a Poly-Prep Column packed with a 1-mL BV of AG1-X2 resin (*see* **Note 5**).
9. Wash the column with 40 mL of water to remove all neutral- or positive-charged contaminants.
10. Elute the column with 20 mL of 4 *M* formic acid, collect the elutant, and dry with CVE. This portion contains all the monophosphate sugars.
11. Dissolve the samples in 1 mL of 0.1 *N* HCl and transfer to a 1.5-mL plastic centrifuge tube. Heat at 100°C for 15 min to hydrolyze phosphate-1-sugars to free sugars and dry with CVE (*see* **Note 6**).

12. Elute the column with 20 mL of 2 *M* PAC, collect the elutant, and dry with CVE. This portion contains all the nucleotide sugars.
13. Dissolve the samples in 1 mL of 0.1 *N* HCl and transfer to a 1.5-mL plastic centrifuge tube, heat at 100°C for 15 min to hydrolyze nucleotide sugars to free sugars, and dry with CVE (*see* **Note 7**).

3.1.2. AMAC Labeling

1. Add 5 μL of AMAC to the tube of acid-treated and dried phosphate sugar or nucleotide sugar sample from **Subheading 3.1.1.** and vortex (*see* **Note 8**).
2. Add 5 μL of $NaBCNH_3$ to the tube and vortex.
3. Incubate at 37°C for 18 h in the dark.
4. Dry with CVE.
5. Dissolve in 30% DMSO. Store at –80°C for later analysis.

3.1.3. Fluorophore-Assisted Carbohydrate Electrophoresis

1. Assemble the gel-casting apparatus and prepare resolving gel solution by mixing 4 mL of resolving gel stock solution, 2 mL of 4X stock resolving gel buffer, 2 mL of water, and 20 μL of 10% APS.
2. Polymerization is initiated by the addition of 20 μL of TEMED and the solution is poured into the casting apparatus to a height of 0.5 cm below the bottom of the teeth of the comb. Immediately overlay the gel solution with 1 cm of water. Polymerization occurs after 15 min.
3. Prepare the stacking gel solution by mixing 1 mL of stacking gel stock solution, 0.5 mL of 4X stacking gel buffer, 0.5 mL of water, and 6 μL of 10% APS. Polymerization is initiated by adding 5 μL of TEMED.
4. Pour off the water overlay and fill the remaining space in the mold with the stacking gel solution. Insert a comb to form sample wells. The stacker should polymerize within 15 min.
5. Dilute the 5X mono running buffer fivefold and cool to 4°C. Pour into the electrophoresis apparatus (anode compartment) connected to a circulating cooler set to the appropriate temperature to maintain the gel at 4°C.
6. Insert the gel sandwich into the apparatus. Add 1X running buffer to the cathode compartment.
7. Take out the desired volume of AMAC-labeled sample (dissolved in 30% DMSO) and mix with an equal volume of 2X loading buffer. Apply 1–2 μL of the mixture to the wells with flat-end tips.
8. Connect to a power supply and set it to a constant current of 15 mA (a voltage in the range of 200–1200 V will result). Run until the thorin I marker dye exits the bottom of the gel, usually in 1 h. Turn off the current, remove the gel, and place in the gel imager with both glass plates still attached. A delay of more than 20 min may result in some diffusion of the fluorescent bands. Whereas AMAC-labeled monophosphate sugars need 1 h to get a good separation on the gel, the AMAC-labeled free mono sugars require an additional 2 h (*see* **Note 6**).

3.2. LLO FACE Analysis

3.2.1. LLO Extraction and Partial Purification

1. Wash the cells (90% confluence, >10^7 cells; equivalent to four 15-cm dishes of fibroblast cells or one 10-cm dish of Chinese hamster ovary cells) in cultured dishes with ice-cold PBS twice, add room temperature methanol to the dish, harvest the cells by scraping, and transfer to a 15-mL glass tube.
2. Sonicate for 5–10 min in a water-bath sonicator and dry under a stream of N_2 gas (*see* **Note 9**).
3. If animal tissues are used, harvested animal tissues should be used immediately or frozen in liquid nitrogen. Fresh or frozen tissues are disrupted into 10 volumes of methanol with a Bronson Homogenizer (polytron) for 30 s at a setting of 6. The methanolic suspension is dried under N_2 gas (*see* **Note 9**).
4. Add 10 mL of CM to the tube and sonicate for 5–10 min with occasional vortexing. Centrifuge at 3000*g* for 10 min at room temperature and discard the supernatant. Repeat once.
5. Resuspend the pellet in 2 mL of CM by sonication and dry under N_2 gas.
6. Add 10 mL of water to the tube, sonicate for 5–10 min with occasional vortexing, centrifuge, and discard the supernatant. Repeat once.
7. Resuspend the pellet in 2 mL of methanol by sonication. Dry under N_2 gas (*see* **Note 9**).
8. Add 10 mL of CMW to the tube and sonicate for 5–10 min with occasional vortexing. Centrifuge and collect the supernatant. LLOs are extracted in this fraction. Repeat two or three times.
9. Load the CMW extract onto a Ply-prep column packed with DEAE-cellulose pre-equilibrated with CMW, using 1 mL BV per 5×10^7 cells or 500 mg wet tissue (*see* **Note 10**).
10. Wash with 10 BV of CMW.
11. Wash with 10 BV of 3 m*M* acetic acid in CMW.
12. Elute with 10 BV of 300 m*M* NH_4OAc in CMW.
13. Add 4.3 BV of chloroform and 1.2 BV of water to the tube and vortex. Centrifuge at 3000*g* for 10 min. Remove the upper phase carefully without disturbing the middle layer (most of the salt is removed by this step). Dry the combined middle layer and lower phase under N_2 gas.
14. Add 2 mL of 0.1 *N* HCl (in 50% isopropanol), vortex, incubate at 50°C for 60 min, and dry under N_2 gas or CVE.
15. Add 1 mL of butanol-saturated water to the tube, vortex, and then add 1 mL of water-saturated butanol. Vortex, centrifuge at 3000*g* for 1 min, and collect the lower phase containing the released OS. Dry with CVE.
16. Resuspend the dried OS/salt mixture in 1 mL of water and add 200 μL (packed volume) of AG50W-X8 (hydrogen form) cation exchange resin. Mix for 5 min, centrifuge at 3000*g* for 1 min, and collect the supernatant.
17. Add 200 μL (packed volume) of AG1-X8 (formate form) anion exchange resin to the tube, mix for 5 min, centrifuge at 3000*g* for 1 min, and collect the supernatant containing the OS released from the LLO.

3.2.2. Derivatization of Released OSs with ANTS and ANDS

1. Dry the OS samples released from the LLOs in a CVE in a 1.5-mL microcentrifuge tube (use a 0.5-mL tube if the sample is lower than 200 pmol).
2. Add 5 µL (or 1 µL if the sample is lower than 200 pmol) of ANTS or ANDS solution. Vortex and centrifuge at 3000*g* briefly (*see* **Note 11**).
3. Add 5 µL (or 1 µL if the sample is lower than 200 pmol) of SCB solution. Vortex, centrifuge at 3000*g* briefly, and react at 37°C for 18 h.
4. Dry the reaction mixture with CVE. Dissolve in the desired volume of water.

3.2.3. OS Profiling FACE

1. Assemble the gel-casting apparatus provided with the gel box and prepare the resolving gel solution by mixing 4 mL of resolving gel stock solution, 1 mL of 8X stock resolving gel buffer, 3 mL of water, and 20 µL of 10% APS.
2. Polymerization is initiated by the addition of 20 µL of TEMED and the solution is poured into the casting apparatus to a height of 0.5 cm below the bottom of the teeth of the comb. Immediately overlay the gel solution with 1 cm of water. Polymerization occurs after 15 min.
3. Prepare the stacking gel solution by mixing 1 mL of stacking gel stock solution, 0.25 mL of 8X stacking resolving gel buffer, 0.75 mL of water, and 6 µL of 10% APS. Polymerization is initiated by adding 5 µL of TEMED.
4. Pour off the water overlay and fill the remaining space in the mold with stacking gel solution. Insert a comb to form sample wells. The stacker should polymerize within 15 min.
5. Dilute the stock electrode buffer 10-fold and cool to 4°C. Pour into the electrophoresis apparatus (anode compartment) connected to a circulating cooler set to the appropriate temperature to maintain the gel at 4°C.
6. Insert the gel sandwich into the apparatus. Add the electrode buffer to the cathode compartment. Mix the sample (dissolved in water) with an equal volume of 2X loading buffer. Apply 2–4 µL of the mixture to the wells with flat-end tips.
7. Connect to a power supply and set it to a constant current of 15 mA (a voltage in the range of 200–1200 V will result). Run until the thorin I marker dye exits the bottom of the gel, usually in 1 h. Turn off the current, remove the gel, and place in the gel imager with both glass plates still attached. A delay of more than 20 min may result in some diffusion of the fluorescent bands.

4. Notes

1. All the fluorophores used are light-sensitive. Exposure to the light for a short period of time is OK.
2. We have used two types of gel boxes for FACE. A specialized gel box is offered by Prozyme, San Leandro, CA. This apparatus consists of two chambers: a lower chamber cooled to –4°C by circulating coolant and an upper chamber contacting the bottom chamber and containing electrolyte, in which gel sandwiches are mounted. The upper chamber reaches a running temperature of 4°C during electrophoresis. This apparatus is convenient because gels can easily be removed and

replaced for periodic inspection with ultraviolet light. A temperature-controlled gel box with a matching gel-casting apparatus (model P8DS, Owl Separation Systems, Portsmouth, NH) may also be used, although periodic inspection of gels during the run with ultraviolet light is more cumbersome.

3. The glass plates (Electroverre, EVR glass) must have very low ultraviolet absorbance; our laboratory has obtained them in bulk by special order from Erie Scientific Company, Portsmouth, NH (cat. no. SMC-2101 for front notched plate, cat. no. SMC-2102 for back plate). Gels are sandwiched between 1.0-mm-thick glass plates and are 10-cm wide, 10-cm high, and formed with 0.5-mm-thick spacers. Combs with eight 8-mm-wide teeth (2 mm between teeth) are typically used to form loading wells, but 12-tooth combs can also be used. Pre-assembled FACE gels can be obtained from a commercial supplier (Prozyme). Gel components can be recycled from used gels. The combs and spacer can also be prepared by any competent machine shop.
4. Imager: Many of the available fluorescence imagers using charge-coupled device cameras are suitable for image acquisition and analysis. We have used the Bio-Rad Fluor-S Multi-imager with a 530DF60 filter and Quantity-One software supplied with the imager.
5. When loading the samples onto a column or during the washing and eluting steps, recovery is optimal if the flow rate is kept lower than 20 mL/h. This can usually be achieved by restricting the flow with a 200-µL flat-end gel-loading tip attached to the bottom of the column.
6. Only sugars with free reducing terminals can be labeled with AMAC. Sugar-6-phosphates have free reducing terminals (the 1 position) and thus can be labeled by AMAC. Sugar-1-phosphates cannot be labeled by AMAC because the 1 position is occupied by phosphate and thus has no free reducing terminal. This problem can be solved by weak acid hydrolysis. Sugar-6-phosphates are very resistant to acid and sugar-1-phosphates are very labile. After weak acid treatment, sugar-6-phosphates will remain intact and sugar-1-phosphates will form free sugar, which can then be labeled by AMAC and thus can be separated on a FACE mono gel. On a mono FACE gel the free sugars correspond to phosphate-1-sugars (*see* **Fig. 1**).
7. Nucleotide sugar does not have a reducing terminal and thus cannot be labeled by AMAC directly. With acid treatment, the nucleotide portion is removed and the free sugar can be labeled.
8. If the sample is less than 200 pmol, use 0.5-mL tubes and reduce the reagent volume from 5 µL to 1 µL.
9. When drying the samples in methanol under nitrogen gas in the first step of LLO extraction, the samples can easily form hardened pellet chunks as the final traces of residual water evaporate. To avoid this, carefully dry the samples until slightly damp with no liquid visible. Then add 5–10 mL of CM to resuspend the pellet by sonication and dry under N_2 gas. This step can be repeated two or three times for large pellets, such as those from animal tissues, to fully drive out the residual water. Samples dried in this way form a loose powder and the LLO recovery is very high.

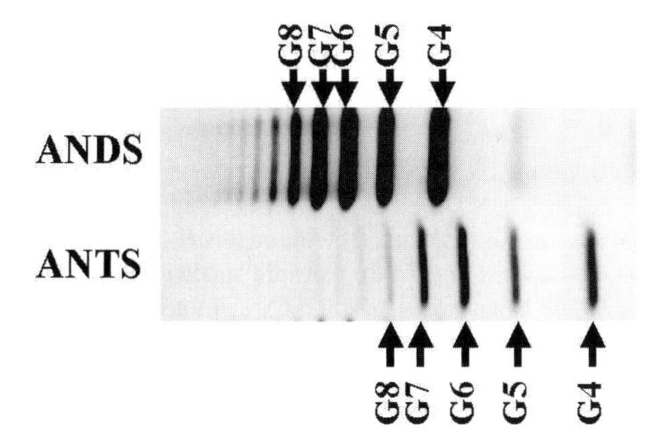

Fig. 4. 8-Aminonophthalene-1,3,6-trisulfonate (ANTS)- and 7-amino-1,3-naphthalene-disulfonic acid (ANDS)-labeled glucose oligosaccharides standard. When ANDS (two negative charges, blue fluorescent, runs slower) and ANTS (three negative charges, green fluorescent, runs faster) were used to label the same amount of glucose oligomers, the intensity of the ANDS signal is five times more than that of ANTS. In this figure, G6 is 50 pmol.

10. When loading the CMW extract onto a DEAE-cellulose column or during the washing and eluting steps, recovery is optimal if the flow rate is kept below 0.5 mL/min. This can usually be achieved by restricting the flow with a 200-µL flat-end gel-loading tip attached to the bottom of the column.
11. ANDS seems more suitable for our imager and gives five times more sensitivity than ANTS (*see* **Fig. 4**).

Acknowledgments

The author would like to thank Professor Mark A. Lehrman for assistance and suggestions with the manuscript. This work was supported by NIH grant GM38545 and Welch grant I-1168.

References

1. Kornfeld, R. and Kornfeld, S. (1985) Assembly of asparagine-linked oligosaccharides. *Annu. Rev. Biochem.* **54,** 631–664.
2. Varki, A., Cummings, R. D., Esko, J. D., Freeze, H. H., Hart, G.W., and Marth, J. (1999) *Essentials of Glycobiology.* Cold Spring Harbor Laboratory Press, Cold Spring Harbor, NY, pp. 85–90.
3. Karaoglu, D., Kelleher, D. J., and Gilmore, R. (2001) Allosteric regulation provides a molecular mechanism for preferential utilization of the fully assembled dolichol-linked oligosaccharide by the yeast oligosaccharyltransferase. *Biochemistry* **40,** 12,193–12,206.

4. Freeze, H. H. (1998) Disorders in protein glycosylation and potential therapy: tip of an iceberg? [Review] *J. Pediatr.* **133,** 593–600.
5. Cho, S. K., Gao, N., Pearce, D. A., Lehrman, M. A., and Hofmann, S. L. (2005) Characterization of lipid-linked oligosaccharide accumulation in mouse models of Batten disease. *Glycobiology* **15,** 637–648.
6. Badet, J. and Jeanloz, R. W. (1988) Isolation and partial purification of lipid-linked oligosaccharides from calf pancreas. *Carbohydr. Res.* **178,** 49–65.
7. Gibbs, B. S. and Coward, J. K. (1999) Dolichylpyrophosphate oligosaccharides: large-scale isolation and evaluation as oligosaccharyltransferase substrates. *Bioorg. Med. Chem.* **7,** 441–447.
8. Chapman, A. E. and Calhoun, J. C. (1988) Effects of glucose starvation and puromycin treatment on lipid-linked oligosaccharide precursors and biosynthetic enzymes in Chinese hamster ovary cells in vivo and in vitro. *Arch. Biochem. Biophys.* **260,** 320–333.
9. Gao, N. and Lehrman, M. A. (2002) Analyses of dolichol pyrophosphate-linked oligosaccharides in cell cultures and tissues by fluorophore-assisted carbohydrate electrophoresis. *Glycobiology* **12,** 353–360.
10. Gao, N. and Lehrman, M. A. (2002) Coupling of the dolichol-P-P-oligosaccharide pathway to translation by perturbation-sensitive regulation of the initiating enzyme, GlcNAc-1-P transferase. *J. Biol. Chem.* **277,** 39,425-39,435.
11. Gao, N. and Lehrman, M. A. (2003) Letter to the Glycoforum: Alternative sources of reagents and supplies for fluorophore-assisted carbohydrate electrophoresis (FACE). *Glycobiology* **13,** 1G–3G.
12. Gill, A., Gao, N., and Lehrman, M. A. (2002) Rapid activation of glycogen phosphorylase by the endoplasmic reticulum unfolded protein response. *J. Biol. Chem.* **277,** 44,747–44,753.

2

Partial Purification of Mannosylphosphorylundecaprenol Synthase From *Micrococcus luteus*

A Useful Enzyme for the Biosynthesis of a Variety of Mannosylphosphorylpolyisoprenol Products

Jeffrey S. Rush and Charles J. Waechter

Summary

Membrane fractions from *Micrococcus luteus* catalyze the transfer of mannose from GDP-mannose to mono- and dimannosyldiacylglycerol, mannosylphosphorylundecaprenol (Man-P-Undec), and a membrane-associated lipomannan. This chapter describes the detergent solubilization, partial purification, and properties of Man-P-Undec synthase. The mobility of the mannosyltransferase activity on sodium dodecyl sulfate-polyacrylamide gel electrophoresis indicates that the enzyme is a polypeptide with a molecular weight of approx 30.7 kDa. Utilizing the broad specificity of the bacterial mannosyltransferase provides a useful approach for the enzymatic synthesis of a wide variety of Man-P-polyisoprenol products.

Key Words: Mannosyltransferase; GDP-mannose, detergent-solubilization; Man-P-polyisoprenols.

1. Introduction

The presence of membrane-bound mannosyltransferases that catalyze the biosynthesis of mono- and dimannosyldiacylglycerol (Man_{1-2}-DAG), mannosylphosphorylundecaprenol (Man-P-Undec), and a membrane-associated lipomannan in *Micrococcus luteus* (formerly *Micrococcus lysodeikticus*) was documented more than 40 yr ago by Lennarz and coworkers ***(1–3)***. In 2004 ***(4)*** structural studies established the structure of the major membrane mannolipid as α-D-mannosyl-(1→3)-α-D-mannosyl-(1→3)-diacylglycerol (Man_2-

From: *Methods in Molecular Biology, vol. 347: Glycobiology Protocols*
Edited by: I. Brockhausen © Humana Press Inc., Totowa, NJ

DAG) by negative-ion electrospray-ionization multistage mass spectrometry (ESI-MS^n). Based on the fragmentation patterns the *sn*-1-position is occupied with a 12-methyltetradecanoyl group, and the *sn*-2 position is acylated with a myristoyl group. Moreover, topological approaches demonstrated that the active sites of the mannosyltransferases catalyzing the transfer of mannose from guanosine 5′-diphosphate-mannose (GDP-Man) to Man_{1-2}-DAG and Man-P-Undec were exposed on the cytoplasmic face of the plasma membrane, whereas the lipid-mediated mannosyltransferases catalyzing the transfer of 48 α-mannosyl units from Man-P-Undec to the membrane-associated lipomannan were oriented toward the exterior face of the cytoplasmic membrane (*see also* Chapter 3). Additional support for the topological model illustrated in **Fig. 1** was obtained from temperature-sensitive mutants selected by a mannose-suicide procedure *(5)*. These enzymatic studies presented direct evidence that Man_2-DAG served as the lipid anchor precursor for the lipomannan after Man_2-DAG and Man-P-Undec diffused transversely (flip-flopped) from the inner leaflet to the external monolayer of the cytoplasmic membrane. In **Fig. 1**, the open circles represent mannose residues derived directly from GDP-Man and the closed circles represent mannose residues transferred to lipomannan via Man-P-Undec.

This chapter describes a procedure for the detergent solubilization and partial purification of *M. luteus* Man-P-Undec synthase (MPUS). The properties and specificity of the mannosyltransferase are characterized, and the enzyme is shown to catalyze the transfer of mannose from GDP-Man to a wide range of polyisoprenyl monophosphate substrates. Thus, MPUS provides a useful reagent for the enzymatic synthesis of many diverse mannosylphosphorylpolyisoprenol substrates functioning as mannosyl donors in the assembly of a variety of complex mannosylated glycoconjugates in prokaryotic and eukaryotic systems. This enzyme has previously been utilized to synthesize Man-P-citronellol, a water-soluble analog of Man-P-dolichol *(6)*. The water-soluble analog is an acceptable substrate for the lipid-mediated mannosyltransferases catalyzing the transfer of mannosyl units from Man-P-dolichol into Man_{6-9}-$GlcNAc_2$-P-P-dolichol, intermediates in the assembly of *N*-linked glycoproteins *(6)*, and has been used in transport-based assays for the protein(s) mediating the transbilayer movement of Man-P-dolichol in the endoplasmic reticulum (ER) of the liver and brain *(7,8)*.

2. Materials

2.1. Optimal Growth Conditions for Lysozyme-Sensitive M. luteus Cells and Preparation of Crude Membrane Fractions

1. *M. luteus* (formerly *M. lysodeikticus*) cells (American Type Culture Collection, Rockville, MD, ATCC no. 4968).

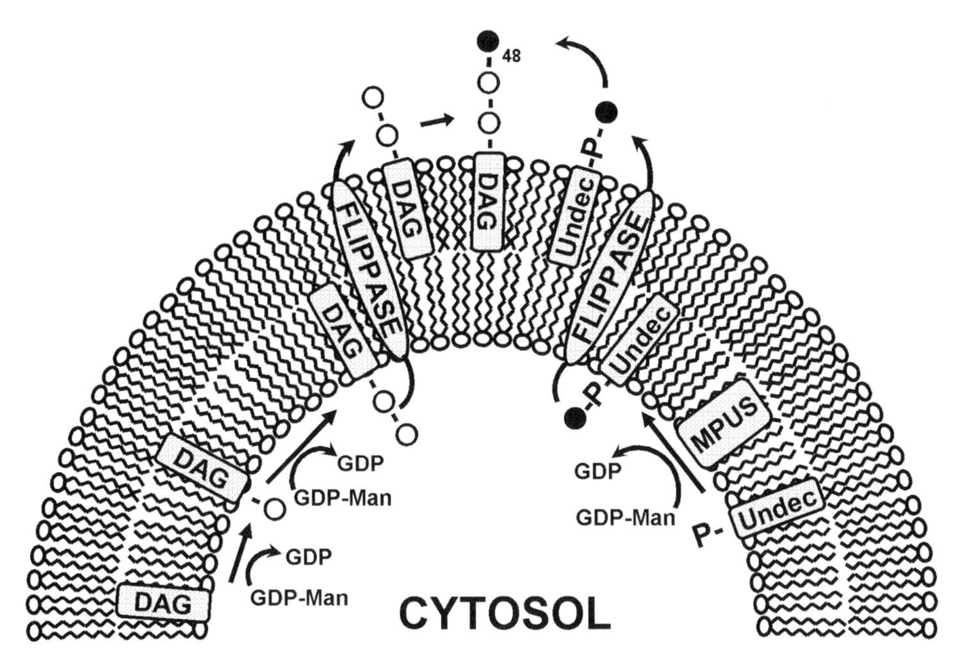

Fig. 1. Proposed model for the synthesis and transbilayer movement of mannolipid intermediates involved in lipomannan assembly in *Micrococcus luteus*. Man-P-Undec and Man_2-DAG are synthesized on the cytosolic monolayer of the bacterial plasma membrane from GDP-mannose and either Undec-P or diacylglycerol, respectively. The glycolipids are then translocated to the exoplasmic monolayer, where Man-P-Undec donates an average of 48 mannosyl units to Man_2-DAG to form the full-length lipomannan. MPUS, Man-P-Undec synthase; DAG, diacylglycerol; Undec-P, undecaprenylphosphate, mannose residues in lipomannan derived directly from GDP-Man (○); mannose residues in lipomannan derived from Man-P-Undec (●).

2. Bacto™Agar (Becton, Dickenson Co., Sparks, MD).
3. Luria broth (LB) medium: 1% Bacto-Peptone (Becton), 0.6% yeast extract (Becton), and 1% NaCl.
4. Sorvall Legend RT refrigerated clinical centrifuge (Kendro Laboratory Products, Newtown, CT).
5. Dulbecco's phosphate-buffered saline (DPBS): 140 m*M* of NaCl, 2.5 m*M* of KCl, 8.1 m*M* of Na_2HPO_4, and 1.5 m*M* of KH_2PO_4 (pH 7.4).
6. 20 m*M* of potassium phosphate (pH 7.2) and 1 *M* of sucrose.
7. Lysozyme (Sigma-Aldrich, St. Louis, MO, cat. no. L-6876).
8. 40 mL Dounce homogenizer (Kontes Glass, Vineland, NJ, cat. no. 885300-0040).
9. Deoxyribonuclease (DNase) and ribonuclease (RNase; Sigma, cat. nos. R-7003 and D-5025).
10. Buffer A: 0.1 *M* of Tris-HCl (pH 7.4), 0.25 *M* of sucrose, 10 m*M* of 2-mercaptoethanol, and 1 m*M* of ethylenediaminetetraacetic acid (EDTA).

2.2. In Vitro Assay for MPUS Activity

1. 15 Ci/mMol of GDP-Man (American Radiolabeled Chemicals, St. Louis, MO) mixed with nonradioactive GDP-Man (Sigma) to a final concentration of 0.2 m*M* and 10–1000 cpm/pmol.
2. 3-[(3-Cholamidopropyl)dimethylammonio]-1-propanesulfonate (CHAPS; Pierce, Rockford, IL).
3. Citronellol (C10 dolichol), geraniol (*E*-C10 polyprenol), and nerol (Z-C10 polyprenol; Sigma).
4. Dolichols (C55 and C95) and polyprenols (C55 and C95; Warszawa, Inc., Warsaw, Poland).
5. 1 m*M* Solution of the desired dolichyl monophosphate or polyisoprenyl monophosphate (for a description of the phosphorylation procedure, *see* **Note 1**) dispersed ultrasonically in 1% (w/v) CHAPS (*see* **Note 2**).
6. 1 *M* of *N*-2-hydroxyethylpiperazine-*N*′-2-ethanesulfonic acid (HEPES)-NaOH, pH 8.0.
7. 1 *M* of $MgCl_2$.
8. 0.1 *M* of 5′-adenosine monophosphate.
9. 0.9% (w/v) NaCl.
10. $CHCl_3/CH_3OH$ (2/1, v/v).
11. $CHCl_3/CH_3OH$/0.9% NaCl (3/48/47, v/v/v).
12. 1% Sodium dodecyl sulfate (SDS).
13. Econo-safe Biodegradeable Counting Cocktail (Research Products International Corp., Mount Prospect, IL).
14. Cellulose chromagrams sheets (Eastman Kodak, Rochester, NY).
15. Bioscan AR-2000 Imaging Scanner (Bioscan, Inc., Washington, DC).
16. Packard TR2100 scintillation spectrometer.

2.3. Detergent-Solubilization and Partial Purification of MPUS Activity

1. Protease inhibitor cocktail, Complete Mini, EDTA-free (Roche Diagnostics GmbH, Mannheim, Germany).
2. Solubilization buffer: 50 m*M* of HEPES-NaOH (pH 8.0), 10 m*M* of 2-ME, 1 m*M* of EDTA, 0.5 m*M* of phenylmethanesulfonyl fluoride (PMSF), and 1X protease inhibitor cocktail.
3. 10% CHAPS.
4. TSK-Gel Toyopearl diethylamino ethyl (DEAE) 650M (Supelco, Bellefonte, PA).
5. Reactive Yellow 86 agarose (Sigma).
6. Hexyl-agarose (Sigma).
7. Column buffer: 10 m*M* of HEPES-NaOH (pH 8.0), 0.25 *M* of sucrose, 20% glycerol, 10 m*M* of 2-ME, 1 m*M* of EDTA, and 0.1% CHAPS.
8. Amicon Ultrafiltration Cell, equipped with a 10,000 MW cut-off YM-5 membrane (Amicon Corp., Lexington, MA).
9. Gel-loading buffer: 50 m*M* of Tris-HCl (pH 6.8), 0.1 *M* of dithiothreitol, 10% glycerol, 2% SDS, and 0.1% bromphenol blue.

3. Methods

M. luteus cells are maintained on LB-agar plates containing 2% agar and LB medium. Cell cultures are initiated from single colonies isolated from the LB-agar plates and grown at 30°C in a shaking incubator in standard LB medium to an OD_{600} of approx 1 (*see* **Note 3**). Crude membrane fractions from *M. luteus* are prepared by hypotonic lysis following lysozyme treatment as described in **Subheading 3.1.**

3.1. Optimal Growth Conditions for Lysozyme-Sensitive M. luteus Cells and Preparation of Crude Membrane Fractions

1. *M. luteus* cells are collected by centrifugation at 500*g*, washed with ice-cold PBS two times, and resuspended in 20 m*M* of potassium phosphate (pH 7.2) and 1 *M* of sucrose to a cell density of approx 200 OD_{600} U/mL.
2. Cells are incubated with 0.25 mg/mL of lysozyme in 20 m*M* of potassium phosphate (pH 7.2) and 1 *M* of sucrose for 30 min at 30°C, and collected by centrifugation at 5000*g* for 10 min.
3. The cell pellet is rapidly suspended into 10 vol of ice-cold 20 m*M* potassium phosphate (pH 7.2) containing 0.5 m*M* of PMSF.
4. The lysate is homogenized with six strokes in a tight-fitting Dounce homogenizer and incubated at 0°C for 30 min with 1 µg/mL of DNAse, 1 µg/mL of RNAse (*see* **Note 4**), and 10 m*M* of $MgCl_2$.
5. Unbroken cells and debris are sedimented by centrifugation at 5000*g* for 20 min and discarded.
6. Crude membranes are recovered from the 5000*g* supernate by centrifugation at 100,000*g* for 60 min.
7. The membrane fraction is resuspended in buffer A and resedimented at 140,000*g* for 30 min in a Beckman TL-100 tabletop ultracentrifuge.
8. Membranes are resuspended in buffer A to a protein concentration of 10–20 mg/mL and stored at –20°C until needed.

3.2. In Vitro Assay for MPUS Activity

1. Reaction mixtures for the determination of MPUS activity contain 50 m*M* of HEPES-NaOH (pH 8.0), 20 m*M* of $MgCl_2$, 0.35% CHAPS, and 5 m*M* of 5′-adenosine monophosphate (*see* **Note 5**), the indicated concentration of the pertinent polyisoprenyl phosphate (dispersed ultrasonically in 1% CHAPS; *see* **Note 2**), enzyme fraction (either crude bacterial membranes or detergent-soluble fraction containing 0.1–1 µg of *M. luteus* membrane protein), and 20 µ*M* of GDP-[^{3}H]Man (10–1000 cpm/pmol) in a total volume of 25 µL.
2. Reactions are incubated in 12-mL glass conical tubes at 30°C and stopped by the addition of 2.5 mL of $CHCl_3/CH_3OH$ (2/1, v/v).
3. The reactions are incubated briefly on ice and centrifuged (low-speed clinical centrifuge) to remove the insoluble material.

4. The organic layer is transferred to a clean 16- × 125-mm glass tube and the insoluble residue is rinsed with an additional 1 mL of $CHCl_3/CH_3OH$ (2/1, v/v). Following centrifugation, the organic layers are combined.
5. Water-soluble reactants and side products are removed by sequential partitioning with 1/5 vol of 0.9% NaCl, followed by 1 mL of $CHCl_3/CH_3OH$/0.9% NaCl (3/48/47, v/v/v) 2 times (*see* **Note 6** and **ref. *9***).
6. The organic phase is transferred to a scintillation vial and dried under a stream of air.
7. Incorporation of [2-^{3}H]mannose into mannolipid is determined by scintillation spectrometry in a Packard TR2100 scintillation spectrometer after the addition of 0.5 mL of 1% SDS and 4 mL of Econo-safe biodegradeable counting cocktail.
8. Enzymatic reactions containing water-soluble isoprenyl phosphates (i.e., neryl-P, geranyl-P, and citronellyl-P) are analyzed chromatographically on cellulose chromagram sheets developed in ethyl acetate/butanol/acetic acid/water (4/3/2.5/4, v/v/v/v; *see* **Note 7**).

3.3. Detergent Solubilization and Partial Purification of MPUS Activity

To identify an effective detergent for the solubilization of MPUS activity, *M. luteus* membranes were incubated for 1 h at 0°C with a variety of detergents and sedimented at 140,000*g* for 30 min. As shown in **Table 1**, analysis of the resulting soluble supernates and pellets indicated that, although MPUS activity was solubilized by all of the detergents tested (except deoxycholate), solubilization with CHAPS resulted in the highest specific activity and most efficient recovery of MPUS activity. An examination of the CHAPS concentration-dependence of solubilization of MPUS (*see* **Fig. 2**) indicated that extraction with 0.5% CHAPS solubilizes approx 75% of the MPUS activity (closed circles), but less than 40% of the membrane protein (open circles).

1. Solubilization mixtures contain *M. luteus* membranes (2 mg/mL of membrane protein), solubilization buffer (*see* **Subheading 2.3., item 2**), and 0.5% CHAPS.
2. Solubilization mixtures are incubated with 0.5% CHAPS for 1 h at 0°C and sedimented at 140,000*g* for 30 min.
3. The supernatant liquid containing MPUS activity is supplemented with 0.25 *M* of sucrose and 20% glycerol from concentrated stock solutions, and immediately purified by ion-exchange chromatography on TSK-Gel Toyopearl DEAE 650M as described in **Subheading 3.4.**

3.4. Chromatography of MPUS on TSK-Gel Toyopearl DEAE 650M

1. CHAPS-soluble MPUS from **Subheading 3.3.** is applied to an 8-mL column of TSK-Gel Toyopearl DEAE 650M equilibrated in column buffer (*see* **Subheading 2.3., item 7**).

Table 1
Comparison of Various Detergents for the Solubilization of MPUS from *Micrococcus luteus* Membranes

		Protein		MPUS activity		
Detergent	Fraction	(mg)	(%)	(nmol/min)	(%)	(nmol/min/mg)
Triton X-100	Homogenate	2		2.44		1.22
	Supernate	1.24	61.4	2.7	100	2.2
	Pellet	0.78		0		0
Tween-20	Homogenate	2		2.9		1.45
	Supernate	0.63	30.4	2.3	77	3.6
	Pellet	1.44		0.68		0.47
Brij 58	Homogenate	2		2.9		1.45
	Supernate	0.96	50.3	0.89	81.8	0.93
	Pellet	0.95		0.2		0.21
CHAPS	Homogenate	2		4.76		2.38
	Supernate	0.77	45.8	3.5	76	4.6
	Pellet	0.91		2.14		1.23
Deoxycholate	Homogenate	2		n.d.		
	Supernate	1.1	59.1	n.d.		
	Pellet	0.76		n.d.		
Nonidet P40	Homogenate	2		2.6		1.3
	Supernate	1.8	81.6	2.74	100	1.52
	Pellet	0.41		0		0

Note: Solubilization mixtures contain 50 m*M* of Tris-HCl (pH 7.4), 0.25 *M* of sucrose, 1 m*M* of 2-mercaptoethanol, 1 m*M* of EDTA, and the indicated detergent (1%, w/v). After 60 min on ice, insoluble material is sedimented by centrifugation at 140,000*g* for 30 min. The supernates and pellets are separated and the pellet is resuspended in solubilization buffer. Assay mixtures contained membrane protein (50 μg), 50 m*M* of Tris-HCl (pH 7.4), 5 m*M* of 5′-adenosine monophosphate, the indicated detergent (0.5%, w/v), 20 μ*M* of GDP-[^{3}H]Man (14.3 cpm/pmol), 100 μ*M* of $Poly_{55}$-P (dispersed ultrasonically in 1% detergent), and 20 m*M* of $MgCl_2$ in a total volume of 0.05 mL. Following incubation for 3 min at 37°C, mannolipid synthesis was assayed as described in **Subheading 3.2.**

n.d., none detected.

2. Following elution with 2 column vols of column buffer, bound proteins are eluted with 60 mL gradient (0–0.5 *M*) of NaCl in column buffer.
3. Fractions of 3 mL are collected and analyzed as shown in **Fig. 3** for protein (open circles) and MPUS (closed circles) activity. MPUS activity elutes from TSK-Gel Toyopearl DEAE 650M in a fairly broad peak of activity centered around 0.35 *M* NaCl.

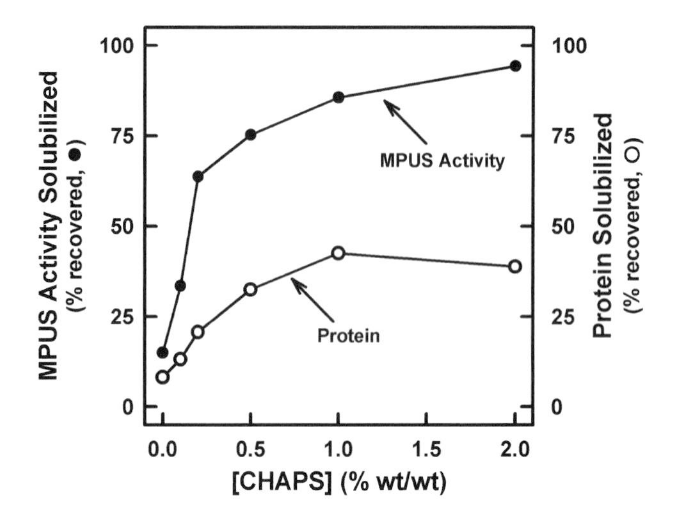

Fig. 2. Effect of CHAPS concentration on the solubilization of *Micrococcus luteus* Man-P-Undec synthase (MPUS) activity. Crude bacterial membranes (2 mg/mL protein) are incubated on ice for 60 min with the indicated concentration of CHAPS and sedimented at 140,000*g*. The supernate and pellet are separated and assayed for protein (○) and activity (●). MPUS assay mixtures contain 50 m*M* HEPES-NaOH (pH 8.0), 20 m*M* of $MgCl_2$, 5 m*M* 5′-adenosine monophosphate, 0.25% CHAPS, 20 μ*M* of GDP-[^{3}H]Man (169 cpm/pmol), and 0.1 m*M* $Poly_{55}$-P (dispersed ultrasonically in 1% CHAPS) in a total volume of 0.025 mL. Following incubation for 10 min at 30°C, mannolipid synthesis is assayed as described in **Subheading 3.2.**

4. Column fractions containing the highest MPUS activity are combined and concentrated by ultrafiltration in an Amicon Ultrafiltration Cell, equipped with a 10,000 MW cut-off YM-5 membrane. The MPUS preparation is freed of NaCl by two rounds of dilution with 10 vol of column buffer followed by reconcentration in the Amicon Ultrafiltration Cell.
5. The partially purified MPUS is concentrated to 5–10 mg/mL (protein) and stored at –80°C (*see* **Note 8**). MPUS can be further purified by chromatography on Reactive Yellow 86 agarose and hexyl-agarose, as described in **Subheading 3.5.**

3.5. Chromatography of TSK-Gel Toyopearl DEAE 650M-Purified MPUS on Reactive Yellow 86 Agarose

1. DEAE-purified MPUS (1 mL, 2.3 mg/mL protein) from **Subheading 3.4.** is applied to a 10-mL column of Reactive Yellow 86 agarose (Sigma) equilibrated in column buffer.
2. The column is eluted with column buffer and 0.5-mL fractions are analyzed as shown in **Fig. 4** for protein (open circles) and MPUS (closed circles). The majority of the "bulk" protein elutes from Reactive Yellow 86 in the exclusion volume, whereas the majority of MPUS is slightly retained.

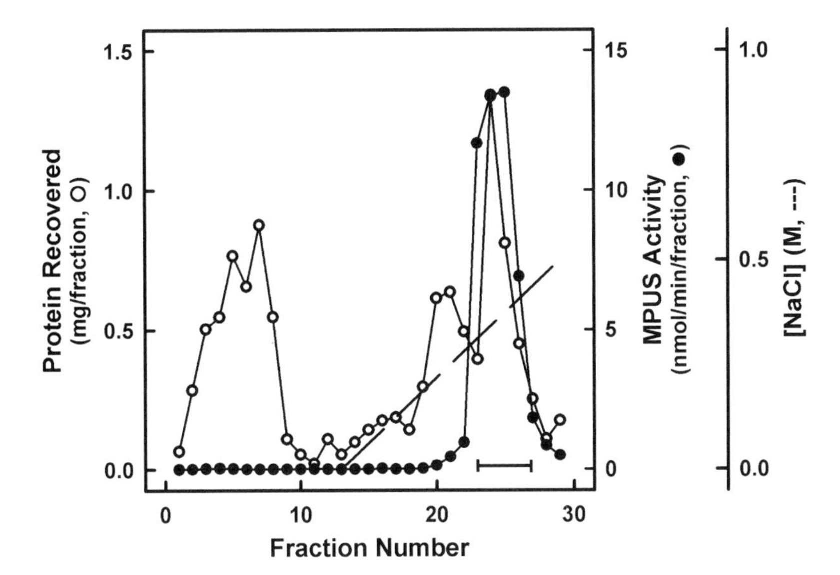

Fig. 3. Chromatographic separation of CHAPS-soluble Man-P-Undec synthase (MPUS) on TSK-Gel Toyopearl DEAE 650M. CHAPS-soluble proteins are chromatographed on an 8-mL column of TSK-Gel Toyopearl diethylamino ethyl 650M and assayed for protein (○) and MPUS activity (●) as described in **Subheading 3.2.** The solid bar indicates the fractions containing MPUS activity used for further purification.

3. Fractions containing MPUS activity are pooled and further purified by chromatography on hexyl-agarose (*see* **Subheading 3.6.**) or concentrated using an Amicon Centricon centrifugal concentrator (Millipore Corp., Bedford, MA) and stored at –20°C. Reactive Yellow 86 chromatography yields a two- to fourfold increase in specific activity of MPUS.

3.6. Chromatography of Reactive Yellow 86-Purified MPUS on Hexyl-Agarose

1. MPUS (~0.5 mg protein) from **Subheading 3.5.** is supplemented with 1 *M* of $(NH_4)_2SO_4$ and applied to a 2-mL column of hexyl-agarose (Sigma), equilibrated in column buffer containing 1 *M* of $(NH_4)_2SO_4$.
2. After elution with 5 column vol of equilibration buffer, the column is eluted with a 10-mL decreasing gradient (1 *M*–0) of $(NH_4)_2SO_4$.
3. 1-mL Fractions are collected and analyzed as shown in **Fig. 5** for protein (open circles) and MPUS (closed circles).
4. MPUS activity binds to hexyl-agarose in the presence of 1 *M* of $(NH_4)_2SO_4$ and elutes during the decreasing salt gradient (*see* **Note 9**).
5. Fractions containing MPUS activity are combined, concentrated using an Amicon Centricon centrifugal concentrator, and stored at –20°C.

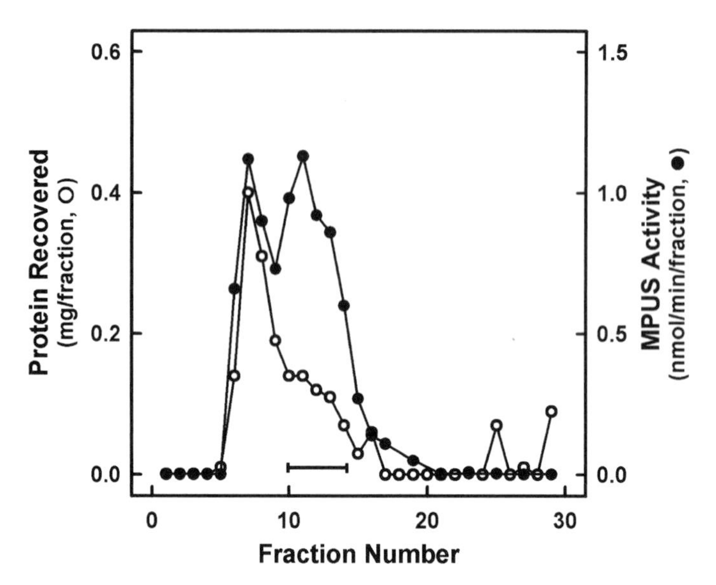

Fig. 4. Chromatography of partially purified Man-P-Undec synthase (MPUS) on Reactive Yellow 86 agarose. Pooled fractions from **Fig. 3** are chromatographed on a 10-mL column of Reactive Yellow 86 agarose and assayed for protein (○) and MPUS activity (●) as described in **Subheading 3.2.** The solid bar indicates the fractions containing MPUS activity used for further purification.

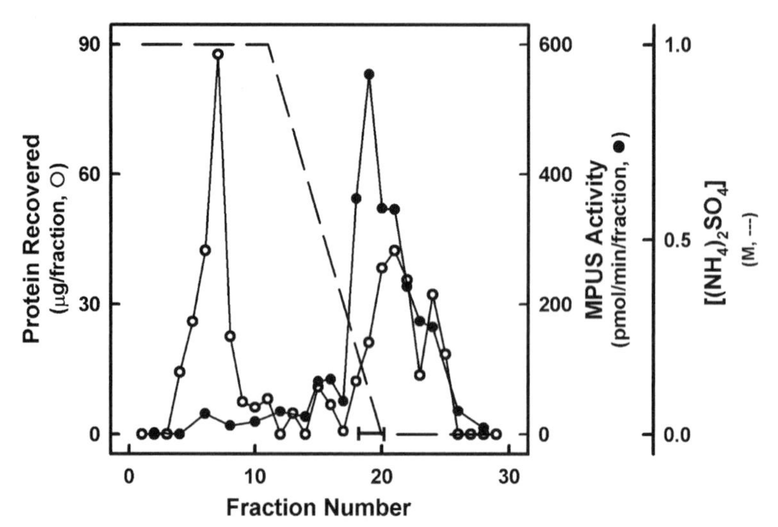

Fig. 5. Chromatography of partially purified Man-P-Undec synthase (MPUS) on hexyl-agarose. Pooled fractions from **Fig. 4** are chromatographed on a 2-mL column of hexyl-agarose and assayed for protein (○) and MPUS activity (●) as described in **Subheading 3.2.** The solid bar indicates the fractions containing MPUS activity used for further purification.

Table 2
Partial Purification of MPUS From *Micrococcus luteus* Membranes

Fraction	Protein (mg)	Protein (%)	MPUS activity (nmol/min/mg)
Membrane homogenate	36.6	100	2.9
CHAPS-soluble extract	28.8	78.7	4.4
TSK-Gel Toyopearl DEAE 650M	8.3	22.7	9.8
Reactive Yellow 86 agarose	4.31	11.8	13.3
Hexyl-agarose	0.08	0.22	137.6

Note: Assay mixtures contain *M. luteus* fraction (1–2 μg of membrane protein), 50 m*M* of HEPES-NaOH (pH 8.0), 5 m*M* of 5′-adenosine monophosphate, 0.3% CHAPS, 20 μ*M* of GDP-[^{3}H]Man (169 cpm/pmol), 100 μ*M* of $Poly_{55}$-P (dispersed ultrasonically in 1% CHAPS), and 20 m*M* of $MgCl_2$ in a total volume of 0.025 mL. Following incubation for 10 min at 30°C, incorporation into mannolipid was determined as described in **Subheading 3.2.**

6. Recoveries of total protein and MPUS activity from a typical purification are presented in **Table 2**. Overall, approx 10.4% of the MPUS activity is recovered in the purified fraction with an enrichment of specific activity of approx 47.5-fold over the initial membrane suspension. Under these conditions, no Man-DAG or Man_2-DAG synthase activity is detected.
7. **Figure 6** compares the time courses of Man-P-Undec synthesis in either crude membranes (panel A) or the fraction purified from TSK-Gel Toyopearl DEAE 650M (panel B) in the presence (closed circles) and absence (open circles) of exogenously added Undec-P. Following purification by ion exchange, there is no detectable mannolipid product after 30 min of incubation unless exogenous polyisoprenyl phosphate acceptor is added (*see* **Note 10**).

3.7. Isoprenyl Monophosphate Specificity

1. To illustrate the utility of partially purified MPUS for the synthesis of Man-P-isoprenols with defined isoprenyl chains, the specificity of MPUS for various isoprenyl monophosphates has been investigated.
2. **Figure 7** shows that MPUS enzymatically mannosylates a variety of polyprenyl phosphates including the two C55 substrates, $Poly_{55}$-P (closed triangles) and Dol_{55}-P (open triangles). In addition, MPUS actively transfers mannosyl units to $Poly_{95}$-P (closed circles) and to Dol_{95}-P (open circles), although at a somewhat slower rate compared to the substrates containing 11 isoprene units. This comparison indicates that MPUS does not require the unsaturated α-isoprene unit present in the naturally occurring substrate for activity, but does prefer the isoprenyl chain length of $Poly_{55}$-P over the longer polyisoprenoids.
3. MPUS will also mannosylate a variety of water-soluble isoprenyl phosphates as shown in **Table 3**, including neryl-P, geranyl-P, and citronellyl-P. This comparison confirms that MPUS does not require an unsaturated α-isoprene. In this

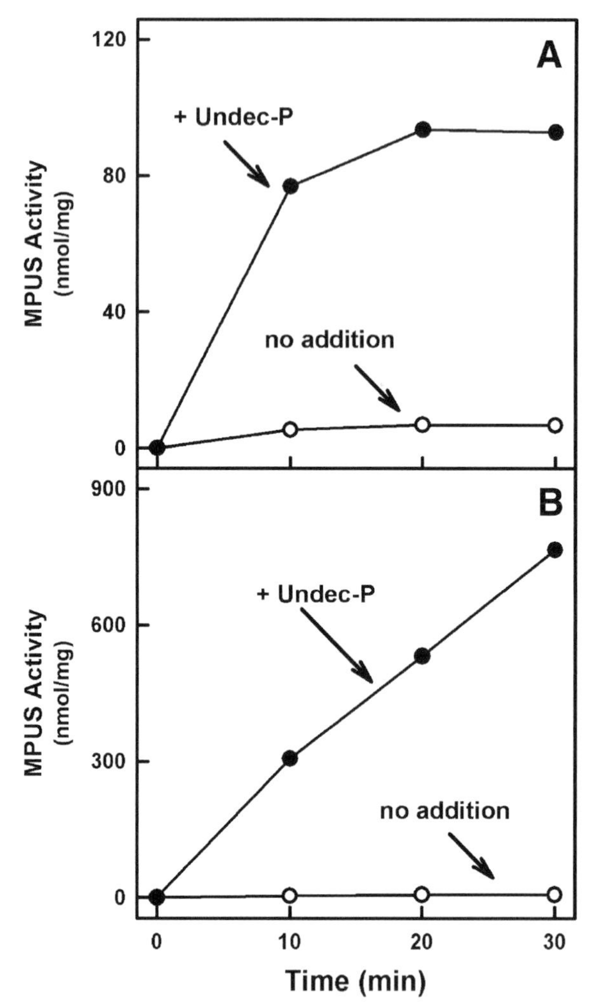

Fig. 6. Dependence of Man-P-Undec synthase (MPUS) on the addition of exogenously added Undec-P. Reaction mixtures and the assay procedure are essentially as described for **Fig. 2** containing either **(A)** *Micrococcus luteus* membrane fraction or **(B)** TSK-Gel Toyopearl DEAE 650M-purified MPUS in either the presence (●) or absence (○) of exogenously added Undec-P.

comparison the concentration of the water-soluble isoprenyl phosphates is 1 m*M*, as the K_m of MPUS for the water-soluble substrates is more than an order of magnitude higher than that for long-chain polyisoprenols.

3.8. Determination of the Molecular Size of M. luteus *MPUS*

To determine the apparent molecular size of *M. luteus* MPUS, partially purified MPUS can be analyzed by SDS-polyacrylamide gel electrophoresis (SDS-PAGE)

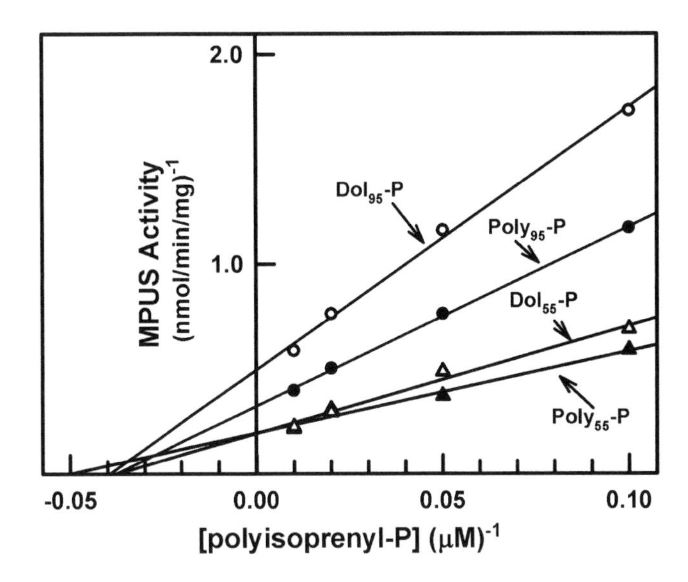

Fig. 7. Effect of chain length and the saturation state of the α-isoprene unit of various polyisoprenyl phosphate substrates on purified Man-P-Undec synthase (MPUS) activity. Partially purified MPUS was assayed with increasing concentrations of either Dol_{95}-P (○), $Poly_{95}$-P (●), Dol_{55}-P (△), or $Poly_{55}$-P (▲) (dispersed ultrasonically in 1% CHAPS) in reaction mixtures as described in **Fig. 2**. Following incubation for 10 min at 30°C, mannolipid synthesis was assayed as described in **Subheading 3.2.** The kinetic data are presented as a Lineweaver-Burk plot.

Table 3
Comparison of Water-Soluble Isoprenyl Phosphates as Substrates for MPUS

Substrate	MPUS activity (nmol/min/mg)
Neryl-P	2.1
Geranyl-P	1.9
Citronellyl-P	2.4

Note: Assay mixtures with water-soluble polyisoprenyl phosphates contained partially purified MPUS (0.27 μg of protein), 50 m*M* of HEPES-NaOH (pH 8.0), 20 m*M* of $MgCl_2$, 5 m*M* 5′-adenosine monophosphate, 0.25% CHAPS (w/v), 200 μ*M* of GDP-[^{3}H]Man (85 cpm/pmol), and 1 m*M* of polyisoprenyl phosphate in a total volume of 0.01 mL. Following incubation for 30 min at 30°C, incorporation into mannolipid was determined as described in **Subheading 3.2.** using the alternative protocol described for water-soluble analogs.

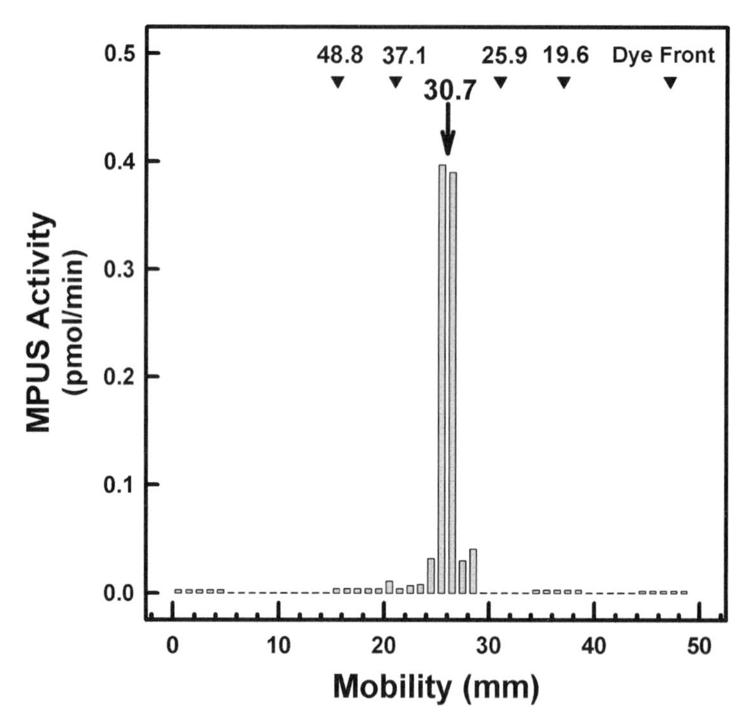

Fig. 8. Determination of apparent molecular weight of Man-P-Undec synthase (MPUS) by SDS-PAGE. DEAE-purified MPUS (~20 μg of membrane protein) is electrophoresed through a 12% discontinuous SDS-polyacrylamide gel according to Laemmli ***(14)***. Following electrophoresis, the gel lane is excised and cut into 1-mm sections. MPUS activity is extracted and assayed as described in **Subheading 3.2.**

on a 12% discontinuous acrylamide gel (ratio of acrylamide to *bis*-acrylamide 29:1) according to Laemmli ***(10)***.

1. CHAPS-soluble *M. luteus* MPUS (40 μg protein) is incubated in gel-loading buffer (*see* **Subheading 2.3., item 8**) at 37°C for 15 min, loaded onto a 12% discontinuous polyacrylamide mini-gel, and electophoresed at 30 mA for 1 h in a water-cooled mini-gel system (Hoeffer Scientific Instruments, San Francisco, CA).
2. Following electrophoresis, the sample lane is excised from the gel and divided into 1-mm slices.
3. The slices are placed in Eppendorf tubes, macerated with a blunt pestle, extracted overnight with 0.05 mL of 10 m*M* HEPES-NaOH (pH 8.0), 0.25 *M* sucrose, 20% glycerol, 10 m*M* 2-mercaptoethanol, 1 m*M* EDTA, and 0.1% CHAPS at 4°C, and the extracts are assayed for MPUS activity (*see* **Fig. 8**).
4. A comparison of the migration of MPUS activity with Benchmark Pre-Stained Protein Ladder molecular-weight markers (Invitrogen) indicated that the apparent molecular weight of the MPUS protein was approx 30.7 kDa (*see* **Note 11**).

4. Notes

1. Dolichols are chemically phosphorylated with phosphorus-oxytrichloride as described by Danilov and Chojnacki ***(11)***; polyprenols are chemically phosphorylated with TBA-phosphate/trichloroacetonitrile in dry acetonitrile as described by Danilov et al. ***(12)***.
2. Polyisoprenyl phosphates are dried in a plastic microcentrifuge tube under a stream of nitrogen and dispersed in 1% CHAPS using a Branson 5200 bath sonicator for 5 min.
3. Extended culture of *M. luteus* above an OD_{600} of 1.0 should be avoided, because this culture condition selects for a pale yellow, lysozyme-resistant variant ***(13)***. We have observed that these bacterial cells no longer contain high-specific-activity MPUS.
4. The lysate is extremely viscous and is incubated with DNAse and RNAse at 0°C until the viscosity is sufficiently reduced to aliquot the liquid into centrifuge bottles.
5. 5′ Adenosine monophosphate stimulates MPUS activity in crude membranes by protecting GDP-Man from nonspecific nucleotide sugar hydrolases ***(14)***. Protection of the nucleotide sugar is not necessary after partial purification by ion exchange.
6. Chloroform/methanol/water mixtures are partitioned by vigorous vortex mixing and incubated for 2 min on ice, followed by a brief centrifugation in a tabletop clinical centrifuge to facilitate phase separation. The aqueous (upper) phase is aspirated to waste and the organic phase is reserved for subsequent partitioning or transfer to a scintillation vial.
7. Following chromatography, the radioactive products are located using a Bioscan AR-2000 Imaging Scanner (Bioscan, Inc., Washington, DC), scraped into a scintillation vial, and analyzed for radioactivity by scintillation spectrometry in a Packard TR2100 scintillation spectrometer after the addition of 0.5 mL of 1% SDS and 4 mL of Econosafe.
8. This preparation is depleted of endogenous polyisoprenyl phosphate acceptor lipids and the competing bacterial mannosyltransferase activities that synthesize mannosyldiacylglycerol and dimannosyldiacylglycerol, and is suitable for the efficient synthesis of Man-P-polyisoprenoids with defined polyisoprenyl chains. The preparation is stable for at least 10 yr at –80°C.
9. Under some circumstances an additional gradient of 0.1–0.5% CHAPS was necessary to efficiently remove MPUS from hexyl-agarose. Octyl-agarose or octyl-sepharose can be substituted for hexyl-agarose for the purification of MPUS.
10. Several properties of MPUS have been examined in detail to define optimal conditions for the synthesis of Man-P-isoprenols. MPUS is optimally active between 0.2% and 0.5% CHAPS and is slightly stimulated by the addition of 10 m*M* of phosphatidylglycerol. MPUS is strongly dependent upon the addition of magnesium ion as shown in **Table 4** and exhibits a broad pH optimum centered around pH 8.0.
11. SDS-PAGE has been used to estimate the molecular size of Man-P-dolichol synthase from archaebacterium *Thermoplasma acidophilum* (42 kDa; *see* **ref. *15***).

Table 4
Effect of Various Divalent Cations on MPUS Activity

Cation	MPUS activity (nmol/min/mg)
None	<0.001
$CaCl_2$	0.02
$MnCl_2$	0.08
$MgCl_2$	2.1

Note: Assay mixtures contain *M. luteus* membrane fraction (0.27 μg of protein), 50 m*M* HEPES-NaOH (pH 8.0), 5 m*M* 5′-adenosine monophosphate, 0.3% CHAPS (w/v), 20 μ*M* GDP-[^{3}H]Man (169 cpm/pmol), 0.5 m*M* EDTA, 100 μ*M* Poly55-P (dispersed ultrasonically in 1% CHAPS), and 10 m*M* of the indicated divalent cation in a total volume of 0.025 mL. Following incubation for 10 min at 30°C, incorporation into mannolipid was determined as described in **Subheading 3.2.**

The apparent molecular weight of 30.7 kDa is significantly different from archael Man-P-dolichol synthase and the Man-P-polyisoprenol synthase of *Mycobacterium tuberculosis* ***(16)***, which apparently exists as a genetic fusion with a putative uncharacterized acyltransferase. However, it is similar in size to the prokaryotic Man-P-polyisoprenol synthases of *Mycobacterium smegmatis* (29 kDa; *see* **ref.** ***16***), *Corynebacterium glutamicum* (30 kDa; *see* **ref.** ***17***), and the eukaryotic DPM1 of *Saccharomyces cerevisiae* (30 kDa; *see* **refs.** ***18*** and ***19***). The DPM1 subunits of *Schizosaccharomyces pombe* (26.6 kDa; *see* **ref.** ***20***) and *Homo sapiens* (28.7 kDa; *see* **ref.** ***21***) lack the C-terminal membrane-anchoring domain present in *S. cerevisiae* and are therefore smaller.

Acknowledgment

The work cited in this article by the authors was supported by NIH Grant GM36065 to Dr. Waechter.

References

1. Lennarz, W. J. and Talamo, B. (1966) The chemical characterization and enzymatic synthesis of mannolipids in *Micrococcus lysodeikticus*. *J. Biol. Chem.* **241,** 2707–2719.
2. Lahav, M., Chiu, T. H., and Lennarz, W. J. (1969) Studies on the biosynthesis of mannan in *Micrococcus lysodeikticus*. *J. Biol. Chem.* **244,** 5890–5898.
3. Scher, M., Lennarz, W. J., and Sweeley, C. C. (1968) The biosynthesis of mannosyl-1-phosphoryl polyisoprenol in *Micrococcus lysodeikticus* and its role in mannan synthesis. *Proc. Natl. Acad. Sci. USA* **59,** 1316–1321.

4. Pakkiri, L. S., Wolucka, B. A., Lubert, E. J., and Waechter, C. J. (2004) Structural and topological studies on the lipid-mediated assembly of a membrane-associated LM in *Micrococcus luteus. Glycobiology* **20,** 8745–8749.
5. Pakkiri, L. S. and Waechter, C. J. (2005) Dimannosyldiacylglycerol serves as a lipid anchor precursor in the assembly of the membrane-associated lipomannan in *Micrococcus luteus. Glycobiology* **15,** 391–302.
6. Rush, J. S., Shelling, J. G., Zingg, N. S., Ray, P. H., and Waechter, C. J. (1993) Mannosylphosphoryldolichol-mediated reactions in oligosaccharide-P-P-dolichol biosynthesis. Recognition of the saturated alpha-isoprene unit of the mannosyl donor by pig brain mannosyltransferase. *J. Biol. Chem.* **268,** 13,110–13,117.
7. Rush, J. S. and Waechter, C. J. (1995) Transmembrane movement of a water-soluble analogue of mannosylphosphoryldolichol is mediated by an endoplasmic reticulum protein. *J. Cell. Biol.* **30,** 529–536.
8. Rush, J. S. and Waechter, C. J. (2004) Functional reconstitution into proteo-liposomes and partial purification of a rat liver ER transport system for a water-soluble analogue of mannosylphosphoryldolichol. *Biochemistry* **43,** 7643–7652.
9. Waechter, C. J. and Scher, M. G. (1981) Methods for studying lipid-mediated glycosyltransferases involved in the assembly of glycoproteins in nervous tissue. In *Research Methods in Neurochemistry, Vol. 5* (Marks, N. and Rodnight, R., eds.), Plenum, New York, NY, pp. 201–233.
10. Laemmli, U. K. (1970) Cleavage of structural proteins during the assembly of the head of bacteriophage T4. *Nature* **227,** 680–685.
11. Danilov, L. L. and Chojnacki, T. (1981) A simple procedure for preparing dolichyl monophosphate by the use of $POCl_3$. *FEBS Lett.* **131,** 310–312.
12. Danilov, L. L., Druzhinina, T. N., Kalinchuk, N. A., Maltev, S. D., and Shibaev, V. N. (1989) Polyprenyl phosphates: synthesis and structure-activity relationship for a biosynthetic system of *Salmonella anatum* O-specific polysaccharide. *Chem. Phys. Lipids* **51,** 191–203.
13. Prasad, A. L. N. and Litwack, G. (1965) Growth and biochemical characteristics of *micrococcus lysodeikticus*, sensitive or resistant to lysozyme. *Biochemistry* **4,** 496–501.
14. Waechter, C. J. and Harford, J. B. (1977) Evidence for the enzymatic transfer of N-acetylglucosamine from UDP-N-acetylglucosamine into dolichol derivatives and glycoproteins by calf brain membranes. *Arch. Biochem. Biophys.* **181,** 185–198.
15. Zhu, B. C. R. and Laine, R. A. (1996) Dolichyl-phosphomannose synthase from the Archae *Thermoplasma acidophilum. Glycobiology* **6,** 811–816.
16. Gurcha, S. S., Baulard, A. R., Kremer, L., et al. (2002) Ppm1, a novel polyprenol monophosphomannose synthase from *Mycobacterium tuberculosis. Biochem. J.* **365,** 441–450.
17. Gibson, K. J. C., Eggeling, L., Maughan, W. N., et al. (2003) Disruption of Cg-Ppm1, a polyprenyl monophosphomannose synthase, and the generation of lipoglycan-less mutants in *Corynebacterium glutamicum. J. Biol. Chem.* **278,** 40,842–40,850.

18. Orlean, P. Albright, C., and Robbins, P. W. (1988) Cloning and sequencing of the yeast gene for dolichol phosphate mannose synthase, an essential protein. *J. Biol. Chem.* **263,** 17,499–17,507.
19. Tomita, S., Inoue, N., Maeda, Y., Ohishi, K., Takeda, J., and Kinoshita, T. (1998) A homologue of *Saccharomyces cerevisiae* Dpm1p is not sufficient for synthesis of dolichol-phosphate-mannose in mammalian cells. *J. Biol. Chem.* **273,** 9249–9254.
20. Colussi, P. A., Taron, C. H., Mack, J. C., and Orlean, P. (1997) Human and *Saccharomyces cerevisiae* dolichol phosphate mannose synthases represent two classes of the enzyme, but both function in *Schizosaccharomyces pombe. Proc. Natl. Acad. Sci. USA* **94,** 7873–7878.
21. Maeda, Y., Tanaka, S., Hino, J., Kangawa, K., and Kinoshita, T. (2000) Human dolichol-phosphate-mannose synthase consists of three subunits, DPM1, DPM2 and DPM3. *EMBO J.* **19,** 2475–2482.

3

Assay of Dolichyl-Phospho-Mannose Synthase Reconstituted in a Lipid Matrix

John S. Schutzbach

Summary

Advances in molecular biology over the last several decades, along with new highly developed methods for protein expression, have enabled investigators to produce and purify large yields of the soluble protein domains of a number of eukaryotic glycosyltransferases and processing glycosidases. The availability of these purified enzymes has in turn allowed determination of the crystal structures of the catalytic domains of some of the proteins, thus providing details of the active site geometry and catalytic mechanisms of the enzymes. It must be remembered, however, that the natural subcellular locations for enzymes involved in glycoprotein and glycolipid synthesis are the membranes of the endoplasmic reticulum and Golgi, where the enzymes exist bound to or inserted in the membrane matrix. Because of technical difficulties, few of the intact enzymes containing their hydrophobic membrane-interactive domains have been purified and studied in a membrane environment, even though the membrane has been shown to have effects on the properties and kinetics of many enzymes. Therefore, a method for the reconstitution of dolichyl-phospho-mannose (Dol-P-Man) synthase in phospholipids and phospholipid membranes will be described in detail. In order to properly characterize membrane glycosyltransferases and glycosidases, it is necessary to investigate the kinetic and catalytic properties of these proteins in a membrane environment. The ultimate goal is to define the topography of the proteins in membranes and also to understand the kinetic and catalytic properties of these enzymes in biological membranes.

Key Words: Dolichyl-phospho-mannose synthase; membrane enzymes; glycosyltransferases; phospholipids.

1. Introduction

A large number of the glycosyltransferases (GTs) and glycosidases (Glcs) that are involved in glycoconjugate biosynthesis and processing, as well as most of the

From: *Methods in Molecular Biology, vol. 347: Glycobiology Protocols*
Edited by: I. Brockhausen © Humana Press Inc., Totowa, NJ

natural substrates for these enzymes, are localized to the two-dimensional matrix of membranes of the endoplasmic reticulum and Golgi. With few exceptions, however, the kinetics and mechanisms of these enzymes have only been studied in solution utilizing soluble protein domains of the enzymes. Furthermore, most of the assays for these enzymes have also utilized low-molecular-weight water-soluble or detergent-solubilized substrates rather than their normal membrane-bound high molecular substrates. If we are to understand the natural properties of these enzymes, however, and to elucidate their mechanisms under in vivo conditions, the kinetic and physical properties of the enzymes must be characterized under more physiological conditions. Perhaps the best rationale for this type of study was given a number of years ago by Efraim Racker in his book on bioenergetics ***(1)***: "The properties of solubilized membrane proteins are different from those of the proteins bound to the membrane." It is therefore important to purify membrane enzymes intact, including their hydrophobic membrane spanning domains, to reconstitute the enzymes in a lipid matrix and to assess the effects of membrane composition and structure on their topology ***(2)***, properties, activity, kinetics, and specificity. This requirement applies to both eukaryotic and prokaryotic GTs and Glcs, although the current protocol is directed to one example of a eukaryotic enzyme. This protocol should be used as a general guideline for studies with other membrane GTs and Glcs. A major obstacle for such studies has always been the isolation and purification of the intact complete enzymes containing their transmembrane domains (*see also* Chapter 2).

As one example, dolichyl-phospho-mannose (Dol-P-Man) synthase catalyzes the reversible transfer of mannose from guanasine 5′-diphosphate-mannose (GDP-Man) to Dol-P. The enzyme has been solubilized from liver microsomes and purified 880-fold over the microsomal fraction in the presence of 0.1% of the nonionic detergent, Nonidet NP-40 ***(3)***. Although crude fractions of the liver enzyme were active when the enzyme was assayed in the presence of detergent, more highly purified fractions were inactive when assayed in detergent solution and the enzyme appeared to be inactivated or denatured upon storage. These purified fractions of liver enzyme were found to be active, however, when detergent-free enzyme was reconstituted with phosphatidylethanolamine (PE) or with phospholipid mixtures of PE and phosphatidylcholine (PC) when the molar proportion of PC was 70% or less.

The mode of activation of the synthase by PE suggested that specific structural features of PE, and/or its unusual macroscopic organization such as its propensity to form a hexagonal phase and to induce the destabilization of bilayer membranes ***(4)***, were involved in the activation. The synthase was found to have a requirement for either hexagonal-phase lipid or for destabilized bilayer structures for optimal activity, and the enzyme was essentially inactive when reconstituted in stable bilayer membranes. The requirement for unique lipid structures was supported by a number of additional studies showing the effects of dolichol on membrane properties and structure. Dolichol was found to promote the destabilization of bilayer structure as well as inducing membrane leakage and enhancing membrane fusion, along with providing the lipid structures necessary for enzyme activity ***(3–12)***.

These studies also provided insight into possible regulatory mechanisms for Dol-P-Man synthase ***(13,14)***. Although the in vivo metabolic regulation of Dol-P-Man synthase has not been elucidated, regulation of the activity of this enzyme could possibly involve covalent modification of the enzyme ***(13)***, or changes in the concentrations of enzyme or of its glycosyl donor or acceptor. Our results suggest, however, that changes in the lipid microenvironment could also regulate Dol-P-Man synthase activity. Diacylglycerol (DAG) and lysophospholipids are present at low concentrations in biological membranes as the result of phospholipid metabolism, and the presence of DAG or lyso-PC has been shown to disturb lipid packing and alter local membrane architecture. The activity of Dol-P-Man synthase reconstituted in phospholipid membranes was shown to be modulated by the incorporation of these compounds into the lipid matrix ***(14)***. DAG-enhanced mannosyl transfer, probably as a consequence of the ability of this conically shaped lipid to physically alter the structure of bilayer membranes, lowered the apparent K_m for Dol-P in PC membranes from 9 μ*M* to 0.3 μ*M* but had no effect on the K_m for GDP-Man. Thus DAG altered the physical properties of the lipid matrix to increase the apparent affinity of the enzyme for Dol-P. The increased affinity of the enzyme for Dol-P in PC/DAG mixtures was likely the result of DAG-induced spreading of phospholipid headgroups, allowing the synthase increased access to the polar phosphate moiety of the polyprenol substrate. In contrast, the incorporation of lyso-PC into either PE dispersion or into PE/PC membranes reduced transferase activity.

The effect of lyso-PC was not owing to micelle formation, but rather to the ability of lysolipids at low concentrations to stabilize the bilayer phase in phospholipid mixtures containing PE. These effects of DAG and lyso-PC on synthase activity suggested that the activity of the enzyme might be modulated by the activity of intracellular phospholipases. Thus when the enzyme was reconstituted in PC vesicles containing 2.4% Dol-P, which supports minimal activity, the rate of mannosyltransfer was increased fivefold by treatment with phospholipase C ***(14)*** and transferase activity was proportional to the quantity of DAG generated *in situ* by phospholipase C. In contrast, when the enzyme was reconstituted in PE/PC membranes that supported maximal transferase activity, mannosyltransfer was inhibited and found to be inversely proportional to the amount of phospholipase A_2 added to the reaction mixture. Because DAG and lyso-PC are generated in vivo by the activity of phospholipases, modulation of Dol-P-Man synthase by these compounds may represent a potential regulatory mechanism for the synthesis of oligosaccharide lipids. The point of this discussion is to emphasize that these results could not be obtained by assays of transferase activity utilizing detergent-solubilized enzyme and substrate. One can only wonder how much information is lacking about the properties of other GTs, because they have not been assayed under more physiological conditions.

Most studies on intact membrane GTs and Glcs, however, have been limited by difficulties in isolating and purifying large amounts of the solubilized membrane enzymes, as well as by real or apparent instability of the purified proteins. The current capability to clone almost all of the GTs and Glcs involved in glycoprotein and glycolipid synthesis and processing, as well as the ability to produce large amounts of these enzymes as recombinant proteins, should greatly simplify the isolation and purification of these intact hydrophobic enzymes and allow studies of the enzymes when reconstituted in phospholipid membranes. In this protocol, an accurate, effective, simple, and rapid assay will be described for a recombinant yeast Dol-P-Man synthase in phospholipid membranes. The rate of mannosyltransfer is determined by separating the hydrophobic product from the water-soluble substrate in a two-phase scintillation cocktail with essentially no background radioactivity. This assay is suitable for many of the enzymes in the dolichol pathway and for other enzymes that transfer sugars to acceptors that are soluble in organic solvents.

2. Materials

2.1. Expression of Dol-P-Man Synthase in Escherichia coli *and Enzyme Purification*

1. Growth of *Escherichia coli.*
   a. 5X M9 medium: 64 g of Na_2HPO_4-$7H_2O$, 15 g of KH_2PO_4, 2.5 g of NaCl, and 5 g of NH_4Cl in 1 L of deionized water (*see* **Note 1**). Divide into five aliquots of 200 mL and sterilize in an autoclave for 20 min at 15 lb/in.^{2} on liquid cycle.
   b. M9ZB medium: 10 g of tryptone (Difco, Sparks, Maryland) and 5 g of NaCl in 800 mL of water are sterilized as above. Add 200 mL of 5X M9 medium, 1 mL of 1 *M* $MgSO_4$ (sterilized separately by filtration through a 0.22-μ filter or in an autoclave), 0.1 mL of 1 *M* $CaCl_2$ (sterilized separately by filtration or in autoclave), and 10 mL of 40% (w/v) glucose (filter-sterilized).
   c. 0.1 mg/mL of ampicillin (filter-sterilized) and 100 m*M* of isopropyl β-thiogalactopyranoside ([IPTG] filter-sterilized; Sigma, St. Louis, MO).
2. Buffer A: 25 m*M* of Na_2HPO_4 and 5 m*M* of $MgCl_2$; adjust pH to 8.0 with 1 *M* NaOH and add 2 mL of mercaptoethanol (ME)/L (Sigma) just before use (*see* **Note 2**).
3. Buffer B: 10 m*M* of Na_2HPO_4, 0.5 m*M* of disodium ethylenediaminetetraacetic acid (Na_2EDTA), 10% glycerol (*see* **Note 3**), 1.0 m*M* of dithiothreitol (DTT), and 0.1% Nonidet P-40 (Sigma; *see* **Note 4**) adjusted to pH 8.0 with 1 *M* NaOH. The glycerol, detergent, and reducing agent should be added just prior to use.

4. Buffer C: 0.1 *M* of Tris-HCl adjusted to pH 7.5 with acetic acid, 10% glycerol, 0.2% ME, and 0.1% (w/v) sodium dodecyl sulfate (SDS). The glycerol, detergent, and reducing agent should be added just prior to use.
5. Hydroxy apatite (Bio-Rad, Richmond, CA), 2.5- × 11-cm column, preferably plastic.
6. Sonifier (Branson model W185 or equivalent).
7. Amicon YM-10 ultrafiltration membranes (Millipore, Billerica, MA).

2.2. Enzyme Assay in Detergent Solution

1. 0.6 mg of Dol-P (American Radiolabeled Chemicals Inc., St. Louis, MO) in $CHCl_3:CH_3OH$ (2:1).
2. GDP-[^{3}H]mannose (50 µCi; 5–15 Ci/mmol, American Radiolabeled Chemicals,) and unlabeled 15 m*M* of GDP-Man (Sigma; *see* **Note 5**).
3. Ultrasonic cleaning bath (Branson model 1510 or equivalent).
4. Components of enzyme Assay I: 0.1 *M* of Tris-HCl/acetate buffer (pH 7.5) containing 0.01 *M* of EDTA. Solutions of 0.1 *M* of $MnCl_2$, 0.1 *M* of $MgCl_2$, 10% Nonidet P-40, and 0.2 *M* of DTT (Sigma).
5. Water-saturated 1-butanol prepared by mixing (shaking) deionized water and 1-butanol in a glass-stoppered bottle, then let the mixture stand until two phases are obtained.
6. Xylene-based scintillation fluid (Scintilene, Fisher Scientific, Pittsburgh, PA).
7. 5-mL Scintillation vials.
8. Scintillation counter.

2.3. Enzyme Assay in Phospholipid Matrices

1. 0.04 mg/mL of Dol-P (American Radiolabeled Chemicals) in $CHCl_3:CH_3OH$ (2:1).
2. Components of enzyme assay II: 25 m*M* of Tris-HCl/acetate (pH 7.5) containing 5 m*M* of $MnCl_2$, 2.5 m*M* of $MgCl_2$, 0.25 m*M* of EDTA, and 5 m*M* of DTT.
3. Ultrasonic cleaning bath.
4. Unsaturated PE, PC, and other lipids from plant sources, usually soy, at 2.0 mg/mL in $CHCl_3$ (Avanti Polar Lipids, Birmingham, Alabama): The unsaturated phospholipids provide optimal activity (*see* **Note 6**).
5. Water-saturated 1-butanol as in **Subheading 2.2., item 5**.
6. Xylene-based scintillation fluid (Scintilene, Fisher Scientific).
7. 5-mL Scintillation vials.
8. Scintillation counter.

3. Methods

3.1. Preparation of Enzymes: Expression of Saccharomyces cerevisiae Dol-P-Man Synthase in E. coli

1. *E. coli* DH5α harboring the plasmid pDM6 (which contains the structural gene [*DPM1*]) for Dol-P-Man synthase is prepared as described in **ref. *15***.

2. This plasmid is then used to prepare the inducible expression vector BL21(DE3)/pDPM1, which contains the structural gene *(DPM1)* for Dol-P-Man synthase ***(16)***. Store as a glycerol stock at –80°C.
3. Prepare one 250-mL flask with 100 mL of M9ZB broth and four 1-L flasks containing 250 mL each of M9ZB broth. Sterilize in an autoclave for 20 min at 15 psi on liquid cycle. Add ampicillin to a final concentration of 100 μg/mL.
4. Insert sterilized platinum wire into glycerol stock of BL21(DE3)/pDPM1 or into a single colony growing on agar.
5. Inoculate the flask containing 100 mL of M9ZB broth with cells of BL21(DE3)/pDPM1 on platinum wire and incubate overnight at 37°C on a water-bath shaker at 150 rpm.
6. Inoculate each of four cultures containing 250 mL of M9ZB broth with 10 mL of the overnight culture. Incubate at 37°C on a water-bath shaker at 200 rpm. After 3 h of incubation, or until an A_{600} of 1.5–1.9, add 2.7 mL of 100 m*M* IPTG to each flask to induce production of the mannosyltransferase. Two hours after induction the bacteria are harvested by centrifugation in screw-capped tubes at 6000*g* for 15 min using a Sorvall GSA rotor or equivalent at 4°C. All additional procedures are carried out at 0–4°C or on ice.
7. Cells are washed by resuspending the pellets in 300 mL of water and centrifuging as in **step 6**, and then are washed with 100 mL of buffer A. After centrifugation, the cell pellet can be frozen overnight or used immediately.

3.2. Purification of Dol-P-Man Synthase

1. Hydroxy apatite is suspended in buffer B and poured into a 2.5- × 11-cm column. The column is washed with 5 column vol of buffer B prior to adding enzyme solution.
2. The cell pellet from **Subheading 3.1., step 7** is thawed (if previously frozen) and suspended in 10 mL of buffer A. The cells are kept at ice-bath temperature and sonically ruptured using a sonifier at a power setting of 4.5 for 3 × 30 s with 1-min intervals to allow for cooling. A particulate fraction is obtained by centrifugation at 38,000*g* in screw-capped tubes for 20 min using a Sorvall SS-34 rotor or equivalent. The precipitate is washed twice with 25 mL of buffer A and then suspended in 5 mL of buffer A.
3. The suspension is diluted with 17.5 mL of buffer B; 2.5 mL of 10% Nonidet P-40 is added, and the mixture is centrifuged at 38,000*g* for 20 min. The supernatant (which often contains significant amounts of activity) and the precipitate are saved. Additional enzyme is solubilized by the addition of 20 mL of buffer C to the precipitate. The precipitate is dispersed by homogenization with a Tekmar homogenizer for 10 s, followed in 10 min by the addition of 2.5 mL of 10% Nonidet P-40. The solution is centrifuged as above and the supernatant is saved. This procedure is repeated once more to yield a third solubilized fraction. Solubilized enzyme fractions containing activity are pooled and assayed (dilute fractions 1:1000 and 1:5000 prior to assay).

4. The pooled solubilized enzyme fractions are applied to a column of hydroxyl apatite (2.5- × 11-cm) previously equilibrated with buffer B. The column is washed with 30 mL of buffer B and then eluted with a linear 400-mL gradient from 0–1 *M* of NaCl in the same buffer. Collect 4.5-mL fractions.
5. Appropriate dilutions (dilute fractions 1:1000 and 1:5000 for assays) of each fraction are assayed for activity, and fractions containing enzyme are pooled and concentrated by ultrafiltration using an Amicon YM-10 membrane. This procedure yields approx 10 mg of purified protein with a specific activity of 8.2 μmol of mannose transferred/min/mg protein (*see* **Note 7**).

3.3. Enzyme Assay in Detergent Solution

This assay is used to determine Dol-P-Man synthase activity in purified fractions and to assess activity in column fractions during purification.

1. The assay mixture is prepared by adding 0.6 mg of Dol-P in $CHCl_3$:CH_3OH (2:1) to a 5-mL polypropylene tube and removing the solvent under a stream of N_2. Add 0.75 mL of 10% Nonidet P-40 and 2.5 mL of 0.1 *M* Tris-HCl/acetate (pH 7.5), vortex thoroughly, and then sonify in a water-bath sonifier for 2 min. Transfer to a larger polypropylene tube. Wash the 5-mL tube an additional two times with 2.5 mL each of 0.1 *M* Tris-HCl buffer and pool washes. Add 1.5 mL of 0.1 *M* $MnCl_2$, 0.75 mL of 0.1 *M* $MgCl_2$, and 0.75 mL of 0.01 *M* EDTA (*see* **Note 8**).
2. Add 0.185 mL of the assay mixture and 0.005 mL of 0.2 *M* DTT to conical 1.5-mL centrifuge tubes. These two solutions can be premixed just prior to enzyme assay in volumes to account for the total number of assays (*see* **Note 9**).
3. Add 0.005 mL of enzyme and incubate at 25°C for 5 min in a water bath.
4. Add 0.005 mL of GDP-[^{3}H]Man solution and mix on a vortex mixer, start stopwatch or timer, and immediately incubate at 25°C.
5. After 5 min, add 0.15 mL of the reaction mixture to a 5-mL scintillation vial containing 0.25 mL of H_2O and 0.45 mL of water-saturated butanol.
6. Mix on a vortex mixer for 10 s and then add 3.6 mL of xylene-based scintillation fluid. Mix again on a vortex mixer for 10 s.
7. Centrifuge the vials for 2 min at 3000*g* in a tabletop centrifuge and then count in a scintillation counter (*see* **Note 10**).
8. Correct total activity by multiplying by 4/3 to account for sample volume counted in scintillation counter. Correct for specific activity of the radioactive GDP-[^{3}H]Man. Express activity as either cpm or μmol of mannose transferred/min/mg protein.

3.4. Enzyme Assay in Phospholipid Matrix

1. 0.05 mL of Dol-P and 0.05 mL of phospholipid(s) are added to 1.5-mL conical polypropylene centrifuge tubes and the solvent is removed under a stream of nitrogen (*see* **Note 11**).
2. 0.19 mL of Tris-HCl buffer is added and the lipids are sonically dispersed by a 30-s immersion in an ultrasonic cleaning bath.

3. 0.005 mL of enzyme (33 ng protein) is added; the mixture is agitated on a vortex mixer and then incubated at 25°C for 5 min in a water bath (*see* **Note 12**).
4. The transferase reaction is initiated by the addition of 0.005 mL of GDP-[^{3}H]Man; mix on a vortex mixer and incubate at 25°C.
5. After 5 min, terminate the reaction by pipetting 0.15-mL aliquots of the reaction mixtures into scintillation vials (1.6 × 5.2 cm) containing 0.45 mL of water-saturated 1-butanol and 0.25 mL of water.
6. Mix on a vortex mixer for 10 s and then add 3.6 mL of xylene-based scintillation fluid. Mix again on a vortex mixer for 10 s.
7. Centrifuge the vials for 2 min at 3000*g* in a tabletop centrifuge and then count in a scintillation counter (*see* **Note 10**).
8. Correct total activity by multiplying by 4/3 to account for sample volume counted in the scintillation counter. Correct for specific activity of the radioactive GDP-[^{3}H]Man. Express activity as either cpm or μmol of mannose transferred/min/mg protein.

4. Notes

1. All chemicals are of the highest quality, or of enzyme grade when available.
2. The reducing agents, either DTT or ME, must be added to buffers just prior to use and should not be stored for more than 24 h in dilute solutions.
3. All solutions are expressed as v/v ratios unless otherwise noted.
4. When working with pure enzymes in low concentrations, all detergents and detergent solutions must be of the highest quality and peroxide-free.
5. Prepare an aqueous solution of GDP-[^{3}H]Man. Take 50 μCi of GDP-[^{3}H]Man that is in ethanol:water (7:3) and remove the solvent under a stream of N_2. Dissolve the GDP-[^{3}H]Man in 1.25 mL of H_2O and add 1.25 mL of 15 m*M* unlabeled GDP-Man. The concentration of unlabeled GDP-Man does not have to be exactly 15 m*M*, but the concentration does have to be known by reading the optical density at 252 nm. The extinction coefficient of GDP-Man at this wavelength is 1.37×10^4. Correct the final concentration of the solution by adding the concentration of radioactive GDP-Man based on the specific activity of the purchased product.
6. Solutions of unsaturated PE from plant sources that are exposed to air have a shelf life of one week or less. These lipids should be stored at –20°C under an atmosphere of inert gas. Phospholipids purchased from suppliers other than Avanti Polar lipids, as well as highly unsaturated phospholipids isolated from animal sources, are often not usable for activation of enzymes in the dolichol pathway because of their instability. Saturated phospholipids do not form bilayers or lipid matrices with the appropriate properties for the optimal activation of Dol-P-Man synthase. For initial studies, be sure to start with fresh solutions of plant PE. By comparison, solutions of unsaturated PC are relatively stable for several weeks, although it is recommended that all lipid solutions are stored under N_2 at –20°C.
7. The enzyme can be stored at –20°C for at least 1 yr and can be stored at 4°C for at least 1 mo. Dol-P-Man synthase is sensitive to oxidation but activity can usually be restored by the addition of fresh reducing agents. If the enzyme appears to

Table 1
Activity of Dol-P-Man Synthase when Reconstituted With Phospholipids

Phospholipid Matrix[a]	Synthase activity[b]	Relative activity
PE	2270	1.0
PC	104	0.05
PE-PC (1:1)	140	0.06
PE-PC-DAG (1:1:0.1)	395	0.17
PE-PC-DAG (13:6.5:1)	612	0.27

[a]Reaction mixtures contained 1 μg of Dol-P and 50 μg of total phospholipid in the indicated ratios. The lipids were sonically dispersed in assay buffer as described in **Subheading 3.3.** 0.6 ng of enzyme was added, the mixtures were incubated for 5 min at 25°C, and the reactions were initiated by the addition of GDP-[^{3}H]Man. PE, phosphatidylethanolamine; PC, phosphatidylcholine; DAG, diacylglycerol.

[b]Activity is expressed as the amount of Dol-P-Man (cpm) formed in 5 min at 25°C.

lose activity, it can usually be reactivated by adding fresh ME to 0.2% alone or with the addition of both ME to 0.2% and SDS to 0.01%.

8. The assay buffer can be stored at 4°C for at least 1 wk. Appropriate volumes (0.185 mL of the mixture per assay) for the number of assays can be mixed with 0.2 *M* of DTT (0.005 mL per assay) prior to setting up assay tubes, but this solution is not stable for more than 1 d.
9. Final concentrations in the enzyme assay are 25 m*M* of Tris-HCl acetate (pH 7.5), 0.25 m*M* of EDTA, 5 m*M* of $MnCl_2$, 2.5 m*M* of $MgCl_2$, 0.25% Nonidet P-40, 4.2 μg of Dol-P, and 5 m*M* of DTT.
10. In control reactions that do not contain enzyme, the radioactivity should range between 50 and 75 cpm or less than 0.1% of the introduced radioactivity. Essentially all of the Dol-P-[^{3}H]Man product partitions into the upper organic phase. Tritium counts in unreacted GDP-[^{3}H]Man or in breakdown products remain in the water phase and are not counted, because the energy of ^{3}H-decay is not sufficient to cross the water/organic phase interface. It is important to note that carbon-14-labeled substrates should not be used in this assay. The same assay procedure can be used to assay dolichyl-P-mannose synthase from mammalian liver, dolichyl-P-glucose synthase, and mannosyltransferase II *(3)*. This procedure may also be applicable for the assay of many of the enzymes involved in the synthesis of lipid-linked oligosaccharides from both eukaryotic and prokaryotic sources.
11. *See* **Table 1** for typical activity of Dol-P-Man synthase reconstituted with phospholipids. The composition of the phospholipid mixtures can be changed or modified in order to determine the effects of lipid composition on enzyme properties and kinetics. However, the total amount of phospholipid should be kept constant.

12. Dol-P-Man synthase readily incorporates into lipid matrices composed of PE or into phospholipid mixtures that form destabilized bilayer structures. The enzyme also incorporates into membranes and lipid vesicles composed of PE/PC/Dol-P mixtures with or without diacylglycerols. Other methods are usually required, however, to incorporate membrane proteins into a lipid–membrane matrix. These methods may include detergent dilution, detergent dialysis, or the use of hydrophobic beads to remove detergents from enzyme–lipid mixtures. The investigator should search the literature, including *Methods in Enzymology* or any of the lipid-oriented journals, to determine the best procedure and protocol for reconstituting other membrane enzymes with lipids.

Acknowledgments

The author would like to acknowledge the significant contributions of John W. Jensen, W. Thomas Forsee, Patsy Hughey, and Janet Zimmerman to the methods and results described in this chapter, as well as the generous support contributed by the National Institutes of Health in research grants CA16777 and GM38643 to John Schutzbach.

References

1. Racker, E. (1976) *A New Look at Mechanisms in Bioenergetics.* Academic, NY, p. 12.
2. Jennings, M. L. (1989) Topography of membrane proteins. *Ann. Rev. Biochem.* **58,** 999–1027.
3. Jensen, J. W. and Schutzbach, J. S. (1985) Activation of dolichyl-phospho-mannose synthase by phospholipids. *Eur. J. Biochem.* **153,** 41–48.
4. Jensen, J. W. and Schutzbach, J. S. (1984) Activation of mannosyltransferase II by nonbilayer phospholipids. *Biochemistry* **23,** 1115–1119.
5. Jensen, J. W. and Schutzbach, J. S. (1986) Characterization of mannosyl-transfer reactions catalyzed by dolichyl-mannosyl-phosphate-synthase. *Carbohydr. Res.* **149,** 199–208.
6. Jensen, J. W. and Schutzbach, J. W. (1988) Modulation of dolichyl-phosphomannose synthase activity by changes in the lipid environment of the enzyme. *Biochemistry* **27,** 6315–6320.
7. Lai, C. S. and Schutzbach, J. S. (1984) Dolichol induces membrane leakage of liposomes composed of phosphatidylethanolamine and phosphatidylcholine. *FEBS Lett.* **169,** 279–282.
8. Lai, C. S. and Schutzbach, J. S. (1986) Localization of dolichols in phospholipid membranes: An ESR spin label study. *FEBS Lett.* **203,** 153–156.
9. Monti, J. A., Christian, S. T., and Schutzbach, J. S. (1987) Effects of dolichol on membrane permeability. *Biochim. Biophys. Acta* **905,** 133–142.
10. Schutzbach, J. S. and Jensen, J. S. (1989) Bilayer membrane destabilization induced by dolichylphosphate. *Chem. Phys. Lipids* **51,** 213–218.
11. Schutzbach, J. S., Jensen, J. W., Lai, C. S., and Monti, J. A. (1987) Membrane structure and mannosyltransferase activities: the effects of dolichols on membranes. *Chemica Scripta* **27,** 109–118.

12. Schutzbach, J. S. (1997) The role of the lipid matrix in the biosynthesis of dolichyl-linked oligosaccharides. *Glycoconj. J.* **14,** 175–182.
13. Banerjee, D. K., Carrasquillo, E. A., Hughey, P., Schutzbach, J. S., Martinez, J. A., and Baksi, K. (2005) *In vitro* phosphorylation by cAMP-dependent protein kinase up-regulates recombinant *Saccharomyces cerevisiae* mannosylphosphodolichol synthase. *J. Biol. Chem.* **290,** 4174–4181.
14. Jensen, J. W. and Schutzbach, J. S. (1989) Phospholipase-induced modulation of dolichyl-phosphomannose synthase activity. *Biochemistry* **28,** 851–855.
15. Orlean, P., Albright, C., and Robbins, P. W. (1988) Cloning and sequencing of the yeast gene for dolichol phosphate mannose synthase, an essential protein. *J. Biol. Chem.* **263,** 17,499–17,507.
16. Schutzbach, J. S., Zimmerman, J. W., and Forsee, W. T. (1993) The purification and characterization of recombinant yeast dolichyl-phosphate-mannose synthase: site-directed mutagenesis of the putative dolichol recognition sequence. *J. Biol. Chem.* **268,** 24,190–24,196.

4

O-Mannosylation in Mammalian Cells

Tamao Endo and Hiroshi Manya

Summary

The *O*-mannosyl glycan is present in a limited number of glycoproteins of brain, nerve, and skeletal muscle. α-Dystroglycan is one of the *O*-mannosylated proteins and is a central component of the dystrophin–glycoprotein complex that has been shown to be related to the onset of muscular dystrophy. We have identified and characterized glycosyltransferases, protein *O*-mannose β1,2-*N*-acetylglucosaminyltransferase (POMGnT1) and protein *O*-mannosyltransferase 1 (POMT1), involved in the biosynthesis of *O*-mannosyl glycans. We subsequently found that loss of function of the *POMGnT1* gene is responsible for muscle–eye–brain disease (MEB). It has also been reported that the *POMT1* gene is responsible for Walker-Warburg syndrome (WWS). MEB and WWS are autosomal recessive disorders characterized by congenital muscular dystrophies with neuronal migration disorders. Therefore, the ability to assay enzyme activities of mammalian *O*-mannosylation would facilitate progress in the identification of other *O*-mannosylated proteins, the elucidation of their functional roles, and the understanding of muscular dystrophies. This protocol describes assay methods for the mammalian POMT and POMGnT.

Key Words: *O*-mannosylation; glycosyltransferase; POMGnT1; POMT1; POMT2; α-dystroglycan; Walker-Warburg syndrome; muscle-eye-brain disease; muscular dystrophy.

1. Introduction

O-mannosylation is a common type of glycosylation in fungi and yeast. These *O*-mannosyl glycans are neutral straight-chain glycans that are composed of one to seven mannose residues. Mammalian *O*-mannosylation is an unusual type of protein glycosylation (*see also* Chapter 5) and is present in a limited number of glycoproteins of brain, nerve, and skeletal muscle ***(1–6)***. We have previously found that the glycans of α-dystroglycan (α-DG) include *O*-mannosyl oligosaccharides, and that a sialyl *O*-mannosyl glycan, Siaα2-3Galβ1-4GlcNAcβ1-2Man, is very different from that of fungi and yeast ***(1)***. Our data also suggest

From: *Methods in Molecular Biology, vol. 347: Glycobiology Protocols*
Edited by: I. Brockhausen © Humana Press Inc., Totowa, NJ

that the sialyl *O*-mannosyl glycan is a laminin-binding ligand of α-DG ***(1)***. α-DG is a central component of the dystrophin–glycoprotein complex (DGC) isolated from skeletal muscle membrane and behaves as a connection between DGC and extracellular matrix molecules, such as laminin, agrin, and neurexin ***(7–9)***. DGC has a crucial role in linking the extracellular basal lamina to the cytoskeletal proteins for stabilization of sarcolemma.

Muscular dystrophies (MDs) are genetic diseases that cause progressive muscle weakness and wasting. Because the causative genes of several MDs have been identified from molecules associated with DGC, it is commonly believed that the dysfunction of DGC causes the development of MDs. Duchenne MD, as a famous case in point, results from mutations of the gene encoding dystrophin in DGC. Recently, scores of reports suggest that aberrant protein glycosylation of α-DG is the primary cause of some forms of congenital MD ***(8–10)***.

Muscle–eye–brain disease (MEB; MIM 253280) and Walker-Warburg syndrome (WWS; MIM 236670) are autosomal recessive disorders characterized by congenital MD, ocular abnormalities, and brain malformation (type II lissencephaly). We previously reported that MEB is caused by mutations in the gene encoding POMGnT1 uridine 5′-diphosphate (UDP)-*N*-acetylglucosamine: protein *O*-mannose β1,2-*N*-acetylglucosaminyltransferase ***(1)***. POMGnT1 is responsible for the formation of the GlcNAcβ1-2Man linkage of *O*-mannosyl glycan ***(11)***. We also demonstrated that protein *O*-mannosyltransferase 1 (POMT1) forms an enzyme complex with POMT2 and is responsible for the catalysis of the first step in *O*-mannosyl glycan synthesis ***(12)***. Mutations in the *POMT1* gene are considered to be the cause of WWS ***(13)***.

The GlcNAcβ1-2Man linkage of *O*-mannosyl glycan is identified only in mammals, and it was impossible to detect POMGnT1 activity by using acceptor substrates such as mannose, mannose–threonine, *p*-nitrophenyl-α-mannose, and mannose-2-aminobenzamide. Therefore, the synthesis of mannosylpeptide as acceptor substrate, derived from the α-DG sequence, enabled us to detect POMGnT1 ***(14)***.

POMT1 encodes a protein that is homologous to members of the family of protein *O*-mannosyltransferases (PMTs) in yeast. In yeast, PMTs catalyze the transfer of a mannosyl residue from dolichyl phosphate mannose (Dol-P-Man) to serine–threonine residues of certain proteins ***(15)***. However, using the same methods as those applied to yeast, POMT activity was not detected in mammalian tissues and cells. This difference between mammals and yeast may depend largely on the specificity of the acceptor peptide sequence and the effect of detergent. We established the method for POMT assay in mammals by using recombinant α-DG expressed in *Escherichia coli* as acceptor substrate and *n*-octyl-β-D-thioglucoside (OTG) as detergent ***(12)***. This protocol describes assay methods for mammalian POMT and POMGnT.

2. Materials

2.1. Preparation of Enzyme Sources

1. *pcDNA3.1-POMGnT1*, *pcDNA3.1-POMT1*, and *pcDNA3.1-POMT2* expression plasmids: Human cDNAs encoding POMGnT1 ***(11)***, POMT1 ***(16)***, and POMT2 ***(17)*** are inserted into mammalian expression vectors, pcDNA3.1/Zeo, or pcDNA3.1/Hygro (Invitrogen Corp., Carlsbad, CA).
2. Dulbecco's modified Eagle's medium (DMEM), fetal bovine serum (FBS), 100X penicillin–streptomycin–glutamine liquid (PC-SM-Gln, 10,000 U/mL of penicillin, 10,000 μg/mL of streptomycin, 29.2 mg/mL of glutamine), Lipofectamine transfection reagent and Plus reagent (Invitrogen).
3. Phosphate-buffered saline (PBS): prepare 10X stock with 1.37 *M* of NaCl, 27 m*M* of KCl, 80 m*M* of Na_2HPO_4, and 14.7 m*M* of KH_2PO_4 (adjust to pH 7.4 with HCl if necessary). Store at room temperature. Prepare working solution by dilution with water (1:9) and store at 4°C.
4. Silicone blade cell scraper (Sumilon, Sumotomo Bakelite Co., Tokyo, Japan).
5. Homogenization buffer: 10 m*M* of Tris-HCl (pH 7.4), 1 m*M* of ethylenediaminetetraacetic acid (EDTA), 250 m*M* of sucrose (SET buffer) with protease inhibitor cocktail (Complete, EDTA-free, Roche Diagnostics, Basel, Switzerland). SET buffer is stored at 4°C. Add protease inhibitor cocktail before use.

2.2. Preparation of Glutathione-S-Transferase-α-Dystroglycan

1. *pGEX-glutathione-S-transferase-α-dystroglycan (GST-α-DG):* Potential *O*-glycosylation sites of α-DG are predicted in the region corresponding to amino acids 313-483 ***(18)***. We amplified this region from mouse brain total ribonucleic acid (RNA) by reverse transcriptase polymerase chain reaction (RT-PCR) using the primer set 5′-GGGAATTCCACGCCACACCTACAC-3′ (sense) and 5′-GGGTC TAGAACTGGTGGTAGTACGGATTCG-3′ (antisense), and subcloned it into the pGEX-4T-3 vector to express the peptide as a GST-fusion protein (Amersham Biosciences Corp., Piscataway, NJ).
2. Luria-Bertani (LB) broth (Invitrogen) supplemented with 50 μg/mL of ampicillin.
3. LB agar plate (1.5% w/v agar) supplemented with 50 μg/mL of ampicillin.
4. Isopropyl-D-thiogalactopyranoside (IPTG; Invitrogen): prepare 1 *M* stock solution in water, sterilize by filtration, and store at –20°C.
5. Ampicillin sodium salt (Nacalai tesque, Kyoto, Japan): prepare 50 mg/mL stock solution in water, sterilize by filtration, and store at –20°C.
6. 1 mL of glutathione–sepharose column (GSTrap; Amersham).
7. Prepare 10 m*M* of reduced glutathione in PBS just before use.
8. 50 m*M* of $(NH_4)HCO_3$: prepare 1 *M* of stock solution (pH 7.0) in water and dilute 100 mL with 1900 mL of water for use.

LIBRARY, UNIVERSITY OF CHESTER

2.3. POMT Assay

1. OTG and 3-[(3-cholamidopropyl)dimethylammonio]-1-propanesulfonate (CHAPS; Dojindo Laboratories, Kumamoto, Japan): prepare 10% (w/v) stock solution in water and store at –20°C.
2. Triton X-100 (Nacalai tesque): prepare 20% (w/v) stock solution in water and store at room temperature.
3. POMT reaction buffer: 10 m*M* of Tris-HCl (pH 8.0), 2 m*M* of 2-mercaptoethanol (2-ME), 10 m*M* of EDTA, and 0.5% of OTG. Store at –20°C.
4. Mannosylphosphoryldolichol95: [Mannose-6-^{3}H] Dol-P-Man (1.48-2.22 TBq/mmol, American Radiolabeled Chemical, Inc., St. Louis, MO). 1.85 MBq of solution in chloroform and methanol is transferred into a screw-cap centrifugal tube and evaporated with a centrifugal evaporator (*see* **Note 1**). Add 1 mL of 20 m*M* Tris-HCl (pH 8.0) and 0.5% CHAPS, and dissolve by sonication with bath-type sonicator in ice-cold water (10 cycles of 15-s pulses with 30-s intervals). Measure radioactivity and then adjust to 40,000 cpm/μL with 20 m*M* of Tris-HCl (pH 8.0) and 0.5% CHAPS. Aliquot and store at –80°C.
5. PBS containing 1% Triton X-100 (1% Triton-PBS). Store at 4°C.
6. 0.5% Triton-*tris* buffer: 20 m*M* of Tris-HCl (pH 7.4) containing 0.5% Triton X-100. The buffer is stored at 4°C.
7. Glutathione–sepharose 4B (Amersham): Prepare a 25% slurry working suspension as follows. Suspension (1 mL, equivalent to 0.75-mL beads) is put in a centrifugal tube. 9 mL of water is added to the suspension and vortexed. After centrifugation at 1000*g* for 1 min the supernatant is removed by aspiration. The beads are rinsed with 10 mL of PBS and collected by centrifugation. 1% Triton-PBS (2.25 mL) is added and stored at 4°C.
8. Liquid scintillation cocktail: 0.4% (w/v) 2,5-Diphenyloxazole (Dojindo), 35% (w/v) polyethylene glycol *p*-isooctylphenyl ether (Nacalai tesque) in toluene.
9. Jack bean-α-mannosidase (Seikagaku Corp., Tokyo, Japan): 0.8 U of enzyme is dissolved in 50 μL of 0.1 *M* ammonium acetate buffer (pH 4.5). The enzyme solution is dried up with a centrifugal evaporator and stored at –20°C. The dried enzyme is dissolved with 50 μL of 1 m*M* $ZnCl_2$ before use.

2.4. POMGnT1 Assay

1. UDP-*N*-acetyl-D-glucosamine (UDP-GlcNAc; Sigma-Aldrich Corp., St. Louis, MO): Prepare a 1-m*M* stock solution in water and store at –20°C.
2. UDP-GlcNAc [glucosamine-6-^{3}H(N)] (UDP-[^{3}H]-GlcNAc, 0.74-1.66 TBq/mmol, PerkinElmer, Inc., Wellesley, MA). Store at –20°C.
3. Benzyl-α-D-mannopyranoside (Sigma-Aldrich): Prepare 100 m*M* of stock solution in 20% ethanol and store at –20°C.
4. Mannosylpeptide (Ac-Ala-Ala-Pro-Thr[Man]-Pro-Val-Ala-Ala-Pro-NH_2; *see* **Note 2**): prepare 2 m*M* of stock solution in water and store at –20°C.
5. POMGnT reaction buffer: 140 m*M* of methanesulfonic acid (MES; adjust pH to 7.0 with NaOH), 2% Triton X-100, 5 m*M* of adenosine 5′-monophosphate (AMP),

200 m*M* of GlcNAc, 10% glycerol, and 10 m*M* of $MnCl_2$. Store at –20°C without $MnCl_2$ ($MnCl_2$ is added just before use).
6. Reverse-phase column for high-performance liquid chromatography (HPLC): Wakopak 5C18-200 column (4.6 × 250 mm, Wako Pure Chemical Industries, Osaka, Japan).
7. 0.1% Trifluoroacetic acid (TFA) in water (Solvent A): Add 1 mL of TFA to 1000 mL of HPLC-grade water and degas with an aspirator before use.
8. 0.1% TFA in acetonitrile (Solvent B): Add 1 mL of TFA to 1000 mL of HPLC-grade acetonitrile and degas by sonication before use.
9. Liquid scintillation cocktail as described in **Subheading 2.3., item 8**.
10. Streptococcal β-*N*-acetylhexosaminidase (HEXaseI, Prozyme, San Leandro, CA): 50 mU of enzyme is dissolved with 50 μL of 0.3 *M* citrate phosphate buffer (pH 5.5) and stored at –20°C.
11. 0.05 *N* of NaOH, 1 *M* of $NaBH_4$, and 4 *N* of acetic acid solution in water.
12. AG-50W-X8 (H^+ form, Bio-Rad Laboratories, Hercules, CA).

2.5. Sodium Dodecyl Sulfate-Polyacrylamide Gel Electrophoresis and Western Blotting

Sodium dodecyl sulfate-polyacrylamide gel electrophoresis (SDS-PAGE) and Western blotting are carried out in accordance with standard methods. Please refer to experimental guidebooks. Some points are described in the following items.

1. 4X Loading buffer (modified Laemmli *[19]* buffer): 250 m*M* of Tris-HCl (do not adjust pH), 8% (w/v) SDS, 40% (w/v) glycerol, 2.84 m*M* of 2-ME, and 0.005% (w/v) bromophenol blue. Store at –20°C.
2. Antibodies: Rabbit antisera specific to the human POMT1, POMT2, and POMGnT1 are produced by using synthetic peptides corresponding to residues 348–362 (YPMIYENGRGSSH) of POMT1, 390–403 (HNTNSDPLDPSFPV) of POMT2, and 649–660 (KEEGAPGAPEQT) of POMGnT1, respectively. Anti-rabbit IgG is conjugated with horseradish peroxidase (HRP; Amersham).
3. Coomassie Brilliant Blue R-250 (CBB): Prepare a 0.1% solution in methanol: acetic acid:water (40:10:50) and store at room temperature.
4. Enhanced chemiluminescent (ECL) reagent kit (Amersham).
5. Amplify fluorographic reagent (Amersham).
6. Hyperfilm ECL and Kodak BioMax MS X-ray film are purchased from Amersham.

3. Methods

The POMT activity is based on the amount of [^{3}H]-mannose transferred from Dol-P-Man to GST-α-DG *(12)*. The reaction product is purified with a glutathione–sepharose column, and the radioactivity of mannosyl GST-α-DG is measured by a liquid scintillation counter. The POMGnT1 activity is based on the amount of [^{3}H]GlcNAc transferred from UDP-GlcNAc to benzyl-α-mannose

(Benzyl-Man; *see* **ref.** ***20***) or mannosylpeptide (Ac-Ala-Ala-Pro-Thr[Man]-Pro-Val-Ala-Ala-Pro-NH_2; *see* **ref.** ***14***). The reaction product is purified with a reverse-phased HPLC and the radioactivity is measured. We also synthesized several mannosylpeptides derived from mucin box sequences of α-DG. These mannosylpeptides are not commercially available but it is possible to use Benzyl-Man, which is commercially available, as a substitute.

POMGnT1 and POMT activities are detected in various mammalian cells and mammalian tissues. This chapter describes the methods that use the microsomal membrane fraction of rat brain and human embryonic kidney 293T (HEK293T) cells as the enzyme source. To demonstrate that the gene products of *POMGnT1*, *POMT1*, and *POMT2* have enzymatic activity, the cells transfected with *POMGnT1* or *POMT1* and *POMT2* are used. Although whole cells instead of membrane fractions may be used as an enzyme source, we recommend using membrane fractions because mammalian tissues and cells have a low specific activity ***(12)***.

3.1. Preparation of Enzyme Sources

3.1.1. Cell Culture and Preparation of Cell Membrane Fraction

1. HEK293T cells are maintained in DMEM supplemented with 10% FBS, 2 m*M* of L-glutamine, and 100 U/mL of penicillin/50 μg/mL of streptomycin at 37°C with 5% CO_2.
2. The expression plasmids of human *pcDNA3.1–POMT1* and *pcDNA3.1–POMT2* are transfected into HEK293T cells using Lipofectamin PLUS reagent according to the manufacturer's instructions.
   a. The day before transfection, plate cells into a 100-mm culture dish with antibiotic-free 10% FBS-DMEM so that they are 60–70% confluent the day of transfection. Avoid antibiotics during transfection.
   b. Dilute 4 μg of DNA with 750 μL of serum-free DMEM, add the 20 μL of Plus reagent, and let stand at room temperature for 15 min (reagent A). In another tube, dilute 30 μL of Lipofectamin reagent with 750 μL of serum-free DMEM (reagent B).
   c. Mix reagent A with reagent B and let stand at room temperature for 15 min (reagent C).
   d. During **step c**, replace the medium on the cells with 5 mL of serum-free DMEM.
   e. Add reagent C to the cells from **step d** and incubate at 37°C with 5% CO_2 for 3 h.
   f. Add 5 mL of 20% FBS-DMEM to the cells from **step e** and culture for 2–3 d.
3. The culture supernatants are removed by aspiration and the cells are rinsed gently with cold PBS. Then 5 mL of cold PBS is added, and the cells are scraped into centrifugal tubes and washed with 10 mL of cold PBS. The cells are collected by centrifugation at 1000*g* for 10 min at 4°C (*see* **Note 3**).
4. The cell pellet is broken with a tip-type sonicator in 500 μL of homogenization buffer (*see* **Note 4**). After centrifugation at 900*g* for 10 min, the supernatant is

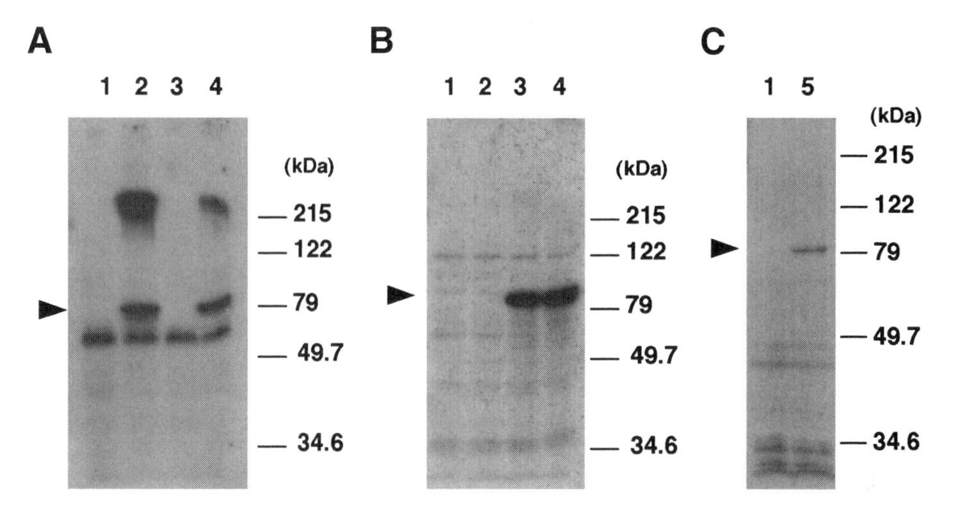

Fig. 1. Western blot analysis of **(A)** POMT1, **(B)** POMT2, and **(C)** POMGnT1 expressed in HEK293T cells. Lanes 1, cells transfected with vector alone; lanes 2, cells transfected with human *POMT1*; lanes 3, cells transfected with human *POMT2*; lanes 4, cells cotransfected with *POMT1* and *POMT2*; lane 5, cells transfected with human *POMGnT1*. The proteins (20 μg of membrane fraction) were subjected to sodium dodecyl sulfate-polyacrylamide gel electrophoresis (10% gel), and the separated proteins were transferred to a polyvinylidene difluorite membrane. The polyvinylidene difluorite membrane was stained with **(A)** anti-POMT1, **(B)** anti-POMT2, or **(C)** anti-POMGnT1 antibody. Arrowheads indicate the positions of the corresponding molecules. Molecular-weight standards are shown on the right. (Reprinted with permission from **ref. *12***. Copyright 2004 by National Academy of Sciences.)

dispensed in halves and subjected to ultracentrifugation at 100,000*g* for 1 h. The precipitates thus obtained are used as microsomal membrane fraction (*see* **Note 5**).
5. Half of the precipitates obtained in **step 4** are used to determine protein concentration and are subjected to Western blotting, and the remainder are used to assess the enzymatic activity.
6. Western blot is performed for detection of products (*see* **Fig. 1**). The microsomal fraction (20 μg) is separated by SDS-PAGE (10% gel) and proteins are transferred to a polyvinylidene difluorite membrane. The membrane, after blocking in PBS containing 5% skim milk and 0.5% Tween-20, is incubated with each antibody and then the membrane is treated with anti-rabbit IgG conjugated with HRP. Proteins bound to an antibody are visualized with ECL.

3.1.2. Preparation of Brain Membrane Fraction

1. The brain is harvested from a newborn rat (F344/N, Nihon SLC, Shizuoka, Japan) and rinsed with cold PBS. For every gram of brain, 9 mL of homogenization

buffer is immediately added and homogenized on ice using a potter's homogenizer at 800 rpm (8 strokes).

2. Nuclei, cellular debris, and connective tissues are removed by centrifugation at 900*g* for 10 min. For preparation of microsomal membranes, the postnuclear supernatant is subjected to ultracentrifugation at 100,000*g* for 1 h. The pellet fraction is aliquoted and stored at –80°C until used.

3.2. Preparation of GST-α-DG

1. BL21(DE3) *E. coli* cells are transformed with *pGEX-GST-α-DG*. Cultures are prepared by growing a single colony overnight in LB broth at 37°C. The overnight culture is then used to inoculate a fresh 50-mL culture, which is grown at 37°C to *A*620 = 0.5. At this point, 1 m*M* of IPTG is added to the culture in order to induce GST-α-DG expression. The induced cells are grown in parallel for an additional 4 h at 37°C, and harvested by centrifugation at 6000*g* for 15 min at 4°C.
2. The cell pellet is suspended in 10 mL of PBS (pH 7.4) and broken with a tip-type sonicator (*see* **Note 6**). The cell supernatant is recovered by ultracentrifugation at 100,000*g* for 1 h.
3. Recombinant GST-α-DG proteins are purified from the supernatant with a GSTrap column in a fast protein liquid chromatography system (Amersham) in the following manner: Pre-equilibrate the GSTrap column with 10 mL of PBS. Load the supernatant onto the column and wash with PBS at a flow rate of 0.2 mL/min. The absorbed recombinant GST-α-DG proteins are eluted with 10 mL of 10 m*M* reduced glutathione in PBS at a flow rate of 1 mL/min.
4. The purified GST-α-DG is dialyzed with 50 m*M* of $(NH_4)HCO_3$, pH 7.0.
5. Protein concentration is determined by bicinchonic acid (BCA) assay (Pierce, Rockford, IL), and the purity of GST-α-DG is checked by SDS-PAGE visualized with CBB (*see* **Subheading 2.5, item 3** and **Fig. 2A**).
6. The GST-α-DG aliquots are dispensed by 10 μg in microcentrifugal tubes, dried up with a centrifugal evaporator, and kept at –80°C.

3.3. POMT Assay

1. The POMT reaction buffer is added to the microsomal membrane fraction at a protein concentration of 4 mg/mL. The fraction is suspended by moderate pipetting and solubilized for 30 min on ice with occasional mild stirring.
2. 20 μL of the solubilized fraction and 2 μL of Dol-P-Man solution (from **Subheading 2.3., item 4**) are added to the dried GST-α-DG (**Subheading 3.2., step 6**), vortexed, and spun down gently. Immediately incubate the reaction mixture at 25°C for 1 h. The reaction is stopped by adding 200 μL of 1% Triton-PBS (*see* **Note 7**).
3. The reaction mixture is centrifuged at 10,000*g* for 10 min. The supernatant is transferred into a screw-cap tube with a packing seal (*see* **Note 8**). Mix 400 μL of 1% Triton-PBS and 40 μL of 25% slurry glutathione–sepharose beads with the supernatant and rotate with a rotary mixer at 4°C for 1 h.
4. After centrifugation at 1000*g* for 1 min, the supernatant is removed by aspiration and the beads are washed three times with 0.5% Triton-tris buffer. 2% SDS is

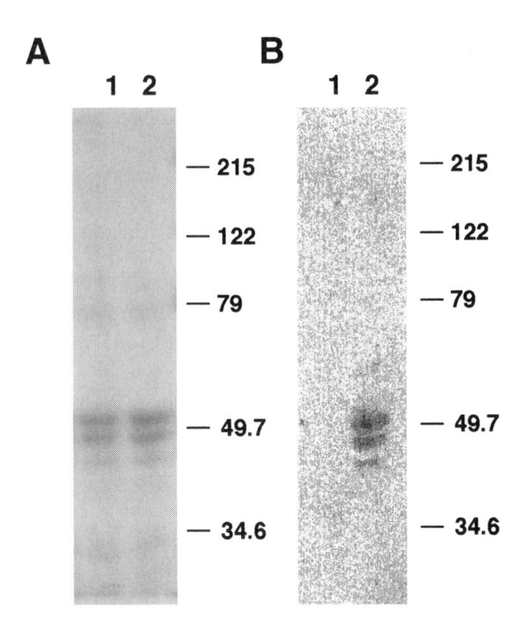

Fig. 2. Incorporation of [^{3}H]-mannose into glutathione-*S*-transferase-α-dystroglycan (GST-α-DG). Dol-P-[^{3}H]Man and GST-α-DG are incubated with HEK293T cell membrane fraction in POMT reaction buffer. After incubation, the products are recovered by the glutathione–sepharose 4B beads and subjected to sodium dodecyl sulfate-polyacrylamide gel electrophoresis (10% gel). GST-α-DG was detected as triplet bands at around 50 kDa by CBB staining **(A)**. Because all bands were stained with the anti-GST antibody (data not shown) the largest molecular-weight band was thought to be the full-length GST-α-DG, and the smaller bands were probably fragments of degraded GST-α-DG. The radioactivity of [^{3}H]-mannose was detected by autoradiography **(B)**, and the radioactivity was incorporated into the GST-α-DG in the presence of both the membrane fraction and an acceptor. Lanes 1, incubation with GST-α-DG but without membrane fraction; lanes 2, incubation with membrane fraction and GST-α-DG. Molecular-weight standards are shown on the right. (Reprinted with permission from **ref.** ***12***. Copyright 2004 by National Academy of Sciences.)

added to the beads and boiled at 100°C for 3 min. The suspension is cooled down to room temperature and mixed with liquid scintillation cocktail. The radioactivity adsorbed by the beads is measured using a liquid scintillation counter (*see* **Fig. 3A**).

5. The incorporation of radioactive mannose into GST-α-DG can be detected by SDS-PAGE and subsequent autoradiography as follows. Instead of 2% SDS in **step 4**, add 20 μL of 2X loading buffer to the beads followed by boiling at 100°C for 3 min. After centrifugation at 1000*g* for 1 min, the supernatant is subjected to SDS-PAGE. Gel is stained with CBB to visualize GST-α-DG, soaked in amplify

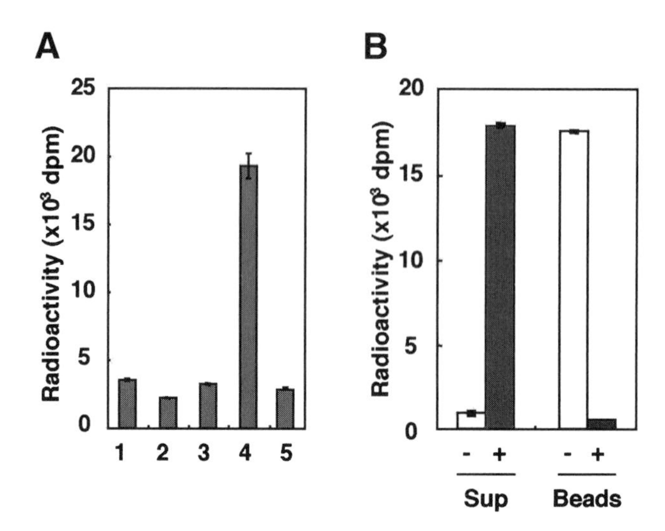

Fig. 3. **(A)** POMT activity of human POMT1 and POMT2 expressed in HEK293T cells. Bar 1, cells transfected with vector alone; bar 2, cells transfected with human *POMT1*; bar 3, cells transfected with human *POMT2*; bar 4, cells cotransfected with *POMT1* and *POMT2*; bar 5, a mixture of the membrane fractions from the *POMT1*-transfected cells and *POMT2*-transfected cells. **(B)** α-Mannosidase digestion of mannosyl-GST-α-DG. Glutathione–sepharose 4B beads bearing [^{3}H]-mannosyl-GST-α-DG were incubated with jack bean-α-mannosidase for 60 h. The radioactivities of the supernatant (Sup) and the beads (Beads) were measured by liquid scintillation counting. Closed bars, active α-mannosidase; open bars, inactive (heat-treated) α-mannosidase. The radioactivity released to the supernatant by active α-mannosidase is shown. (Reprinted with permission from **ref. *12***. Copyright 2004 by National Academy of Sciences.)

fluorographic reagent for 30 min to enhance detection efficiency of tritium, dried with a vacuum gel dryer, and exposed to X-ray film (*see* **Fig. 2B**).

6. The linkage of the mannosyl residue to peptide is determined as follows. Instead of 2% SDS in **step 4**, 50 μL of jack bean-α-mannosidase (0.8 U) is added to the beads and incubated at 37°C. Jack bean-α-mannosidase (0.8 U) is added fresh every 24 h and is incubated for up to 60 h. Inactivated jack bean-α-mannosidase, prepared by heating the enzyme for 5 min at 100°C, is used as a control. After incubation, the radioactivity of the supernatant and the beads is measured using a liquid scintillation counter (*see* **Fig. 3B**).

3.4. POMGnT1 Assay

1. 10 μL of 1 m*M* UDP-GlcNAc, 10 μL of UDP-[^{3}H]GlcNAc (100,000 dpm/nmol), and 10 μL of 2 m*M* mannosylpeptide (or 100 m*M* Benzyl-Man) are mixed in a microcentrifugal tube and dried with a centrifugal evaporator.

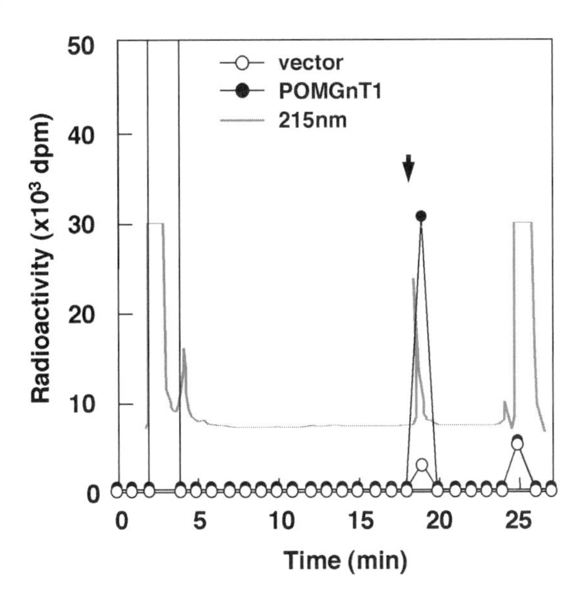

Fig. 4. POMGnT activity of human POMGnT1. UDP-[^{3}H]GlcNAc and mannosylpeptide were reacted with membrane fraction in POMGnT1 reaction buffer and then subjected to reversed-phase high-performance liquid chromatography. The mobile phase consists of (1) 100% A for 5 min, (2) a linear gradient to 75% A, 25% B for 20 min, (3) a linear gradient to 100% B for 1 min, and (4) 100% B for 5 min. The peptide separation was monitored at 214 nm and the radioactivity of each 1-mL fraction was measured by liquid scintillation counting. Arrow indicates the elution position of the mannosylpeptide. Vector (○), cells transfected with vector alone; POMGnT1 (●), cells transfected with human *POMGnT1*.

2. The POMGnT reaction buffer is added to the microsomal membrane fraction at a protein concentration of 2 mg/mL. The fraction is suspended with a bath-type sonicator on ice and solubilized by moderate pipetting until transparent. After centrifugation at 10,000*g* for 10 min, 20 μL of the supernatant is added to the dried substrate (prepared in **step 1**), vortexed gently, and incubated at 37°C for 2 h. The reaction is stopped by boiling at 100°C for 3 min. 180 μL of water is added to the reaction mixture and filtered with a centrifugal filter device.
3. The filtrate is analyzed by reversed-phase HPLC on the following condition: the gradient solvents are aqueous 0.1% TFA (solvent A) and acetonitrile containing 0.1% TFA (solvent B). The mobile phase consists of (1) 100% A for 5 min, (2) a linear gradient to 75% A, 25% B for 20 min, (3) a linear gradient to 100% B for 1 min, and (4) 100% B for 5 min. The peptide separation is monitored by measuring the absorbance at 214 nm, and the radioactivity of each 1-mL fraction is measured by liquid scintillation counting (*see* **Fig. 4**).
4. The reaction product is characterized by two different methods: (1) The product is dried up by an evaporator and then incubated with 50 mU of streptococcal

β-*N*-acetylhexosaminidase at 37°C for 48 h. After incubation, the enzyme is inactivated by boiling at 100°C for 3 min. The enzyme-digested sample is rechromatographed as described in **step 3**. (2) β-Elimination is performed as follows: the product is dissolved in 500 μL of 0.05 *N* NaOH and 1 *M* of $NaBH_4$, and incubated for 18 h at 45°C. After adjusting the pH to 5.0 by adding 4 *N* of acetic acid, the solution is applied to a column containing 1 mL of AG-50W-X8 (H^+ form) and the column is then washed with 10 mL of water. The effluent and the washing are combined and evaporated. After the remaining borate is removed by repeated evaporation with methanol, the residue is analyzed by high-pH anion-exchange chromatography with pulsed amperometric detection (*see* **ref. *14***).

4. Notes

1. Do not dry completely; approx 10 μL of solvent should remain.
2. Synthesis of mannosylpeptide substrate: Mannosylpeptide (Ac-Ala-Ala-Pro-Thr(Man)-Pro-Val-Ala-Ala-Pro-NH_2) is synthesized in a solid-phase manner using 9-fluorenyloxymethylcarbonyl (Fmoc) chemistry. Fmoc-Thr(Man)-OH is synthesized as follows: the reaction of phenyl 2,3,4,6-tetra-*O*-benzyl-1-thio-D-mannopyranoside and *N*-benzyloxycarbonyl-L-threonine benzyl ester (Z-Thr-OBzl) in the presence of *N*-iodosuccinimide and trifluoromethanesulfonic acid give the desired protected mannosyl threonine derivative (Z-Thr(Man(OBzl)$_4$)-OBzl) with a 77% yield. After deprotection of all benzyl groups and the Z group by catalytic hydrogenation, Fmoc-OSu is reacted with the residue to give the desired Fmoc-Thr(Man)-OH with a 75% yield. The product is easily purified by solid-phase extraction using a polymeric adsorbent, such as Dianion HP-20 (Nippon Rensui Co., Tokyo, Japan) or Amberlite XAD-2 (Organo, Tokyo, Japan).

   After the final deprotection from the glycopeptide resin, the crude mannosyl peptide is purified on a C18-preparative reversed-phase column (Inertsil ODS-3, 20X 250 mm, GL Sciences Inc., Tokyo, Japan) eluted by mixing solvent A (0.1% TFA in water) with solvent B (0.1% TFA in acetonitrile) at 45°C at a flow rate of 10 mL/min as follows: 25 min at 5% solvent B, linear gradient to 10 min at 35% solvent B. The glycopeptide separation is monitored continuously by measuring the absorbance at 214 nm. The structure of the product is identified by ^{1}H-NMR, amino acid analysis (6 *M* HCl, 110°C, 24 h), and matrix-assisted laser desorption ionization time-of-flight mass spectrometry. Distilled water is referred to as water in this text.
3. Cell pellets can be stored at –80°C after removal of PBS.
4. Typical sonication conditions to reach semitranslucent cell suspensions are: 10 cycles of 0.6-s pulse with 0.4-s intervals, and these procedures are repeated again.
5. The precipitate can be stored at –80°C after removal of supernatant.
6. Semitranslucent cell suspensions are obtained by 3-s sonication with 3-s intervals for 5–10 min.
7. POMT activity is inactivated in the presence of Triton X-100.

8. When using a screw-cap tube, a packing seal is required to prevent the leakage of radioactivity.

Acknowledgments

This study was supported by a Research Grant for Nervous and Mental Disorders (17A-10) from the Ministry of Health, Labour and Welfare of Japan, and a Grant-in-Aid for Scientific Research on Priority Area (14082209) from the Ministry of Education, Culture, Sports, Science and Technology of Japan.

References

1. Chiba, A., Matsumura, K., Yamada, H., et al. (1997) Structures of sialylated *O*-linked oligosaccharides of bovine peripheral nerve α-dystroglycan. The role of a novel *O*-mannosyl-type oligosaccharide in the binding of α-dystroglycan with laminin. *J. Biol. Chem.* **272,** 2156–2162.
2. Yuen, C. T., Chai, W., Loveless, R. W., Lawson, A. M., Margolis, R. U., and Feizi, T. (1997) Brain contains HNK-1 immunoreactive *O*-glycans of the sulfoglucuronyl lactosamine series that terminate in 2-linked or 2,6-linked hexose (mannose). *J. Biol. Chem.* **272,** 8924–8931.
3. Sasaki, T., Yamada, H., Matsumura, K., Shimizu, T., Kobata, A., and Endo, T. (1998) Detection of *O*-mannosyl glycans in rabbit skeletal muscle α-dystroglycan. *Biochim. Biophys. Acta* **1425,** 599–606.
4. Smalheiser, N. R., Haslam, S. M., Sutton-Smith, M., Morris, H. R., and Dell, A. (1998) Structural analysis of sequences *O*-linked to mannose reveals a novel Lewis *X* structure in cranin (dystroglycan) purified from sheep brain. *J. Biol. Chem.* **273,** 23,698–23,703.
5. Chai, W., Yuen, C. T., Kogelberg, H., et al. (1999) High prevalence of 2-mono- and 2,6-di-substituted manol-terminating sequences among O-glycans released from brain glycopeptides by reductive alkaline hydrolysis. *Eur. J. Biochem.* **263,** 879–888.
6. Endo, T. (1999) *O*-mannosyl glycans in mammals. *Biochim. Biophys. Acta* **1473,** 237–246.
7. Michele, D. E., Barresi, R., Kanagawa, M., et al. (2002) Post-translational disruption of dystroglycan-ligand interactions in congenital muscular dystrophies. *Nature* **418,** 417–422.
8. Michele, D. E. and Campbell, K. P. (2003) Dystrophin-glycoprotein complex: post-translational processing and dystroglycan function. *J. Biol. Chem.* **278,** 15,457–15,460.
9. Montanaro, F. and Carbonetto, S. (2003) Targeting dystroglycan in the brain. *Neuron* **37,** 193–196.
10. Endo, T. (2004) Structure, function and pathology of *O*-mannosyl glycans. *Glycoconj. J.* **21,** 3–7.
11. Yoshida, A., Kobayashi, K., Manya, H., et al. (2001) Muscular dystrophy and neuronal migration disorder caused by mutations in a glycosyltransferase, POMGnT1. *Dev. Cell* **1,** 717–724.

12. Manya, H., Chiba, A., Yoshida, A., et al. (2004) Demonstration of mammalian protein *O*-mannosyltransferase activity: coexpression of POMT1 and POMT2 required for enzymatic activity. *Proc. Natl. Acad. Sci. USA* **101,** 500–505.
13. Beltran-Valero De Bernabe, D., Currier, S., Steinbrecher, A., et al. (2002) Mutations in the *O*-mannosyltransferase gene *POMT1* give rise to the severe neuronal migration disorder Walker-Warburg syndrome. *Am. J. Hum. Genet.* **71,** 1033–1043.
14. Takahashi, S., Sasaki, T., Manya, H., et al. (2001) A new β-1,2-*N*-acetylglucosaminyltransferase that may play a role in the biosynthesis of mammalian *O*-mannosyl glycans. *Glycobiology* **11,** 37–45.
15. Strahl-Bolsinger, S., Gentzsch, M., and Tanner, W. (1999) Protein *O*-mannosylation. *Biochim. Biophys. Acta* **1426,** 297–307.
16. Jurado, L. A., Coloma, A., and Cruces, J. (1999) Identification of a human homolog of the *Drosophila* rotated abdomen gene *(POMT1)* encoding a putative protein *O*-mannosyl-transferase, and assignment to human chromosome 9q34.1. *Genomics* **58,** 171–180.
17. Willer, T., Amselgruber, W., Deutzmann, R., and Strahl, S. (2002) Characterization of POMT2, a novel member of the *PMT* protein *O*-mannosyltransferase family specifically localized to the acrosome of mammalian spermatids. *Glycobiology* **12,** 771–783.
18. Ibraghimov-Beskrovnaya, O., Ervasti, J. M., Leveille, C. J., Slaughter, C. A., Sernett, S. W., and Campbell, K. P. (1992) Primary structure of dystrophin-associated glycoproteins linking dystrophin to the extracellular matrix. *Nature* **355,** 696–702.
19. Laemmli, U. K. (1970) Cleavage of structural proteins during the assembly of the head of bacteriophage T4. *Nature* **227,** 680–685.
20. Zhang, W., Vajsar, J., Cao, P., et al. (2003) Enzymatic diagnostic test for Muscle-Eye-Brain type congenital muscular dystrophy using commercially available reagents. *Clin. Biochem.* **36,** 339–344.

5

Methods for Analysis of Unusual Forms of *O*-Glycosylation

Aleksandra Nita-Lazar and Robert S. Haltiwanger

Summary

The identification of the novel forms of O-linked glycosylation, *O*-fucose, and *O*-glucose requires the development of new methods for their analysis. Here we describe approaches to analyze these novel *O*-glycans. The major method involves metabolic radiolabeling of recombinant glycoproteins expressed in Lec1 Chinese hamster ovary (CHO) cells. The glycoproteins are purified from the media and the stoichiometry of modification is determined by comparing protein levels (by immunoblot) and incorporated radioactivity (by fluorography). The *O*-glycans are subsequently released by alkali-induced β-elimination, and released saccharides are analyzed using a combination of chromatography and exoglycosidase digestion. With these methods, we can determine both stoichiometry and the structure of the glycans on the expressed proteins. We have begun to utilize mass spectrometry in addition to metabolic radiolabeling methods to analyze these structures.

Key Words: *O*-fucose; *O*-glucose; epidermal growth factor repeats; thrombospondin type 1 repeats; oligosaccharide analysis.

1. Introduction

The novel forms of protein *O*-glycosylation, *O*-fucose, and *O*-glucose are notably different from classical *O*-glycans. Unlike mucin-type *O*-glycans, these have relatively short and unbranched structures. Both *O*-fucose and *O*-glucose can exist as either mono- or oligosaccharide (OS) species. *O*-fucose can be elongated to two types of OSs, suggesting that two separate *O*-fucose pathways exist *(1)*. One of the *O*-fucose OSs is the tetrasaccharide Sia-α2,3/6-Gal-β1,4-GlcNAc-β1,3-Fuc-α1-*O*-Ser/Thr. To date this structure has been found exclusively on proteins containing epidermal growth factor (EGF)-like

From: *Methods in Molecular Biology, vol. 347: Glycobiology Protocols*
Edited by: I. Brockhausen © Humana Press Inc., Totowa, NJ

repeats *(2,3)*. The second *O*-fucose OS is the disaccharide Glc-β1,3-Fuc-α1-*O*-Ser/Thr, which was found on proteins containing thrombospondin type 1 repeats (TSRs; *see* **refs. *4*, *5***, and ***5a***). *O*-glucose, detected so far only on proteins containing EGF repeats, can be elongated to the trisaccharide Xyl-α1,3-Xyl-α1,3-Glc-β1-*O*-Ser *(6)*.

The *O*-fucose and *O*-glucose modifications on EGF repeats have generated a great deal of excitement recently owing to the importance of *O*-fucose in Notch signaling *(7–9)*. EGF repeats are small protein domains characterized by six conserved cysteine residues forming three disulfide bonds that provide the distinct fold *(10)*. They are found in many cell-surface and secreted proteins, and many EGF repeats contain the consensus sites for *O*-fucosylation ($C^2X_{4-5}S/TC^3$; *see* **refs. *11*** and ***12***) and/or *O*-glucosylation (C^1XSXPC^2; *see* **ref. *13***). The extracellular domain of Notch1 contains 36 tandem EGF repeats *(14)* with multiple *O*-fucosylation and *O*-glucosylation consensus sites *(15)*. A number of studies have demonstrated that the enzyme responsible for the addition of *O*-fucose to EGF repeats, protein *O*-fucosyltransferase 1 (*O*-FucT-1; *see* **refs. *16*** and ***17***), is essential for proper Notch function in both *Drosophila melanogaster (18,19)* and mice *(20)*. *O*-FucT-1 knockout mice show an embryonic lethal phenotype characteristic of a Notch signaling defect *(20)*. Recent work has shown that *O*-FucT-1 localizes to the endoplasmic reticulum and may play a role in folding and quality control during Notch receptor biosynthesis *(21,22)*. Fringe, a known modulator of Notch function *(23–25)*, is an *O*-fucose-specific β1,3-*N*-acetylglucosaminyltransferase, catalyzing the committed step in *O*-fucose tetrasaccharide biosynthesis *(26,27)*. These and other studies have demonstrated the importance of *O*-fucose saccharides in the Notch signaling pathway (for reviews *see* **ref. *7***).

Like EGF repeats, TSRs are small cysteine-knot motifs containing six conserved cysteine residues forming three disulfide bonds (although in a different pattern than EGF repeats). TSRs are also characterized by conserved Trp, Ser, and Arg residues *(28)*. They are found in a number of cell-surface and secreted proteins and appear to play important roles in protein–protein interactions. They can be modified with the *O*-fucose disaccharide Glc-β1,3-Fuc on the sequence $C_1XX(S/T)C_2G$ (*see* **refs. *4*** and ***5a***). The *O*-fucosylation site in thrombospondin-1 is present at a putative CD-36 binding site, suggesting a possible role for the glycan in regulation of this interaction *(29–31)*. Recent studies in our laboratory have demonstrated that a novel enzyme, *O*-FucT-2, is responsible for the addition of *O*-fucose to TSRs *(31a)*.

We have developed specific methods to analyze both *O*-fucose and *O*-glucose glycans. These methods have been adapted from more traditional methods of *O*-glycan analysis *(32–35)*, although several modifications have been used to enhance radiolabeling of the desired saccharides. For instance, we use Chinese hamster ovary (CHO) cells with mutations in more common glycosylation pathways developed by Dr. Pamela Stanley (Albert Einstein College of Medicine) to minimize incorporation of radiolabel into unwanted structures. For metabolic radiolabelings of *O*-fucose saccharides, we take advantage of Lec1 cells ***(36,37)***. These cells lack GlcNAc transferase 1 and cannot make complex or hybrid-type *N*-glycans. The majority of [^{3}H]-fucose is incorporated into *N*-glycan structures in CHO cells ***(38)***; the use of Lec1 cells eliminates this "background," allowing simple analysis of minor structures such as *O*-fucose saccharides ***(1)***. To enhance radiolabeling of *O*-glucose saccharides we use [^{3}H]-galactose as the precursor. Galactose is rapidly converted to uridine 5′-diphosphate (UDP)-galactose upon entering the cell, which is subsequently epimerized to UDP-glucose ***(39)***. Thus the UDP-glucose pool in the cell becomes radiolabeled without incorporating label into all of the products of glucose metabolism. Lec1 cells can be used for the same reasons described above for the *O*-fucose radiolabelings, although [^{3}H]-galactose can also be incorporated into mucin-type *O*-glycans in Lec1 cells ***(40)***. Alternatively we have used Lec8 cells, which have a defect in the Golgi UDP-galactose transporter ***(41)***. Because the UDP-galactose is inefficiently transported into the Golgi in Lec8 cells, the incorporation of [^{3}H]-galactose is dramatically reduced. We have successfully used both Lec8 and Lec1 cells to study *O*-glucose modifications on EGF repeats (*see* **ref. *15*** and Nita-Lazar and Haltiwanger, in preparation). Radiolabeling experiments can be performed in other cell lines (e.g., HeLa, Cos-1, 293T), although the high level of radiolabel incorporated into *N*-glycans can obscure the signal from these more minor *O*-glycans. This background can be significantly reduced by removal of *N*-glycans with peptide *N*-glycosidase F. Nonetheless, these extra steps reduce yields and complicate the analysis ***(42)***.

Either endogenous or recombinant glycoproteins produced in CHO cells can be metabolically labeled with radioactive monosaccharides ***(1,15)***. We typically transfect (either stably or transiently) plasmids encoding proteins of interest into the appropriate CHO cell line to produce sufficient material for analysis. The constructs contain amino terminal signal sequences to ensure secretion and C-terminal tags (e.g., MYC and His6) for detection and purification purposes ***(26,43)***. The protein is then purified from cell lysates or the media using appropriate antibodies (e.g., against the protein itself or against a tag on a recombinant protein) or Nickel (Ni)-Sepharose if the protein has an His6-tag. At this point, radiochemical purity needs to be established by sodium dodecyl sulfate-polyacrylamide gel electrophoresis (SDS-PAGE) or fluorography. A simple β-elimination followed by acid hydrolysis can be used to compare the extent of the radioactivity directly bound to the protein (yielding an alditol) vs that linked to other sugars (yielding an aldose; *see* **refs. *1*** and ***15***). If some of the radiolabel is not in the form of the alditol, the percentage of alditol can be calculated and used to determine what percentage of total radioactivity was directly linked to the protein (as either *O*-fucose or *O*-glucose). The extent of glycosylation can then be determined by comparing protein levels (using immunoblots) to the level of radioactivity (using fluorography, normalized with percent alditol if necessary ***[43]***). As this is only a relative extent of glycosylation, it can only be used to compare the extent of glycosylation within a specific experiment (e.g., comparison of wild-type and glycosylation site mutants ***[43]***).

The structures of the radioactive *O*-glycans can then be examined after release from the protein by alkali-induced β-elimination. The released OSs are analyzed by size-exclusion chromatography (Superdex) in combination with glycosidase digestions to determine the size and sequence of the OS chain ***(15)***. Linkages between sugars can be confirmed using high-performance anion-exchange chromatography (HPAEC) analysis (Dionex) in combination with standard compounds. We have used this approach successfully to determine the structure of *O*-fucose and *O*-glucose glycans in EGF repeats ***(15)*** and *O*-fucose in TSRs ***(31a)***. A flowchart describing these steps is shown in **Fig. 1**. Recently we have begun to utilize mass spectrometry (MS) to analyze the structures of *O*-fucose and *O*-glucose glycans. Although MS methodologies are extremely sensitive and require no radioactivity, they cannot be used on saccharides of unknown composition or structure. Many excellent reviews describing the use of MS to analyze OS structure have been published in this series and in this book (*see* **refs. *44*** and ***45*** and Chapters 6 and 8).

2. Materials

2.1. Reagents and Tools

1. Radioactive (^{3}H) sugars (L-[6-^{3}H]fucose and D-[6-^{3}H]galactose; American Radiolabeled Chemicals, St. Louis, MO).
2. Lec1 and Lec8 mutant CHO cells (American Type Culture Collection, Manassas, VA).
3. Alpha-minimal essential medium (α-MEM) cell culture media (Invitrogen, Carlsbad, CA) with 10% bovine calf serum (BCS) and 10 m*M* of penicillin/streptomycin.
4. OptiMem I cell culture media (Invitrogen, Carlsbad, CA).
5. Dowex-50, H$^+$-form (Bio-Rad, Hercules, CA).
6. Acetone (cold, kept at –20°C).
7. 1-mL and 10-mL disposable syringes and 18-gage needles.
8. Disposable columns (Bio-Rad).
9. Sep-Pak C18 (Waters, Milford, MA).
10. 100% Methanol.
11. MilliQ water (or water of similar purity).
12. Superdex™peptide gel filtration column (Pharmacia Biotech/GE Healthcare, Piscataway, NJ) connected to the high performance liquid chromatography (HPLC) capable of 0.5 mL/min flow, with refractive index detector for standardization.

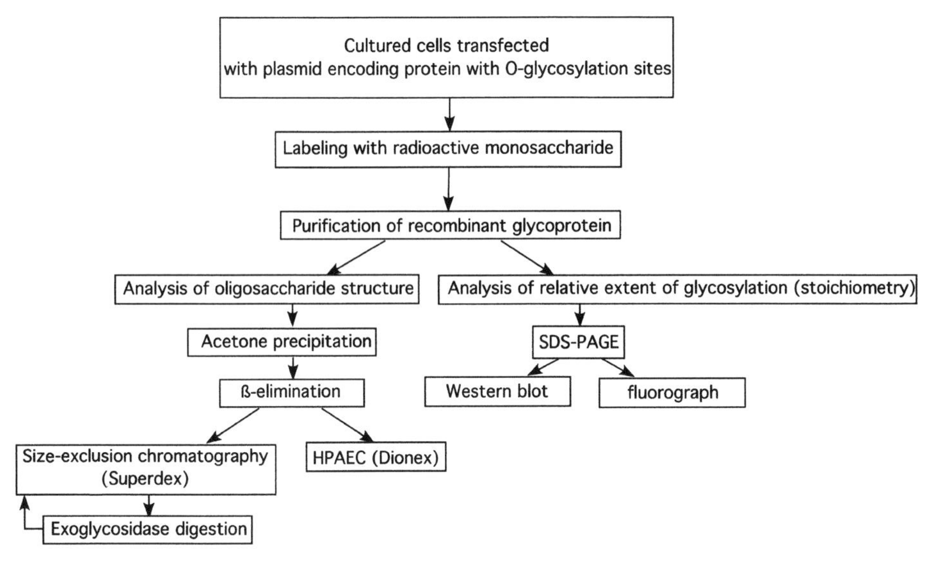

Fig. 1. Flow chart of the methods used in structural analysis of O-linked glycans.

13. Dionex DX300 HPLC system equipped with pulsed amperometric detection (PAD-2 cell) and CarboPac MA-1 column (Dionex Corp., Sunnyvale, CA).
14. Nickel-nitrilotriacetic acid (Ni-NTA) Sepharose beads (Qiagen, Valencia, CA).
15. Nitrocellulose membrane (Bio-Rad).
16. Appropriate antibodies for Western blot protein detection.
17. En^3Hance autoradiography enhancer (Perkin Elmer, Boston, MA). **Caution:** Corrosive!
18. Exoglycosidases: *N*-acetylneuraminidase I (α2,3-specific sialidase) from Glyko, Inc. (San Leandro, CA); β-galactosidase *(Diplococcus pneumoniae)* from Roche Molecular Biochemicals (Indianapolis, IN); β-hexosaminidase (jack bean) from Sigma.
19. Standards: dextran D-4133 (Sigma-Aldrich, St. Louis, MO) hydrolyzed according to **ref. *46***, monosaccharides (Sigma), alditol sugar standards prepared by reduction of the corresponding sugar as in **ref. *47***, and disaccharide standards synthesized as described in **ref. *15***.

2.2. Stock Solutions

1. Radioimmunoprecipitation assay (RIPA) buffer: 50 m*M* of Tris-HCl (pH 8.0), 150 m*M* of NaCl, 1% NP40, 0.5% deoxycholate, and 0.1% sodium dodecyl sulfate (SDS).
2. 0.1 *M* of ethylenediaminetetraacetic acid (EDTA), pH 8.0.
3, 2 *M* of $NaBH_4$ and 100 m*M* of NaOH (β-elimination solution) made fresh.
4. 4 *M* of acetic acid.
5. 50% Slurry of Dowex-50 in 20% methanol.
6. 2 *M* of trifluoroacetic acid (TFA).
7. 1 *M* of NaCl.

3. Methods (*see* Note 1)

3.1. Metabolic Labeling of Cells in Culture

1. Seed one plate (100-mm diameter) with 10^6 of cells stably transfected with the glycoprotein of interest, or transfect the cells transiently before radiolabeling.
2. Grow the cells to approx 50% confluency.
3. Change the medium to 5 mL of α-MEM (*see* **Note 2**) containing 20 μCi/mL of [^{3}H]fucose or [^{3}H]galactose.
4. Incubate the cells for 48 h, collect the media, and purify the radioactively labeled recombinant glycoprotein (*see* **Note 3**).

3.2. Protein Purification (see Note 4)

1. Centrifuge the harvested medium to pellet the cell debris.
2. Add 100 μL of Ni-NTA beads (50% slurry, washed thoroughly with RIPA) per 5 mL of media.
3. Incubate by rotating for 1 h at 4°C.
4. Centrifuge at 5000*g* and discard the supernatant.
5. Wash three times with 1 mL of RIPA buffer. Remove the supernatant with an 18-gage needle using a 1-mL syringe each time.
6. Elute with 250 μL of 100 m*M* EDTA.
7. Centrifuge and collect the supernatant.
8. Count 5 μL in the liquid scintillation counter for quantification purposes.
9. Analyze a portion of the sample by SDS-PAGE-fluorography to check the radiochemical purity:
   a. Load an equal amount of radioactivity (at least 2000 cpm) in each sample onto the 10% polyacrylamide gel.
   b. Perform fluorography using En^{3}Hance (*see* **Note 5**) according to the manufacturer's instructions.
10. Perform analysis of the portion of radiolabel directly linked to protein as *O*-glycan vs other glycans (*see* **Note 6**).

3.3. Analysis of Relative Extent of Glycosylation (Stoichiometry) by Western Blot and Fluorography

1. Load an equal amount of radioactivity (at least 2000 cpm) onto two SDS-polyacrylamide gels of appropriate pore size for the protein being analyzed.
2. With the first gel, perform Western blot and probe with appropriate antibodies.
3. With the second gel, perform fluorography using En^{3}Hance (*see* **Note 5**) according to the manufacturer's instructions.
4. Scan both films (making sure film is in the linear range) and determine a ratio of radioactivity (from fluorography) to protein (from Western blot). Normalize the scan from fluorography using the percentage of alditol (*see* **Note 6**) as 100%.

3.4. Release of O-Glycans by Alkali-Induced β-Elimination

1. Precipitate purified protein (~10,000–25,000 cpm) with acetone using 4 vol of chilled acetone for 1 vol of the protein solution and incubating at least 2 h at –20°C.

2. Pellet the precipitate by centrifugation at maximum microcentrifuge speed at 4°C, remove the supernatant with the 18-gage needle, and air-dry the pellet.
3. Add 500 μL of β-elimination solution to the acetone-precipitated protein pellet (*see* **Note 7**).
4. Incubate at 55–65°C for 18–24 h.
5. Cool and neutralize by slowly adding 4 *M* of acetic acid (dropwise) on ice until the pH is below 6.0 (*see* **Note 8**).
6. Desalt by passing the sample through 3 mL of Dowex-50 (H+) resin (50% slurry) in a disposable Bio-Rad column with an 18-gage needle attached (Dowex-50 should be washed with 15 mL of water before loading the sample). Collect the flow-through in a 50-mL conical tube.
7. Wash with 10 mL of water. Add to the flow-through.
8. Pass the sample through the Sep-Pak C18 column (Sep-Pak must be preconditioned with 10 mL of 100% methanol and then with 10 mL of water) using a 10-mL syringe. Collect the flow-through containing OSs in a 50-mL conical tube.
9. Freeze and lyophilize the flow-through.
10. Remove residual borate by performing methanol dry-down. Add sufficient 100% methanol to the sample to resuspend, mix well, and evaporate in the vacuum centrifuge on the medium heat setting (about 60°C). Repeat twice with 0.5 mL of methanol or until white powder no longer diminishes.
11. Dissolve the sample in 100 μL of water and determine the radioactivity recovered (CPM concentration) using 5 μL of the sample.

3.5. Analysis of OS Structure

3.5.1. Size-Exclusion Chromatography (see **Note 9**)

This is performed as described in **ref. *1***.

1. HPLC-Superdex run conditions: run the sample (about 10,000 cpm) through the Superdex column and flow 0.5 mL/min for 45 min (*see* **Note 10**).
2. Collect 90 fractions (0.5 min/fraction; each fraction is 250 μL).
3. Analyze the radioactivity in aliquots (size depends on total amount of radioactivity) using a liquid scintillation counter.

3.5.2. Exoglycosidase Digestions

1. Pool radioactive chromatographic peaks from size fractionation on Superdex column.
2. Dry the pooled fraction in the vacuum centrifuge.
3. Resuspend in 50 μL of sodium acetate buffer, pH 5.5.
4. Add appropriate exoglycosidases: e.g., *N*-acetyl neuraminidase I (α2,3 specific sialidase; 10 mU), β1,4-galactosidase (5 mU), or β-hexosaminidase (250 mU). Include mock-digested controls (without enzyme).
5. Incubate for 18 h at 37°C.

6. Terminate the reaction by diluting the samples with 50 μL of water and heating for 5 min at 100°C.
7. Subject the samples to chromatography on the Superdex column (*see* **Subheading 3.5.1.**).

3.5.3. Charge Analysis of OSs

This is performed as described in **ref.** ***38***.

1. Dry the OS fraction digested or mock digested with sialidase (*see* **Subheading 3.5.2.**) in the vacuum centrifuge.
2. Resuspend the sample in 500 μL of 2 m*M* Tris-base.
3. Pass through a 330-μL QAE-Sephadex column equilibrated with 2 m*M* of Tris-base.
4. Collect the flow-through (unbound fraction, 0 m*M* of NaCl representing neutral species).
5. Elute with increasing amounts of NaCl, releasing species with negative charges: 20 m*M* of NaCl, 1 negative charge; 70 m*M* of NaCl, 2 negative charges; 100 m*M* of NaCl, 3 negative charges; 140 m*M* of NaCl, 4 negative charges; and 1 *M* of NaCl, 5 or more negative charges. Collect 0.5-mL fractions.
6. Monitor the fractions by scintillation counting.

3.5.4. Acid Hydrolysis

1. Resuspend lyophilized OSs in 2 *M* of TFA.
2. Heat at 100°C for 2 h.
3. Dry in a Speed Vac evaporator.
4. Resuspend in water for HPAEC.

3.5.5. High-Performance Anion-Exchange Chromatography

1. Mix radioactive samples after acid hydrolysis with appropriate standards: 1 nmol for each fucitol, fucose, and glucose for [^{3}H]fucose-labeled sample, or 1 nmol for each galactosaminitol, glucose, glucitol, galactose, and galactitol for [^{3}H]galactose-labeled sample.
2. Subject samples to chromatography on CarboPac MA-1 column at 0.4 mL/min using the following gradients: 0–11 min, 0.1 *M* of NaOH; 11–21 min, 0.1–0.7 *M* of NaOH; 21–40 min, 0.7 *M* of NaOH.
3. Follow the standards by PAD-2 cell.
4. Collect 0.5-min (0.2-mL) fractions.
5. Monitor the fractions for radioactivity by scintillation counting.

4. Notes

1. **Caution:** Remember to work in a space assigned for radioactivity, wear appropriate protection (gloves, lab coat), and properly dispose of radioactive waste.
2. Opti-MEM (Invitrogen) serum-free can be used to avoid contamination with serum proteins, but it can reduce the incorporation of radioactivity.
3. Cell-associated proteins can be purified from the cell lysates using the same method ***(15)***.

4. We use cells expressing His6-tagged recombinant mouse Notch1 fragments (EGF 1–5, 6–10, 11–15, 16–18, 19–23, 24–28, and 29–36) and purify them using Ni-NTA affinity chromatography.
5. **Caution:** This reagent is very corrosive. Wear gloves and execute caution.
6. To check whether the radiolabel (fucose or glucose) is directly attached to the protein in O-linkage (e.g., *O*-fucose or *O*-glucose) or attached through other sugars, perform acid hydrolysis (*see* **Subheading 3.4.**) on a portion of OS immediately after alkali-induced β-elimination (*see* **Subheading 3.4.**), omitting the desalting and methanol blow-down steps; perform HPAEC (*see* **Subheading 3.4.**) in the presence of fucitol or fucose (for glycans labeled with [^{3}H]fucose, radioactivity migrating together with fucitol indicates *O*-fucose) and glucitol, glucose, galactitol, and galactose (for glycans labeled with [^{3}H]galactose, radioactivity migrating together with glucitol indicates *O*-glucose). *See* **ref. *1*** for more details.
7. Physically break up the protein pellet to solubilize.
8. Neutralization is usually complete after approx 10 drops. Add one drop every 5 min and keep the tube lid open to release hydrogen. Vortex before measuring pH with broad-range pH paper.
9. Size markers are generated by mild acid hydrolysis of dextran, according to Kobata ***(46)***.
10. To be certain that the Superdex column does not become contaminated, inject 100 μL of 1 *M* NaCl, collect the fractions, and monitor radioactivity by scintillation counting. If radioactivity is released by this procedure, the column must be washed with 1 *M* of NaOH for 4–5 h and then re-equilibrated with water.

Acknowledgments

The authors would like to thank members of the Haltiwanger laboratory for critical reading of this manuscript. Original research was supported by NIH grant GM 61126.

References

1. Moloney, D. J., Lin, A. I., and Haltiwanger, R. S. (1997) The O-linked fucose glycosylation pathway:Evidence for protein specific elongation of O-linked fucose in Chinese hamster ovary cells. *J. Biol. Chem.* **272,** 19,046–19,050.
2. Harris, R. J., Van Halbeek, H., Glushka, J., et al. (1993) Identification and structural analysis of the tetrasaccharide NeuAcα(2→6)Galβ(1→4)GlcNAcβ (1→3)Fucα1→*O*-linked to serine 61 of human factor IX. *Biochemistry* **32,** 6539–6547.
3. Nishimura, H., Takao, T., Hase, S., Shimonishi, Y., and Iwanaga, S. (1992) Human factor IX has a tetrasaccharide *O*-glycosidically linked to serine 61 through the fucose residue. *J. Biol. Chem.* **267,** 17,520–17,525.
4. Hofsteenge, J., Huwiler, K. G., Macek, B., et al. (2001) C-mannosylation and O-fucosylation of the thrombospondin type 1 module. *J. Biol. Chem.* **276,** 6485–6498.

5. Gonzalez de Peredo, A., Klein, D., Macek, B., Hess, D., Peter-Katalinic, J., and Hofsteenge, J. (2002) C-mannosylation and O-fucosylation of thrombospondin type 1 repeats. *Mol. Cell. Proteomics* **1,** 11–18.

5a. Luo, Y., Koles, K., Vorndam, W., Haltiwanger, R. S., and Panin, V. M. (2006) Protein O-fucosyltransferase 2 adds O-fucose to thrombospondin type 1 repeats. *J. Biol. Chem.* **281,** 9393–9399.

6. Nishimura, H., Kawabata, S., Kisiel, W., et al. (1989) Identification of a disaccharide (Xyl-Glc) and a trisaccharide (Xyl_2-Glc) *O*-glycosidically linked to a serine residue in the first epidermal growth factor-like domain of human factors VII and IX and protein Z and bovine protein Z. *J. Biol. Chem.* **264,** 20,320–20,325.

7. Haltiwanger, R. S. (2002) Regulation of signal transduction pathways in development by glycosylation. *Curr. Opin. Struct. Biol.* **12,** 593–598.

8. Haltiwanger, R. S. and Lowe, J. B. (2004) Role of glycosylation in development. *Annu. Rev. Biochem.* **73,** 491–537.

9. Haines, N. and Irvine, K. D. (2003) Glycosylation regulates notch signaling. *Nat. Rev. Mol. Cell. Biol.* **4,** 786–797.

10. Campbell, I. D. and Bork, P. (1993) Epidermal growth factor-like modules. *Curr. Opin. Struct. Biol.* **3,** 385–392.

11. Panin, V. M., Shao, L., Lei, L., Moloney, D. J., Irvine, K. D., and Haltiwanger, R. S. (2002) Notch ligands are substrates for EGF protein O-fucosyltransferase and fringe. *J. Biol. Chem.* **277,** 29,945–29,952.

12. Shao, L. and Haltiwanger, R. S. (2003) O-fucose modifications of epidermal growth factor-like repeats and thrombospondin type 1 repeats: unusual modifications in unusual places. *Cell Mol. Life Sci.* **60,** 241–250.

13. Harris, R. J. and Spellman, M. W. (1993) O-linked fucose and other post-translational modifications unique to EGF modules. *Glycobiology* **3,** 219–224.

14. Artavanis-Tsakonas, S., Rand, M. D., and Lake, R. J. (1999) Notch signaling: Cell fate control and signal integration in development. *Science* **284,** 770–776.

15. Moloney, D. J., Shair, L., Lu, F. M., et al. (2000) Mammalian Notch1 is modified with two unusual forms of O-linked glycosylation found on Epidermal Growth Factor-like modules. *J. Biol. Chem.* **275,** 9604–9611.

16. Wang, Y. and Spellman, M. W. (1998) Purification and Characterization of a GDP-fucose: Polypeptide fucosyltransferase from Chinese hamster ovary cells. *J. Biol. Chem.* **273,** 8112–8118.

17. Wang, Y., Shao, L., Shi, S., Harris, R. J., et al. (2001) Modification of epidermal growth factor-like repeats with O-Fucose: Molecular cloning of a novel GDP-fucose: Protein O-Fucosyltransferase. *J. Biol. Chem.* **276,** 40,338–40,345.

18. Okajima, T. and Irvine, K. D. (2002) Regulation of notch signaling by O-linked fucose. *Cell* **111,** 893–904.

19. Sasamura, T., Sasaki, N., Miyashita, F., Nakao, S., Ishikawa, H. O., et al. (2003) Neurotic, a novel maternal neurogenic gene, encodes an O-fucosyltransferase that is essential for Notch-Delta interactions. *Development* **130,** 4785–4795.

20. Shi, S. and Stanley, P. (2003) Protein O-fucosyltransferase I is an essential component of Notch signaling pathways. *Proc. Natl. Acad. Sci. USA* **100,** 5234–5239.

21. Okajima, T., Xu, A., Lei, L., and Irvine, K. D. (2005) Chaperone activity of protein O-fucosyltransferase 1 promotes notch receptor folding. *Science* **307,** 1599–1603.
22. Luo, Y. and Haltiwanger, R. S. (2005) O-fucosylation of notch occurs in the endoplasmic reticulum. *J. Biol. Chem.* **280,** 11,289–11,294.
23. Panin, V. M., Papayannopoulos, V., Wilson, R., and Irvine, K. D. (1997) Fringe modulates notch ligand interactions. *Nature* **387,** 908–912.
24. Fleming, R. J., Gu, Y., and Hukriede, N. A. (1997) *Serrate*-mediated activation of *Notch* is specifically blocked by the product of the gene *fringe* in the dorsal compartment of the *Drosophila* wing imaginal disc. *Development* **124,** 2973–2981.
25. Klein, T. and Arias, A. M. (1998) Interactions among Delta, Serrate and Fringe modulate Notch activity during Drosophila wing development. *Development* **125,** 2951–2962.
26. Moloney, D. J., Panin, V. M., Johnston, S. H., et al. (2000) Fringe is a Glycosyltransferase that modifies Notch. *Nature* **406,** 369–375.
27. Bruckner, K., Perez, L., Clausen, H., and Cohen, S. (2000) Glycosyltransferase activity of Fringe modulates Notch-Delta interactions. *Nature* **406,** 411–415.
28. Adams, J. C. and Tucker, R. P. (2000) The thrombospondin type 1 repeat (TSR) superfamily: diverse proteins with related roles in neuronal development. *Develop. Dynamics* **218,** 280–299.
29. Asch, A. S., Silbiger, S., Heimer, E., and Nachman, R. L. (1992) Thrombospondin sequence motif (CSVTCG) is responsible for CD36 binding. *Biochem. Biophys. Res. Commun.* **182,** 1208–1217.
30. Dawson, D. W., Pearce, S. F., Zhong, R., Silverstein, R. L., Frazier, W. A., and Bouck, N. P. (1997) CD36 mediates the In vitro inhibitory effects of thrombospondin-1 on endothelial cells. *J. Cell Biol.* **138,** 707–717.
31. Jimenez, B., Volpert, O. V., Crawford, S. E., Febbraio, M., Silverstein, R. L., and Bouck, N. (2000) Signals leading to apoptosis-dependent inhibition of neovascularization by thrombospondin-1. *Nat. Med.* **6,** 41–48.
31a. Luo, Y., Nita-Lazar, A., and Haltiwanger, R. S. (2006) Two distinct pathways for O-fucosylation of epidermal growth factor-like or thrombospondin type 1 repeats. *J. Biol. Chem.* **281,** 9385–9392.
32. Hounsell, E. F. (1993) A general strategy for glycoprotein oligosaccharide analysis. *Methods Mol. Biol.* **14,** 1–15.
33. Varki, A. (1994) Metabolic radiolabeling of glycoconjugates. *Methods Enzymol.* **230,** 16–32.
34. Davies, M. J. and Hounsell, E. F. (1998) HPLC and HPAEC of oligosaccharides and glycopeptides. *Methods Mol. Biol.* **76,** 79–100.
35. Campbell, B. J. and Rhodes, J. M. (1998) Purification of gastrointestinal mucins and analysis of their O-linked oligosaccharides. *Methods Mol. Biol.* **76,** 161–182.
36. Stanley, P. and Siminovitch, L. (1977) Complementation between mutants of CHO cells resistant to a variety of plant lectins. *Somatic Cell Genetics* **3,** 391–405.
37. Stanley, P. (1992) Glycosylation engineering. *Glycobiology* **2,** 99–107.

38. Lin, A. I., Philipsberg, G. A., and Haltiwanger, R. S. (1994) Core fucosylation of high-mannose-type oligosaccharides in GlcNAc transferase I-deficient (Lec1) CHO cells. *Glycobiology* **4,** 895–901.
39. Frey, P. A. (1996) The Leloir pathway: a mechanistic imperative for three enzymes to change the stereochemical configuration of a single carbon in galactose. *FASEB J.* **10,** 461–470.
40. Misra, A. K., Ujita, M., Fukuda, M., and Hindsgaul, O. (2001) Synthesis and enzymatic evaluation of mucin type core 4 O-glycan. *Carbohydr. Lett.* **4,** 71–76.
41. Stanley, P. (1987) Biochemical characterization of animal cell glycosylation mutants. *Methods Enzymol.* **138,** 443–458.
42. Yan, Y. T., Liu, J. J., Luo, Y. E. C., Haltiwanger, R. S., Abate-Shen, C., and Shen, M. M. (2002) Bi-functional activity of Cripto as a ligand and co-receptor in the Nodal signaling pathway. *Mol. Cell. Biol.* **22,** 4439–4449.
43. Shao, L., Moloney, D. J., and Haltiwanger, R. S. (2003) Fringe modifies O-Fucose on mouse Notch1 at epidermal growth factor-like repeats within the ligand-binding site and the abruptex region. *J. Biol. Chem.* **278,** 7775–7782.
44. Nilsson, B. (1993) Sequence and linkage analysis of N- and O-linked glycans by fast atom bombardment mass spectrometry. *Methods Mol. Biol.* **14,** 35–46.
45. Hansson, G. C. and Karlsson, H. (1993) Gas chromatography and gas chromatography-mass spectrometry of glycoprotein oligosaccharides. *Methods Mol. Biol.* **14,** 47–54.
46. Kobata, A. (1994) Size fractionation of oligosaccharides. *Methods Enzymol.* **230,** 200–208.
47. Haltiwanger, R. S., Holt, G. D., and Hart, G. W. (1990) Enzymatic addition of *O*-GlcNAc to nuclear and cytoplasmic proteins. Identification of a uridine diphospho-*N*-acetylglucosamine:peptide β-*N*-acetylglucosaminyltransferase. *J. Biol. Chem.* **265,** 2563–2568.

6

Structural Analysis of Naturally Occurring Sialic Acids

Johannis P. Kamerling and Gerrit J. Gerwig

Summary

Over the years several methodologies have been developed for the structural analysis of naturally occurring sialic acids (Sias), a family with more than 62 members. Currently there are two primary instrumental approaches: analysis of volatile Sia derivatives by gas-liquid chromatography (GLC) combined with electron-impact mass spectrometry (EI/MS), and analysis of fluorescently labeled Sias by high-performance liquid chromatography (HPLC) eventually coupled with electrospray mass spectrometry (ESI/MS). This chapter presents both approaches in detail. The volatile Sia derivatives are comprised of trimethylsilylated methyl ester derivatives, heptafluorobutylated methyl ester derivatives, or pertrimethylsilylated derivatives. The fluorescent Sia derivatives are prepared by reaction with 1,2-diamino-4,5-methylenedioxybenzene. For the identification of the different Sia derivatives, detailed GLC, HPLC, EI/MS, and ESI/MS data are included.

Key Words: Sialic acid; neuraminic acid; gas–liquid chromatography; high-performance liquid chromatography; mass spectrometry.

1. Introduction of Sialic Acids

Sialic acids (Sias) occur in nature in a great chemical diversity. These biologically important monosaccharides may be present in glycoprotein, glycolipid, and glycosylphosphatidylinositol membrane anchor glycans, and in oligosaccharides and homo- and heteropolysaccharides. They usually occur as terminal units, but examples where Sias have internal positions are well known (*see also* Chapters 7 and 11). Although the free forms have mainly the β-anomeric ring structure (>93%), glycoconjugate-bound Sias occur specifically in the α-anomeric form. In the literature, several recent reviews are available which focus on the biological sources of Sias and their significance in a variety of biological processes ***(1–6)***.

From: *Methods in Molecular Biology, vol. 347: Glycobiology Protocols*
Edited by: I. Brockhausen © Humana Press Inc., Totowa, NJ

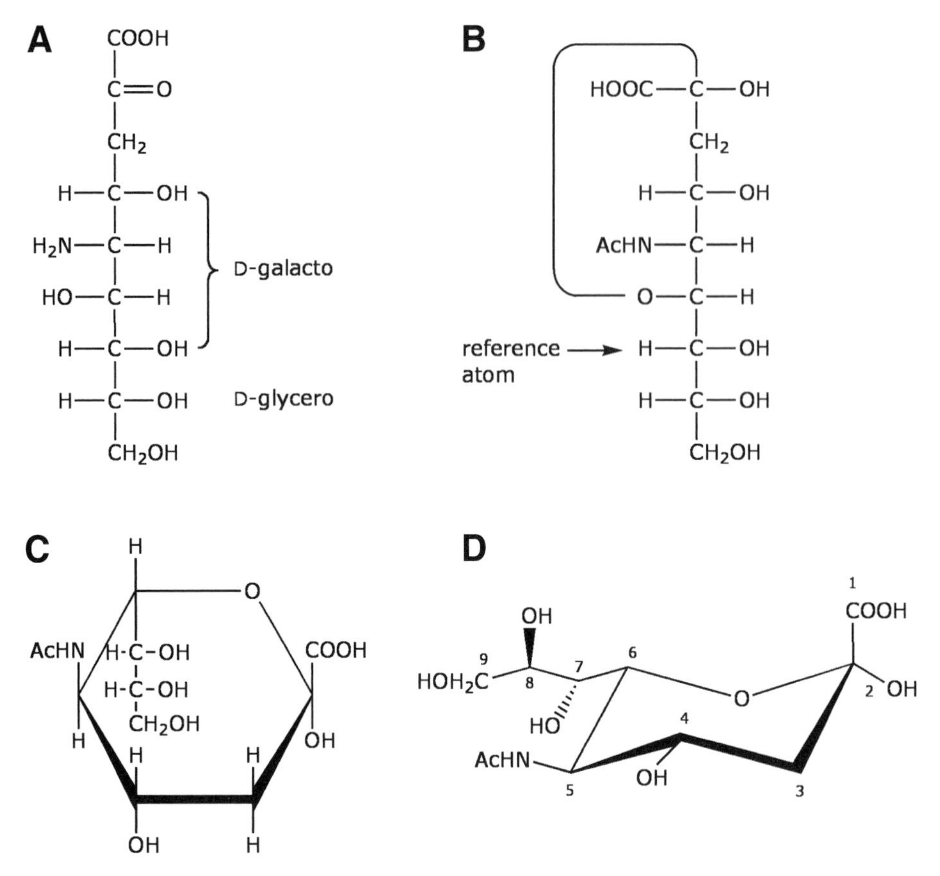

Fig. 1. Chemical structures for simple sialic acids in different views. **(A)** 5-Amino-3,5-dideoxy-D-*glycero*-D-*galacto*-non-2-ulosonic acid (Neu, open chain, Fischer projection formula). **(B)** 5-acetamido-3,5-dideoxy-D-*glycero*-α-D-*galacto*-non-2-ulopyranosonic acid (α-Neu5Ac, Fischer projection formula, note that C-7 is the anomeric reference atom). **(C)** α-Neu5Ac (Haworth formula). **(D)** α-Neu5Ac ($^2C_5$ chair conformation).

In **Table 1**, 62 naturally occurring members of the Sia family currently identified are listed, together with their abbreviations *(3,5,7,8)*. The mother-molecule neuraminic acid, which does not occur in free form in nature owing to its immediate cyclization to form an internal Schiff base, is systematically named 5-amino-3,5-dideoxy-D-*glycero*-D-*galacto*-non-2-ulosonic acid. It is abbreviated as Neu, where the D-notation is implied in the trivial name (*see* **Fig. 1**). Chemically, a Sia is a 2-keto-carboxylic acid, a deoxysugar, and an amino sugar. In general, the amino group is *N*-acetylated (5-acetamido-3,5-dideoxy-D-*glycero*-D-*galacto*-non-2-ulopyranosonic acid; *N*-acetylneuraminic acid; Neu5Ac) or *N*-glycolylated (5-hydroxyacetamido-3,5-dideoxy-D-*glycero*-

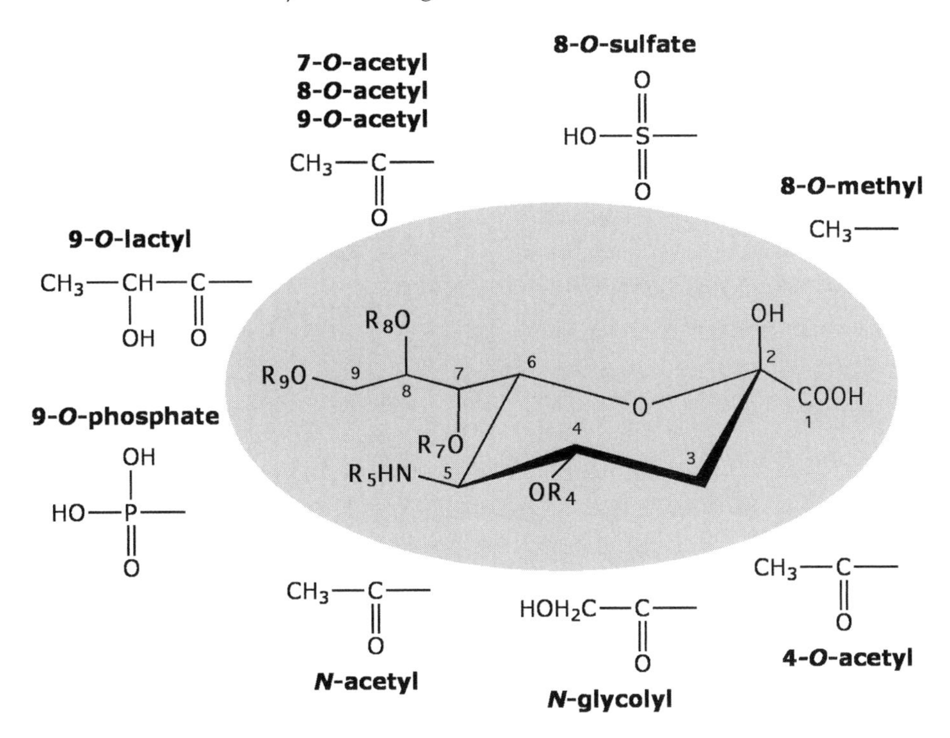

Fig. 2. Schematic overview of the family of naturally occurring sialic acids, as detailed in **Table 1**.

D-*galacto*-non-2-ulopyranosonic acid; *N*-glycolylneuraminic acid; Neu5Gc). The hydroxyl groups may be free, esterified (acetylated, lactylated, sulfated, phosphorylated), or etherified (methylated) (*see* **Table 1**). A schematic overview is given in **Fig. 2**.

The structural analysis of naturally occurring Sias is currently carried out by the following two main approaches: analysis of volatile Sia derivatives by gas-liquid chromatography (GLC) combined with electron-impact mass spectrometry (EI/MS), and analysis of fluorescently labeled Sias by high-performance liquid chromatography (HPLC) eventually coupled with electrospray mass spectrometry (ESI/MS).

Acid hydrolysis and enzymatic hydrolysis are the primary methods applied for the release of Sia from Sia-containing material. Each approach has advantages and disadvantages. Acid hydrolysis always gives rise to some de-*O*-esterification. Moreover, differences in rates of release are influenced by the substitution patterns and the type of glycosidic linkages. In the case of enzymatic hydrolysis with sialidases, linkage specificity as well as a reduced or complete lack of susceptibility must be taken into account. It should be noted that in most cases much lower amounts of Sias are released by sialidases than by acid hydrolysis. This chapter concentrates on acid hydrolysis protocols. In work-up procedures and analyses, pH values lower than 4.0 and higher than 6.0 should be avoided as much as possible to prevent solvolysis of *O*-acetyl groups and migration of *O*-acetyl groups at C-7/C-8 to C-9 ***(9,10)***.

2. Materials

2.1. Release of Sias From Glycoconjugates

2.1.1. Hydrolysis With Acetic Acid

1. Acetic acid (Merck, Darmstadt, Germany).
2. Diethyl ether (Fluka Chemie AG, Buchs, Switzerland).

2.1.2. Hydrolysis With Propionic Acid

1. Propionic acid (Sigma-Aldrich Chemie BV, Zwijndrecht, The Netherlands).

2.2. Preparation of Volatile Sia Derivatives for GLC and GLC-EI/MS Analysis

2.2.1. Preparation and GLC (-EI/MS) Analysis of Trimethylsilylated Methyl Ester Derivatives

1. Phosphor pentoxide (Sigma-Aldrich).
2. Methanol (Biosolve BV, Valkenswaard, The Netherlands).
3. Dowex H^+ (Fluka).
4. Diazomethane in diethyl ether (Fluka), prepared using diazogen 99% (Acros, Geel, Belgium) according to the kit instructions.
5. Pyridine (Sigma-Aldrich).
6. Hexamethyldisilazane (Acros).
7. Trimethylchlorosilane (Merck).
8. MD800/GC8060 GLC-EI/MS instrument (Fisons Instruments/Interscience, Breda, The Netherlands).
9. AT-1 capillary column, 30 m × 0.25 mm (Alltech, Breda, The Netherlands).

2.2.2. Preparation and GLC (-EI/MS) Analysis of Heptafluorobutylated Methyl Ester Derivatives

1. Phosphor pentoxide (Sigma-Aldrich).
2. Methanol (Biosolve).
3. Diazomethane in diethyl ether (Fluka), prepared using diazogen 99% (Acros) according to the kit instructions.
4. Acetonitrile (Biosolve).
5. Heptafluorobutyric anhydride (Sigma-Aldrich).
6. Carlo Erba GC8000/Riber 10-10H GLC-EI/MS instrument (Riber, France).
7. CP-Sil 5CB capillary column, 25 m × 0.32 mm (Chrompack France, Les Ullis, France).

Table 1
Survey of Established Structures of Naturally Occurring Members of the Sialic Acid Family

Sialic acid	Abbreviation
Neuraminic acid	Neu[a]
Neuraminic acid 1,5-lactam	Neu1,5lactam
N-Acetylneuraminic acid	Neu5Ac
5-*N*-Acetyl-4-*O*-acetyl-neuraminic acid	Neu4,5Ac_2
5-*N*-Acetyl-7-*O*-acetyl-neuraminic acid	Neu5,7Ac_2
5-*N*-Acetyl-8-*O*-acetyl-neuraminic acid	Neu5,8Ac_2
5-*N*-Acetyl-9-*O*-acetyl-neuraminic acid	Neu5,9Ac_2
5-*N*-Acetyl-4,9-di-*O*-acetyl-neuraminic acid	Neu4,5,9Ac_3
5-*N*-Acetyl-7,9-di-*O*-acetyl-neuraminic acid	Neu5,7,9Ac_3
5-*N*-Acetyl-8,9-di-*O*-acetyl-neuraminic acid	Neu5,8,9Ac_3
5-*N*-Acetyl-4,7,9-tri-*O*-acetyl-neuraminic acid	Neu4,5,7,9Ac_4
5-*N*-Acetyl-7,8,9-tri-*O*-acetyl-neuraminic acid	Neu5,7,8,9Ac_4
5-*N*-Acetyl-4,7,8,9-tetra-*O*-acetyl-neuraminic acid	Neu4,5,7,8,9Ac_5
5-*N*-Acetyl-9-*O*-L-lactyl-neuraminic acid	Neu5Ac9Lt
5-*N*-Acetyl-4-*O*-acetyl-9-*O*-lactyl-neuraminic acid	Neu4,5$Ac_2$9Lt
5-*N*-Acetyl-7-*O*-acetyl-9-*O*-lactyl-neuraminic acid	Neu5,7$Ac_2$9Lt
5-*N*-Acetyl-8-*O*-methyl-neuraminic acid	Neu5Ac8Me
5-*N*-Acetyl-4-*O*-acetyl-8-*O*-methyl-neuraminic acid	Neu4,5$Ac_2$8Me
5-*N*-Acetyl-9-*O*-acetyl-8-*O*-methyl-neuraminic acid	Neu5,9$Ac_2$8Me
5-*N*-Acetyl-8-*O*-sulpho-neuraminic acid	Neu5Ac8S
5-*N*-Acetyl-4-*O*-acetyl-8-*O*-sulpho-neuraminic acid	Neu4,5$Ac_2$8S
5-*N*-Acetyl-9-*O*-phosphoro-neuraminic acid	Neu5Ac9P[b,c]
5-*N*-Acetyl-2-deoxy-2,3-didehydro-neuraminic acid	Neu2en5Ac[c]
5-*N*-Acetyl-9-*O*-acetyl-2-deoxy-2,3-didehydro-neuraminic acid	Neu2en5,9Ac_2[c]
5-*N*-Acetyl-2-deoxy-2,3-didehydro-9-*O*-lactyl-neuraminic acid	Neu2en5Ac9Lt[c]
5-*N*-Acetyl-2,7-anhydro-neuraminic acid	Neu2,7an5Ac[c]
5-*N*-Acetyl-4,8-anhydro-neuraminic acid	Neu4,8an5Ac[d]
5-*N*-Acetylneuraminic acid 1,7-lactone	Neu5Ac1,7lactone
5-*N*-Acetyl-9-*O*-acetyl-neuraminic acid 1,7-lactone	Neu5,9$Ac_2$1,7lactone
5-*N*-Acetyl-4,9-di-*O*-acetyl-neuraminic acid 1,7-lactone	Neu4,5,9$Ac_3$1,7lactone
N-Glycolylneuraminic acid	Neu5Gc
4-*O*-Acetyl-5-*N*-glycolyl-neuraminic acid	Neu4Ac5Gc
7-*O*-Acetyl-5-*N*-glycolyl-neuraminic acid	Neu7Ac5Gc
8-*O*-Acetyl-5-*N*-glycolyl-neuraminic acid	Neu8Ac5Gc

(continued)

Table 1 *(continued)*

Sialic acid	Abbreviation
9-*O*-Acetyl-5-*N*-glycolyl-neuraminic acid	Neu9Ac5Gc
4,7-Di-*O*-acetyl-5-*N*-glycolyl-neuraminic acid	Neu4,7$Ac_2$5Gc
4,9-Di-*O*-acetyl-5-*N*-glycolyl-neuraminic acid	Neu4,9$Ac_2$5Gc
7,9-Di-*O*-acetyl-5-*N*-glycolyl-neuraminic acid	Neu7,9$Ac_2$5Gc
8,9-Di-*O*-acetyl-5-*N*-glycolyl-neuraminic acid	Neu8,9$Ac_2$5Gc
4,7,9-Tri-*O*-acetyl-5-*N*-glycolyl-neuraminic acid	Neu4,7,9$Ac_3$5Gc
7,8,9-Tri-*O*-acetyl-5-*N*-glycolyl-neuraminic acid	Neu7,8,9$Ac_3$5Gc
4,7,8,9-Tetra-*O*-acetyl-5-*N*-glycolyl-neuraminic acid	Neu4,7,8,9$Ac_4$5Gc
5-*N*-Glycolyl-9-*O*-lactyl-neuraminic acid	Neu5Gc9Lt
4-*O*-Acetyl-5-*N*-Glycolyl-9-*O*-lactyl-neuraminic acid	Neu4Ac5Gc9Lt
7-*O*-Acetyl-5-*N*-Glycolyl-9-*O*-lactyl-neuraminic acid	Neu7Ac5Gc9Lt
8-*O*-Acetyl-5-*N*-Glycolyl-9-*O*-lactyl-neuraminic acid	Neu8Ac5Gc9Lt
4,7-Di-*O*-acetyl-5-*N*-Glycolyl-9-*O*-lactyl-neuraminic acid	Neu4,7$Ac_2$5Gc9Lt
7,8-Di-*O*-acetyl-5-*N*-Glycolyl-9-*O*-lactyl-neuraminic acid	Neu7,8$Ac_2$5Gc9Lt
5-*N*-Glycolyl-8-*O*-methyl-neuraminic acid	Neu5Gc8Me
7-*O*-Acetyl-5-*N*-glycolyl-8-*O*-methyl-neuraminic acid	Neu7Ac5Gc8Me
9-*O*-Acetyl-5-*N*-glycolyl-8-*O*-methyl-neuraminic acid	Neu9Ac5Gc8Me
7,9-Di-*O*-acetyl-5-*N*-glycolyl-8-*O*-methyl-neuraminic acid	Neu7,9$Ac_2$5Gc8Me
5-*N*-Glycolyl-8-*O*-sulpho-neuraminic acid	Neu5Gc8S
5-*N*-Glycolyl-9-*O*-sulpho-neuraminic acid	Neu5Gc9S
N-(*O*-Acetyl)glycolylneuraminic acid	Neu5GcAc
N-(*O*-Methyl)glycolylneuraminic acid	Neu5GcMe
2-Deoxy-2,3-didehydro-5-*N*-glycolyl-neuraminic acid	Neu2en5Gc[c]
9-*O*-Acetyl-2-deoxy-2,3-didehydro-5-*N*-glycolyl-neuraminic acid	Neu2en9Ac5Gc[c]
2-Deoxy-2,3-didehydro-5-*N*-glycolyl-9-*O*-lactyl-neuraminic acid	Neu2en5Gc9Lt[c]
2-Deoxy-2,3-didehydro-5-*N*-glycolyl-8-*O*-methyl-neuraminic acid	Neu2en5Gc8Me[c]
2,7-Anhydro-5-*N*-glycolyl-neuraminic acid	Neu2,7an5Gc[c]
2,7-Anhydro-5-*N*-glycolyl-8-*O*-methyl-neuraminic acid	Neu2,7an5Gc8Me[c]
4,8-Anhydro-5-*N*-glycolyl-neuraminic acid	Neu4,8an5Gc[d]
5-*N*-Glycolylneuraminic acid 1,7-lactone	Neu5Gc1,7lactone
9-*O*-Acetyl-5-*N*-glycolyl-neuraminic acid 1,7-lactone	Neu9Ac5Gc1,7lactone

[a]Present only in bound form.
[b]Biosynthetic intermediate.
[c]Present only in free form.
[d]Not occurring as such in nature, but sometimes isolated owing to hydrolytic conditions. For a survey of typical biological sources of the various sialic acids, *see* **refs.** ***3,5,7,8***.

2.2.3. Preparation and GLC (-EI/MS) Analysis of Pertrimethylsilylated Derivatives

1. Phosphor pentoxide (Sigma-Aldrich).
2. Pyridine (Sigma-Aldrich).
3. Hexamethyldisilazane (Acros).
4. Trimethylchlorosilane (Merck).
5. MD800/GC8060 GLC-EI/MS instrument (Fisons).
6. AT-1 capillary column, 30 m × 0.25 mm (Alltech).

2.3. GLC-EI/MS Data of Trimethylsilylated Sia Derivatives

Materials are as described in **Subheadings 2.2.1.** and **2.2.3.**

2.4. GLC-EI/MS Data of Heptafluorobutylated Sia Methyl Ester Derivatives

Materials are as described in **Subheading 2.2.2.**

2.5. Preparation of Fluorescent Sia Derivatives for HPLC and LC-ESI/MS Analysis

2.5.1. Preparation and HPLC (-ESI/MS) Analysis of 1,2-Diamino-4,5-Methylenedioxybenzene Sia Derivatives

1. Acetic acid (Merck).
2. 1,2-Diamino-4,5-methylenedioxybenzene (DMB) dihydrochloride (Sigma-Aldrich).
3. β-Mercaptoethanol (Sigma-Aldrich).
4. Sodium hydrosulfite (Sigma-Aldrich).
5. Spectroflow 400 HPLC solvent delivery system (ABI Analytical/Kratos Division, Ramsey, NJ).
6. Spectroflow 980 programmable fluorescence detector (ABI).
7. Cosmosil 5C18-AR-II reversed-phase column, 250 × 4.6 mm (Nacalai Tesque, Kyoto, Japan).
8. Acetonitrile (Biosolve).
9. Methanol (Biosolve).
10. Milli-Q water (Millipore BV, Etten-Leur, The Netherlands).
11. API-I simple quadrupole ESI/MS instrument (Perkin-Elmer Sciex Instruments, Thornhill, Canada).
12. Microbore ultrasphere octadecylsilane (ODS) column, 250 × 2 mm (Beckman, Fullerton, CA).

2.6. HPLC-ESI/MS Data of DMB Sia Derivatives

Materials are as described in **Subheading 2.5.1.**

3. Methods

3.1. Release of Sias From Glycoconjugates

Several approaches have been reported for the effective acid hydrolysis of the labile glycosidic linkage between Sia and a neighboring monosaccharide. All these procedures are not optimal for giving the real spectrum of Sias originally present in the sialoglycoconjugate under study, especially in the case of a mixture of (*O*-acylated) *N*-acylneuraminic acids. The three most applied acid hydrolysis systems are formic acid/HCl ***(11)***, acetic acid ***(10)***, and propionic acid ***(12)***. The use of H_2SO_4 is not recommended. The oldest formic acid/HCl protocol comprises a two-step hydrolysis, whereby glycoconjugate probes are incubated with aqueous formic acid (pH 2.0 for 1 h at 70°C), followed by incubation with HCl (pH 1.0 for 1 h at 80°C). After each step the liberated Sias are recovered by centrifugation, ultrafiltration, or dialysis. In the case of a spectrum of (*O*-acylated) *N*-acylneuraminic acids, the supernatant, ultrafiltrate, or diffusate of the formic acid hydrolysis contains the majority of the *O*-acylated *N*-acylneuraminic acids, while that of the HCl hydrolysis contains mostly Neu5Ac and Neu5Gc. In the case of low-molecular-mass substances, isolations can be carried out by gel-permeation chromatography. For purification of Sias, combined cation-anion exchange chromatography is often included. Although these conditions do not lead to significant de-*N*-acylation, de-*O*-acylation has been shown to occur to an extent of approx 30–50%. On the other hand, milder acidic conditions result in incomplete release of Sias. Most research groups currently use acetic acid or propionic acid protocols.

3.1.1. Hydrolysis With Acetic Acid

1. Glycoconjugate probes are incubated with 500 µL of 2 *M* acetic acid for 3 h at 80°C ***(10)***. Because the exact hydrolysis time required can vary with different glycoconjugates, reported incubation times vary between 90 min and 4 h.
2. Solutions are cooled, concentrated *in vacuo* or lyophilized, or directly used in derivatization protocols.
3. Lipid impurities can be removed by diethyl ether extraction. If further purifications are necessary, different types of concentrators are used ***(10,12)***.

3.1.2. Hydrolysis With Propionic Acid

1. Glycoconjugate probes are incubated with 500 µL of 2 *M* propionic acid for 4 h at 80°C (*see* **ref. *12***; 4 *M* propionic acid has also been proposed ***[13]***).
2. Solutions are cooled, then concentrated *in vacuo* or lyophilized, or used directly in derivatization protocols.

3. If further purifications are necessary, different types of concentrators are used. This more recent approach may significantly decrease the *O*-acyl group loss and migration when compared with the use of 2 *M* acetic acid or 0.5 *M* formic acid, while also providing better yields.

3.2. Preparation of Volatile Sia Derivatives for GLC and GLC-EI/MS Analysis

Starting with free Sia pools, mainly present in their β-anomeric form, volatile Sia derivatives are generated using mild derivatization procedures such as esterification with diazomethane followed by trimethylsilylation ***(14)*** or heptafluorobutylation ***(7)***, or direct pertrimethylsilylation ***(11)***.

3.2.1. Preparation and GLC (-EI/MS) Analysis of Trimethylsilylated Methyl Ester Derivatives

1. Lyophilized Sia samples (ng-μg amounts), dried over P_2O_5, are dissolved in anhydrous methanol (200 μL).
2. Dowex H^+ in 80 μL of methanol is added (Sias should be in the COOH form), and the mixture is filtered over cotton wool.
3. Diazomethane in diethyl ether is added until a faint yellow color remains for 5 min at room temperature. The solution is concentrated to dryness using a stream of nitrogen and dried over P_2O_5.
4. The residue is dissolved in 10 μL of trimethylsilylation reagent (pyridine:hexamethyldisilazane:trimethylchlorosilane, 5:1:1), and the mixture is kept for 2 h at room temperature.
5. Analyze 3-μL samples by GLC-EI/MS using an AT-1 capillary column. The temperature program is usually 220°C for 25 min, then 6°C/min to 300°C, and finally 6 min constant at 300°C. The injector temperature is 230°C, the source temperature is 200°C, and the electron voltage is 70 eV.

3.2.2. Preparation and GLC (-EI/MS) Analysis of Heptafluorobutylated Methyl Ester Derivatives (see ***Notes 1** and **2***)

1. Lyophilized Sia samples (ng-μg amounts), dried over P_2O_5, are dissolved in anhydrous 200 μL of methanol, and diazomethane in 200 μL of diethyl ether is added. Samples are left for 4 h at room temperature without stirring.
2. The solution is concentrated to dryness using a stream of nitrogen, and 200 μL of acetonitrile and 25 μL of heptafluorobutyric anhydride are added. The mixture is heated for 5 min at 150°C (sand bath), cooled to room temperature, and concentrated to dryness using a stream of nitrogen.
3. The residue is dissolved in an appropriate amount of acetonitrile, and aliquots are analyzed by GLC-EI/MS using a CP-Sil 5CB capillary column. The temperature program is usually 90°C for 3 min, then 5°C/min to 260°C. The injector temperature is 260°C, the source temperature is 150°C, and the electron voltage is 70 eV.

3.2.3. Preparation and GLC (-EI/MS) Analysis of Pertrimethylsilylated Derivatives

1. Lyophilized Sia samples (ng-µg amounts), dried over P_2O_5, are dissolved in 30 µL of trimethylsilylation reagent (pyridine:hexamethyldisilazane:trimethylchlorosilane, 5:1:1), and the solutions are kept for 2 h at room temperature.
2. 3-µL Samples are analyzed by GLC-EI/MS using an AT-1 column. The temperature program is usually 220°C for 25 min, then 6°C/min to 300°C, and finally 6 min constant at 300°C. The injector temperature is 230°C, the source temperature is 200°C, and the electron voltage is 70 eV.

3.3. GLC-EI/MS Data of Trimethylsilylated Sia Derivatives

A general EI/MS method has been developed for the identification of Sias isolated from biological material and derivatized as volatile trimethylsilylated (TMS) methyl ester derivatives or pertrimethylsilylated derivatives (PTMS). This method has also proved to be useful for the analysis of other isolated Sias, of (partially) *O*-methylated Sia methyl ester methyl glycosides as obtained in methylation analysis, and of synthetic Sia(s) (derivatives) ***(3,14–16)***.

Figure 3 depicts a schematic survey showing the selected fragment ions A-H, which furnish the information (abundances and *m/z* values of the ions) necessary to deduce the complete structure of Sias. As typical examples, **Fig. 4** shows the EI mass spectra of the TMS methyl esters of β-Neu5Ac and β-Neu5Gc, and of the PTMS derivatives of β-Neu5Ac and β-Neu5Gc.

3.3.1 GLC-EI/MS Data of TMS Methyl Ester or PTMS Sia Derivatives

1. Fragments A and B indicate the molecular mass of the Sia derivatives (the molecular ion M is absent), and thus the number and type of substituents.
2. Fragments C–H contain the information concerning the position of the different substituents.
3. Fragments A–H are very useful for ion chromatogram screening of GLC effluents.
4. Fragment A is formed from the molecular ion M by elimination of a methyl group. In TMS (*O*-acylated/*O*-alkylated) *N*-acylneuraminic acid derivatives, the methyl group originates from a trimethylsilyl substituent.
5. Fragment B is obtained by elimination of the C-1 part of the Sia molecule. Eliminations of $OCOCH_3$ in *O*-acetylated Sia derivatives and of NH_2COCH_3 in Neu5Ac derivatives, which in principle give rise to the same *m/z* value as fragment B in the case of $R_1 = CH_3$, contribute little to the abundance of this ion. For *O*-trimethylsilylated *N,O*-acylneuraminic acids (β-anomers) it holds that, when compared with their methyl esters, in their trimethylsilyl esters the intensity of fragment A decreases relative to fragment B.
6. Fragment C is formed by elimination of the C-8,9 part, with localization of the charge on position 7. In general, cleavage occurs between two alkoxylated carbon

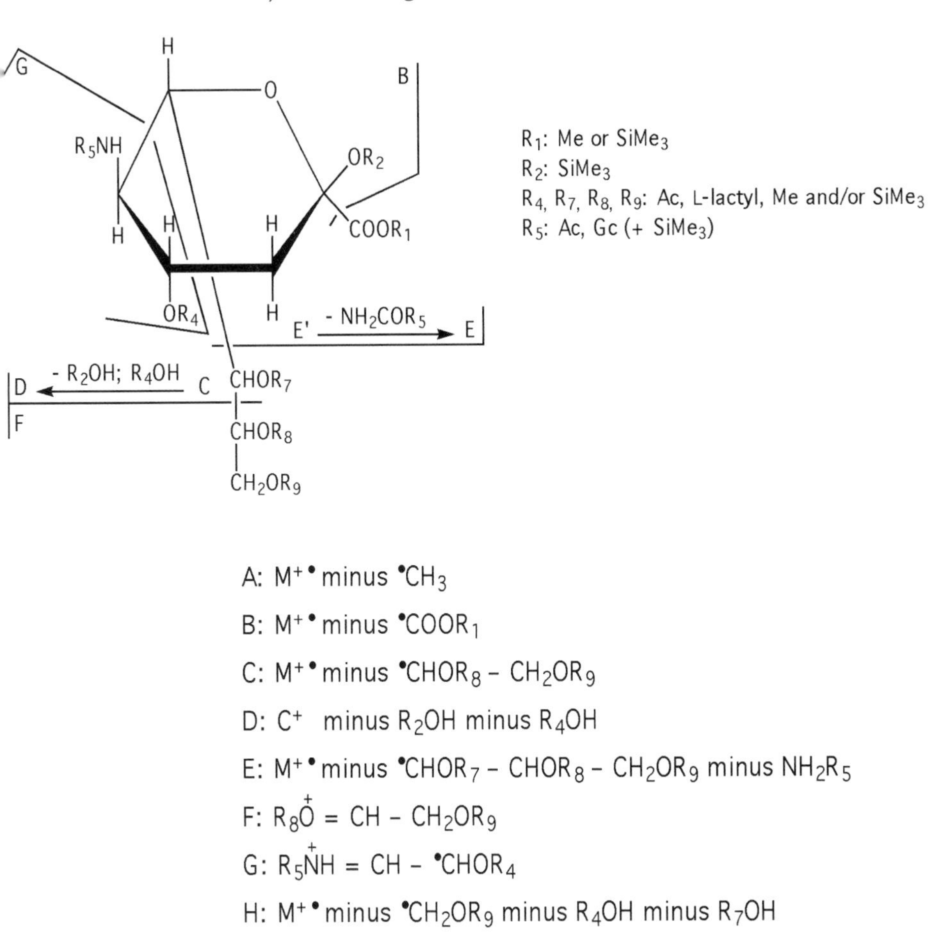

Fig. 3. Survey of the selected fragment ions A–H worked out for the trimethylsilylated methyl ester and the pertrimethylsilylated derivatives of *N*-acylneuraminic acids with *O*-acyl and/or *O*-alkyl substituents (*see* **Tables 2** and **3**).

atoms or between an acetoxylated and an alkoxylated carbon atom, rather than between two acetoxylated carbon atoms. Fragment C only has significant abundance if C-7 bears an ether group. When an ester group is present at C-7, this fragment ion is absent or hardly observable.

7. Fragment D is formed from fragment C by consecutive elimination of R_2OH and R_4OH. It is evident that the occurrence of this fragment ion is dependent on the presence of fragment C.
8. Fragment E is formed by elimination of the side-chain C-7,8,9 and the substituent at C-5. This fragment ion is not seen if an *O*-acyl group is attached to C-4,

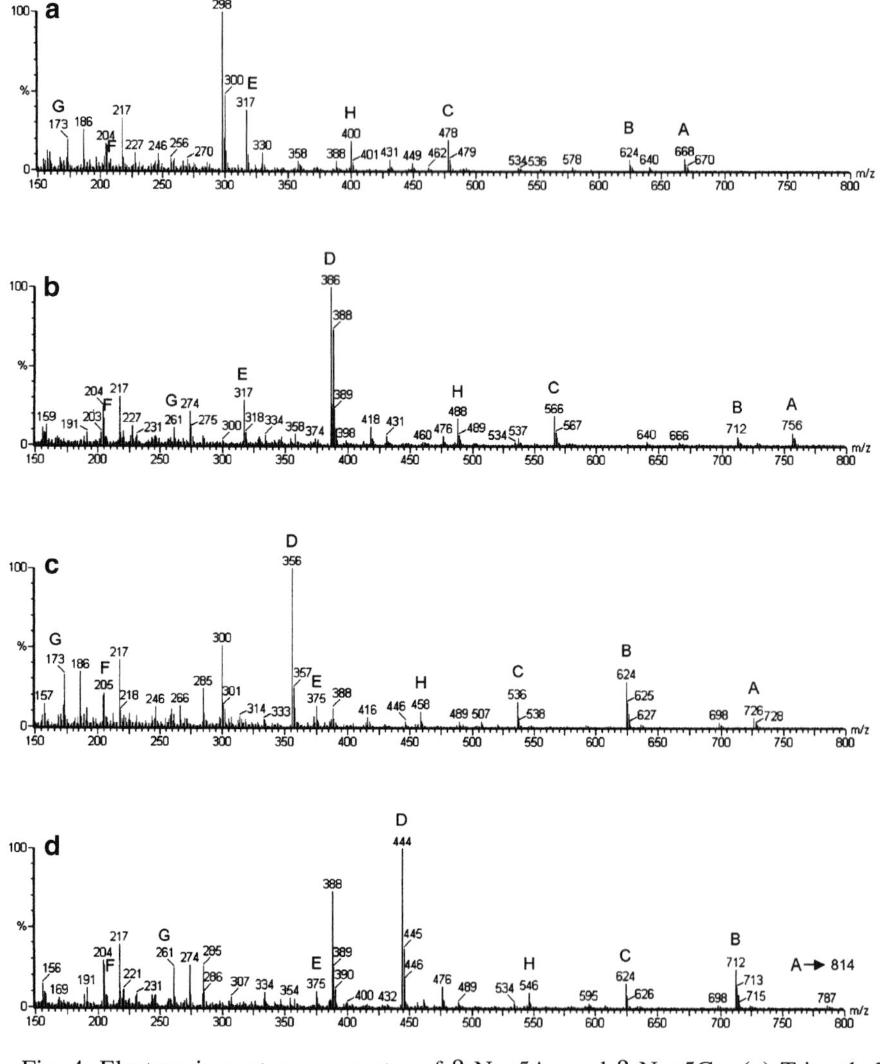

Fig. 4. Electron impact mass spectra of β-Neu5Ac and β-Neu5Gc. **(a)** Trimethylsilylated methyl ester of β-Neu5Ac; **(b)** trimethylsilylated methyl ester of β-Neu5Gc; **(c)** pertrimethylsilylated β-Neu5Ac; **(d)** pertrimethylsilylated β-Neu5Gc.

illustrating that the transition state in the McLafferty rearrangement is not favored when the substituent at C-4 is an ester group rather than an ether group. For *O*-trimethylsilylated *N*,*O*-acylneuraminic acids (β-anomers) it holds that, when compared with their methyl esters, in their trimethylsilyl esters the intensity of fragment E is greatly reduced but still present; instead, an additional fragment derived from fragment E by loss of Me_3SiOH is clearly present.

9. Fragment F contains the C-8,9 part. Based on the same fragmentation rules as mentioned for fragments C and D (**steps 6** and **7**), this ion can be readily formed only if an ether group is connected at C-8.
10. Fragment G consists of the C-4,5 part of the Sia molecule.
11. Fragment H is formed by elimination of the C-9 part of the molecule, followed by elimination of R_4OH and R_7OH; this fragment is only useful in the case of *O*-alkyl substituents.

Table 2 presents the GLC-EI/MS data of a series of naturally occurring Sias, derivatized as TMS methyl esters. **Table 3** presents the GLC-EI/MS data of a series of naturally occurring Sias, derivatized as PTMS derivatives. Sias predominantly occur in the β-anomeric form as previously discussed. However, occasionally the minor amount of α-anomer can be detected separately from the β-anomer as a shoulder. Inspection of the EI/MS data in **Tables 2** and **3** shows that each Sia gives rise to a unique series of A–G fragment ions. The H fragment is also used in case of substitution patterns of *O*-trimethylsilyl and *O*-methyl groups. For EI/MS spectra of members of the Sia family presented in **Tables 2** and **3** but not depicted in **Fig. 4**, *see* **refs.** ***11,14–18***.

3.3.2. Detailed Analyses of GLC-EI/MS Data of TMS Methyl Ester or PTMS Sia Derivatives

1. Fragment G requires more detailed discussion. The occurrence of an *O*-acetyl group instead of an *O*-trimethylsilyl group at C-4 as in the derivatives of Neu4,5Ac_2 (acetyl [AC]), Neu4,5,9Ac_3, Neu4,5$Ac_2$9Lt (lactyl [LT]), and Neu4Ac5Gc (glycolyl [GC]) leads to a negative shift of 30 atomic mass units (amu) for this fragment. Therefore, in the mass spectrum of 4-*O*-acetylated *N*-acetylneuraminic acids the peak at *m/z* 173 is absent: *m/z* 173 shifts to *m/z* 143. However, in all mass spectra a peak at *m/z* 143 with a general formula $C_6H_{11}O_2Si$ is observed. But in the mass spectra of Neu4,5Ac_2, Neu4,5,9Ac_3, and Neu4,5$Ac_2$9Lt, the main contribution to the abundance of *m/z* 143 originates from fragment G ($C_6H_9NO_3$). In the mass spectrum of Neu4Ac5Gc the peak at *m/z* 261 is not observed. For this compound, fragment G shifts to *m/z* 231. Using high-resolution mass spectometry, this fragment ion could not be distinguished from other generally occurring fragment ions in Sia, which also contributes to the intensity of the peak at *m/z* 231.
2. Using the fragment ions A–G, Neu5,8Ac_2 and Neu5,9Ac_2 are distinguished solely on the basis of the intensity of the peak at *m/z* 175 (fragment F). Of course, the mass spectra of these compounds also differ in other aspects. For example, the side chain CH_2OCOCH_3-$CHOSi(CH_3)_3$-$CH{=}O^+Si(CH_3)_3$ in Neu5,9Ac_2 clearly eliminates CH_3COOH, giving rise to the fragment ion at *m/z* 217. In Neu5,8Ac_2, the side chain $CH_2OSi(CH_3)_3$-$CHOCOCH_3$-$CH{=}O^+Si(CH_3)_3$ eliminates CH_3COOH (*m/z* 217) as well as $HOSi(CH_3)_3$ (*m/z* 187).
3. The fragment ion at *m/z* 103 ($CH_2{=}O^+Si(CH_3)_3$) is not characteristic for a primary trimethylsiloxyl group in the Sia derivatives, but can also be formed along other routes.

Table 2
GLC and Characteristic EI/MS Fragment Ions (70 eV) of Trimethylsilylated Methyl Ester Derivatives of Naturally Occurring Sialic Acids (β-Anomers)

		Fragments (*m/z*)							
Sialic acid	R_{Neu5Ac}	A	B	C	D	E	F	G	H
Neu5Ac	1.00	668	624	478	298	317	205	173	400
Neu4,5Ac_2	1.18	638	594	448	298	—	205	143	400
Neu5,7Ac_2	1.04	638	594	—	—	317	205	173	400
Neu5,8Ac_2	1.05	638	594	478	298	317	—	173	—
Neu5,9Ac_2	1.13	638	594	478	298	317	175	173	400
Neu4,5,9Ac_3	1.31	608	564	448	298	—	175	143	400
Neu5,7,9Ac_3	1.14	608	564	—	—	317	175	173	400
Neu5,8,9Ac_3	1.19	608	564	478	298	317	—	173	—
Neu5,7,8,9Ac_4	1.15	578	534	—	—	317	—	173	—
Neu5Ac9Lt	2.55	740	696	478	298	317	277	173	400
Neu4,5$Ac_2$9Lt	3.01	710	666	448	298	—	277	143	400
Neu2en5Ac	1.09	578	—	388	298	227	205	—	—
Neu5Gc	1.81	756	712	566	386	317	205	261	488
Neu4Ac5Gc	2.02	726	682	536	386	—	205	231	488
Neu7Ac5Gc	1.83	726	682	—	—	317	205	261	488
Neu9Ac5Gc	2.04	726	682	566	386	317	175	261	488
Neu7,9$Ac_2$5Gc	2.01	696	652	—	—	317	175	261	488
Neu8,9$Ac_2$5Gc	1.99	696	652	566	386	317	—	261	—
Neu7,8,9$Ac_3$5Gc	1.93	666	622	—	—	317	—	261	—

R_{Neu5Ac} values on 3.8% SE-30 (packed column) at 215°C are given relative to β-Neu5Ac. The R_{Neu5Ac} values are given as a directive to set up an in-house set of GLC values with reference sialic acids. (Data taken from **ref. *3***.)

4. The fragment ions at *m/z* 186 (CH_3CON^+H=CH-CH=CHOSi$(CH_3)_3$ and CH_3CON^+H=CH-C(OSi$(CH_3)_3$)=CH_2) in *N*-acetylneuraminic acids and at *m/z* 274 in *N*-glycolylneuraminic acids only give information about the type of substitution at C-5 (amino group).

3.4. GLC-EI/MS Data of Heptafluorobutylated Sia Methyl Ester Derivatives

In a more recent EI/MS method for the identification of Sias isolated from biological material, heptafluorobutylated (HFB) methyl ester derivatives are used *(7)*. HFB derivatives are rather stable in contrast to the TMS derivatives. It should be noted that in the derivatization procedure the anomeric HO-2 group remains free. The molar ratio between the α- and β-anomers changes slowly if

Table 3
GLC and Characteristic EI/MS Fragment Ions (70 eV) of Pertrimethylsilylated Derivatives of Naturally Occurring Sialic Acids (β-Anomers)

		Fragment ions (*m/z*)							
Sialic acid	R_{Neu5Ac}	A	B	C	D	E	F	G	H
Neu5Ac	1.00	726	624	536	356	375	205	173	458
Neu4,5Ac_2	1.05	696	594	506	356	—	205	143	458
Neu5,9Ac_2	1.02	696	594	536	356	375	175	173	458
Neu5,7,9Ac_3		666	564	—	—	375	175	173	458
Neu5,8,9Ac_3	1.04	666	564	536	356	375	—	173	—
Neu5,7,8,9Ac_4		636	534	—	—	375	—	173	—
Neu5Ac9Lt		798	696	536	356	375	277	173	458
Neu5Ac8Me	0.98	668	566	536	356	375	147	173	400
Neu5,9$Ac_2$8Me	1.00	638	536	536	356	375	117	173	400
Neu2en5Ac	1.01	636	—	446	356	285	205	—	
Neu2,7an5Ac		564	462	374	—	—	205	173	
Neu5Gc	1.19	814	712	624	444	375	205	261	546
Neu4Ac5Gc		784	682	594	444	—	205	231	546
Neu9Ac5Gc	1.21	784	682	624	444	375	175	261	546
Neu7,9$Ac_2$5Gc		754	652	594	414	375	175	261	546
Neu5Gc8Me	1.14	756	654	624	444	375	147	261	488
Neu9Ac5Gc8Me	1.17	726	624	624	444	375	117	261	488
Neu5GcAc	1.21	784	682	594	414	375	205	231	
Neu2en5Gc		724	—	534	444	285	205	—	
Neu2,7an5Gc		652	550	286	—	—	205	261	

R_{Neu5Ac} values on CP-Sil 5 (capillary column), using the program 5 min/140°C; 2°C/min up to 220°C; 15 min/220°C, are given relative to β-Neu5Ac. The R_{Neu5Ac} values are given as a directive to set up an in-house set of GLC values with reference sialic acids. (Data taken from **ref.** *3*.)

samples are kept in the acylation mixtures for a long time, because of mutarotation of the free HO-2 group.

Table 4 presents the GLC-EI/MS data of a series of naturally occurring Sias, derivatized as HFB methyl esters. For EI/MS spectra of members of the Sia family presented in **Table 4**, *see* **ref.** *7*.

1. The EI spectra of the HFB Sia methyl ester derivatives do not show molecular ion peaks.
2. These EI spectra are rather complex owing to the presence of only *O*-acylated groups in most of the naturally occurring Sias (*N*,*O*-acylated neuraminic acids). No use can be made of the preferences in cleavage comparing two neighboring alkoxylated carbon atoms, a neighboring acetoxylated and alkoxylated carbon

Table 4
GLC and Reporter EI/MS Fragment Ions (70 eV) of Heptafluorobutylated Methyl Ester Derivatives of Naturally Occurring Sialic Acids (α,β-Anomers)

Sialic acid	R_{Neu5Ac}'s	Reporter ions (*m/z*)
Neu	1.14/1.15	815-801-765-731-550-534-505-334-294-253
Neu5Ac	0.99/1.00	801-773-757-704-543-490-330-264-238
Neu4,5Ac_2	1.14/1.15	862-833-801-756-704-606-560-493-491-453-320-154-84
Neu5,7Ac_2	1.16/1.16	862-790-773-757-543-493-453-347-279-238
Neu5,8Ac_2	n.d./1.17	862-832-778-776-734-704-605-520-492-307
Neu5,9Ac_2	1.11/1.14	862-813-756-648-620-560-543-519-490-393-347-264-73
Neu4,5,9Ac_3	n.d./1.32	708-648-595-490-335-223-150-84-73
Neu5,7,9Ac_3	1.31/1.33	708-604-494-452-379-238-166-153-73
Neu5,8,9Ac_3	n.d./1.27	708-648-561-351-347-264-238-145-103-73
Neu4,5,7,9Ac_4	n.d./1.96	580-507-281-241-209-205-191-135-73
Neu4,5,7,8,9Ac_5	2.43/2.44	427-368-352-260-213-145-73
Neu5Ac9Lt	1.10/1.11	971-842-776-755-562-514-451-349-238-112
Neu4,5$Ac_2$9Lt	1.45/1.46	962-919-872-748-679-562-451-348-112
Neu5,7$Ac_2$9Lt	n.d./1.32	903-832-562-451-349-238-153-112-73
Neu5Ac8Me	1.09/1.14	833-804-790-693-622-578-364-265
Neu4,5$Ac_2$8Me	n.d./1.47	721-694-678-621-578-424-407
Neu5,9$Ac_2$8Me	n.d./1.47	721-678-620-578-424-96-73
Neu5Ac8S	1.38/1.40	831-688-619-407-365-322-295-122
Neu4,5$Ac_2$8S	1.52/1.53	694-619-534-467-365-295-207-122
Neu5Ac1,7lactone	1.08/1.11	861-841-647-620-533-407-380-350-347-320-252-194-136
Neu5,9$Ac_2$1,7lactone	1.31/1.35	706-604-494-452-450-379-347-306-252-166-136-73
Neu4,5,9$Ac_3$1,7lactone	1.53/1.53	494-450-438-407-394-347-339-254-166-136-83
Neu5Gc	1.25/1.26	859-831-815-761-618-548-404-348-298-294-227
Neu9Ac5Gc	n.d./1.24	903-832-777-647-562-542-349-238-112
Neu4,7$Ac_2$5Gc	n.d./1.48	900-802-706-664-620-559-405-227-211-166-84
Neu4,9$Ac_2$5Gc	n.d./1.37	899-848-703-624-518-366-304-276-238-227-84
Neu7,9$Ac_2$5Gc	n.d./1.40	920-876-847-706-664-620-548-510-406-350-227-153-134-73
Neu8,9$Ac_2$5Gc	n.d./1.30	903-859-744-703-648-594-532-277-145-103
Neu4,7,9$Ac_3$5Gc	n.d./1.31	792-777-703-518-491-304-238-227-73

Table 4 *(continued)*

Sialic acid	R_{Neu5Ac}'s	Reporter ions (*m/z*)
Neu4,7,8,9$Ac_4$5Gc	n.d./1.48	703-624-537-518-304-277-238-227
Neu5Gc9Lt	n.d./1.26	859-857-831-829-815-761-618-548-348-227
Neu4Ac5Gc9Lt	1.25/1.26	901-830-773-686-620-509-407-350-296-227-112
Neu7(8)Ac5Gc9Lt	n.d./1.25	903-861-789-688-619-562-474-408-350-255-195-153-112
Neu4,7$Ac_2$5Gc9Lt	n.d./1.41	961-917-848-677-620-509-407-296-227-153-112-84
Neu7,8$Ac_2$5Gc9Lt	1.40/1.44	920-902-876-748-704-662-632-560-548-255-227-195-112
Neu5Gc1,7lactone	n.d./1.30	859-745-619-519-350-347-277-227-136
Neu9Ac5Gc1,7lactone	n.d./1.41	920-873-815-662-548-388-299-294-227

R_{Neu5Ac} values on CP-Sil 5CB (capillary column), using the program 3 min/90°C; 5°C/min up to 260°C, are given relative to β-Neu5Ac. The R_{Neu5Ac} values are given as a directive to set up an in-house set of GLC values with reference sialic acids. (Data taken from **ref.** *7*.)

atom, and two neighboring acetoxylated carbon atoms, as this is the key of the fragmentation scheme of the TMS derivatives.

3. Characteristic ions from the HFBs are found at *m/z* 69, 119, and 169.
4. Most of the Neu5Ac and Neu5Gc derivatives give rise to peaks at *m/z* 238 and 227, respectively.
5. A peak at *m/z* 84 is specific for Neu4,5,xAc_3 derivatives, a peak at *m/z* 73 for 9-*O*-acetylated Sias, peaks at *m/z* 103 and 145 for 8,9-di-*O*-acetylated Sias, and a peak at *m/z* 136 for 1,7 intramolecular lactones.
6. All 9-*O*-lactyl derivatives contain a peak at *m/z* 112.
7. The peak at *m/z* 505 is specific for Neu, and at *m/z* 122 and 295 for 8-*O*-sulfated Sias.

3.5. Preparation of Fluorescent Sia Derivatives for HPLC and LC-ESI/MS Analysis

By starting from free Sia pools, fluorescent Sia derivatives can be generated by reaction with the fluorogenic reagents DMB (*see* **Fig. 5** and **ref.** ***19***) or *o*-phenylenediamine (OPD; *see* **ref.** ***20***). Most research groups currently use the DMB derivatives for HPLC analysis, and therefore the focus will be on these compounds. A typical HPLC chromatogram is presented in **Fig. 6**.

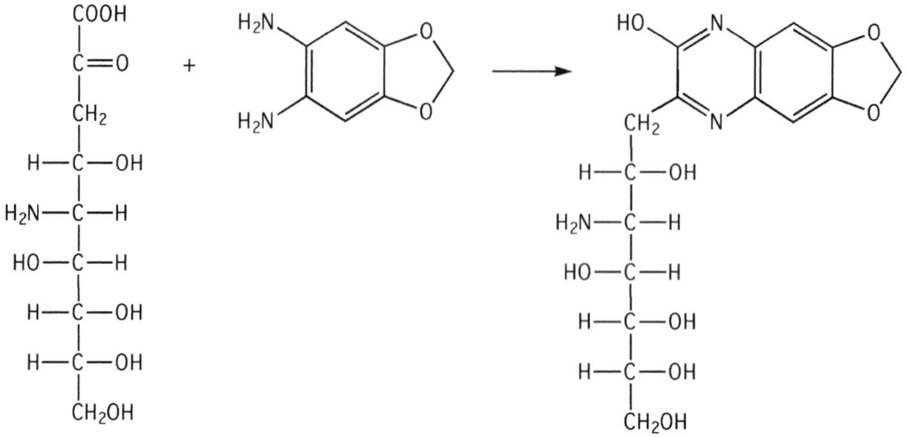

Fig. 5. Conversion of Neu5Ac into the corresponding fluorescent DMB derivative.

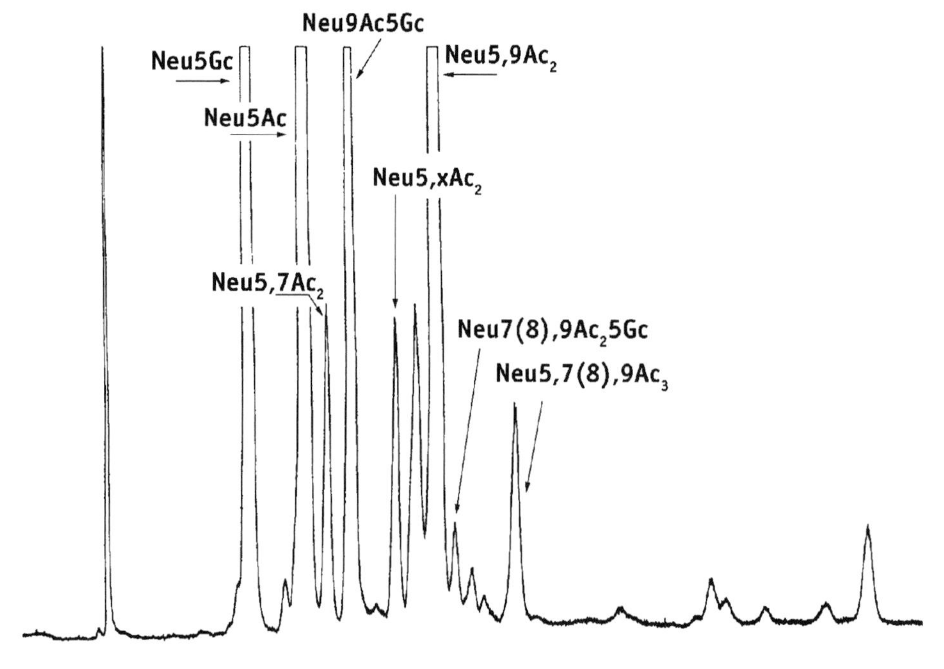

Fig. 6. Typical HPLC pattern of a mixture of fluorescent DMB derivatives of sialic acids on an octadecylsilane column. For elution system details, *see* **Table 5**. (Data taken from **ref.** ***8***.)

3.5.1. Preparation and HPLC (-ESI/MS) Analysis of DMB Sia Derivatives

1. Lyophilized Sia samples (pg-ng amounts) in 200 µL of 2 *M* acetic acid are mixed with 200 µL of 7 m*M* DMB dihydrochloride in 1.4 *M* acetic acid, containing 0.75 *M* of β-mercaptoethanol and 18 m*M* of sodium hydrosulfite. The DMB stock solution can be used for more than 1 wk when stored in a refrigerator. It should be noted that the derivatization can often be carried out directly on the acetic acid hydrolysate of a glycoconjugate without intermediate purification steps.
2. The solution is heated for 2.5 h at 50°C in the dark (1 h at 56°C has also been proposed *[21]*), and after cooling on ice, a fixed volume (2–20 µL) is used for analysis by HPLC.
3. The DMB Sia derivatives are analyzed on a reversed-phase Cosmosil 5C18-AR-II column, using acetonitrile : methanol : water (9 : 7 : 84) as the solvent system at a flow rate of 1 mL/min at room temperature. Authentic Sias are used as reference compounds. Fluorescence detection occurs at an excitation wavelength of 373 nm and an emission wavelength of 448 nm. The DMB derivative of Neu5Ac is used for quantifications as an internal standard.
4. For a combination with ESI/MS, a microbore ultrasphere ODS column has been applied. Here, elutions have been carried out with a linear two-step gradient of acetonitrile : methanol : water (3.85 : 7.00 : 89.15) to 7.15 : 7.00 : 85.85 in 40 min, and then to 11.00 : 7.00 : 82.00 in 10 min at a flow rate of 0.2 mL/min. A survey of HPLC retention times on an ODS column is presented in **Table 5** (*see* **refs.** *8* and *22*).

3.6. HPLC-ESI/MS Data of DMB Sia Derivatives

A sensitive LC-ESI/MS methodology has recently been worked out using the fluorescent DMB Sia derivatives, which is highly suitable for the identification of Sias with free anomeric centers *(8)*. Intermediate purification steps are not always necessary owing to the fluorescent label. Best results can be obtained with a microbore reversed-phase HPLC column (with split for separate fluorescent detection), whereby in the ESI/MS system molecular ion species can be detected on the 5-pmol level, and typical CAD (collisional activation decomposition) fragments on the 10- to 15-pmol level.

Table 5 summarizes ESI/MS data of a series of naturally occurring Sias. Speculative MS fragmentation pathways for several of them have been reported *(8)* (for a series of ESI/MS spectra, *see* **refs.** *8* and *23*). From the information available to date it is clear that a combination of retention time and spectral characteristics allows the identification of the number, type, and position of the various substituents in Sias that can be converted into DMB derivatives.

Table 5
HPLC Relative Retention Times (R_{Neu5Ac}) on an Ultrasphere ODS Column and Characteristic Ions Generated by LC-ESI/MS of DMB Sialic Acid Derivatives

			CAD fragments	
Sialic acid	R_{Neu5Ac}	$[M+H]^+$	$[M+H-H_2O]^+$	Fragments (*m/z*)
Neu5Ac	1.00	426	408	313-295-283-229
$Neu4,5Ac_2$	1.60	468	abs.	408-313-283-229
$Neu5,7Ac_2$	1.08	468	450	313-295-283-229
$Neu5,8Ac_2$	1.31	468	450	313-295-229
$Neu5,9Ac_2$	1.51	468	450	313-295-229
$Neu4,5,9Ac_3$		510	abs.	450-313-295-229
$Neu5,7,9Ac_3$[a]	1.74	510	492	313-295-229
$Neu5,8,9Ac_3$[a]	1.57	510	492	n.d.
$Neu5,7,8,9Ac_4$	2.16	552	534	313-295-229
Neu5Ac9Lt	1.45	498	480	313-295-229
Neu5Ac8S	0.17	506	abs.	426
Neu4,8an5Ac	1.39	408	abs.	313-283-229
Neu5Gc	0.78	442	424	313-295-283-229
Neu4Ac5Gc	1.18	484	abs.	424-313-283-229
Neu7Ac5Gc	0.99	484	466	n.d.
Neu8Ac5Gc	1.04	484	466	313-295-229
Neu9Ac5Gc	1.19	484	466	313-295-229
$Neu7,9Ac_25Gc$[a]	1.66	526	508	313-295-229
$Neu8,9Ac_25Gc$[a]	2.10	526	508	313-295-229
Neu5Gc9Lt	1.14	514	496	n.d.
Neu5Gc8Me	0.98	456	438	313-295-229
Neu7Ac5Gc8Me	1.41	498	480	430-316-313-229
Neu9Ac5Gc8Me	1.90	498	480	316-313-295-229
Neu5Gc8S	0.16	522	abs.	464-442-424-313-229
Neu4,8an5Gc	1.21	424	abs.	313-283-229

[a]Assignments may be interchanged.

Elution with a linear two-step gradient of acetonitrile:methanol:water: 3.85:7.00:89.15 to 7.15:7.00:85.85 in 40 min, and then to 11.00:7.00:82.00 in 10 min, at a flow rate of 0.2 mL/min. The R_{Neu5Ac} values are given as a directive to set up an in-house set of HPLC values with reference sialic acids. (Data taken from **refs.** ***8*** and ***23***.)

abs., absent; n.d., not determined.

1. The molecular mass of the Sia derivative, and thereby the number and type of substituents, is reflected by the pseudomolecular ion $[M+H]^+$ and its sodium adduct $[M+Na]^+$.
2. An important fragment ion is the result of H_2O elimination $[M+H-18]^+$.

3. The ions $[M+H]^+$, $[M+Na]^+$, and $[M+H-18]^+$ are very useful for molecular ion chromatogram screening of HPLC effluents.
4. When focusing on *N,O*-acylneuraminic acids, the $[M+H-18]^+$ ion is suggested to correspond with a 4,8-anhydro ring structure in case both HO-4 and HO-8 are not substituted. Pathways starting from this ring structure fragmentation have been formulated for non-*O*-acylated, 7-*O*-acylated, and 9-*O*-acylated *N*-acylneuraminic acids.
5. Fragment A at *m/z* 313 (elimination of acylamide followed by 2X H_2O, 1X H_2O plus 1X HOAc, or 2X HOAc from the ring structured $[M+H-18]^+$ ion) is present in all Sia derivatives.
6. Fragment D at *m/z* 229 (fragmentation of the ring of the $[M+H-18]^+$ ion; C-3,4,5 containing fragment R-C_3H_3, with R = C-1,2 part) is present in all Sia derivatives.
7. Fragment B (further loss of H_2O from fragment A) at *m/z* 295 is relatively more important compared with fragments A and D when the Sia is substituted at O-9.
8. Fragment C (elimination of C-9 as formaldehyde from the $[M+H-18]^+$ ion) at *m/z* 283 is only present when the Sia is not substituted at O-8 or O-9.
9. It should be noted that the ESI/MS data do not differentiate between 8- and 9-*O*-acylations.
10. 4-*O*-acetylated Sias, missing a $[M+H-18]^+$ ion, are characterized by a major $[M+H-HOAc]^+$ ion.
11. Focusing on *N*-acyl-8-*O*-alkyl-neuraminic acids, it is suggested that etherification of HO-8 (Neu5Ac/5Gc8Me) blocks the cyclization between C-4 and C-8. Furthermore, an ion $[M+H-18]^+$ is present, as well as *m/z* 229 and the fragments arising from the loss of acylamide and two or three molecules of water. The elimination of a formaldehyde part does not occur.

4. Notes

1. It has been stated that in the case of preparing HFB methyl ester derivatives *(7)*, no purification of the liberated Sia pool is needed (in the case of glycoproteins, a short-term centrifugation step is recommended).
2. The report on preparing HFB methyl ester derivatives *(7)* mentions that treatment of standard Neu5Ac with diazomethane reagent up to 15 d at room temperature did not produce Neu5Ac8Me. Experience shows that HO-4 is especially sensitive for *O*-methylation in a diazomethane protocol, and short incubation times are advised.

Acknowledgments

We thank Dr. Roland Schauer (University of Kiel, Germany) and Dr. Hans Vliegenthart (Utrecht University, The Netherlands) for close collaboration in the Sia area for more than 30 yr.

References

1. Schauer, R., ed. (1982) Sialic Acids/Chemistry, Metabolism and Function, Springer-Verlag, Vienna, Austria, *Cell Biology Monographs* **10**.
2. Varki, A. (1992) Diversity in the sialic acids. *Glycobiology* **2,** 25–40.

3. Schauer, R. and Kamerling, J. P. (1997) Chemistry, biochemistry and biology of sialic acids, in *Glycoproteins II* (Montreuil, J., Vliegenthart, J. F. G., and Schachter, H., eds.), Elsevier, Amsterdam, The Netherlands, *New Comprehensive Biochemistry* **29b,** 243–402.
4. Schauer, R. (2000) Achievements and challenges of sialic acid research. *Glycoconjugate J.* **17,** 485–499.
5. Angata, T. and Varki, A. (2002) Chemical diversity in the sialic acids and related α-keto acids: an evolutionary perspective. *Chem. Rev.* **102,** 439–469.
6. Schauer, R. (2004) Sialic acids: fascinating sugars in higher animals and man. *Zoology* **107,** 49–64.
7. Zanetta, J.-P., Pons, A., Iwersen, M., et al. (2001) Diversity of sialic acids revealed using gas chromatography/mass spectrometry of heptafluorobutyrate derivatives. *Glycobiology* **11,** 663–676.
8. Klein, A., Diaz, S., Ferreira, I., Lamblin, G., Roussel, P., and Manzi, A. E. (1997) New sialic acids from biological sources identified by a comprehensive and sensitive approach: liquid chromatography-electrospray ionization-mass spectrometry (LC-ESI-MS) of SIA quinoxalinones. *Glycobiology* **7,** 421–432.
9. Kamerling, J. P., Schauer, R., Shukla, A. K., Stoll, S., van Halbeek, H., and Vliegenthart, J. F. G. (1987) Migration of *O*-acetyl groups in *N*,*O*-acetylneuraminic acids. *Eur. J. Biochem.* **162,** 601–607.
10. Varki, A. and Diaz, S. (1984) The release and purification of sialic acids from glycoconjugates: methods to minimize the loss and migration of *O*-acetyl groups. *Anal. Biochem.* **137,** 236–247.
11. Reuter, G. and Schauer, R. (1994) Determination of sialic acids. *Methods Enzymol.* **230,** 168–199.
12. Mawhinney, T. P. and Chance, D. L. (1994) Hydrolysis of sialic acids and *O*-acetylated sialic acids with propionic acid. *Anal. Biochem.* **223,** 164–167.
13. Chatterjee, M., Kumar Chava, A., Kohla, G., et al. (2003) Identification and characterization of adsorbed serum sialoglycans on *Leishmania donovani* promastigotes. *Glycobiology* **13,** 351–361.
14. Kamerling, J. P., Vliegenthart, J. F. G., Versluis, C., and Schauer, R. (1975) Identification of *O*-acetylated *N*-acylneuraminic acids by mass spectrometry. *Carbohydr. Res.* **41,** 7–17.
15. Kamerling, J. P., Vliegenthart, J. F. G., and Vink, J. (1974) Mass spectrometry of pertrimethylsilyl neuraminic acid derivatives. *Carbohydr. Res.* **33,** 297–306.
16. Kamerling, J. P. and Vliegenthart, J. F. G. (1982) Gas-liquid chromatography and mass spectrometry of sialic acids, in *Sialic Acids/Chemistry, Metabolism and Function* (Schauer, R., ed.), Springer-Verlag, Vienna, Austria, *Cell Biology Monographs* **10,** 95–125.
17. Bergwerff, A. A., Hulleman, S. H. D., Kamerling, J. P., et al. (1992) Nature and biosynthesis of sialic acids in the starfish *Asterias rubens*. Identification of sialo-oligomers and detection of *S*-adenosyl-L-methionine:*N*-acylneuraminate 8-*O*-methyltransferase and CMP-*N*-acetylneuraminate monooxygenase activities. *Biochimie* **74,** 25–38.

18. Kluge, A., Reuter, G., Lee, H., Ruch-Heeger, B., and Schauer, R. (1992) Interaction of rat peritoneal macrophages with homologous sialidase-treated thrombocytes in vitro: biochemical and morphological studies. Detection of N-(O-acetyl) glycoloylneuraminic acid. *Eur. J. Cell Biol.* **59,** 12–20.
19. Hara, S., Yamaguchi, M., Takemori, Y., Furuhata, K., Ogura, H., and Nakamura, M. (1989) Determination of mono-*O*-acetylated *N*-acetylneuraminic acids in human and rat sera by fluorometric high-performance liquid chromatography. *Anal. Biochem.* **179,** 162–166.
20. Anumula, K. R. (1995) Rapid quantitative determination of sialic acids in glycoproteins by high-performance liquid chromatography with a sensitive fluorescence detection. *Anal. Biochem.* **230,** 24–30.
21. Regl, G., Kaser, A., Iwersen, M., et al. (1999) The hemagglutinin-esterase of mouse hepatitis virus strain S is a sialate-4-*O*-acetylesterase. *J. Virol.* **73,** 4721–4727.
22. Manzi, A. E., Diaz, S., and Varki, A. (1990) High-pressure liquid chromatography of sialic acids on a pellicular resin anion-exchange column with pulsed amperometric detection: A comparison with six other systems. *Anal. Biochem.* **188,** 20–32.
23. Klein, A. and Roussel, P. (1998) *O*-Acetylation of sialic acids. *Biochimie* **80,** 49–57.

7

Nonradioactive *Trans*-Sialidase Screening Assay

Silke Schrader and Roland Schauer

Summary

Trans-sialidase (TS; E.C. 3.2.1.18) catalyzes the transfer of preferably α2,3-linked sialic acid to another glycan or glycoconjugate, forming a new α2,3-linkage to galactose or *N*-acetylgalactosamine. In the absence of an appropriate acceptor, TS acts as a sialidase, hydrolytically releasing glycosidically linked sialic acid. Interest in TS has increased rapidly in recent years owing to its great relevance to the pathogenicity of trypanosomes and its possible application in the regiospecific synthesis of sialylated carbohydrates and glycoconjugates. Recently, the authors described a newly developed nonradioactive screening test for monitoring TS activity ***(1)***. In this highly sensitive and specific assay, 4-methylumbelliferyl-β-D-galactoside is used as acceptor substrate and sialyllactose as donor to fluorimetrically detect enzyme activity in the low mU range (~0.1–1 mU/mL possible). The test can be applied to screen a large number of samples quickly and reliably during enzyme purification, for testing inhibitors, and for monitoring TS activity during the production of monoclonal antibodies ***(2)***.

This chapter focuses on the main steps of this assay and gives detailed instructions for performing a nonradioactive TS 96-well-plate fluorescence test. In addition, it describes the controls necessary when starting to monitor an unknown TS and facts to be considered when testing new substrates and inhibitors.

Key Words: *Trans*-sialidase; sialidase; sialic acids; sialyllactose; 4-methylumbelliferyl-β-D-galactoside; nonradioactive screening assay; fluorescence assay; 96-well-plates.

1. Introduction

Sialic acids are a family of about 50 naturally occurring derivatives of the nine-carbon sugar neuraminic acid, and are typically found at the outermost position of sugar chains of glycoconjugates ***(3,4)***. The most widespread form in nature is *N*-acetylneuraminic acid (Neu5Ac, of which the amino group is

From: *Methods in Molecular Biology, vol. 347: Glycobiology Protocols*
Edited by: I. Brockhausen © Humana Press Inc., Totowa, NJ

acetylated), followed by *N*-glycolylneuraminic acid (Neu5Gc) and *O*-acetylated derivatives ***(5)***. Owing to their terminal position, their negative charge and chemical diversity, sialic acids play an important role in cell–cell and cell–molecule interactions ***(6,7)***. The key enzyme of sialic acid catabolism is sialidase (neuraminidase, E.C. 3.2.1.18), an exoglycosidase that hydrolyzes terminal sialic acid residues from glycoproteins, glycolipids, oligosaccharides, and polysaccharides. Sialidases are mainly found in animals of the deuterostomate lineage and in several viruses, bacteria, protozoa, and fungi ***(8)***. In microorganisms, sialidases can play roles in both nutrition and in pathogenicity ***(9)***. An endo-sialidase from phage origin is also known, which hydrolyzes polysialic acids, e.g., colominic acid ***(10)***. In contrast to the "classical" sialidases, *trans*-sialidases (TSs; E.C. 3.2.1.18) catalyze the transfer of preferably α2,3-linked sialic acid directly to terminal β-galactose- or β-*N*-acetylgalactosamine-containing acceptors, giving rise to a new α2,3-linkage ***(11)***. In the absence of an appropriate acceptor this enzyme acts as a sialidase, hydrolytically releasing glycosidically-linked sialic acid (*see* **Fig. 1**).

TS is a GPI-anchored cell surface enzyme that was first described in *Trypanosoma cruzi*, the agent of Chagas disease in South and Central America ***(12)***. It was also found in other trypanosomes ***(13,14)*** and is used to acquire sialic acid from host glycoconjugates for sialylation of parasites plasma membrane glycoproteins ***(15)***. TS was shown to play important roles in the pathogenicity of trypanosomes ***(12,13,16,17)***. This enzyme is considered to be the main virulence factor for infectious African and American trypanosomes, best studied with *T. cruzi*. It strengthens the innate immunity of the parasites by sialylation, masks antigens, protects trypanosomes from proteases, disturbs the host's cytokine network, impairs its immune response, manipulates multiple host cell-signaling pathways, inhibits apoptosis, and facilitates recognition and uptake of trypanosomes by host cells ***(12,13,16–24)***. The occurrence of TS has also been described in *Endotrypanum* species ***(25)***, *Corynebacterium diphtheriae* ***(26)***, and human serum ***(27)***.

Owing to its extensive pathophysiological, pharmacological, and biotechnological significance, interest in research on TS has increased in recent years. Tiralongo et al. described two TS forms with different sialic acid transfer and sialidase activities from *Trypanosoma congolense* (*see* **refs. *2*** and ***28***). In addition, several experiments with amino acid exchanges of the enzyme protein were performed to elucidate the catalytic mechanism of TS reaction ***(29,30)***. Buschiazzo et al. crystallized TS from *T. cruzi* alone and in complex with different sugar ligands and assumed that sialic acid binding triggers a conformational switch activating the enzyme and modulating affinity for the acceptor substrate ***(31)***. The possibility of using TS for the regiospecific synthesis of sialylated carbohydrates and glycoconjugates and the search for potent TS

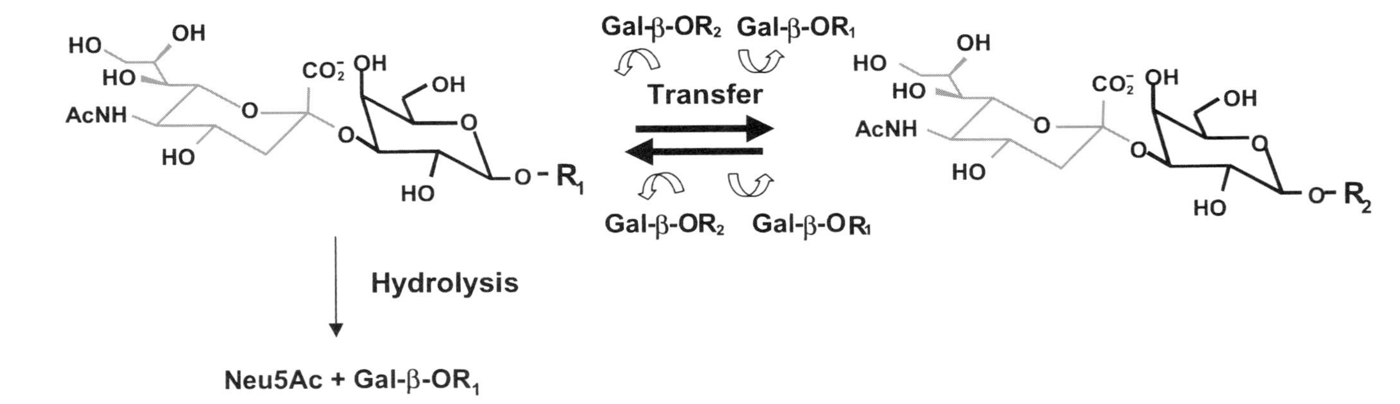

Fig. 1. *Trans*-sialidase acts as a *trans*-glycosidase, catalyzing the transfer of sialic acid from one galactoside, with R_1 as penultimate glycan, to another galactose, bearing R_2 as subterminal glycan, in a reversible reaction. This enzyme also has the property of a sialidase by hydrolyzing the sialic acid–galactose bond in a unidirectional reaction.

inhibitors to reduce trypanosome pathogenicity has led to several approaches and research activities in these areas ***(32,33)***.

Various assays exist for the detection of the activity of this enzyme. Mattos-Guaraldi et al. ***(26)*** developed a peanut lectin hemagglutination assay for testing the TS from *C. diphtheriae*. An enzyme-linked immunoassay using recombinant TS of *T. cruzi* ***(34)*** and a TS inhibition assay ***(35)*** was recently introduced primarily for the diagnostic purpose of Chagas disease. High pH anion-exchange chromatography with pulsed amperometric detection of the reaction products was applied for investigation of the acceptor substrate specificity of lactose derivatives of TS and their inhibitory potency ***(32)***. To facilitate the monitoring of TS activity in various biological samples, a 96-well-plate fluorescence test was developed ***(1)***. The assay can be used to screen TS activity during enzyme purification and the production of monoclonal antibodies, as well as for substrate and inhibitor testing. This TS test, described below, consists of two steps. First, TS is incubated with the substrates sialyllactose and 4-methylumbelliferyl-β-D-galactoside (MUGal). Second, the formed product, 4-methylumbelliferyl-β-D-sialylgalactoside (MUGalNeu5Ac), is separated from the substrate MUGal by use of anion-exchange chromatography, followed by acid hydrolysis of MUGalNeu5Ac and detection of released methylumbelliferone (MU). The separation step is performed in 96-well filter plates. To get optimal results, it is important to adapt this system to the properties of the TS used. In addition to a detailed description of these two steps, this chapter gives instructions to test for possible remaining sialidase activity and considerations in regard to examining potential substrates and inhibitors.

2. Materials

2.1. TS Assay

1. 200 m*M* of *bis/tris* buffer, pH 7.0 (store at room temperature for the detection of *T. congolense* TS); 400 m*M* of piperazine-*N,N′–bis*(2-ethanesulphonate) (PIPES) buffer, pH 7.0 (store at room temperature for the detection of recombinant *T. cruzi* TS [*see* **Note 1**]).
2. 10 m*M* of 3′-sialyllactose (3′-SL; Sigma) or isolated from bovine milk as described by Veh et al. ***(36)***, or 10 m*M* of 3′-sialyllactosamine (3′-SLN) dissolved in water and stored at –20°C (*see* **Note 2**).
3. Dissolve 2 m*M* of MUGal (Sigma) in water, sonicate for 10 min, and store at –20°C. Sonification for 10 min is necessary immediately before use (*see* **Note 2**).
4. Wash 100 mL of Q-Sepharose Fast Flow (Pharmacia) three times with 1 *M* of sodium acetate and soak overnight in 1 *M* of sodium acetate at +4°C, then adjust to a final ratio of gel volume: sodium acetate supernatant of 1 : 1. Store at +4°C.
5. Elution medium: 1 *M* of HCl. Store at room temperature.
6. Stopping media: 6 *M* of NaOH solution and 1 *M* of glycine/NaOH, pH 10.0. Store both at room temperature.

7. Polypropylene 96-well plates (0.5-mL MicroWell™ plates, Nunc, Denmark), which can be sealed with Nunc™ Well Caps.
8. Multi-channel pipet (CappAero™, Dunn Labortechnik GmbH, Germany).
9. UNIFILTER® 800-filter-well plate (GF/D glass fiber filter, Polyfiltronics®, Whatman, UK; *see* **Note 3**).
10. Vacuum manifold (QIAvac 96, Qiagen, Germany).
11. 2-mL Nunc™ 96 DeepWell Plate (Nunc; *see* **Note 4**).
12. Polyolefin sealing tape (Nunc; *see* **Note 4**).
13. Nunc Well Caps (*see* **Note 4**).
14. Black 96-well plates (Microfluor, Dynex Technologies, Chantilly, VA).
15. 96-Well-plate fluorimeter (e.g., TECAN, Deutschland GmbH, Germany).

2.2. MU Calibration Curve and Control of Sialidase Activity

1. Dissolve 4-MU (Sigma-Aldrich) in ddH_2O; prepare a 0.1-m*M* stock solution and store at –20°C, protected from light.
2. Dissolve 10 m*M* of 2′-(4-methylumbelliferyl)-α-D-*N*-acetylneuraminic acid (MUNeu5Ac; Sigma-Aldrich) in ddH_2O and store at –20°C.

3. Methods

The TS assay conditions described are based on the properties of *T. congolense* and recombinant *T. cruzi* TS. Some parameters may change depending on the source and concentration of the TS used: kind, molarity, and pH of the enzyme assay buffer (*see* **Note 1** and **Fig. 2**), incubation temperature and incubation time of the assay, and the amount of enzyme solution applied in the test (*see* **Fig. 3**). The substrates MUGal and 3′-SL have been found to be particularly suitable for the TS from *T. congolense*, as well as for the recombinant TS from *T. cruzi (1)*.

3.1. TS Assay

1. Sonicate the 2-m*M* MUGal stock suspension for 10 min prior to use in the assay in order to obtain a homogeneous dispersion.
2. The assay is performed in polypropylene 96-well plates. Transfer 12.5 µL, 200 m*M* of *bis/tris* buffer, pH 7.0, for testing *T. congolense* TS, or 12.5-µL, 400 m*M* of PIPES buffer, pH 7.0, for testing recombinant *T. cruzi* TS, into each well. Then add 5 µL of 10 m*M* 3′-SL and 20 µL of TS solution (e.g., crude extract or diluted recombinant TS; *see* **Note 5**). As a blank, transfer 20 µL of 50-m*M bis/tris* buffer, pH 7.0 (for *T. congolense* TS) and 20 µL of 100-m*M* PIPES buffer, pH 7.0 (for recombinant *T. cruzi* TS) or 20-µL heat-inactivated enzyme solution (5 min at 95°C) or 20-µL ddH_2O into the wells.
3. The reaction is started by the addition of 12.5 µL of MUGal stock suspension into each well; the plate is shaken, sealed with Nunc Well Caps, and incubated at

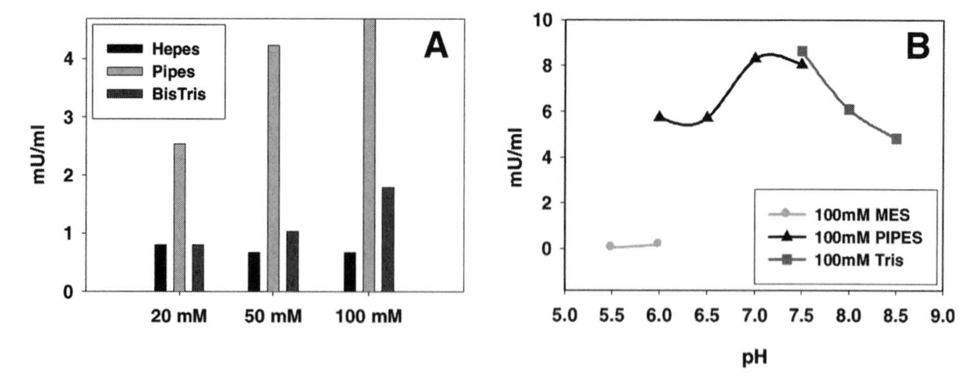

Fig. 2. Dependence of *Trypanosoma cruzi trans*-sialidase activity on **(A)** the type and molarity of buffer and **(B)** pH.

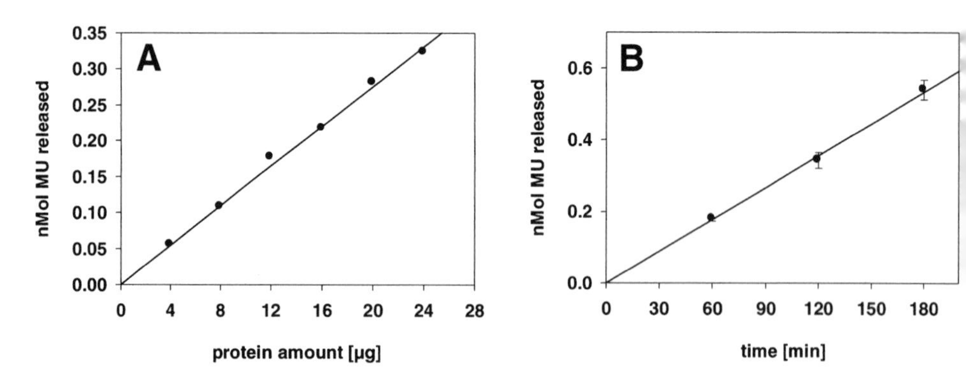

Fig. 3. Dependence of *Trypanosoma congolense trans*-sialidase activity on **(A)** protein amount and **(B)** incubation time.

37°C for 2 h for *T. congolense* TS and at 25°C for 30 min for recombinant *T. cruzi* TS, respectively (*see* **Note 6**).

4. The reaction is stopped by the addition of 350 µL of ice-cold water to each well. Store the plate on ice until applying the filtration step (*see* **Subheading 3.2.**).

3.2. Separation of the TS Product From Substrate Using Filter Plates and Detection of Released MU

1. During incubation time of the TS assay, load a UNIFILTER 800-filter-well plate with 300 µL of Q-Sepharose (acetate form) by transferring 600 µL of Q-Sepharose sodium acetate solution into each well and applying a vacuum with a vacuum manifold. Then wash each mini-column six times with 500 µL of water under vacuum using a multi-channel pipet. The vacuum must be applied slowly and only after the water has been added to each well in order to avoid drying of the columns.

2. Transfer the assay-water mixtures (*see* **Subheading 3.1., step 4**) to the prepared filter plate using a multi-channel pipet, so that each sample corresponds to one mini-column. Apply vacuum again slowly until the mini-column has absorbed the whole sample. Remove excess MUGal from the columns by successive washings with water (6 × 500 µL) under vacuum.
3. After discarding the first 100 µL of the eluent (1 *M* HCl), elute the sialylated product from the columns with 6 × 150 µL of 1 *M* HCl. The eluate is collected in a 2-mL Nunc 96 DeepWell Plate and sealed with polyolefin sealing tape (*see* **Notes 7** and **8**).
4. After hydrolysis of the eluted product in a water bath at 95°C for 45 min, remove the sealing tape immediately and cool the plate on ice for 15 min. Adjust the pH to 10.0 by adding 120 µL of 6 *M* NaOH and 300 µL of 1 *M* glycine/NaOH, pH 10.0 (*see* **Note 9**). Then seal the plate with Nunc Well Caps. After mixing by inversion of the plate two or three times, transfer 300 µL of the reaction mixture into black 96-well plates and quantify the released MU using a 96-well-plate fluorimeter at an excitation and emission wavelength of 365 and 450 nm, respectively (*see* **Note 10**).
5. Fluorescence is taken as a measure for the amount of the product MUGalNeu5Ac. One unit of TS is defined as one µmol of sialylated product formed per minute.

3.3. Performing a Calibration Curve for MU

1. An MU calibration curve must be performed to quantify TS activity. For each MU calibration point, triplicates are run. To allow for possible losses of MU during acid hydrolysis, treat MU similar to the product MUGalNeu5Ac.
2. For each triplicate, transfer 100 µL of elution media (1 *M* of HCl) under vacuum onto the prepared and equilibrated mini-columns of the filter plate and discard (*see* **Subheading 3.2., step 1**). Then collect the applied 6 × 150 µL of 1 *M* HCl under vacuum into a 2-mL Nunc 96 DeepWell Plate.
3. Transfer different amounts of MU into the wells of the 2-mL Nunc 96 DeepWell Plate.
4. After sealing with polyolefin sealing tape, incubate the DeepWell plate in a water bath for 45 min at 95°C. Remove the sealing tape immediately and cool the plate on ice for 15 min. After adjusting the pH to 10.0 and mixing as described above, measure the fluorescence at an excitation and emission wavelength of 365 nm and 450 nm, respectively (*see* **Subheading 3.2., step 4**).
5. The obtained calibration curve can be used to calculate TS activity (*see* **Note 11**).

3.4. Determining Potential Residual Sialidase Activity (see Note 12)

1. Incubate TS with 3′-SL and MUGal as described in **Subheading 3.1.**
2. Stop the reaction with liquid nitrogen.
3. Lyophilize the samples and determine the amount of free *N*-acetylneuraminic acid by fluorimetric high-performance liquid chromatography after derivatization with 1,2-diamino-4,5-methylenedioxybenzene as described by Hara et al. *(37)*.

4. Quantify sialidase activity of TS by incubating 40 μL of enzyme solution with 25 μL of 200-m*M* *bis/tris* buffer, pH 7.0, for testing *T. congolense* TS, or 25 μL of 400-m*M* PIPES buffer, pH 7.0, for testing recombinant *T. cruzi* TS, 25 μL of ddH_2O, and 10 μL of 10-m*M* MUNeu5Ac solution ***(2,38)***. Black 96-well plates must be used for this assay. Cover the plates with a second plate. Incubation time and temperature are the same as for the TS assays. In parallel, incubate different amounts of MU in corresponding buffer (*bis/tris* or PIPES) to perform an MU calibration curve. Stop the reaction by adding 50 μL of 1 *M* glycine/NaOH, pH 10.0, and 150 μL of ddH_2O. Determine fluorescence by shaking the plate at excitation and emission wavelengths of 365 nm and 450 nm, respectively.

3.5. Testing of Potential Substrates and Inhibitors

1. Prepare stock solutions with potential donors, acceptors, and inhibitors by solving the compounds in ddH_2O for testing potential substrates and assay buffer, e.g., 200 m*M* of *bis/tris*, pH 7.0, for testing potential inhibitors, respectively.
2. For testing of potential donors (donor assay), replace 3′-SL with a potential donor (oligosaccharide, glycoprotein, or ganglioside), and run the TS assay as described in **Subheading 3.1.** and quantify as described in **Subheading 3.2.** A TS test with 3′-SL and without any other donor can be run in parallel as a control (3′-SL control assay).
3. For testing of potential donor inhibitors (donor inhibitor assay), add different concentrations of potential donor inhibitors to the TS assay mixture before starting the reaction with MUGal. The final volume of 50 μL and the buffer concentration must be maintained. Run a TS test without potential inhibitors in parallel as a control (3′-SL control assay). Run all assays as described in **Subheading 3.1.** and quantify the product MUGalNeu5Ac as described in **Subheading 3.2.** Inhibitory effects are reflected if fluorescence in the donor inhibitor assay is reduced when compared with the 3′-SL control assay.
4. For testing of potential acceptors, in contrast to the donor testing, an acceptor assay in which the acceptor MUGal is replaced by another acceptor is not possible with the method described here, except another 4-methylumbelliferyl substrate such as 4-methylumbelliferyl-α-D-lactoside or a similar fluorescence compound (e.g., CF_3MU substrates), can be purchased or prepared according to the method described by Engstler et al. ***(39)***. When testing other fluorescence compounds such as CF_3MU substrates, an additional calibration curve with CF_3MU must be created. For further details, *see* **Note 13**.
5. For testing of potential acceptors or acceptor inhibitors (acceptor inhibitor assay), different concentrations of potential acceptors or acceptor inhibitors are added to the TS assay mixture before starting with MUGal (*see* **Subheading 3.1.**). The final volume of 50 μL of this acceptor inhibitor assay and the buffer concentration must be maintained. Run a standard TS test without additional compounds in parallel as a control (3′-SL control assay). Possible inhibitory effects of acceptor substrates appear if fluorescence in the acceptor inhibitor assay is reduced compared with the control assay. However, because this reduction can be caused

both by an acceptor function and by an inhibitor effect, this is not proof and additional experiments must follow. Therefore, the assay described here can be used as a pre-screening for potential acceptors. In case of fluorescence reduction in the inhibitory assay, the products must be separated and further analyzed *(32)* to distinguish between acceptor and inhibitor function. In addition, kinetic studies can give more information about the type of inhibition.

6. Quenching effects can influence all systems that are based on fluorescence measurements. Such quenching effects must be controlled, especially when modifying the enzyme assay (e.g., in the case of testing potential substrates and inhibitors). To test possible quenching in the donor assay, a TS test without the potential donor (assay minus donor) is run in parallel to the standard donor assay (assay plus donor). To measure possible quenching in the inhibitor assay, this assay is run in parallel to the standard assay (assay minus inhibitor; *see* **Table 1**). After the separation step using 96-filter-well plates, performing acid hydrolysis and neutralization (*see* **Subheading 3.2.**), add a defined amount of MU (within the linear range of the MU calibration curve) to 300 µL of each of the parallel TS assays (assay with and without donor or assay with and without inhibitor). Use 300 µL of the reaction mixture of each test without additional MU as controls. After shaking the black 96-well plate, quantify MU using a 96-well-plate fluorimeter at an excitation and emission wavelength of 365 and 450 nm, respectively. Quenching effects exist if the difference between the plus MU and minus MU measurements in the assay with donor or inhibitor are significantly decreased compared to the difference between the plus MU and minus MU measurements of the control assay (assay without donor or assay without inhibitor when testing donors or inhibitors, respectively).

4. Notes

1. The type, molarity, and pH of buffers are useful for monitoring TS of *T. congolense* and recombinant TS of *T. cruzi*, respectively. However, these parameters have to be checked and modified if necessary when quantifying another TS. **Figure 2** shows the type, molarity, and pH of the assay buffer for recombinant *T. cruzi* TS. In this case, 100 m*M* of PIPES, pH 7.0, should be used.
2. Different donor and acceptor substrates can modify the TS assay conditions. Donor substrates like Neu5Acα2,3-*N*-acetyllactosamine (3′-SLN) and fetuin may also be used in this assay *(1)*. A second possible acceptor substrate is 4-methylumbelliferyl-α-D-lactoside (*see* **ref. *1***), which has a better solubility in aqueous solutions than MUGal. Nevertheless, affinity to these substrates depends on the TS tested *(2)*.
3. The 96-filter-well plates must be deep enough to hold 300 µL of packed gel and also provide sufficient volume for the washing and elution media. They must also contain a glass fiber membrane that does not bind the product and has a structure allowing a continuous and moderate flow under vacuum. Resistance to 1 *M* of HCl and reusability are also important properties of the filter plates. Thus, UNIFILTER 800-filter-well plates from Whatman were chosen, which can be reused four or five times.

Table 1
Testing of Possible Quenching Effects in the Donor and Inhibitor Assay

Donor assay				Inhibitor assay (donor or acceptor)			
Assay plus donor (A+D)		Assay minus donor (A–D)		Assay plus inhibitor (A+I)		Assay minus inhibitor (A–I)	
After neutralization:				After neutralization:			
+ MU	– MU	+ MU	– MU	+ MU	– MU	+ MU	– MU
↓	↓	↓	↓	↓	↓	↓	↓
Calculation of difference (+MU) – (–MU): Δ MU(A+D)		Calculation of difference (+MU) – (–MU): Δ MU(A–D)		Calculation of difference (+MU) – (–MU): Δ MU(A+I)		Calculation of difference (+MU) – (–MU): Δ MU(A–I)	
If Δ MU(A+D) < Δ MU(A–D): → Quenching				If Δ MU(A+I) < Δ MU(A–I): → Quenching			

4. The filter-well plate equipment has been tested accurately with regard to special properties. Thus, the 2-mL DeepWell plates consist of polypropylene and are heat-stable, resistant to 1 *M* of HCl, reusable, and deep enough for the volume of the eluate and neutralization medium after acid hydrolysis. In addition, they can be closed with Nunc Well Caps in order to mix the samples after neutralization. The sealing tape chosen is also heat-stable and resistant to 1 *M* of HCl.
5. The amount of used enzyme or enzyme solution (e.g., crude extract) also depends on the sensitivity of the fluorimeter used. In order to measure within a linear range, the determination of fluorescence in dependency on enzyme amount and time is important (**Fig. 3A** shows the dependency of TS activity from *T. congolense* on protein amount). If you must dilute enzyme solutions and observe loss in activity, try adding 0.2% bovine serum albumin to the dilution buffer to stabilize the activity.
6. The incubation time and temperature have been chosen for the TS from *T. congolense* (*see* **Fig. 3B**) and recombinant *T. cruzi* TS ***(1)***, respectively. These parameters must be modified depending on the kind and source of the enzyme used.
7. As shown in **Fig. 4**, the first 100 μL of eluate can be discarded; since the total amount of product (MUGalNeu5Ac) will be eluted in the following 900 μL of 1 *M* of HCl. Successive elution with 6 × 150 μL is important to elute the whole amount of bound MUGalNeu5Ac.
8. The TS assay can be slightly modified. In order to reduce the assay time, the amount of the product MUGalNeu5Ac can also be detected without acid hydrolysis by monitoring its fluorescence at excitation and emission wavelengths of 310–315 and 375 nm, respectively. However, this alternative method is not as sensitive as the method

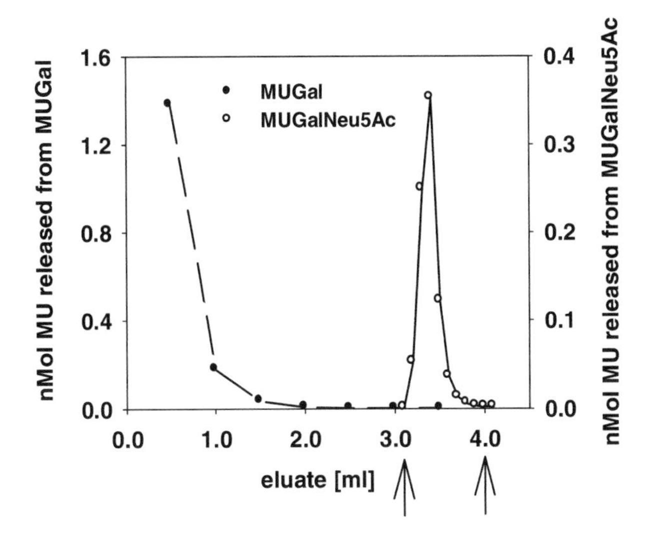

Fig. 4. Elution profile of the acceptor MUGal and the product MUGalNeu5Ac on 300 µL Q-Sepharose packed in wells, detected as *n*mol of MU released after acid hydrolysis and neutralization, during separation with 96-filter-well plates. MUGal was washed from the columns with 500-µL steps of water. MUGalNeu5Ac was eluted with 100-µL steps of 1 *M* HCl as indicated with arrows. (Reprinted from **ref. *1*** with permission from Elsevier.)

involving acid hydrolysis of MUGal, and it requires high enzyme activities or a highly sensitive fluorimeter (*see* **Table 2**). Owing to their identical excitation and emission wavelength maxima, MUGalNeu5Ac and MUGal still must be separated with 96-filter-well plates as described in **Subheading 3.2., steps 1–3** using this alternative method. The product has to be eluted with 1 *M* of ammonium acetate instead of 1 *M* of HCl, and the amount of MUGalNeu5Ac can be detected immediately.

9. The exact amount of 6 *M* of NaOH to be added must be determined after each preparation of NaOH solution. Slightly inaccurate NaOH masses during the preparation of NaOH solution can lead to different pH values after neutralization of the samples. The fluorescence units of measured MU also depend on the pH values in the neutralized samples. The pH values should be at about 10.0. If the pH values do not change within the first determinations, the defined exact volume of added 6 *M* of NaOH solution can be used routinely.
10. Owing to light sensitivity of MU, quantification of released MU should be performed immediately after neutralization of samples.
11. Because MUGalNeu5Ac produced in the TS reaction is not commercially available, known amounts of MUGal were hydrolyzed in 1 *M* of HCl in a water bath at 95°C for 45 min. Hydrolysis of MUGal under these conditions led to the same fluorescence values as for similarly treated MU *(1)*. Therefore, a calibration curve with MU can be performed.

Table 2
Direct Detection of the *Trans*-Sialidase Product MUGalNeu5Ac Without Hydolysis After Elution With 1 *M* of Ammonium Acetate

	+ TS	– TS (control)
MUGalNeu5Ac	231 ± 12.1	4.3 ± 0.4
MU	1594 ± 79.6	50.5 ± 2.2

The values (fluorescence units) are the means of three independent experiments ± standard deviation. The values for the detection of released MU after acid hydrolysis of MUGalNeu5Ac from a corresponding sample are given for comparison. As controls, samples without *trans*-sialidase activity were treated in the same way. (Reprinted from **ref. *1*** with permission from Elsevier.)

12. Tiralongo et al. *(2)* described two TS with different sialic acid transfer and sialidase activities from *T. congolense*. Some authors reported that a part of this sialidase activity could also be present at low acceptor concentrations or when using inefficient acceptors ***(40)***. Therefore, possible remaining sialidase activity can be measured by determination of liberated Neu5Ac using fluorometric high-performance liquid chromatography. Estimation of sialidase activity of TS using MUNeu5Ac as a classical sialidase substrate can give additional information.
13. Agusti et al. ***(32)*** describe a method in which the acceptor specificity of *T. cruzi* TS was measured by high pH anion-exchange chromatography with pulse amperometric detection in combination with different CarboPac columns. This method can be used for further detailed analysis of acceptor specificity.

Acknowledgments

We thank Dr. Evelin Tiralongo of Griffith University, Gold Coast, Australia, for valuable discussions and advice; Dr. Alberto Carlos C. Frasch and Dr. Gastón Paris, Universidad Nacional de Gral., San Martin, Argentina, for recombinant TS from *Trypanosoma cruzi*; and Dr. Teruo Yoshino, International Christian University, Tokyo, for synthesizing 4-methylumbelliferyl-α-D-lactoside. The technical assistance of Marzog El-Madani, Alice Schneider, and Renate Thun is gratefully acknowledged, as well as the financial support of Numico Research (Friedrichsdorf, Germany), Sialic Acids Society (Kiel, Germany), and Fonds der Chemischen Industrie (Frankfurt, Germany).

References

1. Schrader, S., Tiralongo, E., Paris, G., Yoshino, T., and Schauer, R. (2003) A nonradioactive 96-well plate assay for screening of trans-sialidase activity. *Anal. Biochem.* **322,** 139–147.

2. Tiralongo, E., Schrader, S., Lange, H., Lemke, H., Tiralongo, J., and Schauer, R. (2003) Two trans-sialidase forms with different sialic acid transfer and sialidase activities from Trypanosoma congolense. *J. Biol. Chem.* **278,** 23,301–23,310.
3. Schauer, R. (2000) Achievements and challenges of sialic acid research. *Glycoconj. J.* **17,** 485–499.
4. Schauer, R. and Kamerling, J. P. (1997) Chemistry, biochemistry and biology of sialic acids, in *Glycoproteins II* (Montreuil, J., Vliegenthart, J. F. G., and Schachter, H., eds.), Elsevier, Amsterdam, pp. 243–402.
5. Schauer, R. (1988) Sialic acids as antigenic determinants of complex carbohydrates. *Adv. Exp. Med. Biol.* **228,** 47–72.
6. Kelm, S. and Schauer, R. (1997) Sialic acids in molecular and cellular interactions. *Int. Rev. Cytol.* **175,** 137–240.
7. Schauer, R. (2004) Sialic acids: Fascinating sugars in higher animals and man. *Zoology* **107,** 49–64.
8. Roggentin, P., Schauer, R., Hoyer, L. L., and Vimr, E. R. (1993) The sialidase superfamily and its spread by horizontal gene transfer. *Mol. Microbiol.* **9,** 915–921.
9. Corfield, T. (1992) Bacterial sialidases—roles in pathogenicity and nutrition. *Glycobiology* **2,** 509–521.
10. Mühlenhoff, M., Stummeyer, K., Grove, M., Sauerborn, M., and Gerardy-Schahn, R. (2003) Proteolytic processing and oligomerization of bacteriophage-derived endosialidases. *J. Biol. Chem.* **278,** 12,634–12,644.
11. Vandekerckhove, F., Schenkman, S., Pontes de Carvalho, L., et al. (1992) Substrate specificity of the *Trypanosoma cruzi* trans-sialidase. *Glycobiology* **2,** 541–548.
12. Schenkman, S., Jiang, M. S., Hart, G. W., and Nussenzweig, V. (1991) A novel cell surface trans-sialidase of *Trypanosoma cruzi* generates a stage-specific epitope required for invasion of mammalian cells. *Cell* **65,** 1117–1125.
13. Engstler, M., Schauer, R., and Brun, R. (1995) Distribution of developmentally regulated trans-sialidases in the Kinetoplastida and characterization of a shed trans-sialidase activity from procyclic *Trypanosoma congolense. Acta. Trop.* **59,** 117–129.
14. Pontes de Carvalho, L. C., Tomlinson, S., Vandekerckhove, F., et al. (1993) Characterization of a novel trans-sialidase of *Trypanosoma brucei* procyclic trypomastigotes and identification of procyclin as the main sialic acid acceptor. *J. Exp. Med.* **177,** 465–474.
15. Schenkman, S., Ferguson, M. A., Heise, N., de Almeida, M. L., Mortara, R. A., and Yoshida, N. (1993) Mucin-like glycoproteins linked to the membrane by glycosylphosphatidylinositol anchor are the major acceptors of sialic acid in a reaction catalyzed by trans-sialidase in metacyclic forms of *Trypanosoma cruzi. Mol. Biochem. Parasitol.* **59,** 293–304.
16. Fralish, B. H. and Tarleton, R. L. (2003) Genetic immunization with LYT1 or a pool of trans-sialidase genes protects mice from lethal *Trypanosoma cruzi* infection. *Vaccine* **21,** 3070–3080.
17. Calvet, C. M., Meuser, M., Almeida, D., Meirelles, M. N. L., and Pereira, M. C. S. (2004) *Trypanosoma cruzi*-cardiomyocyte interaction: role of fibronectin in the

recognition process and extracellular matrix expression in vitro and in vivo. *Exp. Parasitol.* **107,** 20–30.
18. Nagamune, K., Acosta-Serrano, A., Uemura, H., et al. (2004) Surface sialic acids taken from the host allow trypanosome survival in Tsetse fly vectors. *J. Exp. Med.* **199,** 1445–1450.
19. Todeschini, A. R., Nunes, M. P., Pires, R. S., et al. (2002) Costimulation of host T lymphocytes by a trypanosomal *trans*-sialidase: Involvement of CD43 signaling. *J. Immunol.* **168,** 5192–5198.
20. Mucci, J., Hidalgo, A., Mocetti, E., Argibay, P. F., Leguizamón, M. S., and Campetella, O. (2002) Thymocyte depletion in *Trypanosoma cruzi* infection is mediated by *trans*-sialidase-induced apoptosis on nurse cells complex. *Proc. Natl. Acad. Sci. USA* **99,** 3896–3901.
21. Woronowicz, A., De Vusser, K., Laroy, W., et al. (2004) Trypanosome trans-sialidase targets TrkA tyrosine kinase receptor and induces receptor internatization and activation. *Glycobiology* **14,** 987–998.
22. Chuenkova, M. V. and Pereira, M. A. (2000) A trypanosomal protein synergizes with the cytokines ciliary neurotrophic factor and leukemia inhibitory factor to prevent apoptosis of neuronal cells. *Mol. Biol. Cell* **11,** 1487–1498.
23. Risso, M. G., Garbarino, G. B., Mocetti, E., et al. (2004) Differential expression of a virulence factor, the *trans*-sialidase, by the main *Trypanosoma cruzi* phylogenetic lineages. *J. Infect. Dis.* **189,** 2250–2259.
24. Tribulatti, M. V., Mucci, J., Van Rooijen, N., Leguizamón, M. S., and Campetella, O. (2005) The *trans*-sialidase from *Trypanosoma cruzi* induces thrombocytopenia during acute Chagas' disease by reducing the platelet sialic acid contents. *Infect. Immun.* **73,** 201–207.
25. Medina-Acosta, E., Paul, S., Tomlinson, S., and Pontes-de-Carvalho, L. C. (1994) Combined occurrence of trypanosomal sialidase/trans-sialidase activities and leishmanial metalloproteinase gene homologues in *Endotrypanum sp. Mol. Biochem. Parasitol.* **64,** 273–282.
26. Mattos-Guaraldi, A. L., Formiga, L. C., and Andrade, A. F. (1998) Trans-sialidase activity for sialic acid incorporation on *Corynebacterium diphtheriae. FEMS Microbiol. Lett.* **168,** 167–172.
27. Nikonova, E. Y., Tertov, V. V., Sato, C., Kitajima, K., and Bovin, N. V. (2004) Specificity of human trans-sialidase as probed with gangliosides. *Bioorg. Med. Chem. Lett.* **14,** 5161–5164.
28. Tiralongo, E., Martensen, I., Grötzinger, J., Tiralongo, J., and Schauer, R. (2003) Trans-sialidase-like sequences from *Trypanosoma congolense* conserve most of the critical active site residues found in other trans-sialidases. *Biol. Chem.* **384,** 1203–1213.
29. Paris, G., Cremona, M. L., Amaya, M. F., et al. (2001) Probing molecular function of trypanosomal sialidases: single point mutations can change substrate specificity and increase hydrolytic activity. *Glycobiology* **11,** 305–311.

30. Paris, G., Ratier, L., Amaya, M. F., Nguyen, T., Alzari, P. M., and Frasch, A. C. C. (2005) A sialidase mutant displaying trans-sialidase activity. *J. Mol. Biol.* **345,** 923–934.
31. Buschiazzo, A., Amaya, M. F., Cremona, M. L., Frasch, A. C. C., and Alzari, P. M. (2002) The crystal structure and mode of action of trans-sialidase, a key enzyme in *Trypanosoma cruzi* pathogenesis. *Mol. Cell* **10,** 757–768.
32. Agusti, R., Paris, G., Ratier, L., Frasch, A. C. C., and de Lederkremer, R. M. (2004) Lactose derivatives are inhibitors of *Trypanosoma cruzi* trans-sialidase activity toward conventional substrates in vitro and in vivo. *Glycobiology* **14,** 659–670.
33. Neubacher, B., Schmidt, D., Ziegelmüller, P., and Thiem, J. (2005) Preparation of sialylated oligosaccharides employing recombinant trans-sialidase from *Trypanosoma cruzi*. *Org. Biomol. Chem.* **3,** 1551–1556.
34. Pereira-Chioccola, V. L., Fragata-Filho, A. A., Levy, A. M., Rodrigues, M. M., and Schenkman, S. (2003) Enzyme-linked immunoassay using recombinant transsialidase of *Trypanosoma cruzi* can be employed for monitoring of patients with Chagas' disease after drug treatment. *Clin. Diagn. Lab. Immunol.* **10,** 826–830.
35. Buchovsky, A. S., Campetella, O., Russomando, G., et al. (2001) Trans-sialidase inhibition assay, a highly sensitive and specific diagnostic test for Chagas' disease. *Clin. Diagn. Lab. Immunol.* **8,** 187–189.
36. Veh, R. W., Michalski, J.-C., Corfield, A. P., Sander-Wewer, M., Gies, D., and Schauer, R. (1981) New chromatographic system for the rapid analysis and preparation of colostrum sialyloligosaccharides. *J. Chromatogr.* **212,** 313–322.
37. Hara, S., Yamaguchi, M., Takemori, Y., Furuhata, K., Ogura, H., and Nakamura, M. (1989) Determination of mono-*O*-acetylated *N*-acetylneuraminic acids in human and rat sera by fluorometric high-performance liquid chromatography. *Anal. Biochem.* **179,** 162–166.
38. Warner, T. G. and O'Brien, J. S. (1979) Synthesis of 2′-(4-methylumbelliferyl)-α-D-*N*-acetylneuraminic acid and detection of skin fibroblast neuraminidase in normal humans and in sialidosis. *Biochemistry* **18,** 2783–2787.
39. Engstler, M., Talhouk, J. W., Smith, R. E., and Schauer, R. (1997) Chemical synthesis of 4-trifluoromethylumbelliferyl-α-D-*N*-acetylneuraminic acid glycoside and its use for the fluorometric detection of poorly expressed natural and recombinant sialidases. *Anal. Biochem.* **250,** 176–180.
40. Parodi, A. J., Pollevick, G. D., Mautner, M., Buschiazzo, A., Sanchez, D. O., and Frasch, A. C. C. (1992) Identification of the gene(s) coding for the trans-sialidase of *Trypanosoma cruzi*. *EMBO J.* **11,** 1705–1710.

8

Structural Determination of *O*-Glycans by Tandem Mass Spectrometry

Catherine Robbe, Jean-Claude Michalski, and Calliope Capon

Summary

Nano-electrospray ionization quadrupole time-of-flight mass spectrometry (nanoESI-Q-TOF-MS) provides a sensitive means for mapping and sequencing underivatized *O*-glycans. This chapter describes fragmentation rules of *O*-glycans by ESI-MS/MS and provides a series of diagnostic ions relevant for the determination of the core type, position, and linkage of fucose, sialic acid, and sulphate residues, as well as information on type I or II chains. Positive-ion mode gives information about core type, linkage, and position of fucose residues. Negative-ion mode can be applied for differentiation between isomeric molecules and for analysis of sulphated or sialylated glycans. The current technology successfully determines the sequence of underivatized oligosaccharides in complex mixtures and provides a significant step toward the goal of characterizing all aspects of carbohydrate structure using a single instrument.

Key Words: Mass spectrometry; glycosylation; mucins; blood-group antigens.

1. Introduction

Mucins are very high-molecular-weight glycoproteins, consisting of a protein backbone with hundreds of carbohydrate side chains of varying lengths, sequences, compositions, and anomeric linkages (*see also* Chapter 18). A distinguishing feature of mucins is that they contain O-linked oligosaccharides (OSs); i.e., the sugar chains are attached to the peptide backbone via an *O*-glycosidic linkage between a serine or threonine residue and an *N*-acetylgalactosamine (GalNAc) residue. Mucin OSs are mainly composed of neutral (galactose [Gal], fucose [Fuc], GalNAc, and *N*-acetylglucosamine [GlcNAc]) and acidic monosaccharides (sialic acid or *N*-acetylneuraminic acid [NeuAc]), and can be substituted by other components, such as sulfate groups. **Table 1**

From: *Methods in Molecular Biology, vol. 347: Glycobiology Protocols*
Edited by: I. Brockhausen © Humana Press Inc., Totowa, NJ

Table 1
Oligosaccharide Epitopes on Mucin *O*-Glycans

Epitope	Core structures
Core 1	Galβ1-3GalNAc
Core 2	GlcNAcβ1-6(Galβ1-3)GalNAc
Core 3	GlcNAcβ1-3GalNAc
Core 4	GlcNAcβ1-6(GlcNAcβ1-3)GalNAc
Core 5	GalNAcα1-3GalNAc
Core 6	GlcNAcβ1-6GalNAc
Core 7	GalNAcα1-6GalNAc
Core 8	Galα1-3GalNAc
	Type chain
Type 2	-Galβ1-4GlcNAcβ1-
Type 1	-Galβ1-3GlcNAcβ1-
	Terminal structures
Blood group H	Fucα1-2Galβ1-
Blood group A	GalNAcα1-3(Fucα1-2)Galβ1-
Blood group B	Galα1-3(Fucα1-2)Galβ1-
Lewis a	Fucα1-4(Galβ1-3)GlcNAcβ1-
Lewis b	Fucα1-4(Fucα1-2Galβ1-3)GlcNAcβ1-
Lewis x	Galβ1-4(Fucα1-3)GlcNAcβ1-
Lewis y	Fucα1-2Galβ1-4(Fucα1-3)GlcNAcβ1-
Sda/Cad	GalNAcβ1-4(NeuAcα2-3)Galβ1-3/4GlcNAcβ1-

gives the main structural features of *O*-glycans found in mucins (core, chain type, and antigenic epitopes).

The most current strategies used for structural insight of *O*-glycans are generally based on their release by reductive alkaline β-elimination, followed by high-performance liquid chromatography (HPLC) fractionation of individual OSs and determination of their sequence by a combination of nuclear magnetic resonance (NMR) and mass spectrometry (MS) methods. This procedure is time-consuming and requires an excessive amount of glycoprotein, where quantitative information may be lost owing to several purification steps.

MS has emerged as a key instrumentation to enable accurate qualitative glycosylation analysis ***(1–10)***. MS offers distinct advantages over conventional approaches because of its inherent sensitivity and its capability to obtain structure information through tandem mass spectrometry techniques (MS^n). Both matrix-assisted laser desorption ionization mass spectrometry (MALDI-MS) and electrospray ionization mass spectrometry (ESI-MS) have been successfully employed for OS analysis (*see also* Chapter 6).

1.1. Principle of ESI-MS

ESI is a technique that allows the transfer of ions from solution to gas phase in the mass spectrometer. The analyte is introduced to the source in solution either from a syringe pump or as the eluent flow from liquid chromatography (LC). The analyte solution flow passes through the electrospray needle that has a high potential difference applied to it. This forces the spraying of charged droplets from the needle with a surface charge of the same polarity to the charge on the needle. The charge droplets shrink as the solvent evaporates with the aid of N_2 gas, and it continually disintegrates into smaller highly charged droplets that are capable of producing gas phase ions. These gas phase ions move to the analyzer with the possibility of the ions being subjected to varying pressures and electric fields at the boundary leading to the mass analyzer. The compatibility with low flow rate allows coupling with HPLC (*see* **Note 1**).

In the positive ion mode, predominant ions correspond to $[M+H]^+$ and $[M+Na]^+$ ions, whereas in negative ion mode, most of the ions detected are $[M-H]^-$.

2. Materials

2.1. Release of OS Alditols From Mucin by Alkaline Borohydride Treatment

1. 0.1 *M* of KOH (Sigma, St Louis, MO).
2. 1 *M* of KBH_4 (Sigma).
3. Dowex 50 × 8 cation-exchange (CE) resin (Bio-Rad, Richmond, CA).
4. Dowex 50 × 2 CE resin (Bio-Rad).
5. Methanol.
6. Bio-Gel P2 (Bio-Rad).
7. Ultraviolet detector Uvicord SII (Amersham Biosciences, Uppsala, Sweden).

2.2. Electrospray Tandem Mass Spectrometry

All analyses are performed using a Q-STAR Pulsar quadrupole time-of-flight (Q-q-TOF) mass spectrometer (Applied Biosystems/MDS Sciex, Toronto, Canada) fitted with a nano-electrospray ionization source (Protana, Odense, Denmark).

3. Methods

3.1. Release of OS Alditols From Mucin by Alkaline Borohydride Treatment

1. The mucins are submitted to β-elimination under reductive conditions (0.1 *M* of KOH and 1 *M* of KBH_4; *see* **ref. *11***) at 45°C for 24 h.
2. The reaction is stopped by the addition of Dowex 50 × 8 CE resin (25–50 mesh, H+ form) at 4°C until pH reaches 6.5.
3. After evaporation to dryness, boric acid is distilled as methyl ester in the presence of methanol (three or four times).
4. Total material is submitted to CE chromatography on a Dowex 50 × 2 column (200–400 mesh, H+ form) to remove residual peptides.
5. The solution is further purified by size exclusion chromatography on a column of Bio-Gel P2 (85 × 2 cm id, 400 mesh) equilibrated and eluted with water, at 10 mL/h at room temperature.
6. The OS fractions, detected by ultraviolet absorption at 206 nm, are pooled and lyophilized for structural analysis.

3.2. Electrospray Tandem Mass Spectrometry

1. External calibration is performed prior to each measurement using a 4 pmol/μL solution of taurocholic acid in acetonitrile/water (50/50, v/v) containing 2 m*M* of ammonium acetate.
2. OSs dissolved in water (60 pmol/μL) are acidified by the addition of an equal volume of methanol/0.1% formic acid and sprayed from gold-coated "medium length" borosilicate capillaries (Protana).
3. A potential of 800 V is applied to the capillary tip and the focusing potential set at 100 V, the declustering potential varying between 60 and 110 V.
4. For the recording of conventional mass spectra, TOF data are acquired by accumulation of 10 multiple channel acquisition (MCA) scans over the range *m/z* 400–2500. In the collision-induced dissociation (CID) MS/MS analyses, multiple charged ions are fragmented using nitrogen as collision gas (5.3×10^{-5} torr, where 1 torr = 0.133 kPa), the collision energy varying between 40 and 90 eV to obtain optimal fragmentation. The CID spectra are recorded by the orthogonal TOF analyzer over a range *m/z* 80–2000. Data acquisition is optimized to yield the highest possible resolution and the best signal-to-noise ratio even in the case of low-abundance signals. Typically, the full width at half maximum (FWHM) is 7000 in the measured mass ranges.

3.3. Mass Spectra Interpretation

A systematic nomenclature for labeling fragment ions observed in MS/MS spectra has been introduced by Domon and Costello in 1988 ***(12)***. As shown in **Scheme 1**, A_i, B_i, and C_i are used to designate fragments containing a terminal (nonreducing end) sugar unit, whereas X_j, Y_j, and Z_j represent ions still containing the aglycone or the reducing sugar unit. Subscripts indicate the position

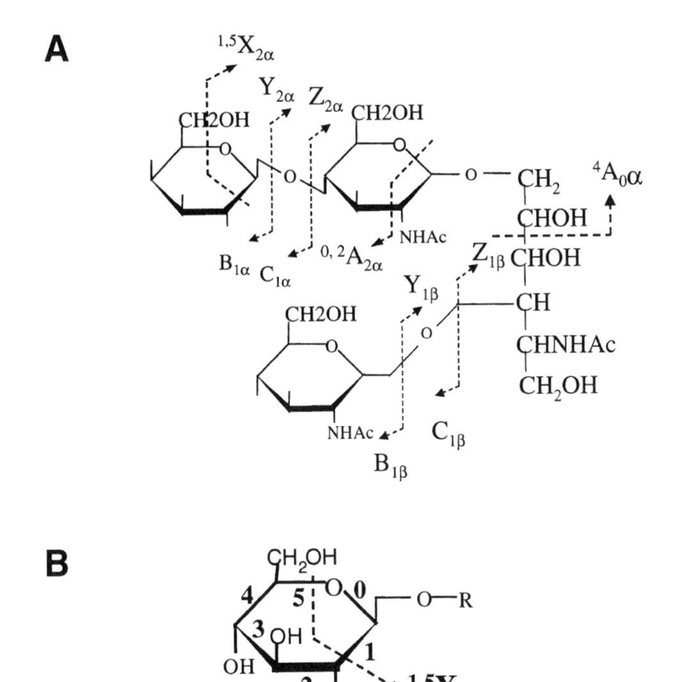

Scheme 1. Fragmentation nomenclature of oligosaccharides.

relative to the termini and superscripts indicate cleavages within carbohydrate rings. Moreover, an α suffix is used to designate cleavages in the 6-linked branch, and a β suffix is used for cleavage in the 3-linked branch from the GalNAcol. In addition, the cleavages within the *N*-acetylgalactosaminitol moiety are denoted as A_0 fragments ***(13)***.

3.3.1. Determination of the Type of Core (see **Note 2**)

Some ions are diagnostic for substituted or unsubstituted GalNAc.

1. In the positive-ion mode, these diagnostic ions are composed of a prominent Y_2 fragment ion accompanied by a smaller Z_2 fragment ion for each *O*-glycan with a singly C-3 or C-6 substituted reducing terminus GalNAcol (*see* **Fig. 1**). When GalNAcol is disubstituted (core 2 or 4), diagnostic ions correspond to two prominent Y_1 fragment ions with their Z_1 ions. Two other fragments are also present in the spectrum, which could be assigned as Z_1-H_2O-type fragment ions.
2. In the negative-ion mode, all diagnostic ions are present in the spectrum but the intensities are different: the intensities of Y_i ions are less than those of Z_i or Z_i-H_2O (*see* **Fig. 2**). Moreover, Karlsson et al. ***(10)*** have demonstrated that one of the most pronounced features found for mucin OS alditols with a singly C-3 substituted GalNAcol is the ion with a loss of 223 Da (loss of the reducing end

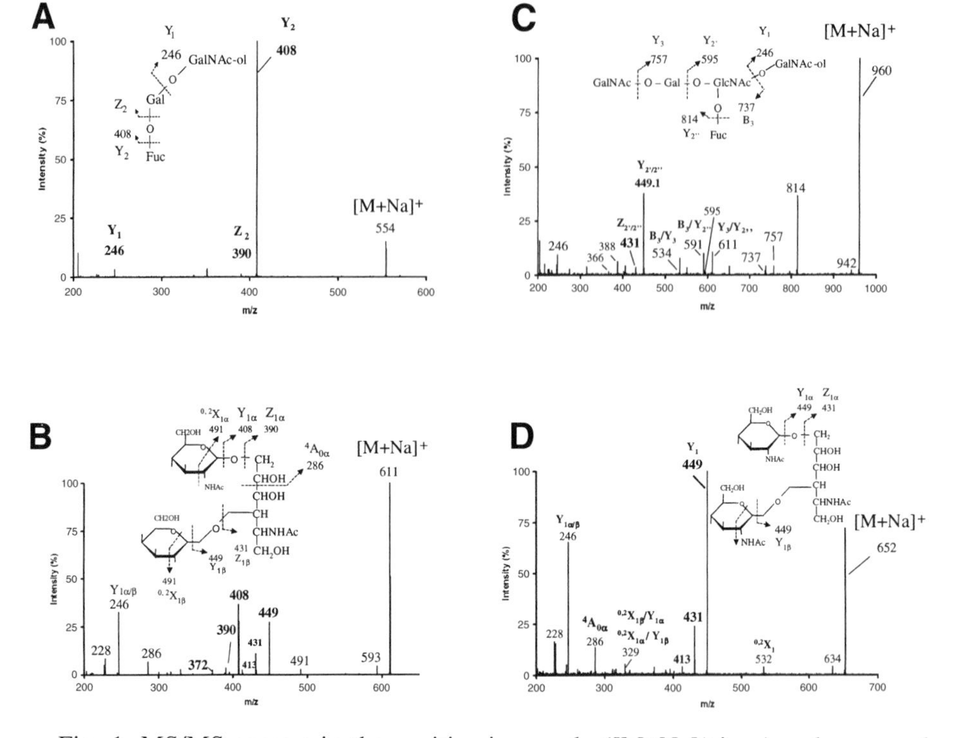

Fig. 1. MS/MS spectra in the positive-ion mode ([M+Na]$^+$ ions) and proposed fragmentation of oligosaccharides (OSs) with different core types: **(A)** Core 1 OS; **(B)** Core 2 OS; **(C)** Core 3 OS; **(D)** Core 4 OS.

HexNAcol) from the pseudomolecular ion. An additional characteristic ion is also recovered in core 1 and 3 structures resulting in the loss of 108 Da ($C_3H_8O_4$) from the [M–H]$^-$ ion. This ion is never found in structures with a disubstituted GalNAcol, nor can it be observed when only the C-6 is elongated.

Fragmentation of core 2 or core 4 glycans gives a ^4A_i-type fragment ion corresponding to a cleavage of the GalNAcol between C-4 and C-5, and allowing the determination of the composition of the substituent on C-6 (*see* **ref. *14*** and **Table 2).**

3.3.2. Discrimination Between Type 1 and Type 2 Chains

Chai et al. ***(1)*** have previously shown for milk OSs that the double glycosidic D-type cleavage of a 3-linked GlcNAc and the saccharide ring fragmentation of the ^{0,2}A type from a 4-linked GlcNAc allowed differentiation of type 1 (Galβ1-3GlcNAc) and type 2 (Galβ1-4GlcNAc) chains. In mucin *O*-glycans no D-type cleavages are found, but fragmentation spectra of

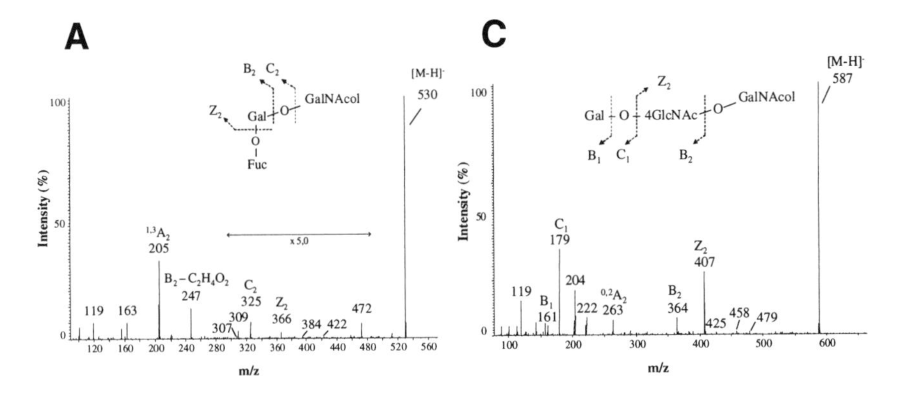

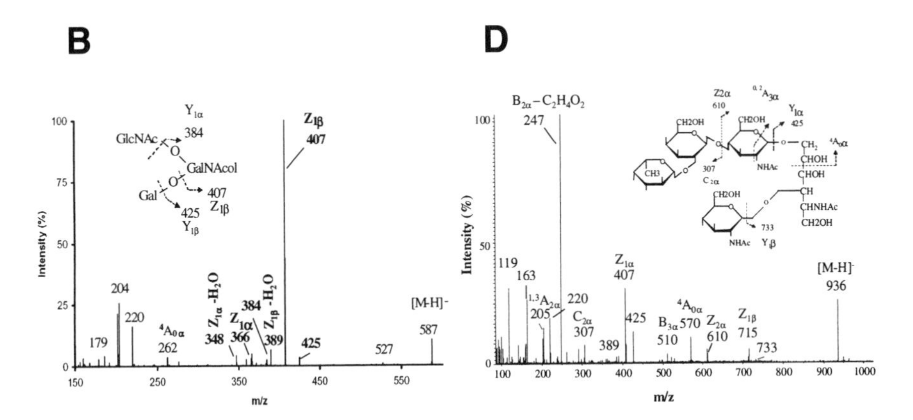

Fig. 2. MS/MS spectra in the negative-ion mode ([M-H]⁻ ions) and proposed fragmentation of oligosaccharides (OSs) with different core types: **(A)** Core 1 OS; **(B)** Core 2 OS; **(C)** Core 3 OS; **(D)** Core 4 OS.

type 2 chain OSs exhibit a specific $^{0,2}A_i$ cleavage with concomitant loss of water from a GlcNAc substituted on C-4. In the small type 2 elongated core 3 structure this fragment dominates the spectrum, but in more complex structures it is of lower intensity (5–10%; *see* **Fig. 2C,D**). In this fragment, the GlcNAc moiety is converted into an enone-type structure, as proposed by Karlsson et al. ***(10)***.

3.3.3. Determination of the Position and Partial Linkage Type of Fucose (see **Note 3***)*

1. In the positive-ion mode, fragmentation of fucosylated OSs is characterized by instability of the fucose linkage, sometimes rendering the assignment of the

Table 2
Diagnostic Ions in Negative-Ion Mode for Sequence Determination of Mucin *O*-Glycans

Epitope	*m/z*	Specific fragmentation	Sequence information
Core 1	384, 366		Galβ1-3GalNAcol
	$[M–H]^-$ – 223		Loss of the GalNAcol
	$[M–H]^-$ – 108		Loss of $C_3H_8O_4$
Core 3	425, 407		GlcNAcβ1-3GalNAcol
	$[M–H]^-$ – 223		Loss of the GalNAcol
	$[M–H]^-$ – 108		Loss of $C_3H_8O_4$
Core 2	384, 366, 348		
	425, 407, 389		
	Molecular mass of C-6 branch – H + 60	$^{4}A_{0\alpha}$ cleavage of GalNAcol	C-6 branch composition
Core 4	425, 407, 389		
	628, 610		
	Molecular mass of C-6 branch – H + 60	$^{4}A_{0\alpha}$ cleavage of GalNAcol	C-6 branch composition
Type 2 chain			
	Molecular mass of the fragment + 102	$^{0,2}A_i$ type cleavage of GlcNAc	
	263, 281		Galβ1-4GlcNAc
	409		Fucα1-2Galβ1-4GlcNAc
	466		GalNAcβ1-4Galβ1-4GlcNAc (blood group Sda/Cad)
Type 1 chain		Predominantly Zi type cleavage	
Sulfated oligosaccharides			
	241	B_i type ion	Gal-SO3

	282	B_i type ion	GlcNAc-SO3
	97		HSO4–
	139	$^{1,3}A_i$, $^{0,4}A_i$, $^{2,4}A_i$	
	199	$^{0,2}A_i$, $^{2,4}X_i$	Sulfate residue on C-4 or C-6 of Gal, GlcNAc or GalNAc
Sialylated oligosaccharides			
	290	B_i type cleavage	NeuAc
	– 221	$^{0,2}X_i$ ("ion j")	
	– 44	loss of CO_2	
	306	$^{4}A_{0\alpha}$– CO_2	NeuAc6 linked
	408, 611	B_i – CO2	NeuAc3 linked
	513	$Y_{1\beta}$	NeuAc6GalNAcol
Blood group			
H	205	$^{1,3}A_2$ cleavage of Gal	
	247	B_2- $C_2H_4O_2$	Fuc2Gal
	307, 325	B_2, C_2 cleavages	
Lewis a/x			
	$[M–H]^-$ – 344	Z_i/Z_i cleavages	Gal(Fuc3/4)GlcNAc[a]
	$[M–H]^-$ – 374	Z_i/Z_i – CH_2O	
Lewis b/y			
	$[M–H]^-$ – 490	Z_i/Z_i cleavages	Fuc2Gal(Fuc3/4)GlcNAc[a]
	$[M–H]^-$ – 520	Z_i/Z_i – CH_2O	
		Characteristic Z_i/Z_i fragment ions	
A Lewis a/x	$[M–H]^-$ – 547, – 577		(GalNAcα1-3)Gal(Fuc3/4)GlcNAc[a]
B Lewis a/x	$[M–H]^-$ – 506, – 536		(Galα1-3)Gal(Fuc3/4)GlcNAc[a]
A Lewis b/y	$[M–H]^-$ – 693, – 723		Fuc2(GalNAcα1-3)Gal(Fuc3/4)GlcNAc[a]
B Lewis b/y	$[M–H]^-$ – 652, – 682		Fuc2(Galα1-3)Gal(Fuc3/4)GlcNAc[a]

[a]Loss of branch substituting GlcNAc.

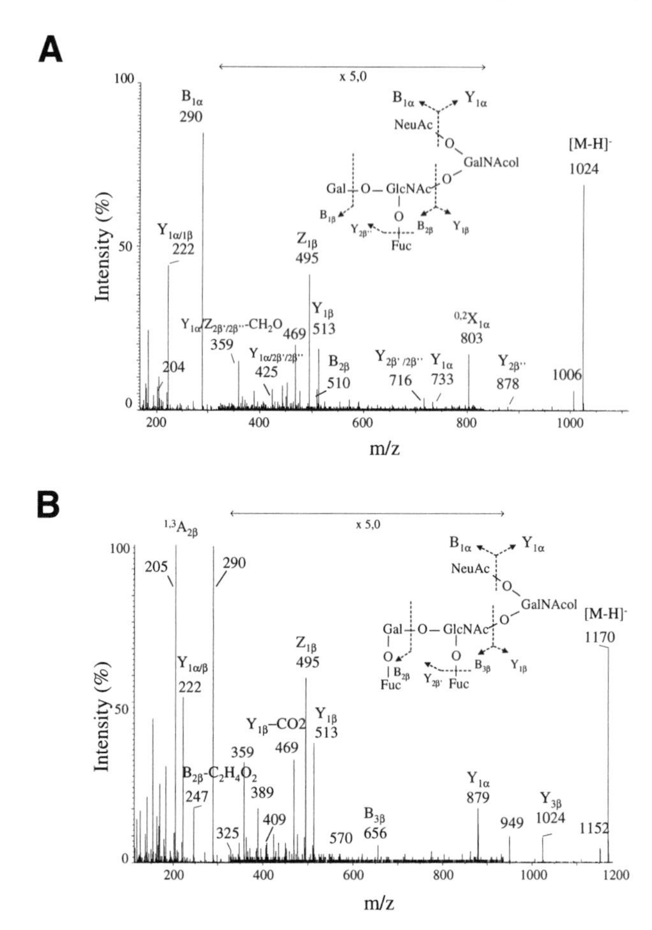

Fig. 3. MS/MS spectra in the negative-ion mode ([M-H]$^-$ ions) and proposed fragmentation of two fucosylated oligosaccharides at *m/z* 1024 and *m/z* 1170.

position of fucose difficult. Each molecule yields a characteristic fragmentation pattern allowing reconstruction of the glycan sequence of this type of residue, but no diagnostic ions can be defined.

2. In the negative-ion mode, when GlcNAc is disubstituted by both a Gal and Fuc residue, its fragmentation leads to the formation of a conjugated diene product formed by dual elimination on the GlcNAc residue (Z_i/Z_i and Z_i/Z_i-CH_2O fragments; *see* **ref. *10*** and **Fig. 3**). The extension of the Lewis a (Le^a) and Lewis x (Le^x) epitopes into Lewis b (Le^b) and Lewis y (Le^y) epitopes (with a Fucα1-2 linked to the nonreducing end of the galactose) does not inhibit the formation of the diene, but as the OS extends the intensity of the diene fragments decreases. These fragment formations are also accompanied by the loss of acetyl and acetate

from parts of the parent ion other than the substituted GlcNAc (Z_i/Z_i-C_2H_2O and Z_i/Zi-$C_2H_4O_2$).

Cross-ring cleavages of blood group H determinants (Fucα1-2Galβ1-) including the fucose and part of the galactose ($^{1,3}A_2$ fragment) could be detected in the fragments of the smaller structures, but this cannot be used unambiguously to assign the fucose to the C-2 of galactose without further confirmation by other linkage isomers. Nonreducing end blood group H determinant fragments also include weak B_2 and C_2 cleavages, accompanied by additional loss of acetate parts (B_2-$C_2H_4O_2$). This particular fragment is characteristic of a C-2 substituted Gal.

No diagnostic cross-ring fragments for the ABO-type epitopes can be defined. However, the presence of these epitopes can be suggested from the primary OS sequence elucidated by the glycosidic cleavages using MS/MS fragmentation. Linkage-specific fragment ions from OSs containing $Le^{b/y}$ extended with blood group A or B could more easily be identified, since C-3 and C-4 disubstituted GlcNAc gives characteristic Z_i/Z_i fragment ions.

3.3.4. Sequencing of Acidic (Sulphated and/or Sialylated) OSs (see **Note 4***)*

3.3.4.1. Sulphated OSs

1. The presence of sulphate is shown by the fragment ions at *m/z* 97 ($HSO4^-$) and 139 in all spectra (*see* **Fig. 4**). As previously shown *(3)*, the fragment ion at *m/z* 199 is a diagnostic ion for locating the sulphate group when this sulphate residue is linked to C-4 or C-6 of GalNAc, Gal, or GlcNAc, but not to the C-3 of Gal or GlcNAc.
2. The location of the sulphate group on a Hex or HexNAc is indicated by the presence of either *m/z* 241 $[\text{Hex-SO3}]^-$ or *m/z* 282 $[\text{HexNAc-SO3}]^-$.
3. Another feature characterizing all sulphated OS spectra is that the first Bi fragment ion observed always corresponds to the fragment with a sulphate group attached to the monosaccharide.

3.3.4.2. Sialylated OSs

In human mucin *O*-glycans, sialic acid can be either α2-3 linked to a Gal residue at the nonreducing end or α2-6 linked to the GalNAcol. Diagnostic ions allow differentiation between these two types of linkage (*see* **Fig. 5**).

1. In all spectra of sialylated OSs, the major ion is a B1 ion at *m/z* 290 corresponding to the residue of sialic acid. Moreover, a fragment ion corresponding to a mass difference of 221 Da relative to the precursor ion is always present at low intensity. It is generated via a $^{0,2}X$-type cleavage of the sialic acid.

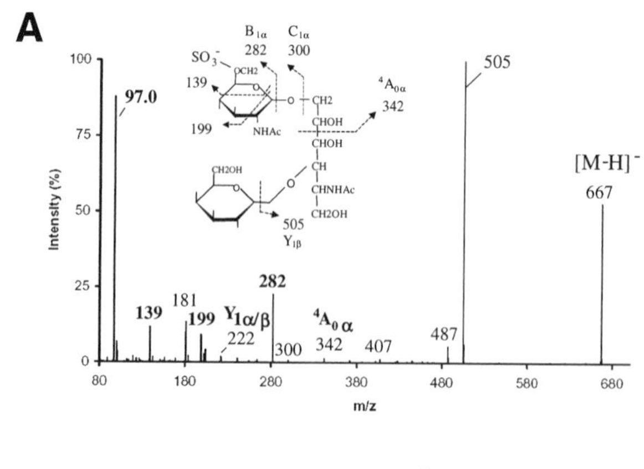

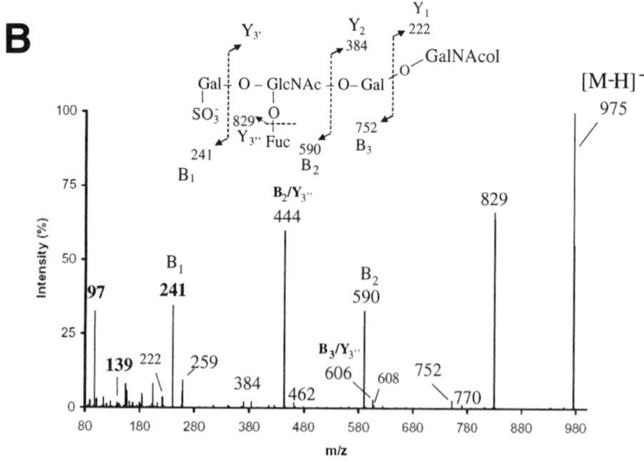

Fig. 4. MS/MS spectra and proposed fragmentation of different sulphated oligosaccharides at *m/z* 667 and *m/z* 975, recorded in the negative-ion mode.

2. The ions at *m/z* 306 and 513 are detected only in the CID spectrum (collision-induced dissociation) of α2-6 sialylated glycans. The fragment ion at *m/z* 306 is generated via a specific ring cleavage and is assigned as a $^4A_{0\alpha}$ ion accompanied by loss of CO_2 of the α2-6 linked sialic acid. The fragment ion at *m/z* 513 can be referred to as the diagnostic ion for a NeuAc residue directly α2-6-linked to the GalNAcol, as it is generated by a cleavage of the glycosidic linkage of the C-3 branch of the GalNAcol, giving a $Y_{1\beta}$ fragment ion.
3. When sialic acid is α2-3 linked to a Gal residue, B_i-type ions accompanied by loss of CO_2 are generated, giving ions at *m/z* 408 (NeuAcα2-3Gal) or 611 ([NeuAcα2-3]GalNAcβ1-4Gal) in the case of Sda/Cad antigens ***(15)***.

3.4. Example: Sequencing of the Ion at m/z 1170

Figure 3B shows the MS/MS spectrum of the [M–H]⁻ ion at *m/z* 1170. Interpretations of the observed fragments are given in the figure.

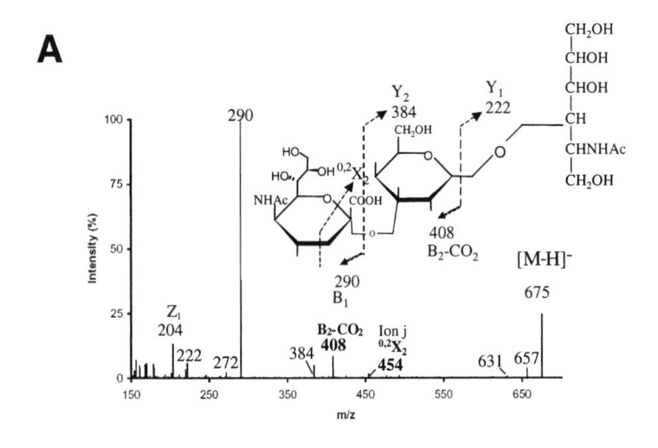

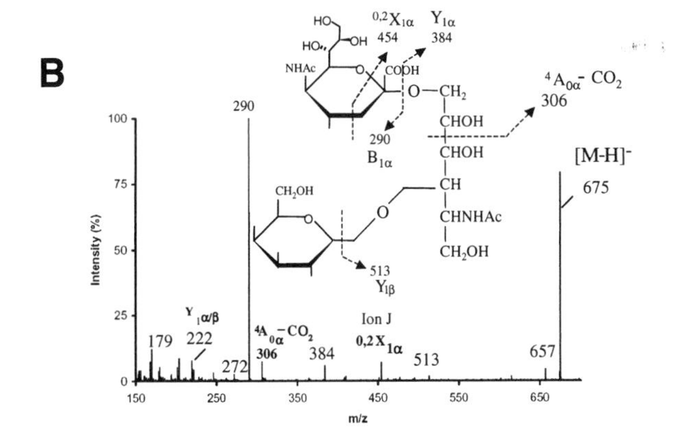

Fig. 5. MS/MS spectra and proposed fragmentation of two sialylated oligosaccharides (OSs) at *m/z* 675, recorded in the negative-ion mode: **(A)** OS with a NeuAcα2-3 linked to a Gal residue, and **(B)** OS with a NeuAcα2-6 linked to a GalNAcol.

1. Ions at *m/z* 290 ($B_{1\alpha}$) and *m/z* 879 ($Y_{1\alpha}$) are owing to the loss of the sialic acid, and the ion at *m/z* 513 indicates that NeuAc is α2-6 linked to the GalNAcol. The specific $^{0,2}X_{1\alpha}$ cleavage of the sialic acid is seen at *m/z* 949.
2. The $Z_{2\beta}/Y_{1\alpha}$ ion at *m/z* 407 indicates that a GlcNAc residue is linked to the GalNAcol. The ions at *m/z* 656 (loss of 223 Da from the desialylated precursor ion) and *m/z* 771 (loss of 108 Da) confirm that GlcNAc is β1-3 linked to the GalNAcol, constituting a core 3 glycan.
3. The $^{0,2}A_3$ ion at *m/z* 409 is diagnostic for a Fucα1-2Galβ1-4 linked to a residue of GlcNAc.
4. The presence of a Le^y antigen (Fucα1-2Galβ1-4[Fucα1-3]GlcNAcβ1) is confirmed by ions at *m/z* 389 and 359, corresponding respectively to a loss of 490 and 520 Da from the desialylated precursor ion.

5. Finally, the blood group H determinant is shown by the ions at *m/z* 205 ($^{1,3}A_{2\beta}$ cleavage of the Gal residue), *m/z* 247 ($B_{2\beta}$-$C_2H_4O_2$), and *m/z* 325 ($C_{2\beta}$).

4. Notes

1. It has been shown that it is now possible to combine online separation techniques such as LC, capillary electrophoresis, and MS, enabling the separation and characterization of OS isomers ***(9)***. Using graphitized carbon columns, ESI LC/MS analysis can be performed with sensitivities down to low fentomole levels and allowing detection of both neutral and negative OSs in the one separation.
2. Eight different core structures of *O*-glycans exist in mammalian mucins. Cores 1 and 8 contain a Gal linked by a β1-3 or α1-3 linkage to the reducing GalNAc respectively, whereas cores 3, 5, 6, and 7 contain a HexNAc (GalNAc or GlcNAc) linked to the GalNAc. Thus, one problem occurs when studying mucin OSs by mass spectrometry: it is very difficult to distinguish between different cores with the same molecular mass (i.e., cores 1 and 8 or cores 3, 5, 6, and 7). In this way, only knowledge of the sample origin (intestine, meconium, bovine, human, etc.) can give some information to differentiate between the different isomers.
3. In mucin *O*-glycans, fucose residue can be α1-2 linked to a Gal residue defining blood group H, Le^b, or Le^y epitopes, α1-3 linked to a GlcNAc residue as found in Le^x or Le^y determinants, or α1-4 linked to a GlcNAc residue constituting Le^a or Le^b antigens.
4. Sialic acid (NeuAc) and sulphate are readily cleaved in the fragmentation process, leaving the remaining fragment uncharged and therefore not detected. Sequencing of OSs with this approach therefore relies on sensitive detection of low-abundance fragment ions.

Acknowledgments

The authors are grateful for the support of the CNRS (Unité Mixte de Recherche CNRS/USTL 8576; Director, Dr. Jean-Claude Michalski) and of the Ministère de la Recherche et de l'Enseignement Supérieur.

References

1. Chai, W., Piskarev, V., and Lawson, A. M. (2001) Negative-ion electrospray mass spectrometry of neutral underivatized oligosaccharides. *Anal. Chem.* **73,** 651–657.
2. Harvey, D. J. (1999) Matrix-assisted laser desorption/ionization mass spectrometry of carbohydrates. *Mass Spectrom. Rev.* **18,** 349–450.
3. Thomsson, K. A., Karlsson, H., and Hansson, G. C. (2000) Sequencing of sulphated oligosaccharides from mucins by liquid chromatography and electrospray ionization tandem mass spectrometry. *Anal. Chem.* **72,** 4543–4549.
4. Wheeler, S. F. and Harvey, D. J. (2000) Negative ion mass spectrometry of sialylated carbohydrates: discrimination of *N*-acetylneuraminic acid linkages by MALDI-TOF and ESI-TOF mass spectrometry. *Anal. Chem.* **72,** 5027–5039.

5. Chai, W., Piskarev, V., and Lawson, A. M. (2002) Branching pattern and sequence analysis of underivatized oligosaccharides by combined MS/MS of singly and doubly charged molecular ions in negative-ion electrospray mass spectrometry. *J. Am. Soc. Mass Spectrom.* **13,** 670–679.
6. Sagi, D., Conradt, H. S., Nimtz, M., and Peter-Katalinic, J. (2002) Sequencing of tri- and tetraantennary *N*-glycans containing sialic acid by negative mode ESI QTOF tandem MS. *J. Am. Soc. Mass Spectrom.* **13,** 1138–1148.
7. Pfenninger, A., Karas, M., Finke, B., and Stahl, B. (2002) Structural analysis of underivatized neutral human milk oligosaccharides in the negative ion mode by nano-electrospray MS^n (Part 1: Methodology). *J. Am. Soc. Mass Spectrom.* **13,** 1331–1340.
8. Robbe, C., Capon, C., Coddeville, B., and Michalski, J. C. (2004) Diagnostic ions for the rapid analysis by nano-electrospray ionization quadrupole time-of-flight mass spectrometry of *O*-glycans from human mucins. *Rapid Commun. Mass Spectrom.* **18,** 412–420.
9. Karlsson, N. G., Wilson, N. L., Wirth, H. J., Dawes, P., Joshi, H., and Packer, N. H. (2004) Negative ion graphitised carbon nano-liquid chromatography/mass spectrometry increases sensitivity for glycoprotein oligosaccharide analysis. *Rapid Commun. Mass Spectrom.* **18,** 2282–2292.
10. Karlsson, N. G., Schulz, B. L., and Packer, N. H. (2004) Structural determination of neutral *O*-linked oligosaccharide alditols by negative ion LC-electrospray-MSn. *J. Am. Soc. Mass Spectrom.* **15,** 659–672.
11. Carlson, D. M. (1968) Structures and immunochemical properties of oligosaccharides isolated from pig submaxillary mucins. *J. Biol. Chem.* **243,** 616–626.
12. Domon, B. and Costello, C. E. (1988) A systematic nomenclature for carbohydrate fragmentations in FAB-MS/MS spectra of glycoconjugates. *Glycoconj. J.* **5,** 397–409.
13. Karlsson, N. G., Karlsson, H., and Hansson, G. C. (1996) Sulphated mucin oligosaccharides from porcine small intestine analysed by four-sector tandem mass spectrometry. *J. Mass Spectrom.* **31,** 560–572.
14. Schulz, B. L., Packer, N. H., and Karlsson, N. G. (2002) Small scale analysis of *O*-linked oligosaccharides from glycoproteins and mucins separated by gel electrophoresis. *Anal. Chem.* **74,** 6088–6097.
15. Robbe, C., Capon, C., Coddeville, B., and Michalski, J. C. (2004) Structural diversity and specific distribution of *O*-glycans in normal human mucins along the intestinal tract. *Biochem. J.* **384,** 307–316.

9

Detailed Structural Analysis of *N*-Glycans Released From Glycoproteins in SDS-PAGE Gel Bands Using HPLC Combined With Exoglycosidase Array Digestions

Louise Royle, Catherine M. Radcliffe, Raymond A. Dwek, and Pauline M. Rudd

Summary

In contrast to the linear sequences of protein and DNA, oligosaccharides are branched structures. In addition, almost all glycoproteins consist of a heterogeneous collection of differently glycosylated variants. Glycan analysis therefore requires high-resolution separation techniques that can provide detailed structural analysis, including both monosaccharide sequence and linkage information. This chapter describes how a combination of high-performance liquid chromatography (HPLC) and exoglycosidase enzyme array digestions can deliver quantitative glycan analysis of sugars released from glycoproteins in sodium dodecyl sulfate-polyacrylamide gel electrophoresis gel bands by matching HPLC elution positions with a database of standard glycans.

Key Words: *N*-glycan analysis; HPLC; exoglycosidase.

1. Introduction

As the roles for posttranslational modification of proteins become more apparent, it is vital to fully characterize their glycosylation. Alterations in protein glycosylation accompany many diseases ***(1–3)***, so analysis of these glycan changes should lead to a better understanding of disease pathology and provide valuable diagnostic biomarkers (*see also* Chapters 6 and 8). It is also necessary to monitor glycosylation during the production of recombinant glycoproteins (GPs) to ensure a consistent therapeutic product. Glycans can be linked to protein either via the amide group of an asparagine residue (*N*-glycans) or to the hydroxyl group of a serine or threonine residue (*O*-glycans). The high-performance liquid

From: *Methods in Molecular Biology, vol. 347: Glycobiology Protocols*
Edited by: I. Brockhausen © Humana Press Inc., Totowa, NJ

chromatography (HPLC)-based analysis of enzymatically released, fluorescently labeled *N*-glycans is covered in the protocols in this chapter.

Peptide-*N*-glycanase F (PNGaseF) is used to remove *N*-glycans ***(4)***, with or without α1-6 linked core fucose. The released glycans are fluorescently labeled for sensitive detection (10 fmol) following HPLC analysis. PNGaseF will not remove any *N*-glycans which have an α1-3 linked core fucose such as that found on many plant derived glycans; this requires the GP to be digested with trypsin before using PNGaseA to remove the glycans ***(5,6)***. There is as yet no generic O-glycanase, making removal of intact, non-reduced *O*-glycans possible only through chemical methods.

As a consequence of the oligosaccharide processing pathway there is extensive diversity of glycan structures. There are a number of different ways in which the monosaccharides can be linked together; e.g., galactose can be linked to GlcNAc at the 2, 3, 4, or 6 position by either an α or β linkage, giving eight possible isomeric structures. To add to this complexity, in contrast to proteins and nucleic acids, glycans are usually branched structures.

Most GPs exist as a heterogeneous population of glycoforms in which a range of different glycans is present at each glycosylation site. For example, there are 32 different structures found on human IgG, which has only a single glycosylation site at Asn 297 on each heavy chain ***(7,8)***. The largest of these is a disialylated, digalactosylated, bisected, core-fucosylated biantennary glycan; the remaining chains are smaller, less-processed glycans lacking some of these monosaccharides. The structures of the glycans depend on the levels of specific glycosyl transferases, the availability of appropriate monosaccharides, and the sugar nucleotide donors (cell-specific glycosylation) as well as the local three-dimensional structure of the protein at the glycosylation site. Thy-1 is an example of a GP in which there is site-specific glycosylation. Each of the three glycosylation sites contain a different range of sugars ***(9)***. Local environmental factors can also play a role, and a GP produced under different fermentation conditions or grown in different cell lines can have quite different glycan structures in each case ***(10)***.

2. Materials

To eliminate as much background contamination as possible from the HPLC analysis of released glycans, high-purity chemicals (including water) are used throughout (*see* **Notes 1** and **2**). The quality of plasticware can also affect the results, as lubricants used in their manufacture can produce peaks in the HPLC profiles (*see* **Notes 3** and **4**). Non-powdered gloves should be used at all stages of GP purification, glycan release, and analysis; otherwise a polysaccharide ladder may be detected by HPLC that will obscure any glycans present in the sample (*see* **Fig. 1**).

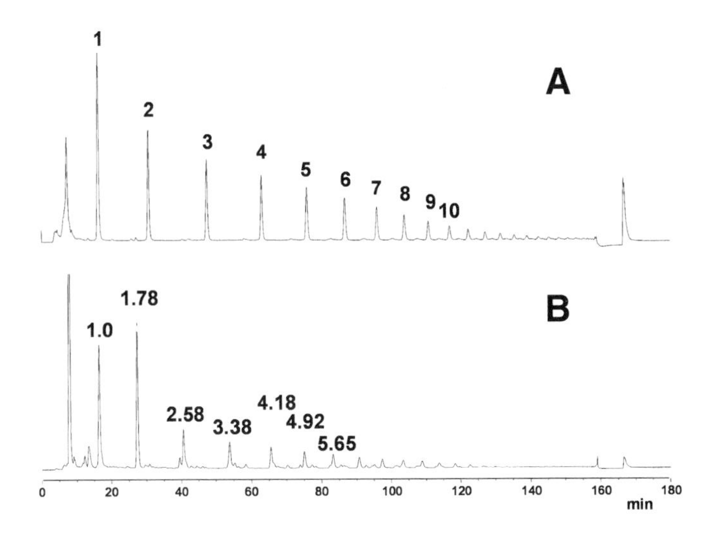

Fig. 1. NP-HPLC profiles of **(A)** dextran ladder; **(B)** contamination from the use of powdered gloves. The numbers on the profiles are glucose unit values.

2.1. Sample Preparation

1. Laemli sample buffer (5X): 0.04 g of Bromophenol blue, 0.625 mL of stacking buffer, pH 6.6 (*see* **Subheading 2.2., item 2**), 1 mL of 10% SDS, and 0.5 mL of glycerol in 2.875 mL of water.
2. 0.5 *M* of dithiothreitol (DTT): dissolve 7.71 mg of DTT in 100 µL of water (*see* **Note 2**) and immediately freeze in single-use (20-µL) aliquots at –20°C.
3. 100 m*M* of iodoacetamide: prepare 18.5 mg in 1 mL of water and immediately freeze in single-use (20-µL) aliquots at –20°C.

2.2. Sodium Dodecyl Sulfate-Polyacrylamide Gel Electrophoresis

1. 30% (w/w) acrylamide: 0.8% (w/v) *bis*-acrylamide stock solution (37.5:1; Protogel ultrapure protein and sequencing electrophoresis grade, gas stabilized; National Diagnostics, Hessle, Hull, UK). **Caution:** this is a neurotoxin when unpolymerized, so care should be taken to avoid exposure.
2. Stacking buffer: 0.5 *M* of Tris-HCl (6 g for 100 mL), adjusted to pH 6.6 with HCl (*see* **Note 5**).
3. Gel buffer: 1.5 *M* of Tris-HCl (18.2 g for 100 mL) adjusted to pH 8.8 with HCl.
4. 10% Sodium dodecyl sulfate (SDS): 1 g of SDS in 10 mL of water.
5. 5X running buffer: 144 g of glycine, 30 g of Tris-HCl, and 10 g of SDS in 2 L of water.
6. Ammonium peroxodisulphate (APS): prepare a 10% solution (1 g in 10 mL of water) and immediately freeze in single-use (200-µL) aliquots at –20°C.
7. *N,N,N,N′*-tetramethyl-ethylenediamine (TEMED; *see* **Note 6**).

8. Water-saturated butanol: Shake equal volumes of water and butanol together in a glass bottle. Use the top layer.
9. Molecular-weight markers: Sigmamarker wide range (molecular weight 6,500–205,000; Sigma-Aldrich Company Ltd., Poole, Dorset, UK). Dissolve according to manufacturer's instructions and store frozen in 10-μL aliquots at –20°C.
10. Coomassie stain: 1.25 g of Coomassie R-250 (brilliant blue), 250 mL of methanol, 50 mL of concentrated acetic acid, and 200 mL of water.
11. Destain 1: 50% (v/v) methanol, 7% (v/v) concentrated acetic acid, 43% water.
12. Destain 2: 5% (v/v) methanol, 7% (v/v) concentrated acetic acid, 88% water.
13. Mini-gel system: XCell *SureLock*™ Mini-Cell (Invitrogen, Paisley, UK).
14. Gel cassettes (Invitrogen).

2.3. N-*Glycan Release and Extraction*

1. PNGaseF buffer: 20 m*M* of $NaHCO_3$, pH 7.0 (0.168 g in 100 mL of H_2O adjusted to pH 7.0 with HCl). Store frozen at –20°C in 10-mL aliquots.
2. PNGaseF stock solution: PNGaseF (Roche Diagnostics GmbH, Mannheim, Germany) made up in H_2O to 1000 U/mL.
3. PTFE membrane filter (Millex-LCR, Millipore, Watford, Herts, UK).
4. 1-mL Plastic syringe (PP/PE syringe, Sigma; cat. no. Z230723).
5. 1% Formic acid in water.

2.4. *Fluorescent 2-AB Labeling and Cleanup*

1. LudgerTag 2-AB labeling kit (Ludger Ltd., Abingdon, Oxon, UK) or Glyko Signal 2-AB Labeling Kit (Prozyme, San Leandro, CA).
2. Whatman 3MM chromatography paper cut into 10- × 3-cm pieces.
3. Acetonitrile.
4. 2.5-mL Plastic syringe (PP/PE syringe, Sigma; cat. no. Z116858).
5. PTFE membrane filter (Millex-LCR, Millipore).

2.5. *NP-HPLC Profiling of 2-AB-Labeled* N-*Glycans*

1. Acetonitrile: E Chromasolv® for HPLC far UV (Riedel-de Haën, Sigma; cat. no. 34888; *see* **Note 7**).
2. 2 *M* Ammonium formate, pH 4.4, normal phase (NP) stock solution: Weigh 184.12 g of formic acid into a 2-L glass beaker. Place the beaker in an ice bath to which salt has been added to take the temperature down to –10°C, add 1 L of water, and stir with a glass rod. Adjust the pH by adding 25% ammonia solution. This causes a rapid rise in temperature, so the ammonia must be added in small amounts. Add 4X 50 mL of ammonia, making sure that the temperature drops between each addition; then add in 5-mL aliquots until pH 4.4 is reached at room temperature. Transfer the solution to a 2-L volumetric flask and make up to 2 L with water. Store this stock solution in a brown Winchester bottle at room temperature.
3. Waters 2695 (Waters Ltd., Elstree, Herts, UK) separations module with a Waters 474 or 2475 fluorescence detector (or another HPLC system which delivers a reproducible shallow gradient and has a fluorescence detector).

4. NP column: TSK Amide-80 250- × 4.6-mm column (Anachem, Luton, UK; cat. no. TSK 13071).
5. 2-Aminobenzamide (2-AB)-dextran ladder (2-AB-glucose homopolymer, Ludger).

2.6. Exoglycosidase Digestions

1. A selection of exoglycosidase enzymes such as those listed in **Table 1** is needed. These can be purchased from a variety of companies such as Prozyme, Ludger, Merck Biosciences (Nottingham, UK), New England Biolabs (Hitchin, Herts, UK), or Sigma. However, the standard and specificity of these enzymes can vary, so it is prudent to check their specificity occasionally against standard *N*-glycans (obtainable from Ludger or Prozyme).
2. Incubation buffers: 10X 50m*M* of sodium acetate, pH 5.5, for mixed enzyme incubations. 5X Incubation buffers at the optimal pH are usually supplied with individual enzymes.
3. Protein-binding filters for removal of enzymes before HPLC (Micropure-EZ, Millipore; cat. no. 42530).

2.7. Weak Anion Exchange-HPLC Profiling of 2-AB-Labeled N-*Glycans*

1. 2 *M* of ammonium formate, pH 9.0. Weak anion exchange (WAX) stock solution: make up in the same way as NP stock solution (*see* **Subheading 2.5., item 2**), but adjust pH to 9.0.
2. Methanol.
3. WAX column: Vydac protein WAX, 7.5- × 50-mm (cat. no. 301 VHP 575).
4. 2-AB Bovine serum fetuin *N*-glycan standard. This is a mixture of mono-, di-, tri-, and tetrasialylated glycans which can be obtained by releasing and labeling the *N*-glycans from bovine serum fetuin (Sigma) as detailed in this chapter. Alternatively, individual *N*-glycans may be obtained from Ludger or Prozyme.

3. Methods

The GP under investigation is first run on SDS-PAGE and visualized by staining with Coomassie blue. The GP can be run as either the reduced or nonreduced GP, although reduction ensures that the protein is fully unfolded and accessible to digestion by PNGaseF, and is therefore the preferred option. Reduction of the GP also has the advantage of separating subunits which can then be analyzed separately (*see* **Fig. 2**). The method described here is based on that published by Küster et al. ***(11)***, with modifications by Radcliffe et al. ***(12)*** that involve reduction and alkylation of the GP before SDS-PAGE, freezing the gel pieces before PNGaseF treatment, and extensive washing of the treated gel pieces to remove the released glycans. Further modifications include more extensive washing of the gel pieces before incubation with PNGaseF, which helps to remove background contamination; removal of the step with AG-50 (H^+ form) resin, which can lead to the loss of some larger glycans (it has been established that desalting is not required); and the addition of an incubation

Table 1
Exoglycosidase Enzymes for Glycan Analysis

Enzyme	Abbreviation	Specificity
Arthrobacter ureafaciens sialidase	Abs	Sialic acid α2-6>3,8
Streptococcus pneumoniae sialidase recombinant in *Escherichia coli*	Nan1	Sialic acid α2-3,8
Bovine testis galactosidase	Btg	Gal β1-3,4>6
S. pneumoniae galactosidase	Spg	Gal β1-4
Coffee bean α galactosidase	Cbag	Gal α1-3,4,6
Bovine kidney fucosidase	Bkf	Fuc β1-6>2>>3,4
Almond meal fucosidase	Amf	Fuc β1-3,4
S. pneumoniae-*N*-acetylhexosaminidase[a]	Sph	GlcNAc β1-2>3,4,6
Jack bean-*N*-acetylhexosaminidase[a]	Jbh	GlcNAc, GalNAc β1-2,3,4,6
Jack bean mannosidase	Jbm	Man α1-2,3,6

[a]The exact linkage specificities of the *N*-acetylhexosaminidases vary according to their concentration as well as between different recombinant species. It is therefore important to check these against standard bisected, tri-, or tetraantennary glycans if any structures other then biantennary *N*-glycans are present.

with 1% formic acid to ensure that all released glycans are converted to aldoses before fluorescence labeling. These modifications ensure maximum recovery and minimum background contamination, enabling publishable profiles to be obtained from as little as 2 μg of GP.

Following release, the glycans are labeled with the fluorophore 2-AB and separated by NP-HPLC. The advantage of using 2-AB over charged labels such as 2-aminoanthranilic acid (2-AA) is that the order of elution on the column is related to the number of sugar residues in the glycan, so that the larger glycans elute later even when sialic acid residues are present (which does not apply when using 2-AA). The elution times of glycans are expressed in glucose units (GU) by reference to a dextran ladder ***(13)***. Each individual glycan structure has a GU value that is directly related to the number and linkage of its constituent monosaccharides. Thus GU values can be used to predict structures, since each monosaccharide in a specific linkage adds a given amount to the GU value of a given glycan. The use of arrays of exoglycosidases in combination with NP-HPLC profiling enables the individual monosaccharides and linkages to be determined ***(13,14)***. In addition, WAX-HPLC can be used to separate glycans

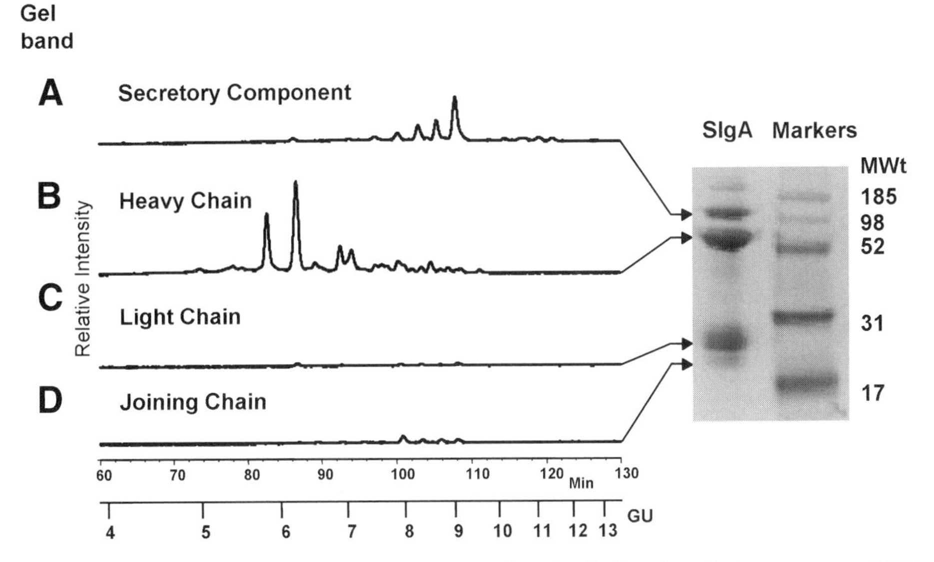

Fig. 2. Human secretory IgA (SIgA) was reduced, alkylated, and then run on a 10% *bis-tris* gel. The *N*-glycans were released by in-gel PNGaseF digestion, 2-AB-labeled, and run on NP-HPLC. The HPLC traces are all on the same scale, so the relative size (glucose unit) and abundance of the *N*-glycans can be compared. (Reproduced from **ref. *10*** with permission from The American Society for Biochemistry and Molecular Biology).

on the basis of charge ***(15)***, which can be very useful when sialic acids or sulphate groups are present.

Fingerprinting the whole undigested glycan pool from NP-HPLC profiles may be sufficient for comparing the glycosylation between batches of cultured GPs where the range of possible glycans is well-established. However, confirmation of any initial assignments by exoglycosidase sequencing is essential in most cases as a number of glycan structures can co-elute, e.g., the hybrid structure Fuc_1 $GlcNAc_2Man_4GalNAc_1Gal_1Neu5Ac_1$ (FcMan4A1G1S1) co-elutes with the complex biantennary $GlcNAc_2Man_3GalNAc_2Gal_2Neu5Ac_1$ (A2G2S1) at GU 7.9 ***(14)***.

3.1. Preparation of Sample

1. Recommended amounts of sample to run are 5–10 μg of GP per well to ensure that clear separated bands are seen. Samples are usually reduced and alkylated before SDS-PAGE to ensure maximum release of glycans by PNGaseF.
2. Reduce the sample by adding 4 μL of 5X sample buffer, 2 μL of 0.5 *M* DTT, and water to make up to a total of 20 μL. Incubate for 10 min at 70°C. (If samples are to be run nonreduced, then incubate without DTT and do not alkylate).
3. Alkylation of reduced samples: add 2 μL of 100 m*M* iodoacetamide to the reduced samples and incubate for 30 min in the dark at room temperature.

Table 2
Reagent Quantities for Two Running Gels

	Percentage Gel					
	6%	10%	12.5%	14%	15%	17.5%
Protogel	2.4 mL	4.0 mL	5.0 mL	5.6 mL	6.0 mL	7.0 mL
Gel buffer	3.0 mL	3.0 mL	3.0 mL	3.0 mL	3.0 mL	3.0 mL
H_2O	6.6 mL	5.0 mL	4.0 mL	3.4 mL	3.0 mL	2.0 mL
SDS (10%)	120 μL	120 μL	120 μL	120 μL	120 μL	120 μL
APS (10%)	120 μL	120 μL	120 μL	120 μL	120 μL	120 μL
TEMED	12 μL	12 μL	12 μL	12 μL	12 μL	12 μL

Table 3
Reagent Quantities for Two Stacking Gels

Protogel	0.665 mL
Stacking buffer	1.25 mL
H_2O	3.05 mL
SDS (10%)	50 μL
APS (10%)	50 μL
TEMED	5 μL

3.2. SDS-PAGE and Preparation of Gel Bands for Glycan Removal

1. These instructions assume the use of an Invitrogen vertical mini gel system with freshly prepared gels (80 × 80 × 1 mm).
2. Choose a percentage gel appropriate to the size of the protein. Ten percent gel is the most commonly used; however, use 6% gel for proteins greater than 120 kDa, 12.5% gel for 70–200 kDa, or 15% gel for small (<70 kDa) proteins. Mix together the solutions for both the running gels (*see* **Table 2**) and for the stacking gels (*see* **Table 3**) in separate plastic tubes but do not add the TEMED until just before pouring the gel. The APS should be taken out of the freezer just before use and added penultimately.
3. Have two empty gel cassettes ready. Add the APS and the TEMED to the running gel mixture and mix well by inverting the tube (do not vortex as this can introduce air bubbles). Fill the gel cassettes with running gel up to the line approx 2 cm from the top edge. Cover with a layer of water-saturated butanol, tap out the bubbles, and leave to set for 15–20 min.
4. Pour off the butanol and rinse the gel top with water, and dry off any remaining droplets with a piece of filter paper.
5. Add APS and TEMED to the stacking gel mixture and mix well by inversion. Fill the top of the cassettes with stacking gel. Insert the comb, ensuring no bubbles remain. Leave to set for 15–20 min.

6. Prepare the running buffer by diluting 200 mL of 5X running buffer with 800 mL of water.
7. Carefully remove the combs and rinse the wells three times with running buffer, shaking out the buffer with each rinse. Peel off the tape from the bottom of the gel plate.
8. Load the gels with the sample using gel-loading pipet tips, starting with the tip at the bottom of the well and lifting it up slowly as the well fills. Load 5 μL of molecular weight markers in one or two wells.
9. Assemble the gel unit according to the manufacturer's instructions with a magnetic stirrer in the bottom. Add a small amount of running buffer to the inner compartment and make sure there are no leaks. Fill the inner compartment, taking care not to disturb the samples in the wells. Fill the outer chamber to about three quarters. Put the lid on and connect the power supply. With the power off, set the voltage to 500 V and the current to 25 mA per gel (50 mA for two gels). Switch on and run until the blue line reaches the bottom of the cassette (~45 min for a 10% gel).
10. Pry the gel cassette apart with a palette knife and discard the top plate. Gently drop each gel into separate plastic boxes containing enough Coomassie stain to cover and leave to stain for about 2 h (or it can be left overnight) on a platform shaker (note that the gel will shrink).
11. Tip out the stain and cover with destain 1. Place on a platform shaker for 5 min.
12. Tip out destain 1 and replace with destain 2. Place on a platform shaker for several hours or overnight until sufficiently destained (*see* **Note 8**).
13. Photograph the gel.
14. On a clean glass plate over a light box, cut out the Coomassie-stained bands from the gel with a clean scalpel. Cut into approx 1-mm^3 pieces and transfer the pieces from each band to 1.5-mL Eppendorf tubes, and freeze for at least 2 h or overnight (*see* **Note 9**).
15. Wash the gel pieces thoroughly by adding 1 mL of acetonitrile, vortexing, and then mixing on a roller mixer for 30 min at room temperature. Pipet off and discard the liquid. Repeat this washing procedure with 1 mL of PNGaseF buffer, 1 mL of acetonitrile, 1 mL of PNGaseF buffer, and 1 mL of acetonitrile.
16. Dry the gel pieces in a vacuum centrifuge.

3.3. N-*Glycan Release and Extraction*

1. The *N*-glycans are released by incubation with PNGaseF. The quantities given below are sufficient for one gel band of 2–3 mm in length, 10–15 mm^3.
2. Prepare the PNGaseF by mixing 3 μL of the stock solution with 27 μL of PNGaseF buffer.
3. Add the PNGaseF solution to the gel pieces and leave for 10–15 min until the gel has reswollen; add more PNGaseF solution if the gel has not fully reswollen. Cover the gel with 1–2 mm of PNGaseF buffer (50–100 μL) and close the lids. Incubate overnight at 37°C.
4. Wash the gel pieces thoroughly to extract the glycans. Vortex the gel pieces and spin down in a benchtop centrifuge. Remove any supernatant and retain in

a 1.5-mL Eppendorf tube. Add 200 μL of H_2O to the gel and sonicate for 30 min, remove the supernatant, and add to the retentate. Repeat this procedure with a further 2X 200 μL of H_2O, 200 μL of acetonitrile, 200 μL of H_2O, and 200 μL of acetonitrile (the acetonitrile washes shrink the gel and help to squeeze out the water). Dry the sample down to about 500 μL.

5. Filter the glycan solution into a 1.5-mL Eppendorf tube through a 13-mm 0.45-μm low-protein-binding PTFE membrane filter using a 1-mL plastic syringe (*see* **Note 4**), followed by 200 μL of H_2O to wash out the syringe and filter.
6. Dry down the filtrate and resuspend in 20 μL 1% formic acid and leave at room temperature for 40 min to ensure all the released glycans are converted to aldoses.
7. Dry down in a vacuum centrifuge to remove the formic acid. Resuspend in a known volume of H_2O. This glycan solution can be stored frozen at –20°C.

3.4. Fluorescent 2-AB-Labeling and Cleanup

1. Dry down the glycan sample in a 200-μL PCR tube. Make sure the sample is completely dry.
2. Prepare some freshly-washed 3MM chromatography strips by rinsing in three changes of water and drying in an oven at 65°C (*see* **Notes 10** and **11**). The paper should be washed and used within 24 h, as this reduces contaminants that leach from the paper.
3. Make up the 2-AB labeling solution per the manufacturer's instructions. This mixture is stable for several weeks if stored in the dark at –20°C in a clean glass vial.
4. Add 5 μL of the 2-AB labeling solution to the tube containing the dry glycans and cap the tube. Vortex and spin down. Incubate at 65°C in a dry oven for 30 min. Re-vortex and spin the sample, then return it to the oven for another 2.5 h.
5. Cool the sample in a freezer for 5 min and write the name of the sample in pencil at the top of the paper strip. Spot the sample onto the center line of the filter paper 1 cm from the bottom. Dry with a hair dryer.
6. Place the paper strip in a 100-mL glass beaker containing about 10 mL of acetonitrile; this should wet the paper but not touch the sample spot. Leave in a fume hood for 1–1.5 h. Any free 2-AB should run with the solvent front leaving 2-AB-labeled glycans at the origin. Remove the paper from the beaker and dry. Check that the 2-AB streak is well away from the sample spot by viewing under a UV light. Use a clean pair of scissors to cut out the fluorescent sample spot (do not draw around the spot as this will contaminate it).
7. Put the paper-sample spot into a 2.5-mL plastic syringe (*see* **Note 4**) fitted with a 13-mm 0.45-μm low-protein-binding PTFE membrane filter. Add 0.5 mL of H_2O to the syringe and leave for 10 min for the glycans to dissolve. Push the solution gently through the filter and collect. Repeat with a further 4X 0.5 mL of H_2O. Dry down the solution and redissolve in 100 μL of H_2O (*see* **Note 12**). This solution of 2-AB-labeled glycans should be stored frozen at –20°C and is stable for at least a year.

3.5. NP-HPLC Profiling of 2-AB-Labeled N-Glycans

1. These instructions assume the use of a Waters 2695 separations module with a Waters 474 or 2475 fluorescence detector. Another HPLC system which delivers

Table 4
NP-HPLC Startup Method

Time (min)	Flow (mL/min)	%A	%B
0	0	20	80
4	1	20	80
8	1	95	5
12	1	95	5
16	1	20	80
26	1	20	80
27	0.4	20	80
40	0.4	20	80
41	0.0	20	80

Run time 30 min. A, 50 m*M* ammonium formate, pH 4.4, normal phase (NP) stock solution; B, aceteonitrile.

a reproducible shallow gradient and has a fluorescence detector can be used. Set the fluorescence detector excitation and emission wavelengths to 330 and 420 nm with a bandwidth of 16 nm, set at maximum sensitivity.

2. Samples are injected in 80% acetonitrile. Take an aliquot of the sample (or standard dextran ladder) and make up to 20 µL with water, then add 80 µL of acetonitrile. It is a good idea to use only a small percentage of your sample in the first run in order to get some idea of how much must be loaded to produce a good trace. Usually about 1–5% of each gel band is sufficient.
3. Make up NP-HPLC running buffer by diluting 50 mL of NP stock solution to 2 L with water (this is solvent A). Solvent B is acetonitrile.
4. Set the HPLC to run the 30-min startup method (*see* **Table 4**), followed by a 180-min run (*see* **Table 5**) with no sample injection. This ensures that the column is well conditioned and helps with run-to-run reproducibility. Run a dextran ladder standard followed by the samples, with the injection volume set to 95 µL. Make sure that a dextran standard is run with each batch of samples.
5. Calibration and allocation of GU: The dextran ladder is used to calibrate the HPLC runs against any day-to-day or system-to-system changes (*see* **Fig. 1**). The GU value is calculated by fitting a fifth order polynomial distribution curve to the dextran ladder (usually glucose 1-15), then using this curve to allocate GU values from retention times (Empower GPC software from Waters can be used to calculate GU values). The GU values for neutral *N*-glycans are very reproducible with standard deviations of less than 0.03 between columns ***(13)***. This allows direct comparison with database values collected from a range of instruments over a long period of time. For sialylated glycans, more variation is found (±0.3 for disialylated biantennary) between columns and systems. It is therefore advisable to run sialylated samples together for direct comparison.

Table 5
NP-HPLC Running Method

Time (min)	Flow (mL/min)	%A	%B
0	0.4	20	80
152	0.4	58	42
155	0.4	100	0
157	1	100	0
162	1	100	0
163	1	20	80
177	1	20	80
178.5	0.4	20	80
200	0.4	20	80
201	0.0	20	80

Run time 180 min. A, 50 m*M* ammonium formate, pH 4.4, normal phase (NP) stock solution; B, aceteonitrile.

6. Lifetime of column: In order to maintain resolution of glycan peaks, it is important to monitor the column and change it when the peak widths at 50% height get above 0.65 min for dextran peaks 3 and 8. This is usually after about 800 runs. You may, however, need to be more stringent if you are measuring poorly-resolved peaks.

3.6. Exoglycosidase Digestions

1. Pipet aliquots of the 2-AB-labeled glycan pool into 200-μL microcentrifuge tubes and dry down. Add 1 μL of 10X incubation buffer, pH 5.5 (or 2 μL of 5X manufacturer's buffer), the required enzyme or array of enzymes, and H_2O to make up to 10 μL. Incubate overnight (16–18 h) at 37°C. A typical set of enzyme digestion arrays is shown in **Fig. 3** for human IgG glycans.
2. Prewash the Micropure-EZ enzyme removers with 200 μL of H_2O in a microcentrifuge at half speed (~7000*g*) for 10 min and discard washings.
3. Apply the digested glycan sample to the middle of the filter and centrifuge at full speed (~14,000*g*) for 2 min. Wash out the digestion tube with 20 μL of H_2O and apply to the filter, then centrifuge for another 2 min. Apply a further 100 μL of H_2O to the filter and centrifuge. Dry down the sample, then redissolve in 20 μL of H_2O ready for injection onto the HPLC.

Fig. 3. *(facing page)* NP-HPLC profiles of 2-AB-labeled human IgG *N*-glycans. The top profile shows undigested whole pool glycans, followed by a series of exoglycosidase digestions (*see* **Table 1** for enzyme abbreviations and specificities).

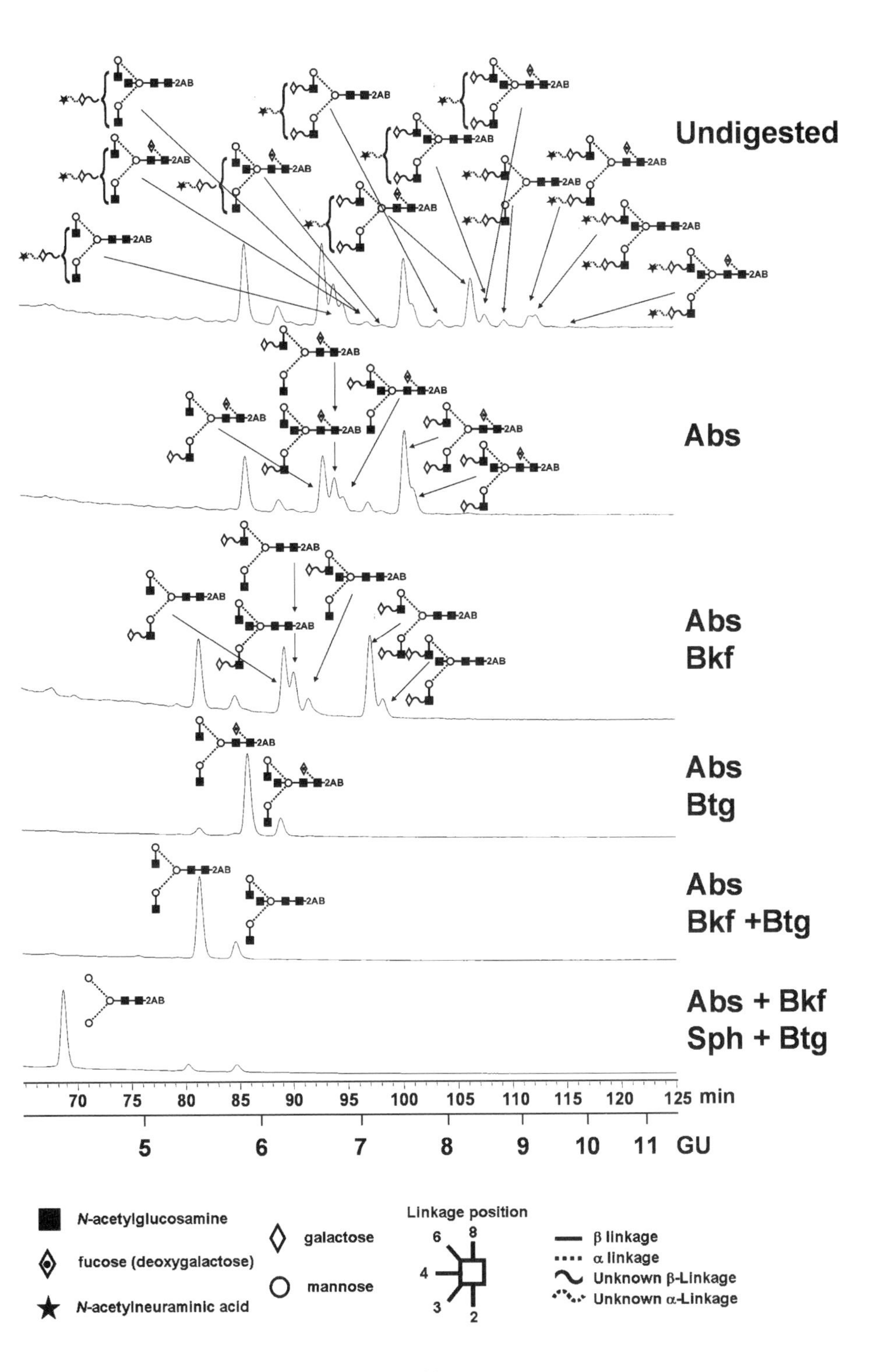
Undigested
Abs
Abs
Bkf
Abs
Btg
Abs
Bkf +Btg
Abs + Bkf
Sph + Btg
2AB
70 75 80 85 90 95 100 105 110 115 120 125 min
5 6 7 8 9 10 11 GU
N-acetylglucosamine
fucose (deoxygalactose)
N-acetylneuraminic acid
galactose
mannose
Linkage position
6 8
4
3 2
β linkage
α linkage
Unknown β-Linkage
Unknown α-Linkage

Table 6
WAX-HPLC Startup Method

Time (min)	Flow (mL/min)	%A	%B
0	0	0	100
5	1	0	100
10	1	0	100
40	1	0	100
41	0	0	100

Run time 30 min. A, 500 m*M* ammonium formate, pH 9.0, weak anion exchange (WAX) stock solution; B, 10% methanol in water.

Table 7
WAX-HPLC Running Method

Time (min)	Flow (mL/min)	%A	%B
0	1	0	100
12	1	5	95
25	1	21	79
50	1	80	20
55	1	100	0
65	1	100	0
66	2	0	100
70	2	0	100
77	2	0	100
78	1	0	100
90	1	0	100
91	0	0	100

Run time 80 min. A, 500 m*M* ammonium formate, pH 9.0, weak anion exchange (WAX) stock solution; B, 10% methanol in water.

3.7. WAX-HPLC Profiling of 2-AB-Labeled N-Glycans

1. Set up equipment as in **Subheading 3.5., step 1**.
2. Inject samples in 100% aqueous solution. Dilute the required amount of sample or fetuin *N*-glycan standard to 100 µL, ready for 95-µL injection.
3. Make up WAX-HPLC running buffer by diluting 250 mL of WAX stock solution to 1 L with water (this is solvent A). Solvent B is 10% methanol in water.
4. Set the HPLC to run the 30-min startup method (*see* **Table 6**), followed by an 80-min run (*see* **Table 7**) with no sample injection. This ensures that the column

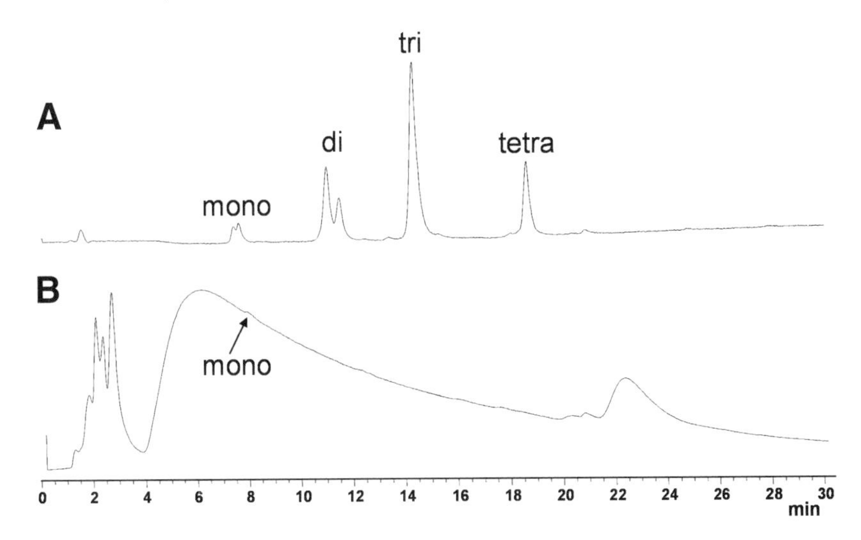

Fig. 4. WAX-HPLC profiles of **(A)** bovine serum fetuin *N*-glycans showing separation of the mono-, di-, tri-, and tetrasialylated glycans. The double peaks of the mono- and disialylated glycans are from the triantennary eluting before the biantennary glycans. **(B)** When excess 2-AB has not been sufficiently removed, the glycan peaks are obscured.

is well-conditioned and helps with run-to-run reproducibility. Run a fetuin *N*-glycan standard followed by the samples.

5. Compare the elution positions of peaks to those of the fetuin standard (*see* **Fig. 4**). Note that the larger triantennary glycans elute before the biantennary glycans with the same charge, and that if excess 2-AB label has not been sufficiently removed, a large "hump" in the baseline which can obscure the glycan peaks is seen when the sample is run on WAX.

3.8. Structural Allocation

1. The GU value for a glycan is directly related to the number and linkage of its constituent monosaccharides; the larger the glycan, the higher its GU value. Thus GU values can be used to predict structures since each monosaccharide in a specific linkage adds a given amount to the GU value of a given glycan (*see* **Table 8**).
2. The preliminary assignment of structures to peaks is made by matching the GU values from the undigested pool with those in a database of known structures (*see* **refs. *13*** and ***14***; *see also* Oxford Glycobiology Institute HPLC database). This will often result in several possible structures for each peak; therefore, exoglycosidase digestion is required to confirm structures. To make a "final" allocation, the NP-HPLC traces and GU values from the exoglycosidase digestions must be interpreted. For example, in **Fig. 2**, the Abs + Bkf + Sph + Btg trace shows that the glycans

Table 8
Incremental GU Values for 2-AB-Labeled *N*-Glycans

Monosaccharide	Linkage	To	GU increment
Mannose	α1-2,3,6	Mannose	0.7–0.9
GlcNAc	β1-2,4,6	α-Mannose	0.5
GlcNAc (bisect)	β1-4	β-Mannose	0.2–0.4
Galactose	α or β1-3,4	GlcNAc or Gal	0.8–0.9
Fucose (core)	α1-6	Core GlcNAc	0.5
Fucose (outer arm)	α1-3,4	GlcNAc	0.8
Fucose (outer arm)	α1-2	Gal	0.5
Neu5Ac	α2-3,6	Gal	0.7–1.2

have digested down to $GlcNAc_2Man_3$ (GU 4.4). The trace above, which omits the *N*-acetylhexosaminidase (Sph), shows two peaks (GU 5.5 and GU 5.8), a biantennary glycan with and without a bisecting GlcNAc. In the trace without the fucosidase (Bkf) the two major peaks are 0.5 GU higher, as they contain core fucose plus minor peaks for these two structures without core fucose. In the trace without galactosidase (Abs + Bkf) the structures (±bisect) have 0, 1, or 2 galactose residues added (note that isomers separate: if galactose is on the upper 6-linked arm it elutes later than if the galactose is on the lower 3-linked arm). By continuing to compare successive enzyme digestions and being guided by the incremental values in **Table 8**, all of the structures can be allocated.

3. As there is one 2-AB fluorescent label per glycan, the fluorescence intensity is directly related to the number of moles of labeled glycans present. Thus, the areas of the HPLC peaks can be used to quantify the amounts of glycans present. Comparison of peak areas within a trace gives the relative proportions of different glycans. For actual quantification of the amounts of glycans, a 2-AB-labeled standard of known amount can be used to generate a calibration curve.
4. Additional information about the number of sialic acids present can be gained from the WAX-HPLC profiles.
5. Digestion with specific sialidases to distinguish between linkages can be carried out and then run on the WAX for comparison.
6. It is also feasible to collect these charge-separated fractions following WAX-HPLC, then run them separately on NP-HPLC with or without further exoglycosidase digestions (*see* **Fig. 5**). The buffer salts used to elute the samples from WAX must be removed before NP-HPLC or exoglycosidase digestion. To do this, dry the collected WAX fractions and redissolve them in 2 mL of H_2O and then redry; repeat two or three times before finally lyophilizing them overnight ***(14)***.

4. Notes

1. High-purity reagents are used throughout, e.g., AnalaR, Ultrapure, or greater than 99%.

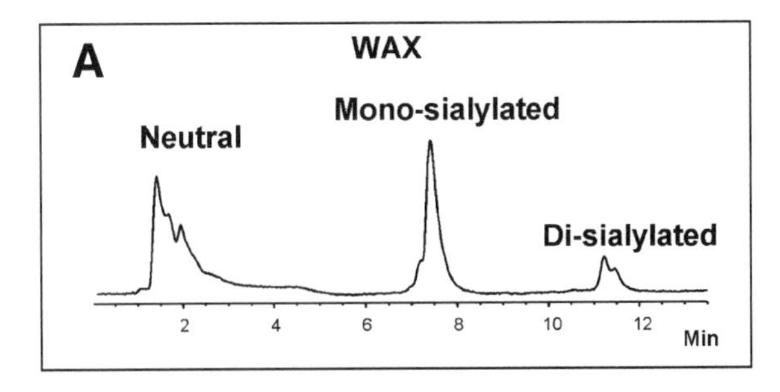

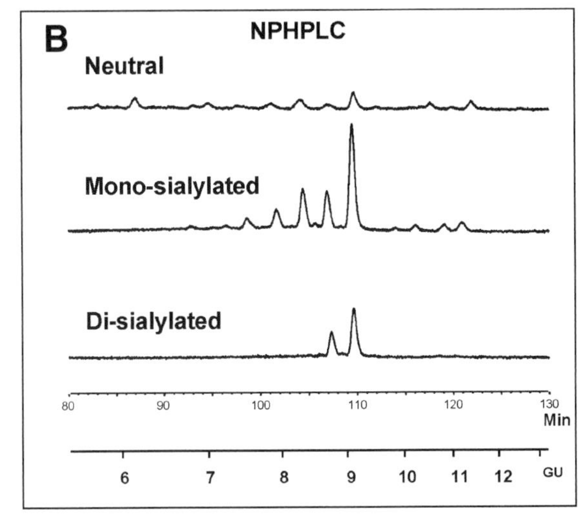

Fig. 5. *N*-glycans from secretory component of human secretory IgA were separated into **(A)** neutral and mono- and disialylated fractions by WAX-HPLC, and then **(B)** fractions were collected and run on NP-HPLC. The majority of the structures were monosialylated. Note that the two disialylated structures co-eluted with monosialylated structures. (Reproduced from **ref. *10*** with permission from The American Society for Biochemistry and Molecular Biology.)

2. Unless otherwise stated, all solutions are prepared in water that has a resistivity of 18 MΩ-cm, is particle-free (>0.22 μm), and has a total organic content of less than ten parts per billion.
3. Sarstedt (Sarstedt Aktiengesellschaft & Co., Nümbrecht, Germany) microtubes are used throughout, as they give very little contamination on the HPLC.
4. It is important to use syringes that do not have any silicone rubber, as this leads to contaminating peaks in the sample.
5. All solutions are stored at room temperature unless otherwise stated.
6. Buy small bottles of TEMED as this may decline in quality after opening, lengthening the time gels take to polymerize.

7. It is important to use acetonitrile with low fluorescence; otherwise, a sloping baseline is seen from the gradient elution.
8. Add a small piece of clean foam rubber packaging material to destain 2, as this greatly speeds up the destain process.
9. Freezing the gel pieces helps break down the matrix a little so that more of the GP is accessible to the PNGaseF.
10. An alternative cleanup method is to use LudgerClean S/GlykoClean S glycan purification cartridges (Ludger or Prozyme) instead of 3MM chromatography paper.
11. When clean surfaces are required, e.g., for drying the paper, use clean aluminium foil as this will not contaminate the sample.
12. When drying volumes of more than 0.5 mL, recovery of the sample can be improved by redissolving the sample in reducing volumes of water, vortexing, and spinning before redrying. For example, starting with 2.5 mL, the sample is redissolved with the following volumes: 1 mL, 0.5 mL, 250 µL, 100 µL, or until the sample reaches the volume required.

Acknowledgment

This work was supported by an endowment from the Oxford Glycobiology Institute.

References

1. Parekh, R. B., Dwek, R. A., Sutton, B. J., et al. (1985) Association of rheumatoid arthritis and primary osteoarthritis with changes in the glycosylation pattern of total serum IgG. *Nature* **316,** 452–457.
2. Block, T. M., Comunale, M. A., Lowman, M., et al. (2005) Use of targeted glycoproteomics to identify serum glycoproteins that correlate with liver cancer in woodchucks and humans. *Proc. Natl. Acad. Sci. USA* **102,** 779–784.
3. Butler, M., Quelhas, D., Critchley, A. J., et al. (2003) Detailed glycan analysis of serum glycoproteins of patients with congenital disorders of glycosylation indicates the specific defective glycan processing step and provides an insight into pathogenesis. *Glycobiology* **13,** 601–622.
4. Kuhn, P., Tarentino, A. L., Plummer, T. H., Jr., and Van Roey, P. (1994) Crystal structure of peptide-N4-(N-acetyl-beta-D-glucosaminyl)asparagine amidase F at 2.2-A resolution. *Biochemistry* **33,** 11,699–11,706.
5. Navazio, L., Miuzzo, M., Royle, L., et al. (2002) Monitoring endoplasmic reticulum-to-Golgi traffic of a plant calreticulin by protein glycosylation analysis. *Biochemistry* **41,** 14,141–14,149.
6. Tekoah, Y., Ko, K., Koprowski, H., et al. (2004) Controlled glycosylation of therapeutic antibodies in plants. *Arch. Biochem. Biophys.* **426,** 266–278.
7. Wormald, M. R., Rudd, P. M., Harvey, D. J., Chang, S. C., Scragg, I. G., and Dwek, R. A. (1997) Variations in oligosaccharide-protein interactions in immunoglobulin G determine the site-specific glycosylation profiles and modulate the dynamic motion of the Fc oligosaccharides. *Biochemistry* **36,** 1370–1380.

8. Takahashi, N., Ishii, I., Ishihara, H., et al. (1987) Comparative structural study of the N-linked oligosaccharides of human normal and pathological immunoglobulin G. *Biochemistry* **26,** 1137–1144.
9. Homans, S. W., Ferguson, M. A., Dwek, R. A., Rademacher, T. W., Anand, R., and Williams, A. F. (1988) Complete structure of the glycosyl phosphatidylinositol membrane anchor of rat brain Thy-1 glycoprotein. *Nature* **333,** 269–272.
10. Parekh, R. B., Dwek, R. A., Rudd, P. M., et al. (1989) N-glycosylation and in vitro enzymatic activity of human recombinant tissue plasminogen activator expressed in Chinese hamster ovary cells and a murine cell line. *Biochemistry* **28,** 7670–7679.
11. Küster, B., Wheeler, S. F., Hunter, A. P., Dwek, R. A., and Harvey, D. J. (1997) Sequencing of N-linked oligosaccharides directly from protein gels: in-gel deglycosylation followed by matrix-assisted laser desorption/ionization mass spectrometry and normal-phase high-performance liquid chromatography. *Anal. Biochem.* **250,** 82–101.
12. Radcliffe, C. M., Diedrich, G., Harvey, D. J., Dwek, R. A., Cresswell, P., and Rudd, P. M. (2002) Identification of specific glycoforms of major histocompatibility complex class I heavy chains suggests that class I peptide loading is an adaptation of the quality control pathway involving calreticulin and ERp57. *J. Biol. Chem.* **277,** 46,415–46,423.
13. Guile, G. R., Rudd, P. M., Wing, D. R., Prime, S. B., and Dwek, R. A. (1996) A rapid high-resolution high-performance liquid chromatographic method for separating glycan mixtures and analyzing oligosaccharide profiles. *Anal. Biochem.* **240,** 210–226.
14. Royle, L., Roos, A., Harvey, D. J., et al. (2003) Secretory IgA N- and O-glycans provide a link between the innate and adaptive immune systems. *J. Biol. Chem.* **278,** 20,140–20,153.
15. Zamze, S., Harvey, D. J., Chen, Y. J., Guile, G. R., Dwek, R. A., and Wing, D. R. (1998) Sialylated N-glycans in adult rat brain tissue—a widespread distribution of disialylated antennae in complex and hybrid structures. *Eur. J. Biochem.* **258,** 243–270.

10

Molecular Modeling of Glycosyltransferases

Anne Imberty, Michaela Wimmerová, Jaroslav Koča, and Christelle Breton

Summary

Glycosyltransferases, the enzymes that build oligosaccharides and glycoconjugates, have received much interest in recent years owing to their biological functions and their potential uses in biotechnology. The analysis of the wealth of sequences that are now available in databases allowed the classification in different families characterized by conserved peptide motifs. Nevertheless, only a limited number of crystal structures is available and molecular modeling appears to be an inescapable tool for rationalization of binding data, engineering of enzyme properties, and design of inhibitors that would be of interest as therapeutic compounds. Because of sequence diversity and limited experimental data, molecular modeling of these enzymes is not straightforward and utilizes the most recent tools, such as fold recognition programs.

Key Words: Glycosyltransferase; nucleotide-sugar; oligosaccharide; molecular modeling; fold recognition; docking; molecular dynamics.

1. Introduction

Glycosyltransferases (GTs) are the enzymes that synthesize oligosaccharides, polysaccharides, and glycoconjugates (*see* Chapters 4, 11, 13, and 16). GTs have been grouped into 76 families on the basis of sequence similarities. Despite the fact that many GTs recognize similar donor or acceptor substrates, there is surprisingly limited sequence identity between different families. Interestingly, only two folds have been observed until now for GTs: fold GT-A, consisting of one $\alpha/\beta/\alpha$ sandwich domain and characterized by the presence of divalent cation in the binding site, and fold GT-B, consisting of two such domains (*[1,2]*; *see* **Fig. 1**).

From: *Methods in Molecular Biology, vol. 347: Glycobiology Protocols*
Edited by: I. Brockhausen © Humana Press Inc., Totowa, NJ

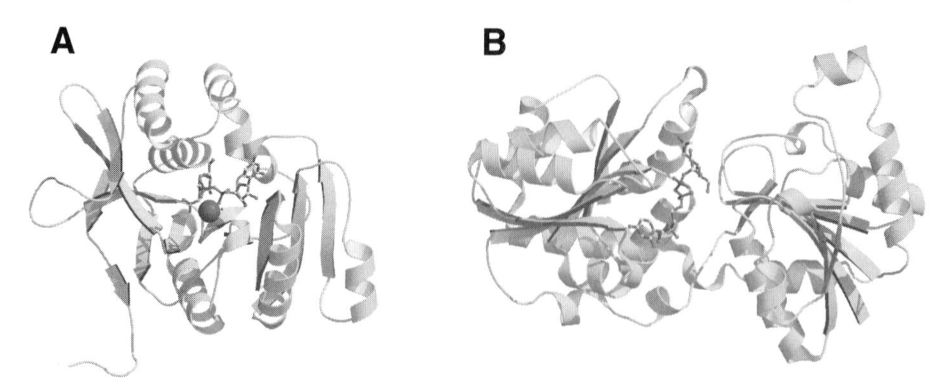

Fig. 1. Graphic representation of the two folds that have been observed in crystal structures of glycosyltransferases. **(A)** Fold GT-A is represented by EXTL2 in complex with UDP-GalNAc and Mn^{2+} (Protein Database code 1OMZ; *see* **ref. *23***). **(B)** Fold GT-B is represented by MurG complexed with UDP-GlcNAc (code 1NLM; *see* **ref. *24***). (This drawing was prepared using the Molscript ***[25]*** and Raster3D ***[26]*** programs.)

Molecular modeling of GTs presents peculiar difficulties. Compared with the hundreds of different enzymes that participate in glycoconjugate synthesis, very few have been crystallized: only 20 of the 82 carbohydrate-active enzymes (CAZY) families ***(3)*** contain at least one GT that has been crystallized. The low degree of sequence similarity within some of the CAZY families and the absence or quasi-absence of similarity between different families represent an insurmountable barrier for classical sequence alignment procedure, which is a prerequisite in a homology-building procedure. Docking of substrates also appears to be a difficult task owing to the flexibility of the nucleotide sugar and the presence of phosphate and divalent cation, for which energy parameters are not always available in modeling software although efforts have recently been made ***(4)***. Furthermore, concerted movements of loops are required for locking the nucleotide sugar in the binding site of the GT-A family, and the enzymes can exist in an "open" or "closed" conformation, therefore complicating the modeling procedure.

Molecular modeling has only recently been used in the field of GTs, but the models have been useful not only for rationalizing experimental data but also for designing directed mutagenesis experiments. Fold recognition coupled with multivariate analysis has been applied to a large number of sequences in order to identify putative GTs ***(5,6)***. Homology modeling, together with docking of substrates, helps in understanding the molecular basis of specificity for human blood group A and B transferases ***(7)*** and for plant glucosyltransferase ***(8)***. Although the time scale of loop opening and closing is too large to be modeled, preliminary molecular dynamics studies allow for identifying the key amino acids involved in the conformational changes ***(9,10)***.

2. Materials

2.1. Hardware

The computer required is either a Silicon Graphics workstation or a personal computer running under Linux. Access to the Internet with reasonable speed access is necessary.

2.2. Computing Programs

1. Visualization of protein structure: Chimera (http://www.cgl.ucsf.edu/chimera/), PyMol (http://pymol.sourceforge.net/), or VMD (http://www.ks.uiuc.edu/Research/vmd/).
2. Fold recognition: Prohit (previously Profit and Proceryon; http://www.proceryon.com/solutions/prohit_pro.html).
3. Homology modeling: Composer within the Sybyl package (http://www.tripos.com/) or Modeller (http://salilab.org/modeller/).
4. Docking: Autodock (http://www.scripps.edu/mb/olson/doc/autodock/).
5. Molecular Dynamics: Amber (http://amber.scripps.edu/).
6. Checking for stereochemical quality of modeled structures: What if (http://swift.cmbi.kun.nl/whatif/).
7. High-quality rendering for publication: Molscript (http://www.avatar.se/molscript/) and Raster3D (http://www.bmsc.washington.edu/raster3d/raster3d.html).

2.3. Internet Access

Addresses of servers that can be used directly on the Internet are given in each protocol.

3. Methods

The protocols described below represent only some possibilities of several ways to proceed (*see* **Note 1**).

3.1. Datamining for Glycosyltransferase Sequences

3.1.1. The Basic Local Alignment Search Tool

The Basic Local Alignment Search Tool (BLAST; http://www.ncbi.nlm.nih.gov/BLAST, http://www.ebi.ac.uk/Tools/similarity.html) provides a method for rapid searching of protein databases. Because the BLAST algorithm detects local as well as global alignments, regions of similarity in unrelated proteins can be detected, and both functional and evolutionary information can be inferred from well-designed queries and alignments.

For large datasets of sequences to be searched (e.g., from whole genomes) multivariate analysis could be helpful in order to discriminate between positive hit and strong background *(5,6)*.

3.1.2. Structure Fold Recognition

Prohit (http://www.proceryon.com/solutions/prohit_pro.html) is one example of the program used in this method.

1. Download structure coordinates of known GTs from Brookhaven protein database and make your own GT library.
2. Prepare a set of protein sequences to be analyzed (e.g., from whole genome).
3. Run the threading method based on fold recognition on the GT library. Several scores for each sequence alignment on a 3D structure of the GT library are calculated.
4. Analyze sequences with the highest scores by additional bioinformatic tools (*see* **Subheading 3.2.**). If not successful, run a multivariate analysis using principal component analysis (PCA; *see* **ref. *11***) to investigate the whole dataset generated by the fold recognition method on all sequences. This method allows grouping, outliers, and trends investigation of your data.
5. Append sequences of known GTs into analysis in order to determine their behavior in PCA. Optimize parameters to get clusters of selected GTs based on their topology. Run the same multivariate analysis on your dataset.
6. Analyze sequences in GT clusters by additional bioinformatic tools (*see* **Subheading 3.2.**).

3.1.3. Analysis of Amino Acid Physicochemical Properties

1. Take a reference set of known GTs and analyze their topology.
2. Analyze physicochemical properties of these GTs and your selected sequences, like hydrophobicity, polarizability, charge, etc.
3. Run multivariate analysis (PCA) to separate individual groups of GTs.
4. Analyze your sequences closed to GT grouping by additional bioinformatic tools.

3.2. Two-Dimensional Analysis of Sequences

1. Search for transmembrane domain(s), if any (TMHMM: http://www.cbs.dtu.dk/services/TMHMM-2.0/).
2. In cases involving a large number of amino acids (>500) consider the possibility of several domains (i.e., bifunctional glycosyltransferase, lectin domain, etc.). DrawHCA (http://psb11.snv.jussieu.fr/hca/index.html/) may be used for identifying regions of low complexity that can correspond to linkers between such domains.
3. Prediction of secondary structures can be done using multiple methods. Use of a consensus approach that compares the results of different algorithms, such as the one proposed on the Network Protein Sequence Analysis site (http://npsa-pbil.ibcp.fr/), is recommended.

3.3. Classification Into Families (Grouping of Distantly-Related Proteins)

1. For each domain of interest, run BLAST (or second-generation program) to identify GTs with a significant sequence similarity. Once a group of related sequences is formed, perform a multiple alignment with several sequences representative of

the family in order to determine how conserved the important amino acids are (*see* **Note 2**).
2. Because fold is more conserved than sequence in GTs, also consider the fold recognition approach to identify distantly related sequences. The 3DPSSM server (http://www.sbg.bio.ic.ac.uk/~3dpssm/index2.html) has been proven to be well adapted in this context (*see* **Note 3**).
3. The DrawHCA program *(12)* is particularly well adapted for aligning sequences of low similarities but sharing structural features (*see* **Fig. 2**).

3.4. Homology Modeling

1. Select two to five crystal structures showing sequence similarities to the target, being careful to choose structures with sufficient differences in sequences.
2. Download the coordinates file (also called pdb file) from Brookhaven protein databank (http://www.rcsb.org/pdb/; *see* **ref.** ***13***).
3. Check the three-dimensional structure by reading the Header (remark part) of the files that have been downloaded and by viewing them with a structure viewer program. Eliminate the ones for which a significant portion of loops are missing because of disorder.
4. Use a program for homology modeling such as Composer *(14)* or Modeller *(15)* for aligning the structures, and for building the structural conserved regions (SCRs) based on the multialignment performed in **Subheading 3.3.**
5. Build each of the missing portions using a loop builder. Do not consider only the defaults conformation provided by the program, but visualize all the possible ones and make a choice based on steric hindrance and similarity with known structure.
6. Optimize the conformations of each of these loops using a simulated annealing approach: minimize only the geometry of the new loop and a few amino acids around while keeping the rest of the protein rigid.
7. Add hydrogen atoms and partial charges in your model, and optimize the position of hydrogen atoms.
8. Check the built structure for steric conflict between amino acids and relieve them by optimizing the side chain without altering the backbone conformation.

3.5. Docking of Nucleotide Sugars and Acceptor Substrates

1. Build the oligosaccharide or nucleotide sugar with a molecular editor or download it from a crystal structure.
2. Add hydrogen atoms if needed and partial charges calculated with semi-empirical methods.
3. If the complex to be modeled is very similar to one already solved by crystallography (e.g., blood group A transferase compared with α3Gal-transferase), position the ligand in the binding site with the same geometry as in the known structure (homology docking).
4. If full approach orientation is needed, use the Autodock program *(16)* considering flexibility at glycosidic linkages and pendent groups (hydroxyl groups, hydroxymethyl, etc.).

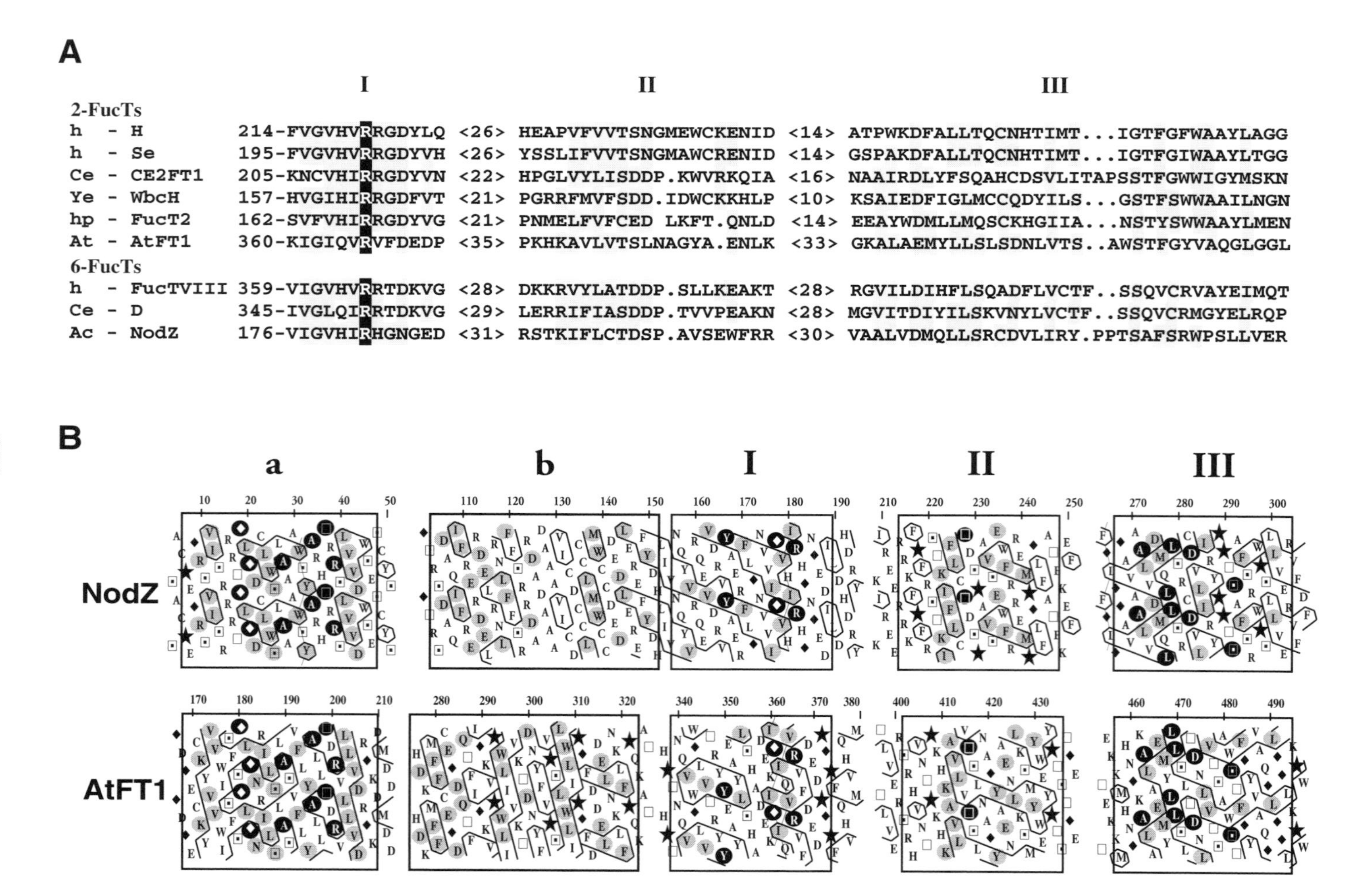
A
I
II
III
2-FucTs
h - H 214-FVGVHVRRGDYLQ <26> HEAPVFVVTSNGMEWCKENID <14> ATPWKDFALLTQCNHTIMT...IGTFGFWAAYLAGG
h - Se 195-FVGVHVRRGDYVH <26> YSSLIFVVTSNGMAWCRENID <14> GSPAKDFALLTQCNHTIMT...IGTFGIWAAYLTGG
Ce - CE2FT1 205-KNCVHIRRGDYVN <22> HPGLVYLISDDP.KWVRKQIA <16> NAAIRDLYFSQAHCDSVLITAPSSTFGWWIGYMSKN
Ye - WbcH 157-HVGIHIRRGDFVT <21> PGRRFMVFSDD.IDWCKKHLP <10> KSAIEDFIGLMCCQDYILS...GSTFSWWAAILNGN
hp - FucT2 162-SVFVHIRRGDYVG <21> PNMELFVFCED LKFT.QNLD <14> EEAYWDMLLMQSCKHGIIA...NSTYSWWAAYLMEN
At - AtFT1 360-KIGIQVRVFDEDP <35> PKHKAVLVTSLNAGYA.ENLK <33> GKALAEMYLLSLSDNLVTS..AWSTFGYVAQGLGGL
6-FucTs
h - FucTVIII 359-VIGVHVRRTDKVG <28> DKKRVYLATDDP.SLLKEAKT <28> RGVILDIHFLSQADFLVCTF..SSQVCRVAYEIMQT
Ce - D 345-IVGLQIRRTDKVG <29> LERRIFIASDDP.TVVPEAKN <28> MGVITDIYILSKVNYLVCTF..SSQVCRMGYELRQP
Ac - NodZ 176-VIGVHIRHGNGED <31> RSTKIFLCTDSP.AVSEWFRR <30> VAALVDMQLLSRCDVLIRY.PPTSAFSRWPSLLVER
B
a
b
I
II
III
NodZ
AtFT1

5. For each docking solution, optimize the position of hydrogen atoms while taking into account electrostatic interactions.
6. Perform a final optimization allowing for minimization of the position of the ligands and the amino acid side chains in the binding site.
7. Visualize the docking solution with the Connolly surface of the protein using the Molcad program (*see* **ref.** ***17*** and **Fig. 3**).

3.6. Molecular Dynamics

1. Select a starting structure. It can either be an X-ray or nuclear magnetic resonance structure. In special cases, a structure obtained by homology modeling may be used.
2. Download the coordinates file (usually called the pdb file) from the Brookhaven protein databank (www.pdb.org; *see* **Note 4**). For example, LgtC galactosyltransferase with UDP-2-deoxy-2-fluorogalactose (UDP-Gal, donor analog) and 4-deoxylactose (acceptor analog) is deposited under the code 1GA8.
3. Check protonation states of histidine residues by WHAT IF software *(18)*.
4. Add all hydrogen atoms using Xleap from the AMBER package (*see* **ref.** ***19*** and **Note 5**).
5. Neutralize the system by adding chloride or sodium counter ions depending on the overall charge of the system.

Fig. 2. *(previous page)* Detection of several conserved regions present in all α2- and α6-fucosyltransferases from prokaryotic and eukaryotic origin, using BLAST and HCA methods. On the basis of sequence similarity, the known α2- and α6-fucosyltransferases have been classified in CAZY in three different families (GT11, GT23, and GT37). Blast analysis coupled to HCA allows the detection of three conserved regions in all these sequences *(27)*. **(A)** Sequence alignment of the three most conserved regions (designated I, II, and III) present in the catalytic domains of all α1,2- and α1,6-fucosyltransferases. The only invariant arginine residue is indicated in white on a black background. Similar amino acids are shaded in gray (groups of similar residues are defined as AILMV, FWY, DENQ, RHK, and CST). Ac, *Azorhizobium caulinodans*; At, *Arabidopsis thaliana*; Ce, *Caenorhabditis elegans*; h, human; hp, *Helicobacter pylori*; Ye, *Yersinia enterocolitica*. **(B)** HCA comparison of NodZ and AtFT1. The protein sequences are written on a duplicated α-helical net and the contour of clusters of hydrophobic residues is drawn. The one-letter code for amino acids is used except for Gly (diamonds), Pro (stars), Ser (squares with solid dots), and Thr (open squares). In addition to the three previously mentioned motifs, NodZ proteins (GT23) display two other regions of similarity with the plant α2-fucosyltransferases (GT37), designated a and b, which are located at the N-terminal side of the catalytic domain. The five most conserved regions have been boxed. Residues that are strictly invariant in all NodZ proteins and plant α2-FucTs are indicated in white on a black background, and those that are highly conserved are shaded in gray. (Reproduced from **ref.** ***27*** with permission from the American Society for Microbiology.)

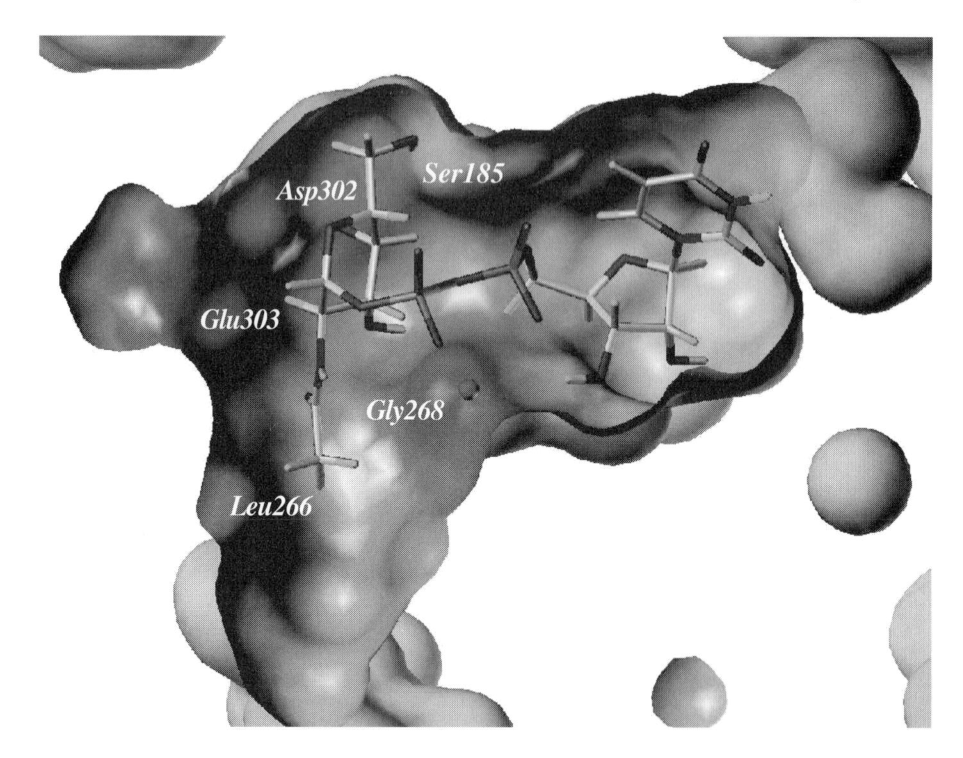

Fig. 3. Docking of UDP-GalNAc in the binding site of human blood group A transferase. (Reproduced from **ref. 7** with permission from Oxford University Press.)

6. Insert the system in a rectangular water box where the layer of water molecules is at least 10 Å.
7. Set all important molecular dynamics (MD) parameters, namely the cut-off value (usually 9–10 Å), constant pressure or constant volume switch, and time step (usually 1–2 fs). Decide whether to use cut-off or PME to treat electrostatic interactions.
8. Run a 1000-step minimization and a 2-ps-long MD during which the protein is frozen and the rest of the system (including counter ions) is allowed to move.
9. Perform several subsequent minimizations during which decreasing force constants (for example, 1000, 500, 125, 25, 0 kcal/mol Å^2) are applied to the backbone atoms to allow the side chains to relax (*see* **Note 6**).
10. Slowly heat the system to 250°K in 10 ps and then to 298.15°K in 40 ps (*see* **Note 7**).
11. Run the production part of the trajectory. The length is dependent on the features you follow and the computational resources you have at hand. Usually the time scale is up to 10 ns, although much longer trajectories are also mentioned in the literature.

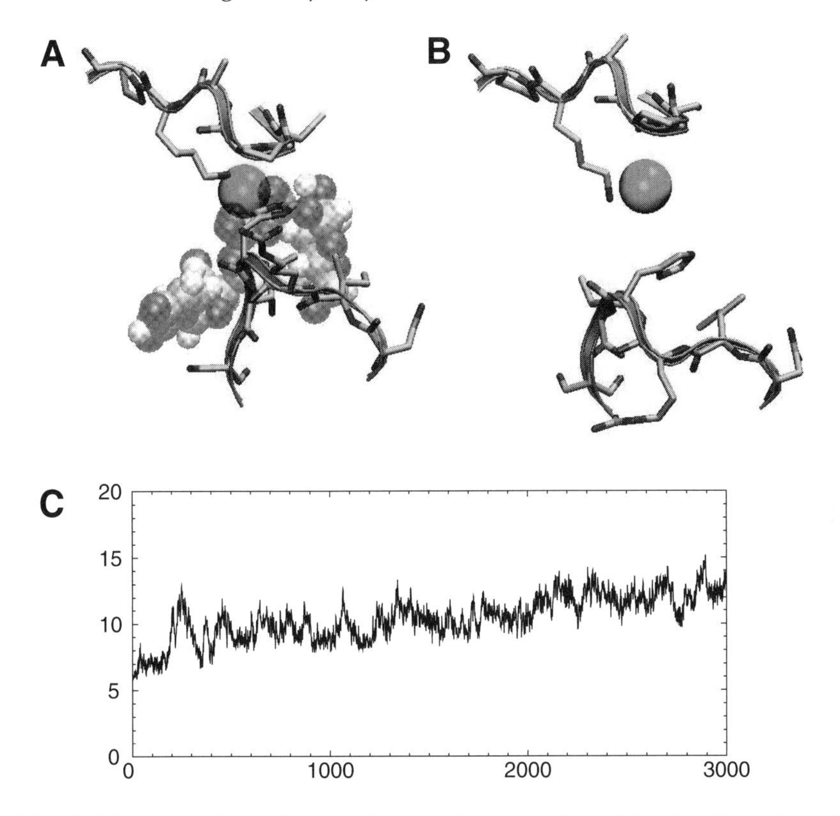

Fig. 4. Molecular dynamics simulation of the opening of the binding site of LgtC (α4-galactosyltransferase from *Neisseria meningitidis*). **(A)** Closed binding site with manganese at the beginning of the simulation (UDP-Gal not present but position indicated in grey). **(B)** Opened binding site at the end of the simulation. **(C)** History of the distance between the α-carbons of His78 and Pro248. (Adapted from **ref. *10*** with permission from Elsevier.)

12. Analyze the resulting trajectories by the PTRAJ module of AMBER, by GROMACS ***(20)***, and by gOpenMol ***(21)*** program packages (*see* **Fig. 4**).
13. Visualize the trajectory, using VMD ***(22)*** or similar software.

4. Notes

1. It is strongly recommended to discuss a particular protocol with an experienced person when entering fields such as homology modeling, docking, or MD, as there are many parameters that can be changed.
2. Some programs may have problems aligning all sequences properly when the sequences are similar only in some short regions. This is because programs such as ClustalW try to find global, not local, alignments. In such cases it may be useful to cut out the similar parts before running alignment.

3. The authors have developed a new fold prediction site, Phyre (http://www.sbg.bio.ic.ac.uk/~phyre/), that may be of interest.
4. Make sure that all heavy atoms required for simulations are present and have the correct connectivity in the structure file. If not, complete the structure very carefully using homology modeling.
5. Make sure that the position of all hydrogens is correct and salt bridges are created. This step may be time-consuming, but it is extremely important as wrongly positioned hydrogen atoms may substantially affect future simulations and generate artifacts.
6. The energy of the system should drop substantially. If this is not the case and energy "oscillates" on high values, it is a sign that something is wrong with the structure. The cause may, for example, be a water molecule unnaturally locked within the protein, or it may even be a single hydrogen atom incorrectly positioned.
7. This is an extremely sensitive part of the simulation. Follow the density of the system and its temperature carefully. Both should continuously reach desired values. If there are jumps, try to perform the procedure more slowly (in a longer time) or use a smaller time step. Jumps would mean "bubbles" in the solvent or local overheating, which could unnaturally change the structure.

References

1. Unligil, U. M. and Rini, J. M. (2000) Glycosyltransferase structure and mechanism. *Curr. Opin. Struct. Biol.* **10,** 510–517.
2. Breton, C., Heissigerova, H., Jeanneau, C., Moravcova, J., and Imberty, A. (2002) Comparative aspects of glycosyltransferases. *Biochemi. Soc. Symp.* **69,** 23–32.
3. Coutinho, P. M., Deleury, E., Davies, G. J., and Henrissat, B. (2003) An evolving hierarchical family classification for glycosyltransferases. *J. Mol. Biol.* **328,** 307–317.
4. Petrova, P., Monteiro, C., Hervé du Penhoat, C., Koča, J., and Imberty, A. (2001) Conformational behavior of nucleotide-sugar in solution: molecular dynamics and NMR study of solvated UDP-glucose in the presence of monovalent cations. *Biopolymers* **58,** 5617–5635.
5. Wimmerova, M., Engelsen, S. B., Bettler, E., Breton, C., and Imberty, A. (2003) Combining fold recognition and exploratory data analysis for searching for glycosyltransferases in the genome of *Mycobacterium tuberculosis. Biochimie* **85,** 691–700.
6. Rosen, M. L., Edman, M., Sjostrom, M., and Wieslander, A. (2004) Recognition of fold and sugar linkage for glycosyltransferases by multivariate sequence analysis. *J. Biol. Chem.* **279,** 38,683–38,692.
7. Heissigerova, H., Breton, C., Moravcova, J., and Imberty, A. (2003) Molecular modeling of glycosyltransferases involved in the biosynthesis of blood group A, blood group B, Forssman and iGb_3 antigens and their interaction with substrates. *Glycobiology* **13,** 377–386.

8. Hans, J., Brandt, W., and Vogt, T. (2004) Site-directed mutagenesis and protein 3D-homology modeling suggest a catalytic mechanism for UDP-glucose-dependent betanidin 5-O-glucosyltransferase from *Dorotheanthus bellidiformis*. *Plant J.* **39,** 319–333.
9. Gunasekaran, K., Ma, B., Ramakrishnan, B., Qasba, P. K., and Nussinov, R. (2003) Interdependence of backbone flexibility, residue conservation, and enzyme function: A case study on beta1,4-galactosyltransferase-I. *Biochemistry* **42,** 3674–3687.
10. Snajdrová, L., Kulhánek, P., Imberty, A., and Koča, J. (2004) Molecular dynamics simulations of glycosyltransferase LgtC. *Carbohydr. Res.* **339,** 995–1006.
11. Hotelling, H. (1933) Analysis of a complex of statistical variables into principal components. *J. Edu. Psych.* **24,** 417–441.
12. Gaboriaud, C., Bissery, V., Benchetrit, T., and Mornon, J. P. (1987) Hydrophobic cluster analysis: an efficient way to compare and analyse amino acid sequences. *FEBS Lett.* **224,** 149–155.
13. Berman, H. M., Westbrook, J., Feng, Z., et al. (2000) The Protein Data Bank. *Nucleic Acids Res.* **28,** 235–242.
14. Blundell, T., Carney, D., Gardner, S., et al. (1988) 18th Sir Hans Krebs lecture. Knowledge-based protein modeling and design. *Eur. J. Biochem.* **172,** 513–520.
15. Marti-Renom, M. A., Stuart, A., Fiser, A. S. R., Melo, F., and Sali, A. (2000) Comparative protein structure modeling of genes and genomes. *Annu. Rev. Biophys. Biomol. Struct.* **29,** 291–325.
16. Morris, G. M., Goodsell, D. S., Halliday, R. S., et al. (1998) Automated docking using a Lamarckian genetic algorithm and and empirical binding free energy function. *J. Comp. Chem.* **19,** 1639–1662.
17. Waldherr-Teschner, M., Goetze, T., Heiden, W., Knoblauch, M., Vollhardt, H., and Brickmann, J. (1992) Advances in Scientific Visualization (Post, F. H. and Hin, A. J. S., eds.), Springer, Heidelberg, pp. 58–67.
18. Vriend, G. (1997) WHATIF, EMBL, Heidelberg.
19. Case, D. A., Darden, T. A., Cheatham, T. E. I., et al. (2004) AMBER, University of California, San Francisco.
20. Berendsen, H. J. C., van der Spoel, D., and van Drunen, R. (1995) GROMACS: A message-passing parallel molecular dynamics implementation. *Comp. Phys. Comm.* **91,** 43–56.
21. Laaksonen, L. (1992) A graphics program for the analysis and display of molecular dynamics trajectories. *J. Mol. Graph.* **10,** 33–34.
22. Humphrey, W., Dalke, A., and Schulten, K. (1996) VMD—Visual Molecular Dynamics. *J. Mol. Graph.* **14,** 33–38.
23. Pedersen, L. C., Dong, J., Taniguchi, F., et al. (2003) Crystal structure of an alpha 1,4-N-acetylhexosaminyltransferase (EXTL2), a member of the exostosin gene family involved in heparan sulfate biosynthesis. *J. Biol. Chem.* **278,** 14,420–14,428.
24. Hu, Y., Chen, L., Ha, S., et al. (2003) Crystal structure of the MurG:UDP-GlcNAc complex reveals common structural principles of a superfamily of glycosyltransferases. *Proc. Natl. Acad. Sci. USA* **100,** 845–849.

25. Kraulis, P. (1991) Molscript: A program to produce both detailed and schematic plots of protein structures. *J. Appl. Crystallogr.* **24,** 946–950.
26. Merrit, E. A. and Murphy, M. E. (1994) Raster3D version 2.0. A program for photorealistic molecular graphics. *Acta Crystallogr.* **D50,** 869–873.
27. Chazalet, V., Uehara, K., Geremia, R. A., and Breton, C. (2001) Identification of essential amino acids in the Azorhizobium caulinodans fucosyltransferase NodZ. *J. Bacteriol.* **183,** 7067–7075.

11

β-Galactoside α2,6-Sialyltransferase and the Sialyl α2,6-Galactosyl-Linkage in Tissues and Cell Lines

Fabio Dall'Olio, Nadia Malagolini, and Mariella Chiricolo

Summary

β-Galactoside α2,6-sialyltransferase (ST6Gal.I) is the principal sialyltransferase responsible for the biosynthesis of the sialyl α2,6-galactosyl linkage. This enzyme and its cognate glycosidic structure are overexpressed in several malignancies and are related to cancer progression. The expression of the enzyme is regulated primarily through the expression of three principal mRNA species differing in the 5′-untranslated exons. The form known as YZ is considered associated with the basal expression of the gene, while forms H and X are specific to the liver and B-lymphocytes, respectively. The authors have studied the expression of ST6Gal.I activity by two different methods using a panel of human cancer cell lines: the expression of α2,6-sialylated sugar chains by the lectin from *Sambucus nigra* (SNA), and the expression of the different mRNA species by RT-PCR using oligonucleotide primers complementary to the isoform-specific regions. Very high levels of ST6Gal.I activity result in high levels of SNA reactivity and are associated with the expression of the H transcript in colon and liver cell lines, and of the X transcript in B cells.

Key Words: Sialyltransferase; sialic acid; *Sambucus nigra* agglutinin; RT-PCR; HPLC; glycosylation.

1. Introduction

Sialic acids are a family of negatively charged sugars that often terminate the oligosaccharide chains of glycoproteins and glycolipids, and can mediate numerous biological phenomena. Sialic acid is most often attached in either α2,3- or α2,6-linkage to a subterminal galactose residue, or in α2,6 to a subterminal

From: *Methods in Molecular Biology, vol. 347: Glycobiology Protocols*
Edited by: I. Brockhausen © Humana Press Inc., Totowa, NJ

N-acetylgalactosamine. The most frequently occurring sialic acid in humans is *N*-acetylneuraminic acid (NeuAc).

β-Galactoside α2,6-sialyltransferase (ST6Gal.I, according to the nomenclature proposed by Tsuji et al. *[1]*) has long been thought to be the only sialyltransferase able to add sialic acid in α2,6-linkage to galactose *(2,3)*. Although a second ST6Gal (ST6Gal.II) has recently been cloned *(4)*, the very strict tissue distribution of ST6Gal.II, together with the fact that it sialylates primarily oligosaccharides, leaves ST6Gal.I as the major sialyltransferase able to elaborate the sialyl α2,6-galactosyl linkage. ST6Gal.I and the cognate sialyl α2,6-galactosyl linkage show an oncodevelopmental pattern of regulation, being overexpressed in a large number of malignancies (*see* **refs. *5–15***; reviewed in **refs. *16*** and ***17***). These modifications have been associated with disease progression and metastasis *(12,18,19)*. ST6Gal.I is primarily regulated at the transcriptional level. Three major types of transcripts differing in the 5′-untranslated regions are expressed in a tissue-specific manner by different promoters *(20)*. One type of transcript expressed by many tissues is thought to represent the basal expression of the gene and contains the 5′-untranslated exons Y and Z (YZ form; *see* **ref. *21***); a second type of transcript, specifically expressed by mature B lymphocytes *(22)*, contains the 5′-untranslated exon X (X form); and the third type of transcript, expressed by liver *(10,23)* and colonic cells (H form; *see* **refs. *8*** and ***9***), lacks exons Y, Z, and X, but contains a short specific nucleotide sequence in front of the first exon. A widely used tool for the detection of the sialyl α2,6-galactosyl linkage is the lectin from *Sambucus nigra* (SNA; *see* **ref. *24***), which allows discrimination of the sialyl α2,6-linkage from other types of sialyl linkages. This chapter describes the measurement of ST6Gal.I activity by two different methods: the detection of α2,6-sialylated sugar chains by SNA dot blot analysis, and the identification of the three major types of ST6Gal.I transcripts by reverse transcription-polymerase chain reaction (RT-PCR) analysis (*see also* Chapter 13).

2. Materials

2.1. Measurement of ST6Gal.I Enzyme Activity With Asialotransferrin as Acceptor

1. Tissue or cell homogenates. Obtain cell lines by ATCC (Rockville, MD) with the exception of SW948FL, which has been selected in our lab *(25)*, and cell lines HepG2 and Louckes, which are a generous gift of Dr. Joseph Lau, Roswell Park Cancer Institute, Buffalo, NY. Homogenize tissues in 10 vol of ice-cold water (*see* **Note 1**) with a potter homogenizer (*see* **Note 2**). Suspend pellets of cell lines, which can be kept frozen at –80°C indefinitely, in 150–300 μL of ice-cold water (for a 75-cm^2 flask). In the case of cell lines, the use of a homogenizer is not generally necessary. Vigorous pipetting up and down is sufficient to ensure

homogenization. Determine the protein concentration according to the Lowry method ***(26)*** and adjust to 5–10 mg/mL. The homogenates must be kept on ice during use. Store at –80°C (*see* **Note 3**).
2. 1 *M* of Na-cacodylate buffer, pH 6.5. Store at 4°C.
3. Triton X-100 (Sigma, St. Louis, MO). Prepare a 10% v/v solution in water. Store at 4°C.
4. 55 mCi/m*M*, 25 μCi/mL of radioactive cytidine 5′-monophosphate *N*-acetylneuraminic acid, CMP-[^{14}C]NeuAc (American Radiolabeled Chemicals, Inc., St. Louis, MO). Store at –20°C.
5. Unlabeled CMP-NeuAc (Sigma). Prepare a solution of 1.5 m*M* in water and store at –20°C.
6. Asialotransferrin (AST). Prepare by mild acid hydrolysis, resulting in the chemical release of sialic acid from human transferrin (Sigma). Dissolve transferrin at a concentration of approx 1 mg/mL in 50 m*M* of H_2SO_4 and incubate for 2 h at 80°C. The solution is then extensively dialyzed against water and lyophilized. Dissolve AST in water at a concentration of 20 mg/mL and store at –20°C.
7. Phosphotungstic acid (Sigma). Prepare 1 L of 1% (w/v) solution in 0.5 *M* of HCl (PTA solution). Store at room temperature.
8. 1 *M* of HCl.
9. Ready Gel liquid scintillation cocktail (Beckman-Coulter Fullerton, CA).

2.2. Measurement of ST6Gal.I Enzyme Activity With N-Acetyllactosamine as Acceptor

1. Use **items 1–5** and **item 9** from **Subheading 2.1.**
2. Dissolve *N*-acetyllactosamine (LacNac; Sigma) in water at a concentration of 50 mg/mL. Store at –20°C.
3. 0.2-μm Syringe filters (Nalgene, Rochester, NY).
4. Acetonitrile (Merck, Darmstadt, Germany).
5. 12.5- × 0.4-cm NH_2-Lichrosorb HPLC column (Merck).
6. Dissolve KH_2PO_4 at a concentration of 15 m*M* in 1 L of water and filter on a 0.45-μm hydrophilic-type HA membrane (Millipore, Billerica, MA).

2.3. Detection of α2,6-Linked Sialic Acid

1. Digoxigenin-3-*O*-methylcarbonyl-ε-aminocapronic acid-*N*-hydroxysuccinimide ester (Roche, Milan, Italy). Dissolve in absolute ethanol at a concentration of 4 mg/100 μL and store at –20°C.
2. 1X Phosphate-buffered saline (PBS): 137 m*M* of NaCl, 2.7 m*M* of KCl, 1.4 m*M* of KH_2PO_4, and 4.3 m*M* of Na_2HPO_4. Adjust pH to 7.5 or 8.5 by the addition of 10 *M* of NaOH (*see* **Note 4**).
3. SNA (Sigma). Dissolve 1 mg in 1 mL of PBS, pH 8.5.
4. Conjugate digoxigenin to SNA (SNA-dig) by the addition of 9 μL of the digoxigenin-3-*O*-methylcarbonyl-ε-aminocapronic acid-*N*-hydroxysuccinimide ester solution to 1 mL of the SNA solution. Incubate for 2 h at room temperature and dialyze extensively against 1X PBS, pH 7.5. Store at –20°C.

5. Peroxidase-conjugated anti-digoxigenin antibodies, 150 U/200 μL Fab fragments (Roche). Prepare aliquots and store at –20°C.
6. Hybond nitrocellulose membrane (Amersham, Little Chalfont, UK).
7. 3MM chromatography paper (Whatman, Maidstone, UK).
8. 1X PBS–Tween: PBS (pH 7.5), 0.1% Tween-20 (Sigma). Prepare 2 L and store at room temperature.
9. PBS–Tween–BSA: add 1 g of bovine serum albumin (BSA; Sigma) to 100 mL of PBS–Tween. Store at 4°C.
10. Maleic acid buffer: 100 m*M* of maleic acid and 150 m*M* of NaCl. Adjust pH to 7.5 with 10 *M* of NaOH.
11. Blocking reagent (Roche). Prepare a 10% (w/v) stock solution in maleic acid buffer by shaking and heating on a heating block. Prepare a working solution by diluting the stock solution in PBS–Tween at a ratio of 1:100.
12. Supersignal West Pico Chemiluminescent substrate (Pierce, Rockford, IL).
13. Exposure cassette (Kodak, Rochester, NY) containing a transparent plastic envelope open on two sides. Fix the envelope to the cassette with adhesive plastic tape.
14. Hyperfilm for chemiluminescent detection (Amersham).

2.4. Detection of the ST6Gal.I Transcripts

1. Prepare RNA from frozen cell pellets by the RNAZolB (Biotecx Laboratories, Houston, TX) and suspend in water (*see* **Note 1**). Store at –80°C.
2. Obtain complementary DNAs (cDNAs) by reverse transcription using the TaKaRa RT-PCR kit version 2.1 (TaKaRa, Shouzo, Japan). Store at –80°C.
3. Oligonucleotide primers for PCR analysis are from MWG (https://ecom.mwgdna.com). Dissolve at 10 μ*M* with water and store at –20°C. The following primers can be used: HLP6, 5′-AAAGGGAGCCGATACCGACC-3′ forward primer complementary to exon Y; EXL.1, 5′-ACATCTCTTCATGTGTATCCT CTG-3′ forward primer complementary to exon X; HepL.1, 5′-GTCTCTTAT TTTTTGCCTTTGCAG-3′ forward primer complementary to the H-specific sequence; EIIL.1, 5′-CATCTTCATTATGATTCACACCAAC-3′ forward primer complementary to exon II (common to all types of transcript); EIVR.1, 5′-TCAT TGTACAAACTGTCTTTGAGGA-3′, reverse primer complementary to exon IV (common to all types of transcript); GAPDHL.1, 5′-GGAGCCAAAAGGGTC ATCATCT-3′ and GAPDHR.1, 5′-ATGCCAGTGAGCTTCCCGTTC-3′ for the detection of the housekeeping glyceraldehyde 3-P dehydrogenase (GAPDH) transcript.
4. Dissolve deoxyadenosine 5′ triphosphate (dATP), deoxycytidine 5′ triphosphate (dCTP), deoxythymidine 5′ triphosphate (dTTP), and deoxyguanosine 5′ triphosphate (dGTP) separately at a concentration of 1.25 m*M* in water. Store at –20°C.
5. Taq polymerase (Eppendorf, Hamburg, Germany).
6. 1X TAE buffer: 40 m*M* of Tris-HCl acetate and 2 m*M* of ethylenediaminetetraacetic acid (EDTA), pH 8.5.

7. Dissolve 10 mg/mL of ethidium bromide (Sigma) in water. **Caution:** this is a mutagen and extreme care must be taken to avoid exposure.
8. Agarose (Bethesda Research Laboratories, Gaithersburg, MD).

3. Methods

3.1. ST6Gal.I Enzyme Activity

The ST6Gal.I enzyme activity can be assayed in whole homogenates of tissues or cell lines. The methods are based on the enzyme-catalyzed transfer of radioactive sialic acid from the donor substrate (CMP-NeuAc) to an acceptor that can be either a glycoprotein or an oligosaccharide. The tissue or cell homogenates generally lack enzyme activities degrading the donor substrate, allowing long incubation times in linear conditions. Two methods will be presented: in the first, the acceptor used is the glycoprotein AST; in the second, the acceptor used is the disaccharide LacNac. The first acceptor allows a rapid and relatively inexpensive determination of the enzyme activity, but is not completely specific in that it can theoretically also serve as acceptor for α2,3-sialyltransferases acting on N-linked chains. The lack of O-linked chains ensures that AST serves as acceptor for the sialyltransferases active only on *N*-linked chains. The second acceptor is highly specific but needs the high-performance liquid chromatography (HPLC) separation of the radioactive products, and is therefore expensive and time-consuming.

3.1.1. Assay for Sialyltransferase Activity Using AST as Acceptor

1. Prepare an AST for each sample to be analyzed by mixing: 4 μL of 1 *M* Na cacodylate buffer (pH 6.5), 2.5 μL of 10% Triton X-100, 6.5 μL of H_2O, 15 μL of 20 mg/mL AST, 1 μL of radioactive CMP-NeuAc, and 1 μL of 1.5 m*M* unlabeled CMP-NeuAc.
2. Prepare an "endogenous" mix, which is identical to the AST mix, but replace the AST with 15 μL of water (total 21.5 μL of water; *see* **Note 5**).
3. Prepare two 1.5-mL minifuge plastic tubes on ice for each sample to be analyzed. Dispense 30 μL of the AST mix in one tube and 30 μL of endogenous mix in the other (*see* **Note 6**).
4. Add up to 20 μL of tissue homogenate (*see* **Note 7**) and water to a final volume of 50 μL.
5. Incubate in a water bath for 3 h at 37°C.
6. At the end of the incubation time, add 1 mL of PTA solution to each tube and centrifuge in a minifuge at maximum speed for 5 min (*see* **Note 8**).
7. Discard the supernatants appropriately, as they contain radioactive waste.
8. Repeat **steps 6** and **7** three times.
9. Add 0.5 mL of 1 *M* HCl to each tube and place in a boiling water bath for 15 min (*see* **Note 9**).

10. Transfer the contents of the tube to a liquid scintillation vial, add 6 mL of liquid scintillation cocktail, and count in a liquid scintillation β-counter.

3.1.2. Assay for Sialyltransferase Activity Using LacNac as Acceptor

1. Prepare a LacNAc mix for each sample to be analyzed by mixing 4 µL of 1 *M* Na cacodylate buffer (pH 6.5), 2.5 µL of 10% Triton X-100, 16.5 µL of H_2O, 5 µL of 50 mg/mL LacNAc, 1 µL of radioactive CMP-NeuAc, and 1 µL of 1.5 m*M* unlabeled CMP-NeuAc.
2. Prepare an endogenous mix identical with that described above in **step 1** (*see* **Note 5**).
3. Set up two 1.5-mL minifuge plastic tubes on ice for each sample. Dispense 30 µL of LacNAc mix in one tube and 30 µL of endogenous mix in the other (*see* **Note 6**).
4. Add up to 20 µL of tissue homogenate (*see* **Note 7**) and water to a final volume of 50 µL.
5. Incubate in a water bath for 3 h at 37°C.
6. At the end of the incubation time, add 0.5 mL of water to each sample and boil the samples for 5 min in a water bath (*see* **Note 10**).
7. Filter through a 0.45-µm membrane filter to prepare for HPLC separation.
8. Equilibrate the NH_2 -Lichrosorb column in 83% acetonitrile/17% 15 m*M* of KH_2PO_4 in isocratic conditions at a flow rate of 1 mL/min.
9. Inject 200 µL of the filtered sample in the HPLC apparatus, and start collecting 2-mL fractions in liquid scintillation vials. After 44 min (fraction 22) change the proportion of the eluents to 78% acetonitrile/22% 15 m*M* of KH_2PO_4 and run isocratically for 30 min.
10. Add 5 mL of liquid scintillation cocktail to each vial and count fractions in a liquid scintillation β-counter (*see* **Fig. 1** and **Note 11**).
11. At the end of the chromatography the column is re-equilibrated to the initial conditions (83% acetonitrile/17% 15 m*M* of KH_2PO_4) for 15 min and is ready for another analysis.

Fig. 1. *(opposite page)* Examples of high-performance liquid chromatography (HPLC) separation of sialyltransferase reaction products obtained using LacNAc as acceptor. In **A**, **B**, and **C**, 250 µg of LacNAc were present in the assay mixture; in **D** the acceptor was omitted. The homogenates used as enzyme sources were from: (A) undifferentiated Caco2; (B,D) differentiated Caco2; (C) HT29. After incubation for 3 h at 37°C the reactions were diluted, boiled, filtered, and subjected to HPLC analysis. The chromatography was run isocratically with 83% acetonitrile/17% 15 m*M* of KH_2PO_4 until fraction 22 (arrowhead), where the proportion of the two eluents was changed to 78% acetonitrile/22% 15 m*M* of KH_2PO_4. The arrows indicate the elution position of (1) authentic NeuAc; (2) α2,3 sialylated LacNAc; and (3) α2,6 sialylated LacNAc. In all four samples a large proportion of radioactivity was recovered as NeuAc since boiling hydrolyzes CMP-NeuAc, releasing free NeuAc. The nature of the small

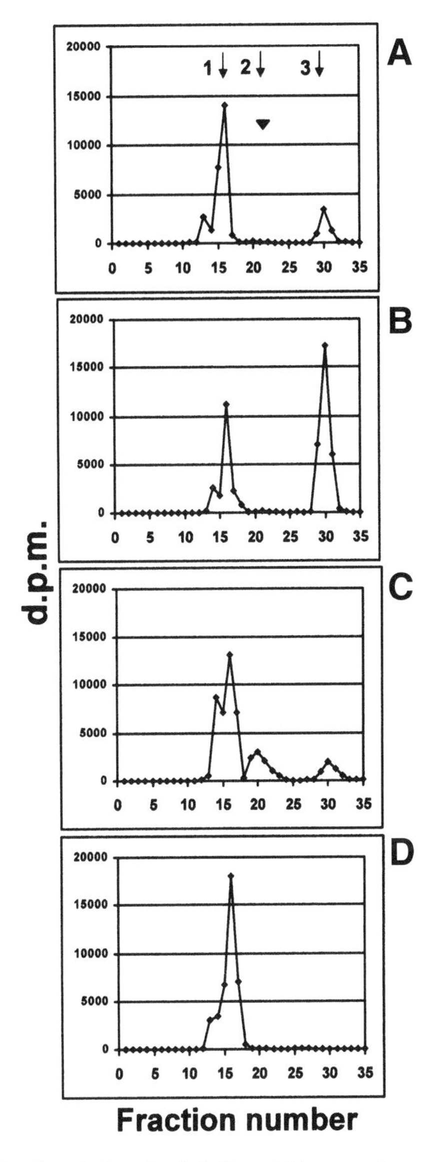

Fig. 1 *(continued).* peak eluted ahead of sialic acid has not been investigated. However, a similar peak is often visible when unlabeled sialic acid is analyzed and detected spectrophotometrically, suggesting that it is not a product of the sialyltransferase action. In Caco2 cells the in vitro differentiation induces a remarkable increase of ST6Gal.I activity, but leaves unchanged the very low levels of α2,3-sialyltransferase activities (*see* **Table 1**). Unlike the majority of colon cancer cells, HT29 cells express relatively high levels of α2,3-sialyltransferase activities resulting in the production of α2,3-sialylated LacNAc, which is eluted just after NeuAc.

Table 1
Sialyltansferase Activity on *N*-Acetyllactosamine and Asialotransferrin of Human Cancer Cell Lines

	Sialic acid (pmoles/mg protein hour) incorporated into		
Cell line	α2,3-sialyl LacNac	α2,6-sialyl LacNac	AST
Caco2 (differentiated)	41	2430	736
Caco2 (undifferentiated)	36	514	156
Colo 205	100	721	216
HT29	811	541	153
SW480	78	180	58
SW620	63	288	86
948FL	95	730	227
SW1417	239	496	155
LoVo	180	297	94
HepG2	18	2135	631
Louckes	0	1285	387
SW48	0	0	0

All the cell lines reported here are from colon carcinoma except HepG2 and Louckes, which are from a hepatocarcinoma and a B-cell lymphoma, respectively. The α2,6-sialyltransferase activity, as measured using LacNac as the acceptor substrate, is usually much higher than that of the α2,3-sialyltransferase activity with the exception of HT29 cells, whose α2,3 activity is higher than the α2,6. It should be noted that the incorporation into AST strictly parallels the α2,6 activity determined with LacNac, not the α2,3 activity, strongly suggesting that AST behaves as a poor acceptor for α2,3-sialyltransferases.

3.2. Detection of α2,6-Sialylated Sugar Chains by Dot Blot Analysis

1. This protocol assumes the use of a Bio-Rad 96-well vacuum manifold.
2. Equilibrate one sheet of Whatman 3MM paper and one of nitrocellulose membrane of the approximate size of the apparatus (*see* **Note 12**) in PBS–Tween for 5 min.
3. Assemble the apparatus with (from top to bottom) the upper part of the manifold, the nitrocellulose membrane, the paper sheet, and the lower part of the manifold (*see* **Note 13**).
4. Prepare 0.1 mg/mL solutions of the samples in 100-μL final volume of PBS–Tween in the first row of a 96-well plate.
5. Prepare serial dilution of the samples in PBS–Tween in 100-μL volumes.
6. Transfer the samples from the wells of the 96-well plate to the wells of the manifold. The use of a multichannel pipet is recommended.
7. Apply the vacuum to the manifold until the entire volume of the samples has been filtered.
8. Remove the vacuum, add 100 μL of PBS–Tween to each well to wash, and reapply the vacuum until the volume has been filtered.

9. Repeat **step 8**.
10. Open the apparatus and briefly wash the blot in PBS–Tween.
11. Incubate the membrane in the working solution of the blocking reagent for 1 h at room temperature on a rocking platform.
12. Remove the blocking reagent solution (*see* **Note 14**).
13. After a brief rinse in PBS–Tween, wash the blot with PBS–Tween for 15 min, then twice for 5 min on a rocking platform.
14. Incubate in 1 μg/mL SNA-Dig dissolved in PBS–Tween–BSA for 1 h at room temperature on a rocking platform.
15. Remove the SNA (*see* **Note 14**).
16. Repeat **step 13**.
17. Incubate with peroxidase-conjugated anti-digoxigenin antibodies diluted 1:20,000 in PBS–Tween–BSA for 1 h at room temperature on a rocking platform.
18. Remove the anti-digoxigenin antibodies solution (*see* **Note 14**).
19. Rinse twice with PBS–Tween and wash once for 15 min and four times for 5 min with PBS–Tween.
20. Prepare the Supersignal substrate working solution by mixing equal volumes of luminol/enhancer solution and the stable peroxide solution. Prepare a sufficient volume to ensure that the blot is completely wet.
21. Move to the darkroom with the Supersignal substrate working solution, the films, and the exposure cassette containing the plastic envelope, scissors, and forceps while the blot is still in the last wash.
22. Incubate the blot with the Supersignal working solution for 5 min.
23. Remove the blot from the substrate working solution and drain the excess solution by placing the corner of the blot on lab paper.
24. Place the blot into the plastic envelope inside the exposure cassette, switch off the light, cover the blot with an autoradiography film, close the cassette, and expose for 30 s.
25. Process the film quickly and after fixing is complete, switch on the light and evaluate whether a different exposure needs to be taken (*see* **Fig. 2** and **Note 15**).

3.3. Detection of the Three Major ST6Gal.I Transcripts by RT-PCR Analysis

The detection of the three major types of ST6Gal.I transcripts (YZ, X, and H) by RT-PCR is based on the use of forward primers specific to the different 5′-untranslated regions, and a common reverse primer complementary to a downstream exon common to the three messenger RNA (mRNA) species.

1. Reverse transcribe 1–5 μg of total RNA using the TaKaRa RT-PCR kit version 2.1, using random 9 mers as primers. Store the cDNA (20-μL final volume) at –80°C.
2. PCR amplify 2 μL of cDNA in a final volume of 50 μL with 1X Eppendorf PCR buffer (1.5 m*M* of $MgCl_2$), 0.25 μ*M* of each primer, 0.2 m*M* of dNTPs, and 0.4 U of Taq polymerase. Amplification conditions are: denaturation, 94°C for 1 min;

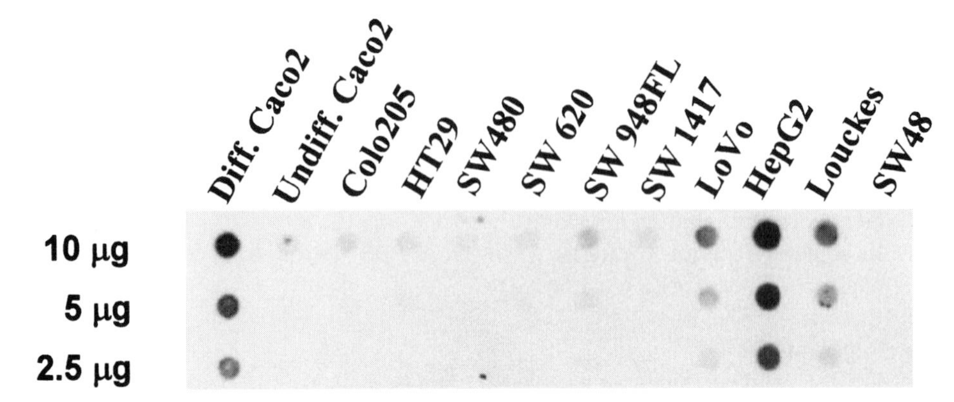

Fig. 2. *Sambucus nigra* (SNA) dot blot analysis of human cancer cells expressing different levels of α2,6-sialyltransferase activity. 10, 5, and 2.5 µg of total cell homogenates of the different cell lines were spotted on a nitrocellulose membrane and stained with SNA-Dig as detailed in text. The three cell lines showing the strongest reactivity are differentiated Caco2, HepG2, and Louckes, which are also the cell lines expressing very high ST6Gal.I activity; SW48 cells, which lack detectable ST6Gal.I activity, are completely unreactive with SNA. However, LoVo cells, which express a low level of ST6Gal.I activity, display a strong SNA reactivity. Together this data indicates that the level of α2,6-sialylation of membrane glycoconjugates is regulated primarily, but not exclusively, by the level of ST6Gal.I activity.

annealing, 60°C for 1 min; extension, 72°C for 1 min for 30–35 cycles, preceded by a single denaturation step of 1 min at 94°C.

3. Cast a 1.2% agarose gel in 1X TAE buffer containing 1 µg/mL of ethidium bromide in the appropriate part of a horizontal electrophoresis apparatus, using a comb that forms wells of at least 30-µL volume.
4. Load 20 µL of each PCR reaction mixture when the gel is solid (*see* **Note 16**).
5. Move the part of the apparatus containing the gel inside the electrophoretic chamber, which is partially filled with 1X TAE buffer.
6. Carefully submerge the gel with 1X TAE buffer until the gel is completely covered by buffer (*see* **Note 17**).
7. Run for approx 45 min at 90 V.
8. Check migration over a UV transilluminator and, when satisfactory, take a photograph with an image acquisition system, such as the Kodak 1D system (*see* **Fig. 3**).

4. Notes

1. The water used in these studies is produced by a Milli-Q Plus apparatus (Millipore) and has a resistivity higher than 15 *M*Ω/cm.
2. Homogenization in water offers the advantage of using the same homogenate for other types of analysis (e.g., other enzymatic determinations, electrophoretic separations, or dot blot analysis).

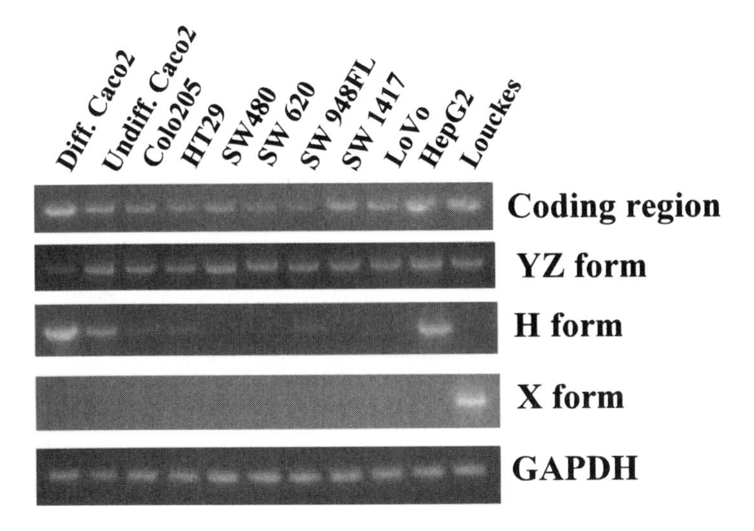

Fig. 3. Expression of the different ST6Gal.I transcripts by a panel of human cancer cells. Cell lines Caco2, COLO205, HT29, SW480, SW620, SW948FL, SW1417, and LoVo are from colorectal cancer; HepG2 are from hepatocarcinoma, whereas Louckes are from a B-cell lymphoma. The amplification of the transcripts was achieved with the following primer pairs: coding region, which is common to all the types of transcript, EIIL.1/EIVR.1; YZ form, HLP6/EIVR.1; X form, EXL.1/EIVR.1; H form, HepL.1/EIVR.1. The housekeeping GAPDH transcript was amplified with primer pair GAPDHL.1/GAPDHR.1. Amplification of the coding region reveals that the overall amount of ST6Gal.I transcripts is similar in the cell lines tested. The YZ transcript is expressed by all the cell lines at a comparable level; the H transcript is expressed only by HepG2 and by some colon cancer cell lines, whereas the X transcript is expressed only by Louckes cells. A comparison with enzyme activity data (*see* **Table 1**) reveals that the three cell lines which express very high levels of activity (differentiated Caco2, HepG2, and Louckes) express, in addition to the YZ transcript, high levels of a non-YZ transcript, which is the H form for colon and liver cells and the X transcript in B-lymphocytes.

3. The homogenates can be stored for years at –80°C without significant loss of activity. Storage at –20°C is not recommended as it causes a progressive loss of activity.
4. It is convenient to prepare 10X stock solutions of PBS, which can be stored at room temperature and used to prepare 1X working solutions.
5. These mixtures can be stored indefinitely at –20°C.
6. The assay without acceptor serves as a control for the incorporation of sialic acid onto endogenous substrates. This incorporation is usually very low, just above the background incorporation (the reaction is blocked immediately after the addition of the homogenate).

7. The optimal amount of proteins ranges between 50 and 150 μg/assay.
8. The samples become cloudy after the addition of PTA owing to the precipitation of proteins. Denatured proteins (including radioactive AST) are separated from unreacted radioactive sugar donor by centrifugation.
9. The linkage between sialic acids and the underlying sugar is labile at high temperature and acid pH. This treatment releases incorporated sialic acid from the precipitated AST.
10. Boiling is necessary to denature proteins and to hydrolyze CMP-NeuAc, yielding NeuAc and CMP. Denatured proteins are more easily removed by the subsequent filtration step. The hydrolysis of CMP-NeuAc reduces the HPLC separation time, since the retention time of NeuAc is much shorter than that of CMP-NeuAc.
11. The first radioactive peak (sialic acid) is eluted around fractions 14–17. The second radioactive peak (when present) is eluted just after sialic acid (fractions 19–23) but is well-separated and is formed by α2,3-sialylated LacNac. The third peak, formed by α2,6-sialylated LacNac, is eluted around fractions 29–31 (7–9 fractions after the eluent proportion has been changed to 78:22).
12. The size of the paper and the nitrocellulose membrane must be large enough to include all 96 wells of the manifold.
13. The screws of the manifold must be very tight to avoid diffusion of the samples.
14. The blocking reagent solution, the SNA working solution, and the antidigoxigenin antibody working solution can be saved for successive uses and stored for at least 1 wk at 4°C.
15. The substrate solution is stable for at least 8 h. This allows repeated exposures to obtain optimal intensity.
16. The dyes generally present in sample buffers (bromophenol blue and/or xylene cyanole) sometimes interfere with the detection of PCR bands; therefore, we routinely omit the addition of a sample buffer in the electrophoretic analysis of PCR products. Under these conditions, the bands yielded by PCR products are sharper. However, this procedure requires that the samples are loaded before the gel is submerged by running buffer.
17. It is crucial to avoid disturbance of the samples during submersion of the gel.

Acknowledgments

Research was supported by funds from PRIN (MIUR, Rome), the University of Bologna (funds for selected research topics), and the Pallotti Legacy for Cancer Research.

References

1. Tsuji, S., Datta, A. K., and Paulson, J. C. (1996) Systematic nomenclature for sialyltransferases. *Glycobiology* **6,** v–vii.
2. Weinstein, J., de Souza e Silva, U., and Paulson, J. C. (1982) Purification of a Gal β 1,4GlcNAc α 2,6 sialyltransferase and a Gal β 1,3(4)GlcNAc α 2,3 sialyltransferase to homogeneity from rat liver. *J. Biol. Chem.* **257,** 13,835–13,844.

3. Weinstein, J., Lee, E. U., McEntee, K., Lai, P. H., and Paulson, J. C. (1987) Primary structure of β-galactoside α 2,6-sialyltransferase. Conversion of membrane-bound enzyme to soluble forms by cleavage of the NH_2-terminal signal anchor. *J. Biol. Chem.* **262,** 17,735–17,743.
4. Takashima, S., Tsuji, S., and Tsujimoto, M. (2002) Characterization of the second type of human β-galactoside α2,6-sialyltransferase (ST6Gal II), which sialylates Galβ1,4GlcNAc structures on oligosaccharides preferentially. Genomic analysis of human sialyltransferase genes. *J. Biol. Chem.* **277,** 45,719–45,728.
5. Dall'Olio, F. and Trere, D. (1993) Expression of α 2,6-sialylated sugar chains in normal and neoplastic colon tissues. detection by digoxigenin-conjugated *Sambucus nigra* agglutinin. *Eur. J. Histochem.* **37,** 257–265.
6. Sata, T., Roth, J., Zuber, C., Stamm, B., and Heitz, P. U. (1991) Expression of α 2,6-linked sialic acid residues in neoplastic but not in normal human colonic mucosa. A lectin-gold cytochemical study with *Sambucus nigra* and *Maackia amurensis* lectins. *Am. J. Pathol.* **139,** 1435–1448.
7. Dall'Olio, F., Malagolini, N., Di Stefano, G., Minni, F., Marrano, D., and Serafini-Cessi, F. (1989) Increased CMP-NeuAc:Galβ1,4GlcNAc-R α 2,6 sialyltransferase activity in human colorectal cancer tissues. *Int. J. Cancer* **44,** 434–439.
8. Dall'Olio, F., Chiricolo, M., and Lau, J. T. (1999) Differential expression of the hepatic transcript of β-galactoside α2,6-sialyltransferase in human colon cancer cell lines. *Int. J. Cancer* **81,** 243–247.
9. Dall'Olio, F., Chiricolo, M., Ceccarelli, C., Minni, F., Marrano, D., and Santini, D. (2000) β-galactoside α2,6 sialyltransferase in human colon cancer: contribution of multiple transcripts to regulation of enzyme activity and reactivity with *Sambucus nigra* agglutinin. *Int. J. Cancer* **88,** 58–65.
10. Dall'Olio, F., Chiricolo, M., D'Errico, A., et al. (2004) Expression of β-galactoside α2,6 sialyltransferase and of α2,6-sialylated glycoconjugates in normal human liver, hepatocarcinoma, and cirrhosis. *Glycobiology* **14,** 39–49.
11. Dalziel, M., Dall'Olio, F., Mungul, A., Piller, V., and Piller, F. (2004) Ras oncogene induces β-galactoside α2,6-sialyltransferase (ST6Gal I) via a RalGEF-mediated signal to its housekeeping promoter. *Eur. J. Biochem.* **271,** 3623–3634.
12. Gessner, P., Riedl, S., Quentmaier, A., and Kemmner, W. (1993) Enhanced activity of CMP-neuAc:Gal β 1-4GlcNAc:α 2,6-sialyltransferase in metastasizing human colorectal tumor tissue and serum of tumor patients. *Cancer Lett.* **75,** 143–149.
13. Lise, M., Belluco, C., Perera, S. P., Patel, R., Thomas, P., and Ganguly, A. (2000) Clinical correlations of α2,6-sialyltransferase expression in colorectal cancer patients. *Hybridoma* **19,** 281–286.
14. Petretti, T., Schulze, B., Schlag, P. M., and Kemmner, W. (1999) Altered mRNA expression of glycosyltransferases in human gastric carcinomas. *Biochim. Biophys. Acta* **1428,** 209–218.
15. Recchi, M. A., Hebbar, M., Hornez, L., Harduin-Lepers, A., Peyrat, J. P., and Delannoy, P. (1998) Multiplex reverse transcription polymerase chain reaction assessment of sialyltransferase expression in human breast cancer. *Cancer Res.* **58,** 4066–4070.

16. Dall'Olio, F. (2000) The sialyl-α2,6-lactosaminyl-structure: biosynthesis and functional role. *Glycoconj. J.* **17,** 669–676.
17. Dall'Olio, F. and Chiricolo, M. (2001) Sialyltransferases in cancer. *Glycoconj. J.* **18,** 841–850.
18. Dall'Olio, F., Malagolini, N., and Serafini-Cessi, F. (1992) Enhanced CMP-NeuAc:Gal β 1,4GlcNAc-R α 2,6 sialyltransferase activity of human colon cancer xenografts in athymic nude mice and of xenograft-derived cell lines. *Int. J. Cancer* **50,** 325–330.
19. Vierbuchen, M. J., Fruechtnicht, W., Brackrock, S., Krause, K. T., and Zienkiewicz, T. J. (1995) Quantitative lectin-histochemical and immunohistochemical studies on the occurrence of α(2,3)- and α(2,6)-linked sialic acid residues in colorectal carcinomas. Relation to clinicopathologic features. *Cancer* **76,** 727–735.
20. Wang, X., Vertino, A., Eddy, R. L., et al. (1993) Chromosome mapping and organization of the human β-galactoside α 2,6-sialyltransferase gene. Differential and cell-type specific usage of upstream exon sequences in B-lymphoblastoid cells. *J. Biol. Chem.* **268,** 4355–4361.
21. Grundmann, U., Nerlich, C., Rein, T., and Zettlmeissl, G. (1990) Complete cDNA sequence encoding human β-galactoside α-2,6-sialyltransferase. *Nucleic Acids Res.* **18,** 667.
22. Stamenkovic, I., Asheim, H. C., Deggerdal, A., Blomhoff, H. K., Smeland, E. B., and Funderud, S. (1990) The B cell antigen CD75 is a cell surface sialytransferase. *J. Exp. Med.* **172,** 641–643.
23. Aas-Eng, D. A., Asheim, H. C., Deggerdal, A., Smeland, E., and Funderud, S. (1995) Characterization of a promoter region supporting transcription of a novel human β-galactoside α-2,6-sialyltransferase transcript in HepG2 cells. *Biochim. Biophys. Acta* **1261,** 166–169.
24. Shibuya, N., Goldstein, I. J., Broekaert, W. F., Nsimba-Lubaki, M., Peeters, B., and Peumans, W. J. (1987) The elderberry (*Sambucus nigra* L.) bark lectin recognizes the Neu5Ac(α 2-6)Gal/GalNAc sequence. *J. Biol. Chem.* **262,** 1596–1601.
25. Dall'Olio, F., Malagolini, N., Di Stefano, G., Ciambella, M., and Serafini-Cessi, F. (1991) α 2,6 sialylation of N-acetyllactosaminic sequences in human colorectal cancer cell lines. Relationship with non-adherent growth. *Int. J. Cancer* **47,** 291–297.
26. Lowry, O. H., Rosebrough, N. J., Farr, A. L., and Randall, R. J. (1951) Protein measurement with the Folin phenol reagent. *J. Biol. Chem.* **193,** 265–275.

12

Visualizing Intracellular Distribution and Activity of Core2 β(1,6)*N*-Acetylglucosaminyltransferase-I in Living Cells

Maëlle Prorok-Hamon, Sylvie Mathieu, and Assou El-Battari

Summary

The core2 β(1,6)-*N*-acetylglucosaminyltransferase-I (C2GnT-I) is expressed by leukocytes and is involved in the synthesis of core2 *O*-glycans that carry sialyl-Lewis x (sLe^x) oligosaccharides. The core2-based sLe^x oligosaccharides (C2-O-sLe^x) have been demonstrated to be physiological selectin ligands that confer high affinity binding. The E-, P-, and L-selectins are adhesion proteins that direct leukocytes in the blood to lymphoid organs and sites of inflammation. They are also thought to be involved in the hematogenous dissemination of carcinoma cells expressing sialyl-Lewis glycans. Therefore, accumulation of data on structure–function relationships of this particular enzyme may represent an important part of investigations into pathologies involving selectins, such us inflammatory disorders and cancer progression. In this regard, studies of the intracellular distribution of C2GnT-I and its interaction with cognate substrates in vivo, as well as the knowledge of posttranslational modification (i.e., glycosylation, oligomerization, and proteolytic processing), may greatly aid in designing potential enzyme inhibitors. C2GnT-I fused to the green fluorescent protein is expressed to allow examination of the protein in living cells and to ease studies on structure–function relationships in vivo and in vitro.

Key Words: Core2 β(1,6)-*N*-acetylglucosaminyltransferase-I (C2GnT-I); leukosialin (CD43); T305 antibody; enhanced green fluorescent protein (EGFP); cloning; Golgi; Sep-Pak C18 columns; O-linked oligosaccharide liquid chromatography.

1. Introduction

E- and P-selectins are adhesion molecules that appear on activated endothelials and platelets to bind to leukocytes harboring sialyl-Lewis x (sLe^x, NeuAcα2, 3Galβ1,4[Fucα1,3]GlcNAc) antigens *(1)*. L-Selectin is another member of this

From: *Methods in Molecular Biology, vol. 347: Glycobiology Protocols*
Edited by: I. Brockhausen © Humana Press Inc., Totowa, NJ

family, but it is constitutively expressed on leukocytes to mediate leukocyte–leukocyte and leukocyte–high endothelial venule (HEV) interactions *(2)*. In addition to their natural function in leukocyte trafficking, the three selectins are involved in several inflammation reactions and may interact with tumor cells bearing sLe antigens, thus initiating adhesion to the endothelium and subsequent migration into the underlying connective tissue *(3)*. It has been shown repeatedly that, depending on the carrier protein considered, the core2-based oligosaccharides (OSs) carrying sLe^x antigen (C2-O-Le^x) represent the "preferred" ligands for all three selectins *(4)*. Therefore, the enzyme core2 β(1,6)-*N*-acetyglucosaminyltransferase-I (C2GnT-I) represents a critical component in human diseases such as inflammation and cancer.

It is important to have powerful and complementary tools to measure the in vivo and in vitro catalytic activities of C2GnT-I. This chapter presents different ways to assay for C2GnT-I activity. The in vivo activity is tested using T305 immunostaining on intact cells, while visualizing the Golgi localization of the enhanced green fluorescent protein (EGFP)-fused C2GnT-I. The in vivo activity is also assayed by metabolic labeling and liquid chromatography of OSs synthesized in the presence (or the absence) of the enzyme. The in vitro activity of C2GnT-I is easily determined owing to the availability of the disaccharide acceptor Gal β(1,3)GalNAcα-*p*-nitrophenyl and its property to be adsorbed on C-18 matrix (Sep-Pak columns). Finally, the protein can be detected by immunoblotting using either the anti-C2GnT-I polyclonal antibody or the anti-EGFP monoclonal antibody.

2. Materials

2.1. Cells and Media

1. Chinese hamster ovary (CHO) cells (CHO-K1; ATCC, Rockville, MD).
2. Ham F-12 medium (Cambrex, Verviers, Belgium) supplemented with 10% fetal calf serum (FCS) and antibiotics, 100 U/mL of penicillin and 100 μg/mL of streptomycin (Complete Ham F-12 medium), and Opti-MEM® medium (Invitrogen, Paisley, Scotland).
3. Phosphate-buffered saline (PBS).
4. Hank's-based enzyme-free cell dissociation buffer (Invitrogen).
5. Trypsin/ethylenediaminetetraacetic acid (EDTA) solution: 0.25% trypsin and 1 m*M* EDTA (Eurobio, Courteboef, France).
6. Tissue culture dishes and plastic tubes (Becton Dickinson, Franklin Lakes, NJ).

2.2. DNAs, Plasmids, Cloning, and Transfection

1. C2GnT-I (Genebank accession no. NM 001490) and leukosialin (sialophorin, CD43; Genebank accession no. NM 003123), both in pcDNA1 vector (Invitrogen),

were kindly provided by Professor Minoru Fukuda (The Burnham Institute, La Jolla, CA; *see also* Chapter 13).

2. Oligonucleotide primers (Genset, Every, France). Forward (sense) primer for C2GnT-I: 5′-*cgtcgt***ggatcc**ATGCTGAGGACGTTGCTGCGA-3′. *BamH*1 site, bold letters; extranucleotides, italicized letters; the C2GnT-I sequence, capital letters, covers between start codon and codon 7. Reverse (antisense) primer for C2GnT-I: 5′-*CACCATGGTGGCG***accggt**GTGTTTTAATGTCTCCAA-3′. *Age*1 site, bold letters; EGFP complementary sequence, italicized letters, covering codon –5 to codon +2 (the start codon is codon +1). The complementary sequence to C2GnT-I, capital letters, covers codon 423 to codon 428; the stop codon, codon 429, is deleted.
3. Taq polymerase, restriction enzymes, and T4-DNA ligase (Promega, Madison, WI).
4. Deionized autoclaved water (*see* **Note 1**).
5. Vectors pEGFP-N1 (Clontech, Palo Alto, CA) and pcDNA-3.1(+) (Invitrogen).
6. DNA gel extraction kit and plasmid (mini and maxi) preparation kits (Qiagen, Hilden, Germany).
7. Ultracompetent *Escherichia coli* Top 10F′ cells.
8. Super optimal catabolite (SOC) medium: 2% Bacto tryptone, 0.5% Bacto yeast extract, 10 m*M* of NaCl, 2.5 m*M* of KCl, 10 m*M* of $MgCl_2$, 10 m*M* of $MgSO_4$, and 20 m*M* of glucose (Invitrogen).
9. Luria-Bertani broth (LB): To make 1 L solution: 10 g of Bacto tryptone (Difco, Detroit, MI), 5 g of Bacto yeast extract (Difco), and 5 g of NaCl in water. Adjust the pH to 7.0. Autoclave and store at room temperature.
10. LB agar supplemented with 100 μg/mL ampicillin (stock solution 50 mg/mL in water) in plastic Petri dishes (Falcon, BD Labware, Franklin Lakes, NJ).
11. LipofectAMINE Plus reagent and neomycin (G418; Invitrogen).

2.3. Immunofluorescence

1. Falcon eight-well glass chamber slides (Becton Dickinson).
2. Antibody binding buffer: PBS containing 1% bovine serum albumin (BSA).
3. Anti-CD43 monoclonal antibody T305 (IgG) and anti-C2GnT-I poyclonal antibody (kindly provided by M. Fukuda), and anti-EGFP monoclonal antibody JL-8 (IgG, Clontech).
4. Rhodamine isothiocyanate (RITC)-labeled goat anti-mouse IgG (Sigma).
5. Fluorescence-activated cell sorter (FACSCalibur analyzer; Becton Dickinson).
6. Zeiss Axiovert 200 (Zeiss, Göttingen, Germany) fluorescence microscope. This inverted microscope is equipped with a sensitive polychrome camera (AxioCamMRC, Zeiss) and a filter set for fluorescent probes such as EGFP and GFP variants blue fluorescent protein (BFP), cyan fluorescent protein (CFP), yellow fluorescent protein (YFP), as well as fluorescein isothiocyanate (FITC), rhodamine, and 4′,6′-diamidino-2-phenylindole (DAPI). The MetaMorph software (v. 6.0) is used for image acquisition and image analysis. This equipment is useful for imaging the dynamic of fluorescent proteins in living cells.

2.4. Sodium Dodecyl Sulfate-Polyacrylamide Gel Electrophoresis

1. 30% Acrylamide/*bis*-acrylamide solution (29:1; Sigma), *N,N,N,N′*-tetramethylethylenediamine (TEMED; Biorad, Hercules, CA,), and 66% (w/v) sucrose. Store at 4°C.
2. 10% Ammonium persulfate (w/v) in water (freeze immediately at –20°C in single-use 500-μL aliquots).
3. 10% Tris-HCl (w/v) sodium dodecyl sulfate (SDS) in water. Store at room temperature.
4. Separating buffer: 3 *M* of Tris-HCl (pH 8.8) and 1 *M* of stacking buffer (pH 6.8). Store at room temperature.
5. 5X Laemmli ***(5)*** SDS-sample buffer: 0.5 *M* of Tris-HCl (pH 6.8), 7.5% (w/v) SDS, 25 m*M* of EDTA, 5 *M* of sucrose, and 0.05% (w/v) of bromophenol blue. Store at –20°C.
6. 10X Running buffer: 0.25 *M* Tris-HCl, 2 *M* of glycine, and 1% SDS. Store at room temperature.
7. Mini Protean-II protein electrophoresis device and All Blue prestained molecular weight markers (Bio-Rad).

2.5. Western Immunoblotting

1. 1X Transfer buffer: 48 m*M* of Tris-HCl (no need to adjust pH), 39 m*M* of glycine, and 20% methanol. Semi-dry transfer cell (Bio-Rad).
2. Polyvinylidene difluorite (PVDF) membrane (Amersham, Buckinghamshire, UK) and 3MM chromatography paper (Whatman, Madison, UK).
3. Ponceau red solution (0.1% Ponceau S in 5% acetic acid; Sigma).
4. Secondary antibody: Horseradish peroxidase-labeled goat anti-mouse IgG (Santa Cruz Biotechnology, Santa Cruz, CA).
5. Tris-buffered saline containing Tween-20 (TBS-T): 25 m*M* of Tris-HCl (pH 7.4), 0.137 *M* of NaCl, and 0.1% Tween-20. Blocking buffer and primary/secondary dilution buffer: TBS-T containing 5% (w/v) nonfat dry milk.
6. Enhanced chemiluminescent (Supersignal) reagent (Pierce, Rockford, IL).

2.6. In Vitro Enzyme Assay for C2GnT–I

1. 5X Assay buffer: 250 m*M* of *N*-morpholino-ethanesulfonic acid (MES; Sigma), pH 7.0, containing 500 m*M* of *N*-acetylglucosamine (GlcNAc; Sigma). Set pH with 1 *M* of NaOH. Store at 4°C.
2. 5X Donor solution: 5 m*M* of uridine 5′-diphosphate (UDP)-GlcNAc (Sigma). Store in aliquots at –20°C. UDP-[6-^{3}H]GlcNAc (36 Ci/mmol, 0.1 mCi/mL; Perkin Elmer, Boston MA).
3. 10X Acceptor solution: 10 m*M* of Galβ1,3GalNAcα-*O*-paranitrophenyl (Toronto Research Chemicals, Downsview ON, Canada). Dissolve in water and store at –20°C.
4. Sep-Pak columns (Sep-Pak C18 cartridges, Waters Corp, Milford, MA).
5. Vacuum dryer (Speedvac, Farmingdale, NY).

6. Liquid scintillation cocktail photon correlation spectroscopy (PCS; Amersham, Arlington Heights, IL).
7. LKB/Wallac 1214 beta-counter (Perkin Elmer).

2.7. Characterization of O-Linked OSs

1. Culture cells under standard conditions until 50–60% confluency.
2. Radiolabeled D-[6-^{3}H(N)]-Glucosamine.HCl (24 Ci/mmol, 1 mCi/mL; Perkin Elmer), tritiated OS standards (sialylated core1 and core2 OSs, kindly provided by M. Fukuda).
3. 1- × 100-cm Sephadex G-50 column (Pharmacia, Uppsala, Sweden). Pronase and $NaBH_4$ (Sigma).
4. 1.2- × 120-cm Biogel P-4 column (Bio-Rad) and 1- × 50-cm Biogel P-2 column (Bio-Rad).
5. Fraction collector (Redirac, LKB). Peristalic pump (Microplex S, LKB) and 5-mL glass tubes. Kimax glass tube equipped with plug ends and rubber stoppers (Boreal Laboratories, Ontario, Canada).
6. Liquid chromatography (LC) buffer: 0.1 *M* of NH_4HCO_3. Store at room temperature.

3. Methods

To study the structure–function relationship of C2GnT-I in vivo, choose a cell line that does not naturally express this enzyme activity (such as CHO cells) and fuse the protein with the EGFP to allow constant monitoring in living cells. The presence of EGFP also offers a number of other advantages, including the possibility of establishing a direct relationship between the fluorescence intensity and the activity of the enzyme (in vivo and in vitro), establishing a visual relationship between the intracellular distribution of the enzyme and its activity (toward specific acceptors) by confocal microscopy, facilitating sorting a stably transfected polyclonal cell population by fluorescence-activated cell sorter (FACS), and detecting proteins on blots with highly sensitive commercially available anti-EGFP antibodies.

3.1. Construction of EGFP-Tagged C2GnT-I

A flowchart of this construction is presented in **Fig. 1A**.

1. Construction of pcDNA3/EGFP plasmid: The vector pEGFP-N1 encodes a variant of the wild-type GFP ***(6,7)*** which has been optimized for brighter fluorescence with an excitation maximum of 488 nm and an emission maximum of 507 nm ***(8)***. It also carries a kanamycin resistance gene for selection and propagation in *E. coli.* In terms of bacterial transformation/propagation and eukaryotic cell transfection/selection; however, cloning in pEGFP-N1 is less efficient than in pcDNA3.1(+). Therefore, transfer the EGFP gene cassette from pEGFP-N1 to pcDNA3.1(+) after digestion of both plasmids with *EcoR*1 and *Not*1, as previously reported ***(9)***. In brief, the *EcoR*1/*Not*1 EGFP cDNA is ligated into

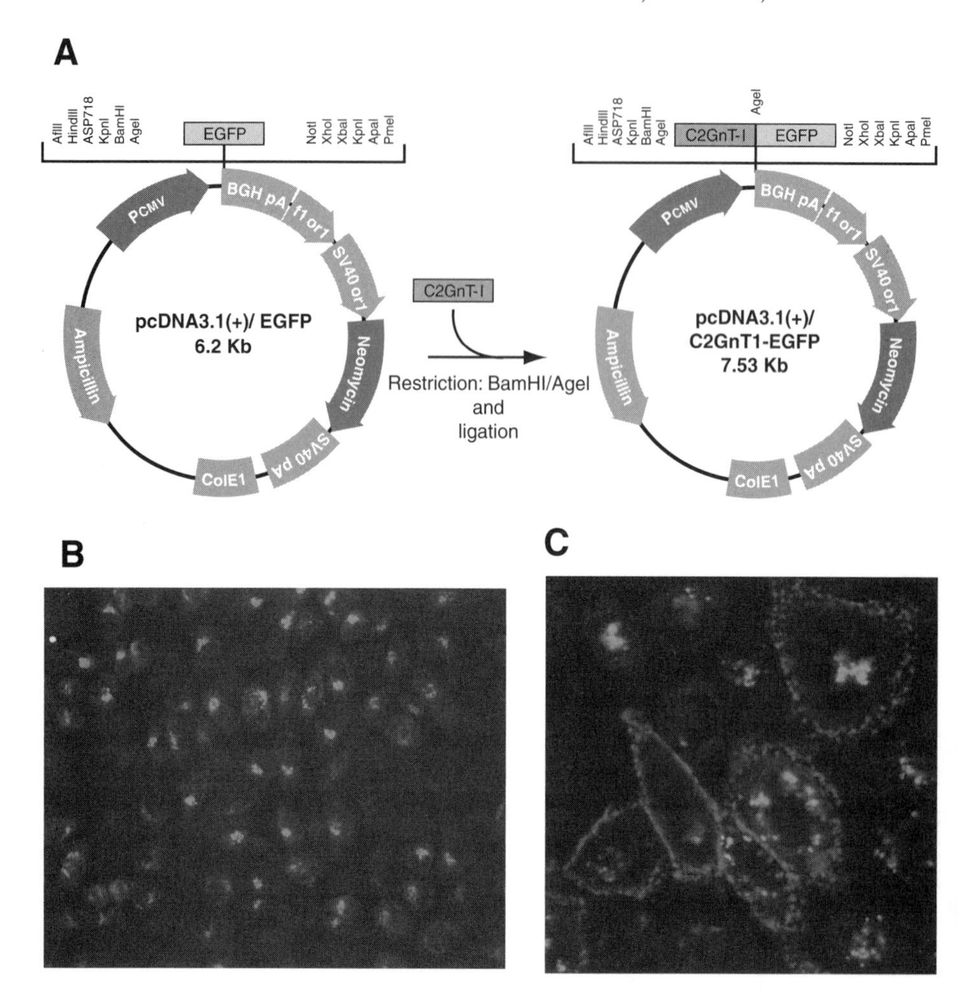

Fig. 1. **(A)** The strategy used for the construction of the expression vector pcDNA3/C2GnT-I-enhanced green fluorescent protein (EGFP) to express an EGFP-conjugated form of C2GnT-I. **(B)** C2GnT-I-EGFP-mediated Golgi staining. Chinese hamster ovary cells were transfected by pcDNA3/C2GnT-I-EGFP construct and neomycin (G418)-resistant transfectants were sorted by fluorescence-activated cell sorter, and a highly-expressing cell population was isolated and is shown here. **(C)** Simultaneous imaging of intracellular distribution of C2GnT-I-EGFP and cell-surface expression of T305 epitope. HighC2 cells **(B)** were transiently transfected with pcDNAI/CD43 construct. They were then incubated with the T305 MAb ***(10)***, stained by rhodamine-conjugated secondary antibody, and examined by fluorescence microscopy using a dual-color (EGFP/RITC) ×40 objective.

*EcoR*1/*Not*1-restricted pcDNA3.1(+) and transformed in Top 10F′ cells. The obtained vector of pcDNA3/EGFP is used to subclone C2GnT-I. Further details on molecular biology handling follow.

2. Construction of C2GnT-I cDNA to be cloned in pcDNA3/EGFP: Amplify the full-length cDNA for C2GnT-I by polymerase chain reaction (PCR) using 0.5 μg of C2GnT-I/pcDNA1 ***(10)*** as a template and the sense and antisense primers listed above (0.2 nmol each). The PCR mixture (final volume 50 μL) also contains 10 nmol each of deoxyribonucleotides (dNTP: dATP, dCTP, dGTP, and dTTP), 5 μL of 10X PCR buffer, and water (up to 50 μL). The PCR protocol involves an initial denaturation step of 1 min 30 s at 94°C, followed by 14 cycles of 3 steps including 1 min 30 s at 94°C, 2 min at 52°C, and 2 min at 68°C. The last cycle is followed by a final extension step of 5 min at 68°C. After the completion of the reaction, the PCR mixture is subjected to 1% (v/w) agarose gel electrophoresis in Tris-acetate-EDTA (TAE) buffer. After ethidium bromide staining (*see* **Note 2**), cut and extract the gel band containing the amplified C2GnT-I DNA fragment using the gel extraction kit. Then estimate the concentration of the purified C2GnT-I DNA by spectrophotometry at 260 nm, and incubate 1 μg of DNA overnight at 37°C in 10 μL with 10X 10 U of *BamH*1 and 1 mL of *BamH*1 buffer (*see* **Note 3**). Perform the same reaction in parallel with 0.2 μg of pcDNA3/EGFP. Add 20 U of *Age*1 together with 10X 2 μL of the corresponding buffer and water to the *BamH*1 reactions, and bring volumes to 20 μL with water. Incubate for an additional 4 h at 37°C. The restricted fragments (C2GnT-I and pcDNA3/EGFP) are then recovered after 1% agarose gel electrophoresis as indicated earlier and mixed in a 5:1 ratio (with respect to the amounts estimated from DNA intensities on the gel, *see* **Note 4**) and incubated overnight at 4°C with 5 U of T4 DNA ligase and 1.5 μL of 10X ligation buffer, in a final volume of 15 μL. Include a "self-ligation" control using water instead of insert (C2GnT-I DNA).
3. Bacterial transformation: Use 5 μL of ligation mixture to transform 50 μL of competent Top 10F′ cells and incubate on ice for 30 min. Transfer the vials to a 42°C bath for a heat-shock of 45 s. Adjust the volume to 500 μL with SOC medium and incubate bacterial suspensions under agitation for 1 h at 37°C. Spread 250-μL aliquots of each bacterial suspension onto LB agar using autoclaved glass beads, and incubate upside-down Petri dishes overnight at 37°C.
4. Pick ampicillin-resistant *E. coli* colonies by sterile pipet tips to inoculate 2 mL of LB media in 15-mL round-bottom polystyrene tubes containing 100 μg/mL of ampicillin. Incubate with vigorous shaking for 5–7 h.
5. To check for the presence of the C2GnT-I insert, plasmid DNAs are prepared by alkaline lysis using the miniprepartion kit from Qiagen according to the manufacturer's instructions. The presence of C23GnT-1 insert is confirmed either by enzyme restriction *BamH*1/*Age*1 or PCR, as specified earlier.

3.2. Transfection of CHO Cells With pcDNA3/C2GnT-I-EGFP

1. Harvest CHO cells by Trypsin/EDTA and plate in a 6-well plate (5×10^4 cells/well) in complete Ham F12 medium 1 d before transfection.

2. Cells should have reached 50–60% confluency by the day of transfection. Transfer 100 μL of Opti-MEM to a 5-mL round-bottom polystyrene tube and add 1 μg of DNA (miniprep quality; *see* **Note 5**) and 6 μL of reagent Plus, and gently tap the tube five to eight times. Incubate at room temperature for 15 min.
3. Dilute 4 μL of lipofectamine with 100 μL of Opti-MEM in another tube, and then add the mixture dropwise to the first tube. Leave at room tempertaure for 15–20 min for liposomes/DNA complex to form.
4. Remove the media from the culture dishes and wash the cell monolayers three times with 1 mL of Opti-MEM. Add 800 μL of Opti-MEM/well and add the 200 μL of DNA/Lipofectamine mixture dropwise. Incubate for 3 h at 37°C (*see* **Note 6**).
5. Replace the transfection medium with complete Ham F-12 medium and incubate the cells for an additional 48 h.

3.3. Visualization and Isolation of C2GnT-I-Expressing Cells

1. The plasmid carrying the chimeric DNA C2GnT-I-EGFP encodes the full-length C2GnT-I, C-terminally conjugated to the full-length EGFP protein through the fusion peptide Pro-Pro-Leu-Ala-Thr, derived from the multicloning site of the pcDNA3/EGFP vector. This results in the fusion protein C2GnT-I-***His***-Pro-Pro-Leu-Ala-Thr-**Met**-EGFP (where ***His*** is the penultimate amino acid of C2GnT-I and **Met** is the start codon of EGFP).
2. It is possible to detect a typical Golgi staining using an inverted fluorescence microscope equipped with an EGFP filter (presumably EGFP-tagged C2GnT-I) as soon as 5 h posttransfection (although only a few cells exhibit such a staining). Thirty to fifty percent of cells exhibit a strong Golgi staining 24 h later.
3. The pcDNA3/C2GnT-EGFP vector carries the neomycin (G418) resistance gene that allows isolation of drug-resistant clones. Harvest the cells 48 h after transfection by trypsin/EDTA, resuspend in 1 mL of complete Ham F-12 medium (per well), and transfer 200 μL to a 150-mm dish containing 20 mL of medium supplemented with 1 mg/mL of G418. Change the medium every 2 d during the first 2-wk selection period, then every 3 to 4 d as soon as individual clones start forming. It is easy to monitor clone formation throughout the selection period owing to EGFP fluorescence. Positive clones represent 5–10% of total drug-resistant clones after 4 wk. At this stage, positive clones may be picked up using cloning cylinders. However, the clones harvested by this technique are often still heterogenous and a second round of cloning using limiting dilution, for example, may be needed. Again, the EGFP fluorescence can be used to sort/purify a heterogenous polyclonal cell population by FACS. In addition, use of this technique makes it possible not only to isolate fluorescent vs nonfluorescent cells but also to sort cells with respect to their fluorescence intensity, thus isolating for specific goals (i.e., low-, moderate-, and high-expressor cell lines).
4. Harvest the G418-resistant transfectants as described in **step 3** from 150-mm dishes and pour into 25-cm^2 flasks in complete Ham F-12 medium containing 200 μg/mL of G418, and grow cells until confluency. Harvest the cells, pellet at

600 rpm for 5 min, and gently resuspend as 10^6 cells/mL in 0.2 μm of sterile-filtered PBS/BSA. Apply cell suspension to a FACSCalibur analyzer equipped with an argon ion laser, and detect EGFP-expressing cells by an excitation/emission of 488/507 nm. A cell population with high EGFP fluorescence (superior to 10^3 log fluorescence intensity) is collected and recovered in complete medium containing 200 μg/mL G418 and 20% FCS. Fluorescence microscopic examination of this cell line is presented in **Fig. 1B**. Note the typical Golgi staining. These cells, which highly express C2GnT-I-EGFP homogeneously, will be used for the following studies and are referred to as HighC2 cells.

3.4. Western Immunoblotting of Intact C2GnT-I and the EGFP-Conjugated Protein

Once the HighC2 cell line has been established, it is important to structurally analyze the changes that resulted from fusion with EGFP. To this end, use either the anti-EGFP MAb JL-8 to detect the C2GnTI-EGFP from HighC2 cell extracts, or the anti-C2GnT-I polyclonal antibody *(11)* to detect the intact enzyme from the CHO/C2 clone *(12)*.

1. Prepare cells for SDS-polyacrylamide gel electrophoresis (SDS-PAGE). Rinse cell monolayers twice with PBS, harvest using a dissociation solution, and pellet. Pellets are directly resuspended in a 2X SDS sample buffer (*see* **Note 7**) in a ratio of 200 μL of sample buffer per one 25-cm^2 flask of fully confluent cells.
2. Prepare 3-mm-thick SDS gel by mixing 2.6 mL of 30% acrylamide/*bis*-acrylamide, 1.4 mL of separating buffer, 1.2 mL of 66% sucrose, and 5.1 mL of water. Add 104 μL of 10% SDS, 60 μL of 10% ammonium persulfate, and 6 μL of TEMED. Pour the gel (leaving space for a stacking gel) and overlay with water. The gel should polymerize in 20–30 min.
3. Prepare stacking gel by mixing 700 μL of 30% acrylamide/*bis*-acrylamide, 700 μL of stacking buffer, 600 μL of 66% sucrose, and 3.2 mL of water. Add 52 μL of 10% SDS, 30 μL of ammonium persulfate, and 3 μL of TEMED. The gel should polymerize within 30 min.
4. Prepare the running buffer by diluting 100 mL of the 10X solution in 900 mL of water.
5. Once the stacking gel has polymerized, carefully remove the comb and wash the wells with water.
6. Add the running buffer to the upper and lower chambers.
7. Load 50–75 μL per well. Include one well for prestained molecular-weight markers (*see* **Note 8**).
8. Run the gel at 120 V (constant voltage) until the dye reaches the bottom of the gel.
9. Remove the stacking gel and soak in the transfer buffer, together with two pieces of 3MM chromatography paper cut to size (slightly larger than the gel).
10. Cut a PVDF membrane and pre-wet for 10 s in 100% methanol. Wash in water for 5 min and incubate in the transfer buffer for 10 min (*see* **Note 9**).

11. Place the 3MM chromatography paper sheets, the PVDF membrane, and the gel in the semi-dry electroblotting cassette, according to the manufacturer's instructions.
12. Transfer for 1 h at constant amperage (250 mA; corresponding voltage 8V).
13. Remove the membrane after transfer, briefly rinse in PBS, and stain with Ponceau red to visualize proteins. At this stage the blot can be photographed (or scanned). Mark the lanes of interest with a pencil, as the red color is quickly removed by the blocking buffer. The blot may be used immediately or stored frozen in a plastic bag until use.
14. Incubate the membrane with 10 mL of blocking buffer for 1 h at room temperature on a rocking platform. Then cut in two pieces, to be incubated separately with the appropriate primary antibodies.
15. Discard the blocking buffer and incubate the part of the membrane containing proteins from HighC2 cells with 1:500 dilution (in the blocking buffer) of the anti-EGFP mAb JL-8. The other piece, which contains proteins from CHO/C2 cells, is incubated with 1:2000 dilution of the anti-C2GnT-I polyclonal antibody *(11)*. Incubations are carried out for 2 h at room temperature.
16. Remove the primary antibodies and wash the membranes three times with 10 mL of TBS-T (*see* **Note 10**).
17. The secondary antibodies are freshly prepared for each experiment at 1:10,000 dilution in the blocking buffer, and incubation is continued for 1 h at room temperature on a rocking platform. The secondary antibodies are then removed and the membrane washed as in **step 16**.
18. Incubate the blots with chemiluminescent reagents according to the manufacturer's instructions and expose to an X-ray film (in an X-ray film cassette) for a suitable exposure time.
19. An example of the results obtained with anti-EGFP MAb and anti-C2GnT-I polyclonal antibody (pAb) is shown in **Fig. 2**. In this experiment we show that the intact and the EGFP-fused proteins can be detected as a cell-associated (Cell) or secreted form (Med).

3.5. Assay for C2GnT-I Catalytic Activity In Vivo

1. The T305 monoclonal antibody reacts exclusively with the sialylated Core2-branched OSs carried by the mucin protein CD43 *(12)*. For this reason it represents a valuable tool to assay for C2GnT-I activity in vivo. Seed HighC2 cells in six-well plates and transfect as indicated in **Subheading 3.2.** with 1 μg of pcDNA1/CD43 construct. Harvest cells 48 h posttransfection with trypsin/EDTA, resuspend in 2 mL of complete medium, and seed 200 μL of cell suspension in an eight-well glass chamber slide. Culture for another day before analysis.
2. All incubations are carried out on ice. Rinse cell monolayers in eight-well chamber slides three times with cold PBS and incubate with PBS/BSA for 5 min to block nonspecific sites. Increasing dilutions of T305 MAb (from 1:50 to 1:200 in PBS/BSA) are then added onto cell monolayers, and incubation is continued for 1 h on ice. Then rinse the cells three times with PBS and incubate with a 1:200

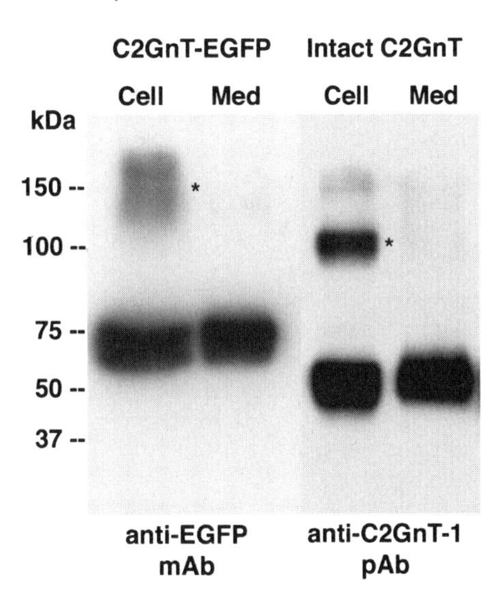

Fig. 2. Immunoblotting of C2GnT-I. Cells (Cell) and media (Med) were subjected to sodium dodecyl sulfate-polyacrylamide gel electrophoresis under nonreducing conditions (in the absence of β-mercaptoethanol) and assayed for C2GnT-I using either the anti-EGFP MAb **(left panel)** or the anti-C2GnT-I polyclonal antibody **(right panel)**. Both the intact C2GnT-I and the chimeric C2GnT-I-EGFP form dimers (*) and are secreted. Note that only monomers are released in the culture medium

dilution (in PBS/BSA) of the secondary antibody RITC-labeled goat anti-mouse IgG. After three rinses with PBS, examine fluorescence using dual color (EGFP/RITC) 40× objective with no bleed-through between the two channels (*see* **Note 11**). The T305 staining of living HighC2 cells transiently transfected with CD43 DNA is shown in **Fig. 1C.**

3.6. Assay of C2GnT-I In Vitro Enzyme Activity

1. In a microfuge (1.5-mL Eppendorf) mix 10 μL of "cold" donor, 5 m*M* of UDP-GlcNAC, and 5 μL of radioactive donor UDP-[6-^{3}H]GlcNAc (1 μL of a 0.1-mCi/mL solution yields 91,000 cpm as determined by liquid scintillation spectrometry using an LKB beta counter). Dry under vacuum in a Speedvac (*see also* Chapter 14).
2. Add per assay: 10 μL of 5X assay buffer, 5 μL of acceptor or water (control), 10 μL of water, and 25 μL of enzyme source (10 mg/mL). Incubate at 37°C for 1 h.
3. Dilute reactions to 5 mL with water and apply to a Sep-Pak column that has been rinsed once with 2 mL of 100% methanol and three times with 5 mL of water. Wash the column two times with 10 mL of water and elute the bound material with 2 mL of 100% methanol.

Table 1
In Vitro C2GnT-I Activity

Cells	C2GnT-I Activity
CHO-K1	Not detected
CHO/C2	86.0 nmol/mg protein/h
HighC2	74.8 nmol/mg protein/h

Cell extracts were assayed for C2GnT-I activity using the synthetic oligosaccharide Galβ1,3GalNAc-*O*-*p*-nitrophenyl as an acceptor and UDP-GlcNAc/UDP-[6-^{3}H]GlcNAc as a donor. The activities were determined as nmols of GlcNAc transferred per mg protein and per hour. Note that EGFP tagging does not significantly alter the C2GnT-I, in vitro.

4. Count aliquots of 500 µL after adding 5 mL of scintillation liquid per vial.
5. **Table 1** shows one representative assay of transferase activity of EGFP-tagged and intact enzyme (in nmol donor transferred/mg protein/h). Note that fusing C2GnT-I with EGFP has no significant effect on the catalytic activity of the enzyme.

3.7. Analysis of O-Linked OSs From HighC2 Cell Line

This is a slightly modified protocol from that of Skrincosky et al. ***(11)***.

1. Prepare cells (50–60% confluency) in two 150-mm culture dishes. Add 10 µCi/mL of radiolabeled D-[6-^{3}H(N)]-glucosamine to 10 mL of complete Ham F-12 medium and sterile filtrate over a 0.2-µm filter. Replace the culture media by 5 mL/dish of radioactive medium. Culture the cells overnight.
2. Discard the radioactive medium and wash the cells once with PBS. Detach the cells by the nonenzymatic cell dissociation solution for 15–20 min at 37°C and centrifuge in conical or round-bottom glass tubes for 5 min at 1000*g*. At this stage, pellets can be stored frozen at –20°C.
3. Add 10 vol of chloroform/methanol (2/1, v/v) to the cell pellets and vortex vigorously. Add methanol to a final ratio with chloroform of 1/1 and centrifuge for 15 min at 865*g*.
4. Discard the supernatant and carefully dry the pellets under a stream of nitrogen.
5. Dissolve the pellets in 1 mL of 0.1 *M* Tris-HCl (pH 8.0), 1 m*M* of $CaCl_2$, add pronase to a final concentration of 2 mg/mL (*see* **Note 12**), and incubate samples for 24 h at 60°C in a toluene atmosphere to prevent bacterial growth.
6. Boil the samples for 10 min to denature the remaining pronase activity and centrifuge for 10 min at 5000*g*. Discard the pellets.
7. Apply the supernatant to the Sephadex G-50 column that has been equilibrated in the LC buffer for at least 24 h before the experiment. Collect 1-mL fractions and count one aliquot of each fraction, as in **Subheading 3.5., step 4**.

8. Pool the V_0 fractions (*see* **Note 13**), lyophilize, and desalt on a Biogel P-2 column that has been equilibrated in water. Pool the V_0 fractions containing total glycopeptides (N- and O-linked) and lyophilize again.
9. Weigh 37.8 mg of $NaBH_4$ in a Kimax glass tube. Resuspend the lyophilized glycopeptides in 1 mL of 50 m*M* NaOH, transfer this solution to the Kimax tube containing $NaBH_4$, and incubate for 16 h at 45°C in a loosely capped tube.
10. Cool at room temperature and neutralize by adding 1 *M* of acetic acid in methanol until the bubbling disappears. Dry samples under a nitrogen stream and dissolve in 1 mL of LC buffer. Centrifuge at 865*g* to remove unsoluble particles and apply to the Sephadex G-50 column as before. Pool fractions between V_0 and V_t, representing the released O-linked OSs. Lyophylize, desalt on a Biogel P-2 column, lyophilize again, and resuspend in 1 mL of LC buffer.
11. Apply to a Biogel P-4, collect fractions, and count aliquots.

A representative chromatographic OS profile is shown in **Fig. 3**. Note that, as expected, the core2 OSs (*see* **Fig. 3C**) are much more enriched in HighC2 cells (fraction 30–40; *see* **Fig. 3B**) than in parental CHO cells (used as a control).

4. Notes

1. Deionized autoclaved water is used for all molecular biology experiments. It is referred to as "water" throughout this chapter.
2. **Caution:** Ethidium bromide is a highly toxic chemical that can be inhaled or absorbed through the skin. It is important to avoid any direct contact with the chemical. Strictly follow the safety instructions provided by the manufacturer. On the other hand, be aware that UV-generated damage to DNA stained with ethidium bromide reduces the transformation efficiency. One minute of exposure to 312 nm of UV can reduce the transformation efficiency by more than 90%. Alternatively, it is possible to use a less hazardous chemical for the detection of DNA, such as the "REDTAQ DNA Polymerase" produced by Sigma.
3. When performing double restriction reactions, it is important to start with the enzyme having the lowest salt concentration in order to optimize the hydrolysis while avoiding the so-called "star activity." For example, in the case of *Age*1 and *BamH*1 (both from Promega), start with *BamH*1 because its buffer contains 100 m*M* of NaCl, whereas *Age*1 buffer has 150 m*M* of KCl.
4. If 1 μL of each purified insert (C2GnT-I) and vector (pcDNA3/EGFP) have equivalent intensities, mix 10 μL of insert with 2 μL of vector and add 1.5 μL of T4 DNA ligase (3 U/μL) and 10X 1.5 μL of buffer.
5. Because the EGFP fluorescence allows us to easily monitor transfection efficiency, we find that DNA of Qiagen "miniprep" quality has a higher transfection efficiency than an equivalent amount of Qiagen DNA "maxiprep" quality.
6. It is possible to increase the transfection incubation time for longer periods (overnight), but this may cause cell death owing to the toxicity of the DNA/

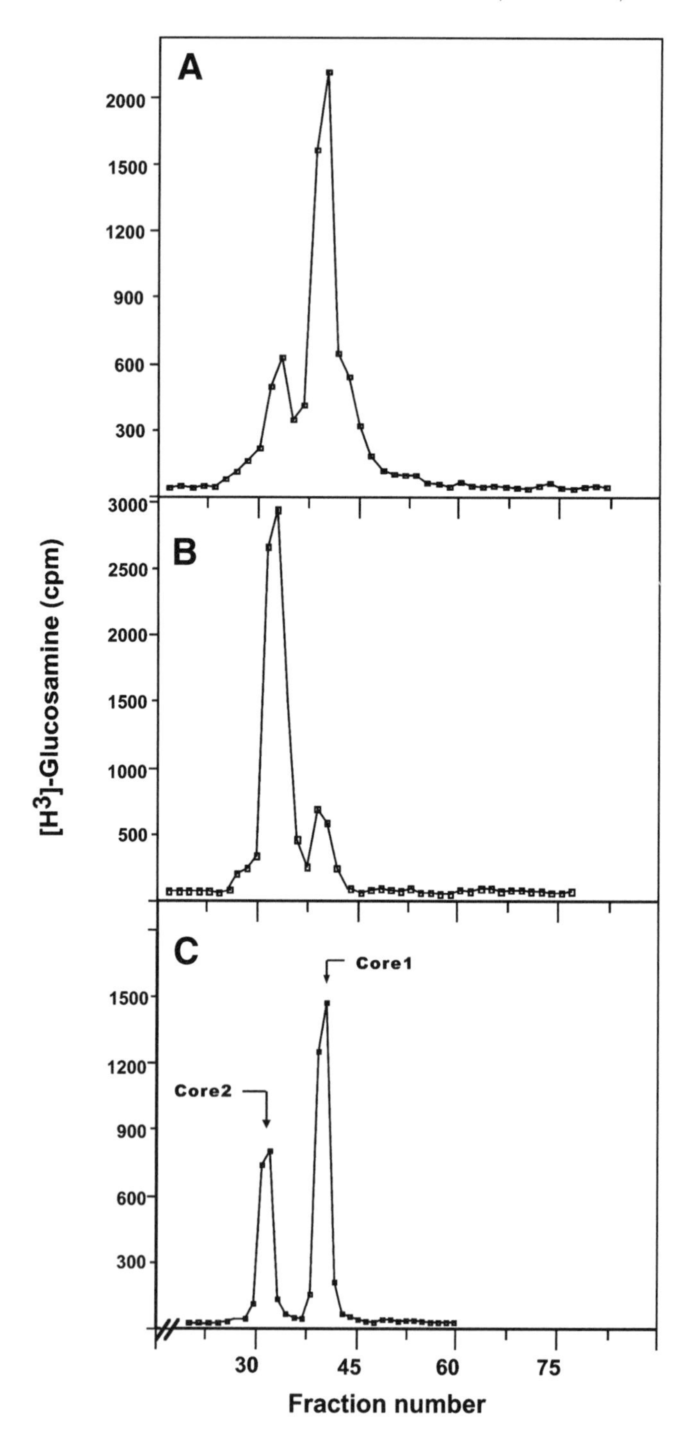
A
B
C
Core1
Core2
2000
1500
1200
900
600
300
3000
2500
1000
500
30
45
60
75
Fraction number
[H3]-Glucosamine (cpm)

liposome mixture. Do not use too much penicillin/streptomycin as antibiotics to decrease transfection efficiency.

7. It is possible that mixing cells directly with SDS sample buffer will result in a viscous solution that is very hard to load into wells (owing to DNA/SDS interaction). If this happens, briefly sonicate the extract for 1 or 2 s while avoiding foam formation.
8. The authors prefer 3-mm-thick gel because a standard micropipet and regular tips can be used to load samples in the bottom of the wells.
9. The PVDF membrane must be kept wet at all times. Should it dry out, re-wet in methanol and rinse in water as indicated.
10. The primary antibodies can be kept frozen and reused two to three times without significant loss of activity.
11. The use of the chamber slides and the inverted fluorescence microscope does not necessarily fix cells before immunostaining in observations less than 15 min under the microscope. In fact, because cells are still alive, there might be internalization of antigen–antibody complexes. To avoid this problem, return the cells to ice for 5 min or change the medium to cold PBS before another microscopic examination. Because the cells are viable it is possible to return the cells to culture for another purpose, provided that all media used throughout this experiment are 0.2-µm sterile-filtered.
12. Pronase preparation may have contaminating glycosidase activities. Therefore, incubate a stock solution of pronase (e.g, 10 mg/mL in water) for 1 h at 37°C before use.
13. V_0 (void volume) and V_t (total volume) are determined by calibrating the column with Dextran Blue and phenol red dyes, respectively.

Acknowledgments

We would like to thank Professor Minoru Fukuda (The Burnham Institute) for providing us with C2GnT-I and CD43 constructs, as well as the C2GNT-I polyclonal antibody and core1- and core2-radiolabeled oligosacchardes. We are also grateful to Mrs. Carmona Sylvie for her technical assistance.

Fig. 3. *(previous page)* Fractionation of O-linked oligosaccharides derived from HighC2 cells, compared to those derived from the parental Chinese hamster ovary (CHO) cells. The [^{3}H]glucosamine-labeled glycopeptides of the whole cell lysates were subjected to alkaline hydrolysis treatment to release *O*-glycans. The released *O*-glycans were isolated by Sephadex G-50 gel filtration and applied to a column of Bio-Gel P-4. Shown are the Biogel P-4 profiles of oligosaccharides derived from CHO-K1 cells **(A)** and HighC2 cells **(B)**. The positions of tritiated standard core2- and core1-oligosaccharides are shown in **(C)**. Note the peak inversion core1/core2 obtained in the presence of C2GnT-I.

References

1. Lowe, J. B. (2003) Glycan-dependent leukocyte adhesion and recruitment in inflammation. *Curr. Opin. Cell Biol.* **15,** 531–538.
2. Rosen, S. D. (2004) Ligands for L-selectin: homing, inflammation, and beyond. *Annu. Rev. Immunol.* **22,** 129–156.
3. Kannagi, R., Izawa, M., Koike, T., Miyazaki, K., and Kimura, N. (2004) Carbohydrate-mediated cell adhesion in cancer metastasis and angiogenesis. *Cancer Sci.* **95,** 377–384.
4. Beum, P. V. and Cheng, P. W. (2001) Biosynthesis and function of beta 1,6 branched mucin-type glycans. *Adv. Exp. Med. Biol.* **491,** 279–312.
5. Laemmli, U. K. (1970) Cleavage of structural proteins during the assembly of the head of bacteriophage T4. *Nature* **227,** 680–685.
6. Prasher, D. C., Eckenrode, V. K, Ward, W. W, Prendergast, F. G, and Cormier, M. J. (1992) Primary structure of the Aequorea victoria green fluorescent protein. *Gene* **111,** 229–233.
7. Chalfie, M., Tu, Y., Euskirchen, G., Ward, W. W, and Prasher, D. C. (1994) Green fluorescent protein as a marker for gene expression. *Science* **263,** 802–805.
8. Cormack, B., Valdivia, R. H, and Falkow, S. (1996) FACS-optimized mutants of the green fluorescent protein (GFP). *Gene* **173,** 33–38.
9. Zerfaoui, M., Fukuda, M., Sbarra, V., Lombardo, D., and El-Battari, A. (2000) Alpha(1,2)-fucosylation prevents sialyl Lewis x expression and E-selectin-mediated adhesion of fucosyltransferase VII-transfected cells. *Eur. J. Biochem.* **267,** 53–61.
10. Bierhuizen, M. F. and Fukuda, M. (1992) Expression cloning of a cDNA encoding UDP-GlcNAc:Gal beta 1-3-GalNAc-R (GlcNAc to GalNAc) beta 1-6GlcNAc transferase by gene transfer into CHO cells expressing polyoma large tumor antigen. *Proc. Natl. Acad. Sci. USA* **89,** 9326–9330.
11. Skrincosky, D., Kain, R., El-Battari, A., Exner, M., Kerjaschki, D., and Fukuda, M. (1997) Altered Golgi localization of core 2 beta-1,6-*N*-acetylglucosaminyltransferase leads to decreased synthesis of branched O-glycans. *J. Biol. Chem.* **272,** 22,695–22,702.
12. Panicot, L., Mas, E., Pasqualini, E., et al. (1999) The formation of the oncofetal J28 glycotope involves core-2 beta6-*N*-acetylglucosaminyltransferase and alpha3/4-fucosyltransferase activities. *Glycobiology* **9,** 35–46.

13

Gene Expression Analysis of Glycosylation-Related Genes by Real-Time Polymerase Chain Reaction

Juan J. García-Vallejo, Sonja I. Gringhuis, Willem van Dijk, and Irma van Die

Summary

Glycan molecules covalently linked to proteins or lipids control vital properties of cells, such as signaling, adherence, and migration through the body. The biosynthesis of such glycans depends on the concerted action of many endoplasmic reticulum and Golgi enzymes, a process that is tightly ordered and regulated. To understand the function of glycoconjugates in cellular interactions, it is crucial to investigate the regulation of expression of the genes encoding the "glycosylation-related" genes, encompassing large families of glycosyltransferases, glycosidases, and sulfotransferases. This chapter describes an easy, flexible, and reliable method of quantitative real-time polymerase chain reaction to measure the expression levels of 80 human glycosylation-related genes that primarily encode common enzymes involved in N- and O-linked protein glycosylation and/or glycolipids. Designing and including additional primer sets to detect more genes can easily extend the system. In order to allow the normalization of gene expression data obtained by real-time polymerase chain reaction within different cells, tissues, or under different experimental conditions, a protocol is included to detect genes suitable for use as endogenous reference genes.

Key Words: mRNA isolation; cDNA synthesis; real-time PCR; quantitative PCR; glycosyltransferases; glycosylation-related genes; gene expression analysis; normalization.

1. Introduction

Oligosaccharide (OS) chains covalently linked to glycoproteins and glycolipids often confer specific biological functions to the molecules carrying them. To understand the function of such glycoproteins or glycolipids in biological processes, it is crucial to determine the factors that control the regulation of expression of their OS chains. Unlike the process of protein synthesis, in which

From: *Methods in Molecular Biology, vol. 347: Glycobiology Protocols*
Edited by: I. Brockhausen © Humana Press Inc., Totowa, NJ

the sequence of amino acids is directly deduced from a nucleic acid matrix, glycosylation is a nontemplate process and includes the formation of branches in the OS chains.

The synthesis of OS chains on glycoconjugates is mediated by the concerted action of many glycosylation-related genes, including glycosyltransferases (GTs), glycosidases, sulfotransferases, and accessory molecules that are localized in the endoplasmic reticulum and Golgi. GTs typically catalyze the transfer of a monosaccharide from a nucleotide-sugar to a growing acceptor-OS attached to a carrier (in some cases directly to the carrier molecule), during transport of the glycoconjugate through the secretory pathway. Monosaccharides are attached sequentially, one at a time, through glycosidic linkages at nonreducing ends and at branching points. The OS product of one reaction is the substrate in the next reaction, and the array of glycosylation-related genes expressed in a particular cell is an important factor in the control of OS synthesis. In addition, the potential synthesis of a particular OS depends on the localization of the enzymes in the Golgi, the acceptor- and branch-specificity of the enzymes, and on the presence of enzymes that compete for the same substrate *(1)*.

Glycosylation-related enzymes are the primary products of their respective genes, and each of these genes may be regulated separately. To investigate regulation of glycosylation, quantification of gene expression of a broad array of glycosylation-related genes is therefore an essential approach. This chapter describes an easy, flexible, and reliable method of quantitative real-time polymerase chain reaction (PCR) to measure messenger ribonucleic acid (mRNA) levels from approx 80 human glycosylation-related genes that primarily encode enzymes involved in N- and O-linked protein glycosylation and/or glycolipid synthesis in one experiment. The quantification of gene expression by real-time PCR has revolutionized the field of gene expression analysis. This method has important advantages over complementary deoxyribonucleic acid (cDNA) microarray technology, especially for expression analysis of glycosylation-related genes. The high sensitivity obtained by real-time PCR allows the detection of relatively small differences (factor 1.5–2) in mRNA expression levels. Minor differences in GT gene expression levels can result in a major shift in OS synthesis, owing to a combination of other factors such as expression levels of other GT genes that compete for the same acceptor OS, and acceptor- and branch-specificity of the enzymes.

Good normalization and calibration methods are essential to allow a reliable quantitative determination of small differences in gene expression and facilitate data exchange and comparison. A simple protocol to detect genes that are stably expressed in the cell lines of choice, or under the experimental conditions used, and thus can be employed as endogenous reference genes to normalize the glycosyltransferase gene expression data obtained by real-time PCR *(2)* is described.

2. Materials

2.1. Sample Preparation

1. Phosphate-buffered saline (PBS): 8.2 g/L of NaCl, 3.1 g/L of $Na_2HPO_4 \cdot 12H_2O$, and 0.3 g/L of $NaH_2PO_4 \cdot 2H_2O$ in nuclease-free water (pH 7.4).
2. Syringe.
3. 21-Gage needle.
4. Mortar and pestle.
5. Liquid nitrogen.
6. Sterile disposable general plastic ware.
7. Disposable cell scraper (Greiner Bio-One, Frickenhausen, Germany).
8. Tabletop cooled microcentrifuge (Heraeus, Osterode, Germany).

2.2. mRNA Isolation

1. The mRNA isolation procedure described in this protocol has been adapted to the mRNA Capture Kit (Roche, Rotkreuz, Switzerland). All the reagents, streptavidin-coated tubes, and caps are included in the kit.
2. Incubator (37°C).

2.3. cDNA Synthesis

1. The cDNA synthesis procedure described in this protocol has been adapted to the Reverse Transcription System (Promega, Madison, WI). All the reagents are included in the kit.
2. ABI-9600 Thermalcycler (Applied Biosystems, Foster City, CA).

2.4. Real-Time PCR

1. ABI 7900HT Real-Time PCR instrument (Applied Biosystems).
2. SYBR Green I Mastermix (Applied Biosystems).
3. Real-time PCR optical plates and optical cover sheets (Applied Biosystems) and filter tips (Greiner).
4. Dissolve oligonucleotides (Invitrogen) in sterile nuclease-free 100 m*M* of Tris/HCl (pH 8.0) and store at –20°C. From this solution, make a working solution by diluting the primers 1:10 in sterile nuclease-free 100 m*M* of Tris/HCl (pH 8.0) and store at –20°C.

3. Methods

3.1. Primer Design

Primers are designed using the computer program Primer Express 2.0 (Applied Biosystems). The general strategy is to choose exon-spanning primers within the coding sequence with an optimal length of 20 nucleotides, a melting temperature between 58 and 60°C, and a limited number of G and C in the 3′-end of the oligonucleotides. The amplicon length varies between 50 and 150 bp. The panels of genes chosen, with their GenBank accession numbers and references, are listed in **Table 1**. Care should be taken in design to ensure the specificity of the primers (*see* **Notes 1**, **2**, and **5**).

Table 1
Primers for Glycosylation-Related Genes

Gene	Linkage	Gene ID	Other names	Primers	Ref.
A. Galactosyltransferases					
β*3GALT1*	β1,3	8708		Fwd: tggaaaatggcctacagtttgtg Rev: gattctgtgcatttcttctggagag	***6***
β*3GALT2*		8707		Fwd: ttacctctcatcagttccagcctag Rev: tggcacaggcattgtgctta	***6***
β*3GALT4*		8705		Fwd: tatgtgctgtcagcgtctgct Rev: acaaagacatcctctaatgggagaa	***6***
β*3GALT5*		10317		Fwd: tgtctccaagagcgtccca Rev: aagaggcatacggagaagcg	***7***
β*3GALT6*		126792	GalT II	Fwd: gacaccgaataccggtccc Rev: gtgcttctccagcatgtcctc	***8***
Core 1 β*3GALT*		56913		Fwd: catccctttgtgccagaacacc Rev: gcaagatcagagcagcaaccag	***9***
β*4GALT*	α1,4	53947	GB3 Synthase, CD77 Synthase	Fwd: tgcatgcgggacttcgt Rev: tgcggatggaacaccactt	***10***
β*4GALT1*	β1,4	2683		Fwd: aacttgacctcggtcccagtgc Rev: ggccgcccatcttcacatttg	***11***
β*4GALT2*		8704		Fwd: cacttcctcgtggccgtcatc Rev: gttgggccgggagcagttg	***12***
β*4GALT3*		8703		Fwd: tagtgggtcctgtgtcggtgtc Rev: aggcacaatgatggctgttcg	***12***
β*4GALT4*		8702		Fwd: gggcttcaacctgactttccac Rev: ttgaatggcacccacgaagtag	***13***
β*4GALT5*		9334		Fwd: tctctcgtcctcgctgctgtac Rev: ccgaagcacctgctcataaacc	***14***

β4GALT6		9331		Fwd: agaagcagcggctggaattt Rev: gcatcgcacggttaaaggtt	***15***
β4GALT7		11285		Fwd: aggtggaccacttcaggttca Rev: agtccgtgctgttgctgct	***16***
B. *N*-Acetylglucosaminyltransferases					
MGAT1	β1,2	4245	GnT-I	Fwd: ggtgattcccatcctggtcat Rev: tcagccgagggccgata	***17***
MGAT2		4247	GnT-II	Fwd: tccggttctgcaggtgttc Rev: caatctctagggtcactacctggaa	***18***
IGNT	β1,3	11041		Fwd: tgattttgaggtcctgaacgaa Rev: cgatataggatcttattgtgctgatttc	***19***
β3GNT-2		10678	β3GnT-1	Fwd: gctggacctcatcgggataa Rev: tgcatagggtgggtagaggc	***20***
β3GNT-3		10331		Fwd: gtgggacttccacgactcctt Rev: gcaccttgtctcctgccact	***20***
β3GNT-4		79369		Fwd: cgtcaccgtctcttcttgacc Rev: tgacttgatggccaggagc	***20***
β3GNT-5		84002	LC3 Synthase	Fwd: gtggtgcccctcccattag Rev: gctccggctgtgtagtcagg	***21***
β3GNT-6		11041	Core 3 GlcNAcT	Fwd: tccctgactgccaagactctg Rev: cgcctggaggaaccactg	***22***
β3GNT-7		93010		Fwd: tcaaaggcgacgatgacgt Rev: tttgttgtctttcctgcgaatg	***23***
MGAT3	β1,4	4248	GnT-III	Fwd: gcacttcttcaagaccctgtccta Rev: agaaaaagctggacaccaggtta	***24***
MGAT4A		11320	GnT-IVA	Fwd: ctgtggaagttttgccttttaagag Rev: tgaaatgggattgagacttgga	***25***

(continued)

Table 1 *(continued)*

Gene	Linkage	Gene ID	Other names	Primers	Ref.
MGAT4B		11282	GnT-IVB	Fwd: cggaggacaagctcttcaaca Rev: agggcctccttgtctgactga	***26***
MGAT5A	β1,6	4249	GnT-V	Fwd: gatgtgcttttctgaatcccaagt Rev: gccgcccgatgaaaact	***27***
MGAT5B		146664	GnT-IX	Fwd: gagaagcggctcatcaaagg Rev: gctctcgtagtacacggtgcc	***28***
C2GNT1		2650	C2GnT-L	Fwd: aggacgttgctgcgaaggagac Rev: cccagcaagctccaagtgtctg	***29***
IGNT1A		2651	GCNT2	Fwd: acagcgttgaaaccgcctc Rev: tcatgcagaacaaagttggca	***30***
C2GNT2		9245	C2GnT-M	Fwd: gctggtcaagtggcagggt Rev: tggccaacaggtgatggtt	***31,32***
C2GNT3		51301	GCNT3	Fwd: gctggtcaagtggcagggt Rev: tggccaacaggtgatggtt	***33***
C. Fucosyltransferase					
FUT1	α1,2	2523	H, HH, HSC	Fwd: gcaggccatggactggtt Rev: cctgggaggtgtcgatgttt	***35***
FUT2		2524	Secretor	Fwd: ctcgctacagctccctcatctt Rev: cgtgggaggtgtcaatgttct	***36***
FUT3	α1,3/4	2525	LE, Les	Fwd: ccagtgggtcctcccga Rev: gccatgtccatagcaggatca	***37***
FUT4	α1,3	2526	ELFT, FCT3A, FUC-TIV	Fwd: gagctacgctgtccacatcacc Rev: cagctggccaagttccgtatg	***38***
FUT5		2527	FUC-TV	Fwd: gtcccgagacgatgccact Rev: ccggtgacaggttccactg	***39***

FUT6		2528		Fwd: atcccactgtgtaccctaatgg Rev: tgccaggcaccatctctgag	***40***
FUT7		2529		Fwd: tccgcgtgcgactgttc Rev: accctcaaggtcctcatagacttg	***41***
FUT9		10690		Fwd: caaatcccatgcagttctgatc Rev: gtggcctagcttgctgaggta	***42***
FUT10	α1,3[b]	84750		Fwd: tgggaggttaggccaatgtg Rev: tccggttgatggtgaagaaac	***48***
FUT8	α1,6	2530		Fwd: tcttcatccccgtcctcca Rev: gagacacccaccacactgca	***44***
D. Sialyltransferases					
ST3GAL I	α2,3	6482	SIAT4A	Fwd: gggcagacagcaaagggaa Rev: ggccgtcacgttagactcaaa	***45***
ST3GAL II		6483	SIAT4B	Fwd: gcctccgactggtttgaca Rev: gtccggtggaagatccatgt	***46***
ST3GAL III[c]		6487	SIAT6	Fwd: gggtcacgaattgacgactatg Rev: gtgatgcgcagtgtcgtttt	***47***
ST3GAL IV		6484	SIAT4C	Fwd: ataagaagcgggtgcgaaaggg Rev: tccgtggctgttgcattggc	***48***
GM3 Synthase		8869	SIAT9, ST3Gal V	Fwd: atcggtgtcattgccgttgt Rev: ttcatagcagccatgcattga	***49***
ST3GAL VI		10402	SIAT10	Fwd: ggccatattcctgagtgctgtc Rev: agctggctttgataaacaaggc	***50***
ST6GAL I	α2,6	6480	SIAT1	Fwd: catccaagcgcaagactgacg Rev: tgtgccctggttgagatgcttc	***51***
ST6GAL II		84620	ST6Gal II	Fwd: agaggtgcacgtgtatgaatatatcc Rev: cgtcgtagtacagctcgtggtagt	***52***

(continued)

Table 1 *(continued)*

Gene	Linkage	Gene ID	Other names	Primers	Ref.
ST6GALNAC I		55808	SIAT7A	Fwd: acatgggccaggagatagaca Rev: atgagagctccgctcaatcg	***53***
ST6GALNAC II		10610	SIAT7B	Fwd: tcaccaagtcatcgcctcc Rev: ttggcactctctgagccgt	***54***
ST6GALNAC III		256435	SIAT7C	Fwd: catgattcgagttgtgtccca Rev: tcatattgcggaaaggtccc	***55***
ST6GALNAC IV		27090	SIAT7D	Fwd: gttcaccatgatcctcgcg Rev: tgacactcatctagccggcc	***56***
ST8SIA I	α2,8	6489	SIAT8A	Fwd: gcgatgcaatctccctcct Rev: ttgccgaattatgctgggat	***57–59***
ST8SIA II		8128	SIAT8B, STX	Fwd: ccatgaacccctcggtcat Rev: gaaggcagggatccacagg	***60***
ST8SIA III		51046	SIAT8C	Fwd: gtacatgttccacgcggga	***61***
ST8SIA IV		7903	SIAT8D, PST	Fwd: gagcaccaggagacgcaact Rev: gagccagcctttcgaatgatt	***62,63***
ST8SIA V		29906	SIAT8E	Fwd: cactgtgaaccccagcatca Rev: accgacgcgttctcgtaca	***64***
ST8SIA VI		338596	SIAT8F	Fwd: tatgtttccagtgtcccagcc	***65***
E. Sulfotransferases					
GST0	6-O	9469	CHST3, C6ST	Fwd: cagcctgaagatgagaagcaaatac Rev: cggtgctgttggcatctg	***67***
GST1		8534	CHST1, KSGal6ST, C6ST	Fwd: ccacgtccagaacacgctcatc Rev: cggcggcttgatgtagttctcc	***68***
GST2		9435	CHST2, GlcNAc6ST	Fwd: gctctggctcgtcgttcttc Rev: agagaggtcgcagcggtaaag	***69***

GST3		10164	CHST4, HEC-GlcNAc6ST	Fwd: gcctggaggcctattaagcacg Rev: ccaagggctcagaatttgcatc	***70***
I-GST-4		23563	CHST5, I-GlcNAc6ST	Fwd: ctctatctttttgtgcgacatgga Rev: ggctcgttgcccagttga	***71***
GST5		56548	CHST7, C6ST-2	Fwd: cacccggacgttttctacttg Rev: aagagcgaacgcagcatgt	***72***
GP3ST1	3-O	64090	Gal3ST-2	Fwd: tggccttcaggctcaactcc Rev: agggtgcggttgaaatgctc	***73***
GP3ST2		89792	Gal3ST-3	Fwd: agctggtaccccaagctgttc Rev: ggcaaagcgaaacaggatgt	***74***
GP3ST3		79690	Gal3ST-4	Fwd: aggcttctgaccccaaatacatc Rev: ccaaaaccttagctgacccaaac	***75***
F. Miscellanea					
β*3GALNACT-1*	β1,3	8706	β3GalT3 GB4 Synthase	Fwd: gagccgatttacagacaagactttc Rev: gcctgcctggctttcaca	***6***
β*3GALNACT-2*	β1,3	148789	MGC39558	Fwd: tggatgggcctaatttttggt Rev: tccaactcctgccactttcc	***76***
α*3GALNACT-1*	α1,3	26301	GB5 Synthase	Fwd: gcaggtggccagggtatatg Rev: agcaccttggacggcttgt	***77***
β*4GALNACT-2*	β1,4	124872	Sd[a]/Cad Synthase	Fwd: ctggctgccctagagaagacc Rev: cacctttatgcggcacattg	***78***
β*4GALNACT-3*		283358		Fwd: atctggtggatgggcttcct Rev: agatgaacaaaccggagtccc	***79***
β*4GALNACT-4*		338707		Fwd: tggtcaaggacttcccgatc Rev: aggcgagtgtagtcgttggg	***80***
ER MAN I	α1,2	9695		Fwd: cagggccgacagctactatga Rev: acgtagtcttccagcagctgtgt	***81***

(continued)

Table 1 *(continued)*

Gene	Linkage	Gene ID	Other names	Primers	Ref.
GOLGI MAN IA		4121	Man_9-mannosidase, MAN1A1	Fwd: `ggagatccaaagagacatcctactg` Rev: `tggtcctgggccacctt`	***82***
GOLGI MAN IB		10905	α1,2-Mannosidase IB, MAN1A2	Fwd: `ctaaaaaccccggagtcttcct` Rev: `attttatttctcagacgttcttcctctt`	***83***
GOLGI MAN IC		57134	α1,2-Mannosidase IC, MAN1C1	Fwd: `tcaaggagatgatgcagtttgc` Rev: `ttgttagtggacggagttcgttt`	***84***
GOLGI MAN II	α1,3/6	4124	MAN2A1	Fwd: `tggtaatgccaagccttatgttt` Rev: `taacatggtcaaaaaagcaagtcact`	***85***
GOLGI MAN III		4122	MAN2A2	Fwd: `ttctggaccaataccggaaga` Rev: `gtcatatcggaagtcatctccaaga`	***86***

[a]Three different isoforms of this gene exist (IGnT1, IGnT2, and IGnT3), owing to multiple variable first exons ***(30,34)***.

[b]The activity of this enzyme (and FUT11) has not yet been demonstrated. According to computer modeling they are expected to be α1,3-fucosyltransferases.

[c]There are 19 transcripts variants that result in seven different isoforms ***(66)***.

3.2. Sample Preparation

1. Lyse the cells/tissue in lysis buffer:
   a. From adherent cell cultures: Wash cells twice in cold PBS. Add sufficient lysis buffer to cover the cells. Incubate 3–5 min at 4°C. Scrape with the help of a rubber policeman and carefully collect into microcentrifuge tubes. The amount of lysis buffer necessary to completely lyse the cells depends on the size of the dishes/flasks and the amount of cells. Typically, 500 μL is sufficient for a 10-cm^2 dish of confluent primary human endothelial cells.
   b. From cells in suspension: Wash cells twice in cold PBS. Resuspend the cells in 20–40 μL of PBS. Add sufficient lysis buffer and mix well with a pipet. The amount of lysis buffer necessary to completely lyse the cells depends on the amount of cells. Typically, 400 μL is sufficient for a cell culture of 10^6 human monocytes.
   c. From tissue: Snap-freeze up to 20 mg of the tissue in liquid nitrogen and grind with the help of a precooled mortar until reduced to a fine powder. Collect the tissue into a microcentrifuge tube and add sufficient lysis buffer. Mix well with a pipet. Typically, 200 μL is sufficient for 1 mg of liver tissue.
2. In order to completely lyse the cells/tissue and to shear genomic DNA, it is convenient to pass the sample through a 21-gage needle. In the case of the tissue, it is also convenient to spin down the sample at 11,000*g* for 30 s and continue with the supernatant.
3. Store lysates at –80°C up to 6 mo.

3.3. mRNA Isolation (see Note 4)

1. Prepare the biotin-oligo$(dT)_{20}$ working solution by diluting 1:20 the stock of biotin-oligo$(dT)_{20}$ in nuclease-free water.
2. Add 4 μL of the biotin-oligo$(dT)_{20}$ working solution to 50 μL of the cell/tissue lysate.
3. Incubate for 5 min at 37°C.
4. Pass the mix to a streptavidin-coated tube and close the tube.
5. Incubate for 5 min at 37°C.
6. Discard the mix and wash the tube twice with washing buffer at room temperature.
7. Proceed immediately to cDNA synthesis.

3.4. cDNA Synthesis

1. Prepare the retrotranscriptase mix started in **Subheading 3.3.**: $MgCl_2$ (6 μL of 25 m*M* stock per sample), cDNA synthesis buffer (3 μL of a 10X concentrated stock per sample), deoxynucleotide 5′-triphosphate mixture (2 μL of 10 m*M* stock per sample), ribonuclease (RNase) inhibitor (0.75 μL per sample), avian myeloblastosis virus reverse transcriptase (0.75 μL per sample), random hexanucleotides (0.7 μL per sample), and 16.8 μL of nuclease-free water per sample. Mix well.
2. Add 30 μL of the retrotranscriptase mix to each tube.
3. Incubate for 10 min at room temperature.

4. Incubate for 75 min at 42°C.
5. Incubate for 5 min at 99°C.
6. Transfer the contents of the tube (while still at 99°C) to a microcentrifuge tube. Add 40 μL of nuclease-free water to the streptavidin-coated tube, mix, and transfer to the microcentrifuge tube together with the rest of the cDNA. Mix well.
7. Incubate for 5 min on ice. The cDNA samples can be stored at –20°C for up to 2 mo.
8. Proceed to real-time PCR.

3.5. Real-Time PCR (see *Note 3*)

1. The real-time PCR mix contains (per sample, and based on a final volume reaction of 8 μL): 4 μL of 2X concentrated SYBR® Green I master mix (including a reaction buffer, deoxynucleotide 5′-triphosphates, SYBR Green I, and the Taq polymerase), 0.2 μL of the forward primer working solution (*see* **Tables 1** and **2**), 0.2 μL of the reverse primer working solution (*see* **Tables 1** and **2**), and 1.6 μL of nuclease-free water. The reaction should be scaled up according to the number of samples used.
2. Pipet 6 μL of the real-time PCR mix per reaction well in a 96-well optical plate. There should be a nontemplate control reaction well for every set of primers used.
3. Add 2 μL of nuclease-free water to each nontemplate control reaction well.
4. Add 2 μL of the appropriate cDNA sample to each reaction well.
5. Seal the plate with an adhesive optical cover sheet. Make sure that the sides are well-sealed in order to avoid evaporation during the real-time PCR.
6. Centrifuge the plate for 1 min at 250*g* to mix the cDNA with the primers and SYBR Green Master mixture.
7. Place the plate in the real-time PCR instrument and set up the run according to the manufacturer's instructions. A typical real-time PCR run consists of 2 min at 50°C, followed by 10 min at 95°C, then 40 cycles of 15 s at 95°C and 1 min at 60°C. The fluorescence monitoring occurs at the end of each cycle. At the end of every run, a dissociation curve analysis should be performed. Dissociation curve analysis is performed by increasing the temperature of the block from 60 to 95°C at a rate of 1.75°C/min while continuously monitoring fluorescence.

3.6. Analysis

1. *The constant (Ct) and normalized amount of target values.* The Ct value is defined as the number of PCR cycles where the fluorescence signal exceeds the detection threshold value, which is fixed at 10 times the standard deviation of the fluorescence during the first 15 cycles and typically corresponds to 0.2 relative fluorescence units. This threshold should be set constant throughout the study and corresponds to the log linear range of the amplification curve. Another parameter is obtained from the Ct values, the ΔCt, which corresponds to the difference between the Ct of the target gene and the Ct of the endogenous reference gene (ERG) ($\Delta Ct = Ct_{Target} - Ct_{Reference}$). From this value it is possible

Table 2
Primers for Potential Endogenous Human Reference Genes

Gene	Gene ID	Primers	Ref.
GAPDH	2597	Fwd: `aggtcatccctgagctgaacgg`	***87***
		Rev: `cgcctgcttcaccaccttcttg`	
ELF-1α	1915	Fwd: `tcgggcaagtccaccact`	***88***
		Rev: `ccaagacccaggcatacttga`	
MLN51	22794	Fwd: `acctccagtcccagaaacca`	***89***
		Rev: `tccaattctgttctgctatttagttgt`	
UBCH5B	7322	Fwd: `gtactcttgtccatctgttctctgttg`	***90***
		Rev: `gtccattcccgagctattctgtt`	
β-*ACTIN*	60	Fwd: `gctcctcctgagcgcaag`	***91***
		Rev: `catctgctggaaggtggaca`	
CYES	7525	Fwd: `caggtatggtgaaccgtgaagtac`	***92***
		Rev: `tcaattcatggagggattctgg`	
HPRT	3251	Fwd: `tcgagcaagacgttcagtcctg`	***93***
		Rev: `tcgagcaagacgttcagtcctg`	
HUMPSY	5694	Fwd: `acctgatggcgggaatcat`	***94***
		Rev: `atcataccccccataggcact`	
L32	6161	Fwd: `caacattggttatggaagcaaca`	***95***
		Rev: `tgacgttgtggaccaggaact`	
PPMM	5372	Fwd: `aagcgtggaaccttcatcga`	***96***
		Rev: `tcccggatcttctctttcttgtc`	

to obtain the normalized amount of target values (Nt), which corresponds to $2^{-\Delta Ct}$ *(3)*. The normalized amount of target reflects the relative amount of target transcripts with respect to the expression of the ERG. *GAPDH* is a widely-used ERG. However, the expression levels of *GAPDH* might be regulated in some cell systems. Therefore, prior to the gene expression analysis of glycosylation-related genes, a suitable endogenous reference gene should be determined in each particular experimental model *(2)*.

2. *Selecting an ERG.* The procedure to select a suitable ERG has been described in detail *(2)*. The statistical parameters employed are mean, median, standard deviation (SD), and coefficient of variation of the mean (CV), calculated as CV = 100 SD/ mean, which allows the comparison of the variability in the measurements. Basically, a number of potentially suitable ERGs are selected and primers for these genes are designed (*see* **Table 2**). To determine a suitable ERG for a particular experiment, cDNA is synthesized from RNA samples obtained under every condition of interest for the study in which the ERG is going to be used. The samples are assayed for the 10 unrelated ERGs (*see* **Table 2**). Subsequently, the first of the genes is set as potential ERG and the Nt of all the other genes is calculated. The CV is calculated on this set of data. The process is repeated for all of the ERGs.

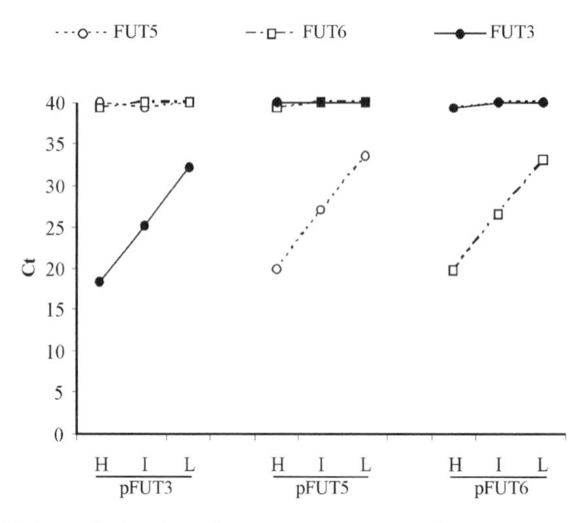

Fig. 1. Specificity of the FUT3, FUT5, and FUT6 primers. The concentration of plasmid DNA used is indicated by the letters H (10^6 copies/µL), I (10^4 copies/µL), and L (10^2 copies/µL). The combination of a dotted line with open circles denotes the reactions. The continuous line (solid circles) represents the Cts obtained with the FUT3 primers, the dotted line (open circles) represents the Cts obtained with the FUT5 primers, and the discontinuous line (open squares) shows the Cts obtained with the FUT6 primers.

Finally, a matrix of CVs that defines high- and low-variability gene clusters is obtained *(2)*. The final ERG is chosen based on the median of the CV of the different potential ERGs.

3. *Dissociation curve analysis.* The dissociation curve analysis should always be performed after the PCR. This process determines the existence of nonspecific amplification (i.e., primer-dimers, miss-priming, and so forth). In the dissociation curve analysis, the plot of the first derivative of the decrease in fluorescence with respect to temperature should show one single peak at the melting temperature predicted for the amplicon produced by each primer set.

4. Notes

1. Primer specificity should be tested by searching the complementary sequence in the whole human genome. Only one sequence corresponding to the gene of interest should be obtained. The dissociation curve analysis should provide a single peak, and a single band of the expected size should be obtained after resolving the PCR products in a DNA agarose gel. An ultimate control for the specificity of the reaction would be to sequence the PCR products. The primers described in this chapter are all tested in this manner.
2. Some glycosylation-related genes share a large sequence similarity, especially in the sequences coding for the catalytic domains. The primers have been designed

Table 3
Fucosyltransferases Gene Expression Profile in Eosinophils According to the Expression of Le^X

Fucosyltransferases		Le^X-negative eosinophils		Le^X-positive eosinophils	
		Av	SD	Av	SD
α1,2	FUT1	10.681	1.047	0.060	0.031
	FUT2	BDL		0.176	0.101
α1/3,4	FUT3	BDL		BDL	
	FUT4	BDL		4.533	2.508
	FUT5	BDL		BDL	
	FUT6	BDL		BDL	
	FUT7	0.004	0.002	0.003	0.001
	FUT9	5.770	3.344	0.018	0.006
α1,6	FUT8	0.023	0.013	0.002	0.001

Results are expressed as the relative abundance (using *GAPDH* as endogenous reference) and constitute the average of triplicate measurements. The experiment was performed twice with similar results. BDL, Below Detection Limit; Av, average; SD, standard deviation). Lewis X (Le^X)-positive and -negative populations were sorted (MoFlo High Performance Cell Sorter, Dako, Glostrup, Denmark) based on the binding of the antibody FH6. The results indicate that the expression of the Le^X antigen in eosinophils is dependent on the expression of FUT4.

in unique sequences, preferably within the coding region. In the case of FUT3, FUT5, and FUT6, the specificity of the primers has been tested using plasmids containing these cDNAs (*see* **Fig. 1**).

3. The method described has been optimized for a certain combination of kits/instruments (as detailed in **Heading 2.**). When adapting this method to other kits/instruments, consider the reoptimization of the real-time PCR reactions.
4. It is very important to use RNase-free glass and plastic material throughout. Working surfaces should also be decontaminated from RNases, and sterile gloves should be used at all times.
5. The authors have used the real-time PCR approach described in this chapter to investigate expression levels of glycosylation-dependent genes in several studies *(2,4,5)*. The collection of primers can easily be extended to suit other research purposes. An important aspect that should be carefully controlled is the specificity of the oligonucleotide primers. For example, the genes encoding FUT3, FUT5, and FUT6 are very similar. **Figure 1** shows that with the primers described here, the expression levels of these enzymes can be determined unambiguously. **Table 3** shows a typical result that can be obtained. As a first approach to understanding the regulation of LewisX (Le^X) synthesis in different subsets of human eosinophils, eosinophils should be FACS-sorted according to the expression of the Le^X glycan antigen using anti-CD15 (Le^X) monoclonal antibody. The expression levels of the

different fucosyltransferase genes in the sorted cell populations were determined, and are shown in **Table 3**. The results indicate that the expression of the Le^X antigen in eosinophils is dependent on expression of FUT4.

Acknowledgment

We thank Dr. J. B. Lowe (Department of Pathology, Howard Hughes Medical Institute, University of Michigan, Ann Arbor, MI) for the kind gift of the plasmids containing the cDNAs for FUT3, FUT4, and FUT5.

References

1. van den Eijnden, D. H. (2000) On the origin of oligosaccharide species. Glycosyltransferases in action. In *Carbohydrates in Chemistry and Biology* (Ernst, B., Hart, G. W., and Sinaÿ, P., eds.), Wiley-VCH, Weinheim, pp. 589–624.
2. Garcia-Vallejo, J. J., Van het Hof, B., Robben, J., et al. (2004) Approach for defining endogenous reference genes in gene expression experiments. *Anal. Biochem.* **329,** 293–299.
3. Livak, K. J. and Schmittgen, T. D. (2001) Analysis of relative gene expression data using real-time quantitative PCR and the 2-CT method. *Methods* **25,** 402–408.
4. Garcia-Vallejo, J. J., van Dijk, W., Van het Hof, B., et al. (2005) Activation of human endothelial cells by tumor necrosis factor-a results in profound changes in the expression of glycosylation-related genes. *J. Cell. Physiol.* **206,** 203–210.
5. Garcia-Vallejo, J. J., van Dijk, W., Van Die, I., and Gringhuis, S. I. (2005) TNFα upregulates the expression of β1,4-Galactosyltransferase I in primary human endothelial cells by mRNA stabilization. *J. Biol. Chem.* **280,** 12,676–12,682.
6. Amado, M., Almeida, R., Carneiro, F., et al. (1998) A family of human β3-galactosyltransferases. Characterization of four members of a UDP-galactose:β-N-acetylglucosamine/β-N-acetyl-galactosamine β1,3-galactosyltransferase family. *J. Biol. Chem.* **273,** 12,770–12,778.
7. Zhou, D., Berger, E. G., and Hennet, T. (1999) Molecular cloning of a human UDP-galactose:GlcNAcβ1,3GalNAc β1, 3 galactosyltransferase gene encoding an O-linked core3-elongation enzyme. *Eur. J. Biochem.* **263,** 571–576.
8. Bai, X., Zhou, D., Brown, J. R., Crawford, B. E., Hennet, T., and Esko, J. D. (2001) Biosynthesis of the linkage region of glycosaminoglycans: cloning and activity of galactosyltransferase II, the sixth member of the β1,3-galactosyltransferase family (β3GalT6). *J. Biol. Chem.* **276,** 48,189–48,195.
9. Ju, T., Brewer, K., D'Souza, A., Cummings, R. D., and Canfield, W. M. (2002) Cloning and expression of human core 1 β1,3-galactosyltransferase. *J. Biol. Chem.* **277,** 178–186.
10. Kojima, Y., Fukumoto, S., Furukawa, K., et al. (2000) Molecular cloning of globotriaosylceramide/CD77 synthase, a glycosyltransferase that initiates the synthesis of globo series glycosphingolipids. *J. Biol. Chem.* **275,** 15,152–15,156.
11. Masri, K. A., Appert, H. E., and Fukuda, M. N. (1988) Identification of the full-length coding sequence for human galactosyltransferase (β-N-acetylglucosaminide: β1,4-galactosyltransferase). *Biochem. Biophys. Res. Commun.* **157,** 657–663.

12. Almeida, R., Amado, M., David, L., et al. (1997) A family of human β4-galactosyltransferases. Cloning and expression of two novel UDP-galactose:β-N-acetylglucosamine β1, 4-galactosyltransferases, β4Gal-T2 and β4Gal-T3. *J. Biol. Chem.* **272,** 31,979–31,991.
13. Schwientek, T., Almeida, R., Levery, S. B., Holmes, E. H., Bennett, E., and Clausen, H. (1998) Cloning of a novel member of the UDP-galactose:β-N-acetylglucosamine β1,4-galactosyltransferase family, β4Gal-T4, involved in glycosphingolipid biosynthesis. *J. Biol. Chem.* **273,** 29,331–29,340.
14. Sato, T., Furukawa, K., Bakker, H., Van den Eijnden, D. H., and Van Die, I. (1998) Molecular cloning of a human cDNA encoding β1,4-galactosyltransferase with 37% identity to mammalian UDP-Gal:GlcNAc β1,4-galactosyltransferase. *Proc. Natl. Acad. Sci. USA* **95,** 472–477.
15. Takizawa, M., Nomura, T., Wakisaka, E., et al. (1999) cDNA cloning and expression of human lactosylceramide synthase. *Biochim. Biophys. Acta* **1438,** 301–304.
16. Okajima, T., Yoshida, K., Kondo, T., and Furukawa, K. (1999) Human homolog of *Caenorhabditis elegans* sqv-3 gene is galactosyltransferase I involved in the biosynthesis of the glycosaminoglycan-protein linkage region of proteoglycans. *J. Biol. Chem.* **274,** 22,915–22,918.
17. Kumar, R., Yang, J., Larsen, R. D., and Stanley, P. (1990) Cloning and expression of N-acetylglucosaminyltransferase I, the medial Golgi transferase that initiates complex N-linked carbohydrate formation. *Proc. Natl. Acad. Sci. USA* **87,** 9948–9952.
18. Tan, J., D'Agostaro, A. F., Bendiak, B., et al. (1995) The human UDP-N-acetylglucosamine: α-6-D-mannoside-β-1,2- N-acetylglucosaminyltransferase II gene (MGAT2). Cloning of genomic DNA, localization to chromosome 14q21, expression in insect cells and purification of the recombinant protein. *Eur. J. Biochem.* **231,** 317–328.
19. Sasaki, K., Kurata-Miura, K., Ujita, M., et al. (1997) Expression cloning of cDNA encoding a human β1,3-N-acetylglucosaminyltransferase that is essential for poly-N-acetyllactosamine synthesis. *Proc. Natl. Acad. Sci. USA* **94,** 14,294–14,299.
20. Shiraishi, N., Natsume, A., Togayachi, A., et al. (2001) Identification and characterization of three novel β1,3-N-acetylglucosaminyltransferases structurally related to the β1,3-galactosyltransferase family. *J. Biol. Chem.* **276,** 3498–3507.
21. Henion, T. R., Zhou, D., Wolfer, D. P., Jungalwala, F. B., and Hennet, T. (2001) Cloning of a mouse β1,3 N-acetylglucosaminyltransferase GlcNAc(β1,3) Gal(β1,4)Glc-ceramide synthase gene encoding the key regulator of lacto-series glycolipid biosynthesis. *J. Biol. Chem.* **276,** 30,261–30,269.
22. Iwai, T., Inaba, N., Naundorf, A., et al. (2002) Molecular cloning and characterization of a novel UDP-GlcNAc:GalNAc-peptide β1,3-N-acetylglucosaminyltransferase (β3Gn-T6), an enzyme synthesizing the core 3 structure of O-glycans. *J. Biol. Chem.* **277,** 12,802–12,809.
23. Kataoka, K. and Huh, N. H. (2002) A novel β1,3-N-acetylglucosaminyltransferase involved in invasion of cancer cells as assayed in vitro. *Biochem. Biophys. Res. Commun.* **294,** 843–848.

24. Ihara, Y., Nishikawa, A., Tohma, T., Soejima, H., Niikawa, N., and Taniguchi, N. (1993) cDNA cloning, expression, and chromosomal localization of human N-acetylglucosaminyltransferase III (GnT-III). *J. Biochem. (Tokyo)* **113,** 692–698.
25. Yoshida, A., Minowa, M. T., Takamatsu, S., et al. (1999) Tissue specific expression and chromosomal mapping of a human UDP-N-acetylglucosamine:α1,3-d-mannoside β1, 4-N-acetylglucosaminyltransferase. *Glycobiology* **9,** 303–310.
26. Yoshida, A., Minowa, M. T., Takamatsu, S., Hara, T., Ikenaga, H., and Takeuchi, M. (1998) A novel second isoenzyme of the human UDP-N-acetylglucosamine:α1,3-D-mannoside β1,4-N-acetylglucosaminyltransferase family: cDNA cloning, expression, and chromosomal assignment. *Glycoconj. J.* **15,** 1115–1123.
27. Saito, H., Nishikawa, A., Gu, J., et al. (1994) cDNA cloning and chromosomal mapping of human N-acetylglucosaminyltransferase V. *Biochem. Biophys. Res. Commun.* **198,** 318–327.
28. Inamori, K., Endo, T., Ide, Y., et al. (2003) Molecular cloning and characterization of human GnT-IX, a novel β1,6-N-acetylglucosaminyltransferase that is specifically expressed in the brain. *J. Biol. Chem.* **278,** 43,102–43,109.
29. Bierhuizen, M. F. and Fukuda, M. (1992) Expression cloning of a cDNA encoding UDP-GlcNAc:Gal β1-3-GalNAc-R (GlcNAc to GalNAc) β1-6GlcNAc transferase by gene transfer into CHO cells expressing polyoma large tumor antigen. *Proc. Natl. Acad. Sci. USA* **89,** 9326–9330.
30. Inaba, N., Hiruma, T., Togayachi, A., et al. (2003) A novel I-branching β1,6-N-acetylglucosaminyltransferase involved in human blood group I antigen expression. *Blood* **101,** 2870–2876.
31. Yeh, J. C., Ong, E., and Fukuda, M. (1999) Molecular cloning and expression of a novel β1,6-N-acetylglucosaminyltransferase that forms core 2, core 4, and I branches. *J. Biol. Chem.* **274,** 3215–3221.
32. Schwientek, T., Nomoto, M., Levery, S. B., et al. (1999) Control of *O*-glycan branch formation. Molecular cloning of human cDNA encoding a novel β1,6-N-acetylglucosaminyltransferase forming core 2 and core 4. *J. Biol. Chem.* **274,** 4504–4512.
33. Schwientek, T., Yeh, J. C., Levery, S. B., et al. (2000) Control of *O*-glycan branch formation. Molecular cloning and characterization of a novel thymus-associated core 2 β1, 6-N-acetylglucosaminyltransferase. *J. Biol. Chem.* **275,** 11,106–11,113.
34. Zhang, T., Haws, P., and Wu, Q. (2004) Multiple variable first exons: A mechanism for cell- and tissue-specific gene regulation. *Genome Res.* **14,** 79–89.
35. Larsen, R. D., Ernst, L. K., Nair, R. P., and Lowe, J. B. (1990) Molecular cloning, sequence, and expression of a human GDP-L-fucose:β-D-galactoside 2-α-L-fucosyltransferase cDNA that can form the H blood group antigen. *Proc. Natl. Acad. Sci. USA* **87,** 6674–6678.
36. Kelly, R. J., Rouquier, S., Giorgi, D., Lennon, G. G., and Lowe, J. B. (1995) Sequence and expression of a candidate for the human Secretor blood group α1,2-fucosyltransferase gene (FUT2). Homozygosity for an enzyme-inactivating nonsense mutation commonly correlates with the non-secretor phenotype. *J. Biol. Chem.* **270,** 4640–4649.

37. Kukowska-Latallo, J. F., Larsen, R. D., Nair, R. P., and Lowe, J. B. (1990) A cloned human cDNA determines expression of a mouse stage-specific embryonic antigen and the Lewis blood group α (1,3/1,4)fucosyltransferase. *Genes Dev.* **4,** 1288–1303.
38. Goelz, S. E., Hession, C., Goff, D., et al. (1990) ELFT: a gene that directs the expression of an ELAM-1 ligand. *Cell* **63,** 1349–1356.
39. Weston, B. W., Nair, R. P., Larsen, R. D., and Lowe, J. B. (1992) Isolation of a novel human α (1,3)fucosyltransferase gene and molecular comparison to the human Lewis blood group α (1,3/1,4)fucosyltransferase gene. Syntenic, homologous, nonallelic genes encoding enzymes with distinct acceptor substrate specificities. *J. Biol. Chem.* **267,** 4152–4160.
40. Weston, B. W., Smith, P. L., Kelly, R. J., and Lowe, J. B. (1992) Molecular cloning of a fourth member of a human α1,3-fucosyltransferase gene family. Multiple homologous sequences that determine expression of the Lewis x, sialyl Lewis x, and difucosyl sialyl Lewis x epitopes. *J. Biol. Chem.* **267,** 24,575–24,584.
41. Sasaki, K., Kurata, K., Funayama, K., et al. (1994) Expression cloning of a novel α1,3-fucosyltransferase that is involved in biosynthesis of the sialyl Lewis x carbohydrate determinants in leukocytes. *J. Biol. Chem.* **269,** 14,730–14,737.
42. Kaneko, M., Kudo, T., Iwasaki, H., et al. (1999) Assignment of the human α1,3-fucosyltransferase IX gene (FUT9) to chromosome band 6q16 by in situ hybridization. *Cytogenet. Cell Genet.* **86,** 329–330.
43. Baboval, T. and Smith, F. I. (2002) Comparison of human and mouse Fuc-TX and Fuc-TXI genes, and expression studies in the mouse. *Mamm. Genome* **13,** 538–541.
44. Yanagidani, S., Uozumi, N., Ihara, Y., Miyoshi, E., Yamaguchi, N., and Taniguchi, N. (1997) Purification and cDNA cloning of GDP-L-Fuc:N-acetyl-β-D-glucosaminide:α1-6 fucosyltransferase (α1-6 FucT) from human gastric cancer MKN45 cells. *J. Biochem. (Tokyo)* **121,** 626–632.
45. Shang, J., Qiu, R., Wang, J., et al. (1999) Molecular cloning and expression of Galβ1,3GalNAc α2, 3-sialyltransferase from human fetal liver. *Eur. J. Biochem.* **265,** 580–588.
46. Kim, Y. J., Kim, K. S., Kim, S. H., et al. (1996) Molecular cloning and expression of human Gal β1,3GalNAc α2,3-sialytransferase (hST3Gal II). *Biochem. Biophys. Res. Commun.* **228,** 324–327.
47. Kitagawa, H. and Paulson, J. C. (1993) Cloning and expression of human Gal β1,3(4)GlcNAc α2,3-sialyltransferase. *Biochem. Biophys. Res. Commun.* **194,** 375–382.
48. Sasaki, K., Watanabe, E., Kawashima, K., et al. (1993) Expression cloning of a novel Gal β (1-3/1-4) GlcNAc α2,3-sialyltransferase using lectin resistance selection. *J. Biol. Chem.* **268,** 22,782–22,787.
49. Ishii, A., Ohta, M., Watanabe, Y., et al. (1998) Expression cloning and functional characterization of human cDNA for ganglioside GM3 synthase. *J. Biol. Chem.* **273,** 31,652–31,655.
50. Okajima, T., Fukumoto, S., Miyazaki, H., et al. (1999) Molecular cloning of a novel α2,3-sialyltransferase (ST3Gal VI) that sialylates type II lactosamine structures on glycoproteins and glycolipids. *J. Biol. Chem.* **274,** 11,479–11,486.

51. Grundmann, U., Nerlich, C., Rein, T., and Zettlmeissl, G. (1990) Complete cDNA sequence encoding human β-galactoside α-2,6-sialyltransferase. *Nucleic Acids Res.* **18,** 667.
52. Takashima, S., Tsuji, S., and Tsujimoto, M. (2002) Characterization of the second type of human β-galactoside α2,6-sialyltransferase (ST6Gal II), which sialylates Galβ1,4GlcNAc structures on oligosaccharides preferentially. Genomic analysis of human sialyltransferase genes. *J. Biol. Chem.* **277,** 45,719–45,728.
53. Ikehara, Y., Kojima, N., Kurosawa, N., et al. (1999) Cloning and expression of a human gene encoding an N-acetylgalactosamine-α2,6-sialyltransferase (ST6GalNAc I): a candidate for synthesis of cancer-associated sialyl-Tn antigens. *Glycobiology* **9,** 1213–1224.
54. Samyn-Petit, B., Krzewinski-Recchi, M. A., Steelant, W. F., Delannoy, P., and Harduin-Lepers, A. (2000) Molecular cloning and functional expression of human ST6GalNAc II. Molecular expression in various human cultured cells. *Biochim. Biophys. Acta* **1474,** 201–211.
55. Sjoberg, E. R., Kitagawa, H., Glushka, J., van Halbeek, H., and Paulson, J. C. (1996) Molecular cloning of a developmentally regulated N-acetylgalactosamine α2,6-sialyltransferase specific for sialylated glycoconjugates. *J. Biol. Chem.* **271,** 7450–7459.
56. Harduin-Lepers, A., Stokes, D. C., Steelant, W. F., et al. (2000) Cloning, expression and gene organization of a human Neu5Ac α2-3Gal β1-3GalNAc α2,6-sialyltransferase: hST6GalNAcIV. *Biochem. J.* **352 Pt 1,** 37–48.
57. Haraguchi, M., Yamashiro, S., Yamamoto, A., et al. (1994) Isolation of GD3 synthase gene by expression cloning of GM3 α-2,8-sialyltransferase cDNA using anti-GD2 monoclonal antibody. *Proc. Natl. Acad. Sci. USA* **91,** 10,455–10,459.
58. Nara, K., Watanabe, Y., Maruyama, K., Kasahara, K., Nagai, Y., and Sanai, Y. (1994) Expression cloning of a CMP-NeuAc:NeuAc α2-3Gal β1-4Glc β1-1′Cer α2,8-sialyltransferase (GD3 synthase) from human melanoma cells. *Proc. Natl. Acad. Sci. USA* **91,** 7952–7956.
59. Sasaki, K., Kurata, K., Kojima, N., et al. (1994) Expression cloning of a GM3-specific α-2,8-sialyltransferase (GD3 synthase). *J. Biol. Chem.* **269,** 15,950–15,956.
60. Kitagawa, H. and Paulson, J. C. (1994) Differential expression of five sialyltransferase genes in human tissues. *J. Biol. Chem.* **269,** 17,872–17,878.
61. Lee, Y. C., Kim, Y. J., Lee, K. Y., et al. (1998) Cloning and expression of cDNA for a human Sia α2,3Gal β1,4GlcNAc α2,8-sialyltransferase (hST8Sia III). *Arch. Biochem. Biophys.* **360,** 41–46.
62. Eckhardt, M., Muhlenhoff, M., Bethe, A., Koopman, J., Frosch, M., and Gerardy-Schahn, R. (1995) Molecular characterization of eukaryotic polysialyltransferase-1. *Nature* **373,** 715–718.
63. Nakayama, J., Fukuda, M. N., Fredette, B., Ranscht, B., and Fukuda, M. (1995) Expression cloning of a human polysialyltransferase that forms the polysialylated neural cell adhesion molecule present in embryonic brain. *Proc. Natl. Acad. Sci. USA* **92,** 7031–7035.

64. Kim, Y. J., Kim, K. S., Do, S., Kim, C. H., Kim, S. K., and Lee, Y. C. (1997) Molecular cloning and expression of human α2,8-sialyltransferase (hST8Sia V). *Biochem. Biophys. Res. Commun.* **235,** 327–330.
65. Takashima, S., Ishida, H. K., Inazu, T., et al. (2002) Molecular cloning and expression of a sixth type of α2,8-sialyltransferase (ST8Sia VI) that sialylates *O*-glycans. *J. Biol. Chem.* **277,** 24,030–24,038.
66. Grahn, A., Barkhordar, G. S., and Larson, G. (2002) Cloning and sequencing of nineteen transcript isoforms of the human α2,3-sialyltransferase gene, ST3Gal III; its genomic organisation and expression in human tissues. *Glycoconj. J.* **19,** 197–210.
67. Fukuta, M., Kobayashi, Y., Uchimura, K., Kimata, K., and Habuchi, O. (1998) Molecular cloning and expression of human chondroitin 6-sulfotransferase. *Biochim. Biophys. Acta* **1399,** 57–61.
68. Fukuta, M., Inazawa, J., Torii, T., Tsuzuki, K., Shimada, E., and Habuchi, O. (1997) Molecular cloning and characterization of human keratan sulfate Gal-6-sulfotransferase. *J. Biol. Chem.* **272,** 32,321–32,328.
69. Uchimura, K., Muramatsu, H., Kaname, T., et al. (1998) Human N-acetylglucosamine-6-O-sulfotransferase involved in the biosynthesis of 6-sulfo sialyl Lewis X: molecular cloning, chromosomal mapping, and expression in various organs and tumor cells. *J. Biochem. (Tokyo)* **124,** 670–678.
70. Bistrup, A., Bhakta, S., Lee, J. K., et al. (1999) Sulfotransferases of two specificities function in the reconstitution of high endothelial cell ligands for L-selectin. *J. Cell Biol.* **145,** 899–910.
71. Lee, J. K., Bhakta, S., Rosen, S. D., and Hemmerich, S. (1999) Cloning and characterization of a mammalian N-acetylglucosamine-6-sulfotransferase that is highly restricted to intestinal tissue. *Biochem. Biophys. Res. Commun.* **263,** 543–549.
72. Kitagawa, H., Fujita, M., Ito, N., and Sugahara, K. (2000) Molecular cloning and expression of a novel chondroitin 6-*O*-sulfotransferase. *J. Biol. Chem.* **275,** 21,075–21,080.
73. Honke, K., Tsuda, M., Koyota, S., et al. (2001) Molecular cloning and characterization of a human β-Gal-3′-sulfotransferase that acts on both type 1 and type 2 (Gal β1-3/1-4GlcNAc-R) oligosaccharides. *J. Biol. Chem.* **276,** 267–274.
74. Suzuki, A., Hiraoka, N., Suzuki, M., et al. (2001) Molecular cloning and expression of a novel human β-Gal-3-*O*-sulfotransferase that acts preferentially on N-acetyllactosamine in N- and O-glycans. *J. Biol. Chem.* **276,** 24,388–24,395.
75. Seko, A., Hara-Kuge, S., and Yamashita, K. (2001) Molecular cloning and characterization of a novel human galactose 3-*O*-sulfotransferase that transfers sulfate to gal β1,3GalNAc residue in *O*-glycans. *J. Biol. Chem.* **276,** 25,697–25,704.
76. Hiruma, T., Togayachi, A., Okamura, K., et al. (2004) A novel human β1,3-N-acetylgalactosaminyltransferase that synthesizes a unique carbohydrate structure, GalNAcβ1-3GlcNAc. *J. Biol. Chem.* **279,** 14,087–14,095.
77. Haslam, D. B. and Baenziger, J. U. (1996) Expression cloning of Forssman glycolipid synthetase: a novel member of the histo-blood group ABO gene family. *Proc. Natl. Acad. Sci. USA* **93,** 10,697–10,702.

78. Montiel, M. D., Krzewinski-Recchi, M. A., Delannoy, P., and Harduin-Lepers, A. (2003) Molecular cloning, gene organization and expression of the human UDP-GalNAc:Neu5Acα2-3Galβ-R β1,4-N-acetylgalactosaminyltransferase responsible for the biosynthesis of the blood group Sda/Cad antigen: evidence for an unusual extended cytoplasmic domain. *Biochem. J.* **373,** 369–379.
79. Sato, T., Gotoh, M., Kiyohara, K., et al. (2003) Molecular cloning and characterization of a novel human β1,4-N-acetylgalactosaminyltransferase, β4GalNAc-T3, responsible for the synthesis of N,N′-diacetyllactosediamine, GalNAc β1-4GlcNAc. *J. Biol. Chem.* **278,** 47,534–47,544.
80. Gotoh, M., Sato, T., Kiyohara, K., et al. (2004) Molecular cloning and characterization of β1,4-N-acetylgalactosaminyltransferases IV synthesizing N,N′-diacetyllactosediamine. *FEBS Lett.* **562,** 134–140.
81. Hosokawa, N., Wada, I., Hasegawa, K., et al. (2001) A novel ER α-mannosidase-like protein accelerates ER-associated degradation. *EMBO Rep.* **2,** 415–422.
82. Bause, E., Bieberich, E., Rolfs, A., Volker, C., and Schmidt, B. (1993) Molecular cloning and primary structure of Man9-mannosidase from human kidney. *Eur. J. Biochem.* **217,** 535–540.
83. Tremblay, L. O., Campbell, D. N., and Herscovics, A. (1998) Molecular cloning, chromosomal mapping and tissue-specific expression of a novel human α1,2-mannosidase gene involved in N-glycan maturation. *Glycobiology* **8,** 585–595.
84. Tremblay, L. O. and Herscovics, A. (2000) Characterization of a cDNA encoding a novel human Golgi α1, 2-mannosidase (IC) involved in N-glycan biosynthesis. *J. Biol. Chem.* **275,** 31,655–31,660.
85. Moremen, K. W. and Robbins, P. W. (1991) Isolation, characterization, and expression of cDNAs encoding murine α-mannosidase II, a Golgi enzyme that controls conversion of high mannose to complex N-glycans. *J. Cell Biol.* **115,** 1521–1534.
86. Misago, M., Liao, Y. F., Kudo, S., et al. (1995) Molecular cloning and expression of cDNAs encoding human α-mannosidase II and a previously unrecognized α-mannosidase IIx isozyme. *Proc. Natl. Acad. Sci. USA* **92,** 11,766–11,770.
87. Arcari, P., Martinelli, R., and Salvatore, F. (1984) The complete sequence of a full length cDNA for human liver glyceraldehyde-3-phosphate dehydrogenase: evidence for multiple mRNA species. *Nucleic Acids Res.* **12,** 9179–9189.
88. Uetsuki, T., Naito, A., Nagata, S., and Kaziro, Y. (1989) Isolation and characterization of the human chromosomal gene for polypeptide chain elongation factor-1 α. *J. Biol. Chem.* **264,** 5791–5798.
89. Tomasetto, C., Regnier, C., Moog-Lutz, C., et al. (1995) Identification of four novel human genes amplified and overexpressed in breast carcinoma and localized to the q11-q21.3 region of chromosome 17. *Genomics* **28,** 367–376.
90. Jensen, J. P., Bates, P. W., Yang, M., Vierstra, R. D., and Weissman, A. M. (1995) Identification of a family of closely related human ubiquitin conjugating enzymes. *J. Biol. Chem.* **270,** 30,408–30,414.
91. Nakajima-Iijima, S., Hamada, H., Reddy, P., and Kakunaga, T. (1985) Molecular structure of the human cytoplasmic β-actin gene: interspecies homology of sequences in the introns. *Proc. Natl. Acad. Sci. USA* **82,** 6133–6137.

92. Sukegawa, J., Semba, K., Yamanashi, Y., et al. (1987) Characterization of cDNA clones for the human c-yes gene. *Mol. Cell Biol.* **7,** 41–47.
93. Jolly, D. J., Okayama, H., Berg, P., Esty, A. C., et al. (1983) Isolation and characterization of a full-length expressible cDNA for human hypoxanthine phosphoribosyl transferase. *Proc. Natl. Acad. Sci. USA* **80,** 477–481.
94. Akiyama, K., Yokota, K., Kagawa, S., et al. (1994) cDNA cloning and interferon gamma down-regulation of proteasomal subunits X and Y. *Science* **265,** 1231–1234.
95. Young, J. A. and Trowsdale, J. (1985) A processed pseudogene in an intron of the HLA-DP β1 chain gene is a member of the ribosomal protein L32 gene family. *Nucleic Acids Res.* **13,** 8883–8891.
96. Hansen, S. H., Frank, S. R., and Casanova, J. E. (1997) Cloning and characterization of human phosphomannomutase, a mammalian homologue of yeast SEC53. *Glycobiology* **7,** 829–834.

14

Analysis of the Glycodynamics of Primary Osteoblasts and Bone Cancer Cells

Inka Brockhausen, Xiaojing Yang, and Mark Harrison

Summary

Bone cells produce a variety of glycoproteins that contribute to bone health, and function in cell adhesion, stabilizing the extracellular matrix, promoting growth and differentiation, and the induction of apoptosis. Some of these processes appear to be disturbed in arthritis. In this chapter, in vitro studies aimed at an understanding of the biological effects of inflammatory stimuli in the bone of arthritis patients are described. The glycodynamics of cells can be studied using primary cultures of osteoblasts or bone cancer cell cultures, to examine the relationship between the biosynthesis of cell-surface glycoproteins and inflammatory stimuli affecting cell growth and cell death. Cell-surface carbohydrates are assessed by lectin staining of cells, and the potential of cells to synthesize glycoproteins is determined by glycosyltransferase assays. These parameters are then related to [^{3}H]thymidine incorporation as a measure of cell proliferation, and to flow cytometry of terminal deoxynucleotidyl transferase dUTP-mediated nick end labeling (TUNEL) and annexin V-stained cells as a measure of apoptosis. These in vitro studies are aimed at an understanding of the role of glycosylation in the bone of arthritis patients, but they can also be applied to other diseases.

Key Words: Bone cells; osteoblasts; osteosarcoma cells; glycoproteins; lectins; glycosyltransferases; apoptosis; cell proliferation; flow cytometry.

1. Introduction

Bone cells produce a variety of proteoglycans and glycoproteins (GPs) that have Ser/Thr-O-linked and Asn-N-linked carbohydrate chains covalently attached to the protein backbone. GPs produced by bone cells function in cell adhesion, as cell-surface receptors, and in stabilizing the extracellular matrix *(1,2)*. These GPs are essential for the maintenance of structurally and

From: *Methods in Molecular Biology, vol. 347: Glycobiology Protocols*
Edited by: I. Brockhausen © Humana Press Inc., Totowa, NJ

functionally healthy bone. The assembly of healthy bone depends on the action of osteoblasts in the production and mineralization of bone matrix, as well as on osteoclasts in remodeling of the bone ***(3–5)***.

Very little is known about the structures, biosynthesis, and functions of the glycan chains of bone GPs, and how these parameters are altered in response to cytokines and growth factors produced at sites of inflammation. Glycan chains of GPs have been shown to be involved in protein expression, conformation, and epitope exposure ***(6–8)***. Glycan chains are critical for the functions of cell-surface GPs and are involved in growth-factor receptor functions ***(9–11)***. Glycosylation has also been shown to be closely linked to cellular and tissue apoptosis ***(12,13)***.

The assembly of the glycan chains of GPs is controlled by the relative activities of many glycosyltransferases (GTs), as well as by the presence and structures of the protein acceptor substrates. The various steps in the assembly pathways are in a dynamic balance that can shift upon external stimulation. The cytokine tissue growth factor-β (TGF-β) has an important role in promoting bone cell growth, whereas tumor necrosis factor-α (TNF-α), which is increased at sites of inflammation, has multiple functions including the induction of apoptosis. It has been shown that cells subjected to cytokine stimulation exhibit altered activities and messenger ribonucleic acid expression of the enzymes that synthesize GP-bound carbohydrate chains ***(14–16)***. Cellular glycosylation and GP biosynthesis are in a dynamic state, and changes under the influence of cytokines or in disease. This phenomenon is called "glycodynamics."

The authors have established in vitro models of osteoblastic cells to determine the molecular actions and effects of growth factors in inducing apoptosis. These agents have been shown to have an impact on the glycodynamics in these cells. Human osteoblasts were found to be similar to non-mucin-producing cells such as endothelial cells, in that they have GT activities that synthesize complex *O*- and *N*-glycans ***(14)***.

In the protocol presented in this chapter, bone cells are grown in cultures and treated with cytokines, growth factors, apoptotic, or inflammatory agents. The cellular parameters of growth and differentiation, as well as apoptosis, are then measured; this is related to cell-surface staining of carbohydrates using biotinylated lectins. Assays of the activities of GTs involved in the biosynthesis of cell-surface GPs are used to help explain the basis for the changes in carbohydrate structures.

This protocol uses bone cells from two different sources. Both of these are of osteoblastic origin resembling the cell type that builds bone in vivo. Primary human osteoblasts can be obtained by digestion of extracellular bone matrix with collagenase, using bone obtained at the time of joint replacement surgery for osteoarthritis. Bone cells can also be isolated from the outgrowth of bone chips, which takes slightly longer. In addition, they can be grown from the mononuclear fraction of bone marrow cells after differentiation with dexamethasone (DEX); however, this usually results in mixed cell cultures. DEX is required to keep primary cells at a differentiated state, but this is not required for stable bone cancer cell cultures. Primary cells are used only at passage 1 to avoid subsequent possible alterations in the state of growth and differentiation.

Since primary osteoblasts grow slowly in culture (especially those isolated from patients with osteoarthritis), osteoblastic cancer cells have been shown to be a useful model. These osteosarcoma-derived cells grow rapidly in adherent cultures, have the appearance of fibroblastic, spindle-shaped osteoblasts, and express similar patterns of carbohydrates and GT activities ***(17)***.

Useful markers of the osteoblastic phenotype are alkaline phosphatase (non-specific), collagen I (highly specific), and osteocalcin (highly specific). All of the cell types used respond to TGF-β, TNF-α, and many other growth factors and cytokines, and can be stimulated to undergo apoptosis.

These study models can contribute to our understanding of the action of inflammatory agents and help to define a role for glycans in growth, differentiation, and apoptosis of bone cells. These methods also allow the studies of specific glycosylation inhibitors, and may be useful for the evaluation of the pathology underlying arthritis and to identify therapeutic targets. These models can be applied to other cell types and can help to elucidate the pathology of inflammatory and other diseases.

2. Materials

2.1. Cell Cultures

2.1.1. Primary Cultures of Human Osteoblasts

1. Obtain fresh human bone chunks and bone marrow from patient samples, generally after knee replacement or hip replacement surgery. Bone is obtained from the metaphyseal portion of the bone, avoiding areas of fibrocartilagenous cyst formation. Other bone samples, especially from younger people, are useful for comparison.
2. Culture medium: minimal essential medium (α-MEM) and fetal bovine serum (FBS; GIBCO/Invitrogen, Carlsbad, CA), DEX (Sigma, St. Louis, MO), stock solution of 10,000 U of penicillin G sodium salt and 10,000 μg of streptomycin sulfate (Invitrogen), 10-cm^2 dishes, 75-cm^2 flasks, and a cell scraper.
3. Trypan Blue and 0.25% Trypsin/1 m*M* of ethylenediaminetetraacetic acid (EDTA; Invitrogen).
4. Collagenase I (Sigma), phosphate-buffered saline (PBS), and 70% ethanol.
5. Histopaque-1077 layer (density 1.077 g/mL; Sigma).

2.1.2. Human Osteosarcoma Cells SJSA-1

1. Human osteosarcoma cells SJSA-1 (American Type Culture Collection, ATCC CRL-2098).
2. Medium: Roswell Park Memorial Institute Medium 1640 (GIBCO), supplemented with 100 U of penicillin/100 μg of streptomycin/mL, and 10% of FBS.

2.1.3. Human Osteosarcoma Cells MG63

1. Human osteosarcoma cells MG63 (ATCC, CRL-1427).
2. Medium: Dulbecco's modified Eagle's medium (DMEM) with 2% glutamine (ATCC) supplemented with 100 U of penicillin/100 μg of streptomycin/mL and 10% of FBS.

2.2. Identification of the Osteoblastic Phenotype

2.2.1. Morphology

1. Inverted light microscope, Trypan Blue (GIBCO), 96-well plates, glass slides and cover slips, BSA, and PBS.

2.2.2. Collagen I Expression by Immunocytochemistry

1. Six-well plates and a light microscope.
2. Polyclonal goat anti-collagen I IgG antibody C18, pre-immune serum, and secondary antibody bovine anti-goat IgG-horseradish peroxidase (HRP; Santa Cruz Biotechnology, Santa Cruz, CA).
3. Tris-HCl (pH 7.6) and H_2O_2.
4. HRP substrate: diaminobenzidine (Sigma).

2.2.3. Osteocalcin Expression by Immunocytochemistry

1. Mouse monoclonal anti-bovine/human osteocalcin (LF-32) antibody (Biodesign International, Saco, ME) and secondary antibody donkey anti-mouse IgG-alkaline phosphatase (ALP; Santa Cruz Biotechnology).
2. ALP substrate: *p*-nitrophenyl phosphate (pNPP; Sigma).

2.2.4. Cytochemistry of Alkaline Phosphatase

1. PBS, formaldehyde, ethanol, BSA, 0.1% Triton X-100 in PBS, and six-well plates.
2. Substrate: naphthyl-AS-BC phosphate and Fast Blue RR salt (Sigma).

2.2.5. Colorimetric Assay of Alkaline Phosphatase Activity in Cell Lysates

1. Harvest buffer: 10 m*M* of Tris-HCl (pH 7.4), 0.2% NP-40, and 2 m*M* of phenylmethylsulfonyl fluoride (PMSF;, Sigma; *see* **Note 1**).
2. pNPP substrate (Sigma) and *tris*-buffered saline (TBS): 50 m*M* of Tris-HCl (pH 9.0).
3. Plate reader (μQuant; Bio-Tek Instruments, Winooski, VT) and 24- and 96-well plates.

2.3. Treatment of Cells With Growth Factors and Apoptosis Inducers

2.3.1. Tumor Necrosis Factor-α

1. Human recombinant TNF-α (R&D Systems, Minneapolis, MN), HCl, and BSA.

2.3.2. Transforming Growth Factor-β

1. Human recombinant TGF-β1 (R&D Systems), HCl, and BSA.

2.3.3. Anti-Fas

1. Mouse anti-Fas (CD95/Apo-1) IgG 3 antibody clone 2R2 (Roche Diagnostics, Mannheim, Germany; *see* **Note 2**). Store at –20°C in small aliquots.

2.4. Cell Proliferation Assay by [^{3}H]Thymidine Incorporation

1. 1 mCi/mL [^{3}H]Thymidine (Perkin-Elmer, Boston, MA) stock solution and TGF-β or other growth factor.
2. Hemocytometer and Zf model cell counter (Coulter, Mississauga, Ontario, Canada).
3. Scintillation counter (e.g., Beckman model LS 6500) and Ready-Solve scintillation fluid (Beckman, Fullerton, CA).
4. Methanol/acetic acid, 3/1 (v/v); 5% trichloroacetic acid, methanol, 1 *N* of NaOH, and 1 *N* of HCl.

2.5. Assay for Cell Differentiation (Alkaline Phosphatase Activity)

See **Subheading 2.2.5.**

2.6. Apoptosis Assays

2.6.1. Terminal Deoxynucleotidyl Transferase dUTP-Mediated Nick-End Labeling Staining for Flow Cytometry

1. Terminal deoxynucleotidyl transferase (TdT)-Frag-EL deoxyribonucleic acid (DNA) fragmentation detection kit: equilibration buffer TdT enzyme, Fluorescein-FragEL TdT labeling mixture, proteinase K in 10 m*M* of Tris-HCl (pH 8.0), and Fluorescein-FragEL mounting media (Oncogene, San Diego, CA).
2. TBS: 20 m*M* of Tris-HCl (pH 7.6) and 140 m*M* of NaCl.
3. Fixing and permeabilization reagents: formaldehyde, 80% ethanol, PBS, and Tris-HCl (pH 8.0).
4. Deoxyribonuclease I (New England Biolabs, Ipswich, MA), Mg-sulfate in TBS, and microcentrifuge.

2.6.2. Terminal Deoxynucleotidyl Transferase dUTP-Mediated Nick-End Labeling Staining for Fluorescence Microscopy

1. TdT-Frag-EL DNA fragmentation detection kit.
2. Formaldehyde, ethanol, PBS, and TBS.

2.6.3. Annexin-V Staining

1. Annexin-V-biotin Apoptosis Detection kit: annexin V-biotin conjugate in 50 m*M* of Tris-HCl (pH 7.4), 0.1 *M* of NaCl, 1% of BSA, 0.02% of Na-azide, propidium iodide, and media binding reagent (Oncogene).
2. Binding buffer: 10 m*M* of *N*-2-hydroxyethylpiperazine-*N'*-ethanesulfonic acid (pH 7.4), 150 m*M* of NaCl, 2.5 m*M* of $CaCl_2$, 1 m*M* of $MgCl_2$, and 4% BSA.
3. Fluorescein-Streptavidin conjugate (Sigma) and PBS.

2.6.4. Fluorescence Microscopy

1. Fluorescence microscope (Axiovert S100, Zeiss) and standard fluorescein filter for 494 nm.
2. Cytospin centrifugation system (Shandon Southern, Sewickley, PA).
3. TBS, 20 µg/mL of Proteinase K, and mounting media.

2.6.5. Flow Cytometry

1. EPICS XL™ flow cytometer with computer and 488-nm argon ion laser source (Beckman Coulter, Miami, FL).

2.7. Lectin Staining of Cell Surfaces

1. 96-Well plates and biotinylated lectins (Sigma and Vector Laboratories, Burlingame, CA).
2. 1 m*M* of $CaCl_2$, 1 m*M* of $MgCl_2$, 0.1 *M* of sodium cacodylate buffer (pH 7.2), and graded ethanol solutions.
3. Fixing reagent: freshly prepared 2% paraformaldehyde and 2.5% glutaraldehyde in 0.1 *M* cacodylate buffer (pH 7.2).
4. Washing reagent (PBS-G): PBS containing 0.2% gelatin.
5. Avidin-ALP conjugate, diluted 1:1000 (Sigma), 1 mg/mL of pNPP in 0.5 m*M* of $MgCl_2$ (pH 9.5), and a microplate reader.

2.8. GT Assays of Cell Homogenates

2.8.1. Preparation of Cell Homogenates

1. Trypsin/EDTA (TE) or cell scraper.
2. 0.25 *M* of sucrose and a glass hand homogenizer with teflon pistle.
3. 4 mg/mL of BSA standard solution and Protein Assay reagent (Bio-Rad, Hercules, CA).

2.8.2. GT Donor Substrates

1. Nonradioactive nucleotide sugars: uridine 5'-diphosphate *N*-acetylglucosamine (UDPα-GlcNAc), UDPα-Gal, UDPα-*N*-acetylgalactosamine (GalNAc), and cytidine 5'-monophosphate (CMPβ-sialic acid; Sigma).
2. Radioactive sugar nucleotides (Amersham, St. Louis, MO and Perkin-Elmer, Woodbridge, Ontario, Canada).

2.8.3. Glycosyltransferase Acceptor Substrates

1. Polypeptide GalNAc-transferase: Thr and Pro containing peptide (5-m*M* solution in water) with mucin tandem repeat sequence (e.g., TVTPTPTPTG; prepared by peptide synthesis).
2. GTs of the *O*-glycosylation pathways: 5-m*M* solutions of GalNAcα-benzyl, Galβ1-3GalNAcα-benzyl, GlcNAcβ-benzyl, and Galβ1-4GlcNAc (Sigma).
3. GlcNAc-transferases of the *N*-glycosylation pathways: 5-m*M* solution of mannose α1-6 (Manα1-6 [Manα1-3]) Manβ-octyl (Toronto Research Chemicals, Toronto, Ontario, Canada).
4. Lyophilizer, water bath, rotor-evaporator, matrix-assisted laser desorption ionization mass spectrometer, and a 500-MHz Proton nuclear magnetic resonance (NMR) spectrometer.
5. High-performance liquid chromatography (HPLC) system: two pumps, gradient controller, injector, ultraviolet (UV) spectrometer, recorder, 4-mm 0.45-μm syringe filters, 25- and 100-μL injection syringes (Waters, Milford, CA), analytical and preparative C18 columns, amine columns and corresponding pre-columns (Phenomenex, Torrance, CA), acetonitrile (HPLC grade), filtered double-distilled water, 15 m*M* of K-phosphate buffer (pH 5.4), and a fraction collector.

2.8.4. GT Assays

1. AG1x8, Cl^--form (Bio-Rad), Sep-Pak C18 cartridges (Waters), and a tabletop vacuum centrifuge.
2. HPLC as described in **Subheading 2.8.3.** with analytical columns only.
3. Radioactive and nonradioactive standard solutions for HPLC (substrates and products, monosaccharides, oligosaccharides, and peptides).
4. Scintillation counter, 7-mL scintillation vials (Fisher Scientific), and Ready-Solve® scintillation fluid (Beckman).

3. Methods

3.1. Cell Cultures

3.1.1. Primary Cultures of Human Osteoblasts

As a source for osteoblasts, bone samples are obtained from surgical knee or hip replacement of patients with osteoarthritis (*see* **Note 3**). Osteoblasts can be conveniently isolated from marrow cells, but are then often contaminated by other cell types ***(18)***. Primary cells are only used from passage 1 to avoid mixtures of cells at different stages of differentiation. Cancer cells can be used at higher passage numbers. Before starting the cultures, it should be calculated how many cells will be needed in the experiments. For better comparison, all cells should be derived from the same patient with the same procedure, and at the same passage number. One hundred million cells may be required, which may be difficult to obtain. Therefore, the preliminary growth conditions are important.

3.1.1.1. Osteoblasts Cultured by Outgrowth From Bone (see **Note 4**)

1. Clean fresh bone sample with PBS and cut into small chips, then culture in a 10-cm^2 culture dish in α-MEM containing 20% FBS, 0.1 μg of DEX/mL, 100 U of penicillin/mL, and 100 μg of streptomycin/mL. Culture chips for 2–3 wk, but change medium every 3 d. Check under the microscope until confluency.
2. Once sufficient cells have grown out of the bone (at confluency), release cells and tissue pieces by scraping. Remove tissue pieces by centrifugation at 1000*g* for 4 s. Then centrifuge the supernatant at 1000*g* for 10 min and suspend the cell pellet in medium, and use for cell cultures in a humidified atmosphere with 5% CO_2 at 37°C. Initially 20% is present in the medium, but is gradually reduced to 10% within 2–3 wk.
3. Change the medium every 4 d until cells are confluent (in 4–8 wk).
4. At confluency, passage cells with TE (passage 0) and then grow to confluency for experiments in larger dishes or plates. Only cells from passage 1 are used for experiments.

3.1.1.2. Osteoblast Cultures After Collagenase Digestion of Bone Tissue

Collagenase digestion will yield osteoblasts faster than the outgrowth method, and the overall yield is higher. However, cells may possibly be damaged by collagenase.

1. Incubate fresh, cleaned bone tissue pieces overnight at 37°C in 10 mL of α-MEM medium containing 160 U of collagenase I/mL.
2. Centrifuge suspensions of tissue pieces and cells at 1000*g* for 4 s. Collect cells from the supernatant by centrifugation at 1000*g* for 10 min and wash once with PBS.
3. Culture cells first in 10-cm^2 dishes and then in 75-cm^2 flasks in α-MEM medium containing 20% FBS and 0.1 μg/mL of DEX. Gradually reduce the concentration of FBS with every second change of medium to 10%.

3.1.1.3. Osteoblasts From Bone Marrow Cells

Cell cultures from human bone marrow stromal cells are rich in osteoblast precursor cells. The fraction of osteoblastic cells can be increased by DEX treatment. Osteoblasts will be the major cell population when using the following procedure. Combinations of morphological inspection, ALP, and osteocalcin expression are good indicators of the proportion of osteoblasts in the cell cultures.

1. Layer 3 mL of bone marrow obtained from surgery onto 3 mL of Histopaque-1077 and centrifuge at 400*g* for 30 min at room temperature. Red cells will be at the bottom and plasma on top. The opaque layer of low-density mononuclear cells (including stromal cells) is near the upper middle layer.
2. Discard the top layer, carefully aspirate the opaque mononuclear cell layer, and transfer to a clean centrifuge tube.

3. Add 10 mL of PBS and centrifuge cells at 400*g* for 10 min at room temperature. Wash the pellet three times with 5 mL of PBS.
4. To increase the proportion of osteoblasts, cells are cultured with 0.1 μ*M* of DEX in the medium α-MEM supplemented with glutamine, 100 U of penicillin/mL, 100 μg of streptomycin/mL, and 10% FBS (*see* **Note 5**).

3.1.2. Human Osteosarcoma Cells SJSA-1

Osteosarcoma cells are stable, fast-growing cells of osteoblastic origin.

1. Grow osteosarcoma cells SJSA-1 in RPM1 1640 medium containing 10% FBS. Under the microscope, cells appear fibroblastic and spindle-shaped upon adherence to the flask. Cells grow very fast.
2. Change the medium every 2–3 d. Cells are confluent and should be passaged after 7–10 d by seeding 40,000–100,000 cells into 75-cm^2 flasks.
3. After confluency, remove the cells from the plate with TE and passage continually.

3.1.3. Human Osteosarcoma Cells MG63

1. Grow human osteosarcoma cells MG63 in DMEM supplemented with 10% FBS, 100 U of penicillin/mL, and 100 μg of streptomycin/mL. Cells have a spindle-shaped, fibroblastic appearance and grow quickly.
2. Harvest and culture the cells as described in **Subheading 3.1.2.**

3.2. Identification of the Osteoblastic Phenotype

3.2.1. Morphology

Osteoblastic cells are identified under the light microscope, and after attachment should have the appearance of spindle-shaped, mononucleated, fibroblastic osteoblasts (*see* **Fig. 1**). The phenotype of these osteoblasts in culture is different from their phenotype in bone tissues. Cell numbers and viability are measured after appropriate dilution with PBS and Trypan Blue reagent using a hemocytometer.

3.2.2. Collagen I Expression by Cytochemistry

1. Seed cells in 1 mL of medium (50,000 cells/well) into each well of 24-well plates on coverslips. Culture cells until pre-confluency, wash with PBS, and treat with 3% BSA in PBS for 15 min to saturate nonspecific binding sites.
2. Fix cells with 1 mL of 3.7% formaldehyde in PBS and dehydrate in 1 mL of 70% ethanol.
3. Wash cells and expose to 3% H_2O_2 for 30 min.
4. Wash cells in PBS and treat with 500 μL of a 1:200 dilution of goat anti-collagen I antibody. Treat control cells with pre-immune serum. Remove the antibody solution after incubation for 1 h at room temperature and wash the cells with PBS three times for 10 min.

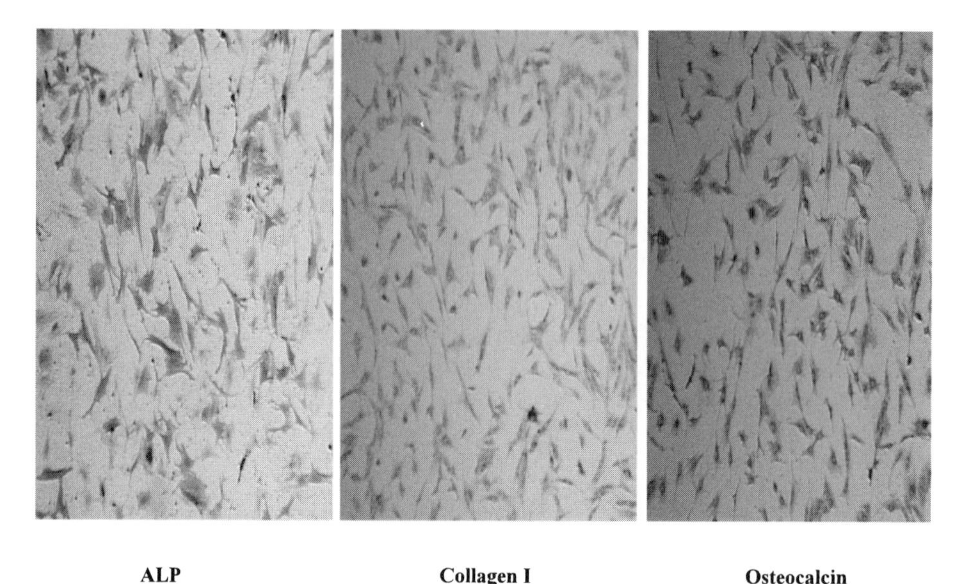

Fig. 1. Osteoblastic cells: morphology of primary osteoblasts visualized by light microscopy. Osteoblastic cells were stained with anti-mouse IgG-alkaline phosphatase (ALP) substrate, anti-collagen I antibody, and anti-osteocalcin antibody as described in **Subheading 3.2.4.** (ALP), **Subheading 3.2.2.** (collagen), and **Subheading 3.2.3.** (osteocalcin).

5. A secondary antibody, bovine anti-goat antibody (500 μL of a 1:100 dilution) linked to HRP is added for 1 h. After washing with PBS, treat cells with HRP substrate diaminobenzidine (final concentration 1 mg/mL in 500 μL of 10 m*M* Tris-HCl, pH 7.6) and 0.05% H_2O_2 for 20 min.
6. Examine stained cells under the light microscope (*see* **Fig. 1**). Osteoblasts give a clear brown color, while controls lacking the specific antibody have a light yellow color.

3.2.3. Osteocalcin Expression

1. To measure osteocalcin expression by cytochemistry, treat cells as described in **Subheading 3.2.2.** but change the antibody to anti-bovine/human osteocalcin (LF-32) antiserum (*see* **Note 6** and **Fig. 1**).

3.2.4. Cytochemistry of Alkaline Phosphatase Activity

1. Grow osteoblasts on six-well plates.
2. Wash the cells at about 50% confluency with 3 mL of PBS and fix with 3 mL of 3.7% formaldehyde in PBS at room temperature for 20 min.

3. Remove the supernatants and wash the cells in PBS three times. Add naphthyl AS-BC phosphate and Fast Blue RR salt, except for control cells.
4. Incubate cells for 10 min at 37°C and remove the supernatant.
5. Visualize the stained cells under the light microscope (*see* **Fig. 1**). Osteoblasts appear as dark blue cells, whereas controls lacking naphthyl AS-BC give a light blue background.

3.2.5. Colorimetric Assays of Alkaline Phosphatase Activity

1. Seed osteoblasts in 24-well plates (50,000 cells/well) and grow to 90% confluency.
2. Wash the cells with PBS and then scrape into a 200-μL ice-cold harvest buffer for lysis at 4°C for 20 min.
3. Add 100-μL aliquots of each lysate to the wells of 96-well plates, and add 100 μL of pNPP ALP substrate solution (1.33 mg/mL in 50 m*M* Tris-HCl, pH 9.0) to each well. Measure each sample in triplicate determinations. Allow the reaction to proceed for 1 h at 37°C.
4. Measure the absorbance of the *p*-nitrophenolate ion at 405 nm with a plate reader. Construct a standard curve using pNPP (concentrations from 1 μ*M* to 1 m*M*) in the same buffer (*see* **Note 7**). Measure the protein concentration by the Bradford/Bio-Rad method. Express ALP activity as nmol nitrophenolate released/min/μg protein.

3.3. Treatment of Cells With Growth Factors, Cytokines, or Apoptosis Inducers

Before treatment with growth factors or cytokines, it is necessary to check the literature for responsiveness of cells to these agents. To amplify the effect of growth factors, reduce the FBS concentration in the medium to 2% before treatment for several days. This may not be necessary to induce apoptosis. The ideal incubation time and concentrations of each cytokine need to be established by time-course and dose–response curves for each cell line (*see* **Note 8**).

3.3.1. Treatment With TNF-α

Grow primary osteoblasts before growth factor treatments in medium with reduced (2%) FBS concentration for 48 h. Grow osteoblastic cancer cells in serum-free medium for 48 h. TNF-α treatment is expected to increase the apoptotic cell population by 2–15%.

1. Depending on the experiments following treatment, grow cells in 96-well plates or in 75-cm^2 flasks. Add TNF-α to the cell medium to a final concentration of 40 ng/mL. Keep cells in culture for 48 h.
2. Treat control cells with BSA to a final concentration of 0.0004% in the cell medium.

3.3.2. Treatment With TGF-β

Primary osteoblast cells are grown before growth factor treatments in 2% FBS for 48 h. Grow osteoblastic cancer cells in serum-free medium.

1. Add TGF-β to the cell medium to a final concentration of 10 ng/mL. Keep cells in culture for 48 h.
2. Treat control cells with HCl and BSA to a final concentration of 20 μ*M* of HCl and 0.001% BSA in the cell medium.

3.3.3. Treatment With Anti-Fas

Anti-Fas can induce apoptosis in many cell types. Negative-control cells lack anti-Fas. Positive-control cells can be treated with an apoptotic agent known to be effective in the specific cell type used.

1. Remove the medium from 75-cm^2 flasks of confluent osteoblastic cells. Add 4 mL of fresh medium containing 4 μg of anti-Fas IgM or IgG antibody. The IgG and IgM antibodies induce apoptosis at a level of 0.2 to 1 μg/mL of cell medium.
2. Incubate cells for 18 h at 37°C.
3. After the induction of apoptosis, cells are examined for their apoptotic phenotype by morphology, Trypan Blue exclusion, and adherence to plates.
4. Detach cells from the plates with TE and dehydrate and prepare for flow cytometry or for fluorescence microscopy as described in **Subheading 3.6.** (*see* **Note 9**).

3.4. Cell Proliferation Assay by [^{3}H]Thymidine Incorporation

Cell proliferation (DNA synthesis) is a sensitive way to assess the effect of growth factors. The increase in cell numbers can be estimated with the hemocytometer in parallel with proliferation assays, or more accurately with a cell counter. Cell growth (increase in cell numbers) may not be apparent in primary human bone cells because cell division is relatively slow (*see* **Fig. 2**). Controls do not contain growth factor.

1. Seed osteoblasts at a density of 2500 cells/well in 96-well plates (four wells are used per treatment). More cells may be needed for primary osteoblasts that grow slowly.
2. Grow cells to 30–50% confluency before growth factor treatment; this is usually achieved in less than 48 h. Culture cells in 100 μL of medium containing 1 ng of TGF-β or other growth factors for 48 h.
3. Add 1 μL of 1 mCi [^{3}H]thymidine to each well to a final concentration of 1 μCi/mL and incubate cells for 24 h.
4. Remove supernatants and fix cells in each well with 200 μL of methanol : acetic acid (3 : 1) for 10 min at room temperature. Remove supernatants and add 200 μL of 100% methanol, then wash once with 5% TCA for 5 min and three times with methanol.

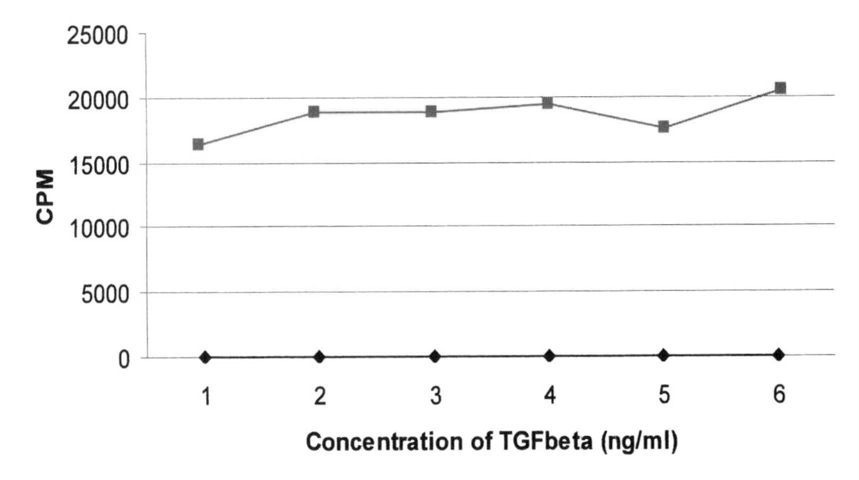

Fig. 2. Cell proliferation measured by [^{3}H]thymidine incorporation of SJSA-1 cells, in response to transforming growth factor-β (TGF-β) 1. Cell proliferation was measured after incubation of SJSA-1 osteosarcoma cells for 48 h with varying concentrations of TGF-β 1 in the cell culture medium, as described in **Subheading 3.4**.

5. Remove supernatants and air-dry the plates for 5–10 min at room temperature. Plates may be stored at room temperature or processed.
6. Lyse cells in each well with 200 μL of 1 *N* NaOH for 5 min. Transfer 180 μL of the lysates to scintillation vials containing 180 μL of 1 *N* HCl to neutralize the base. Add 5 mL of scintillation fluid and determine the radioactivity by scintillation counting. Results are expressed as percentage of controls, and statistical analysis is done by the *t*-test.

3.5. Cell Differentiation Assay by ALP Activity

Osteoblasts acquire ALP upon differentiation; therefore, ALP activity is a good measure of osteoblast differentiation. Cells are incubated in 24-well plates with medium containing cytokines or growth factors, or no addition in controls (12 samples per experimental group). The ALP activity is assayed spectrophotometrically as described in **Subheading 3.2.5.**

3.6. Apoptosis Assays

After treatment of cells with cytokines, anti-Fas, or other agents to induce apoptosis, apoptosis in osteoblastic cells can be measured by several methods, with the terminal deoxynucleotidyl transferase dUTP-mediated nick end labeling (TUNEL) assay being an efficient and reliable method. The TUNEL assay by flow cytometry will determine the percentage of cells at the later (DNA fragmentation) stages of apoptosis (usually a 1–8% increase). Annexin-V assays will indicate the percentage of cells at early and late stages of apoptosis by flow cytometry. Cells can also be examined with a fluorescence microscope. This method is not quantitative, but it is faster and gives a good view of cell shapes, clumping, and distribution of apoptotic cells and nuclei, which can be superimposed with specific dyes.

3.6.1. TUNEL Staining and Flow Cytometry

1. Cells can be treated with apoptosis inducers in small or large dishes. To assess the degree of apoptosis, follow the instructions of the FragEL DNA fragmentation detection kit. Have duplicates (at a minimum) of all treatment groups. In addition, two samples of cells not treated with an apoptosis inducer are used for positive controls and two for negative controls.
2. Fixation: centrifuge one million cells in suspension at 1000*g* for 5 min at room temperature and fix with 1 mL of 4% formaldehyde in PBS for 10 min at room temperature.
3. Centrifuge the cell suspension at 1000*g* for 5 min at 4°C to remove the fixative, and dehydrate the cells in 1 mL of 80% ethanol. This cell suspension can be stored at 4°C for several months.
4. Transfer 1 mL (1 million) of fixed dehydrated cells to a microcentrifuge tube. Centrifuge at 1000*g* for 5 min at room temperature. Remove ethanol.
5. Rehydration: resuspend cells in 200 μL of TBS and incubate for 10–15 min at room temperature. Centrifuge cells at 1000*g* for 5 min. Remove TBS.
6. Permeabilization: resuspend cells in 100 μL of 20 μg/mL proteinase K solution in 10 m*M* of Tris-HCl (pH 8.0). Incubate for 5 min at room temperature, but not longer. Centrifuge cells at 1000*g* for 5 min and remove supernatant.
7. Suspend cells in 100 μL of 1X TdT equilibration buffer. Incubate for 30 min at room temperature, then centrifuge at 1000*g* for 5 min and remove buffer.
8. Negative control: substitute TBS for the TdT reagent.
9. Positive control: incubate 1 million untreated cells with 0.4 U/μL of DNase I in TBS and 1 m*M* of $MgSO_4$ for 30 min.
10. Labeling: prepare TdT labeling mixture by adding 57 μL of Fluorescein-FragEL TdT labeling reaction mixture to 3 μL of TdT enzyme and vortexing. Add 60 μL of the labeling mixture to the cell pellet (1 million cells). Incubate in the dark at 37°C for 1–1.5 h.
11. Terminate the reaction by centrifuging the mixture at 1000*g* for 5 min, removing the reaction mixture, and washing the cells twice with 200 μL of TBS. Resuspend the cells in 500 μL of TBS and analyze the same day by flow cytometry.
12. Flow cytometry: follow the instructions for the flow cytometer (*see* **Fig. 3**). Run standards (negative and positive controls). Run duplicate samples for each treatment group.

3.6.2. TUNEL Staining and Fluorescence Microscopy

The green fluorescence of TUNEL-stained cells can also be detected by fluorescence microscopy. This method is not quantitative, but it gives a good view of cell shapes and the distribution of apoptotic cells.

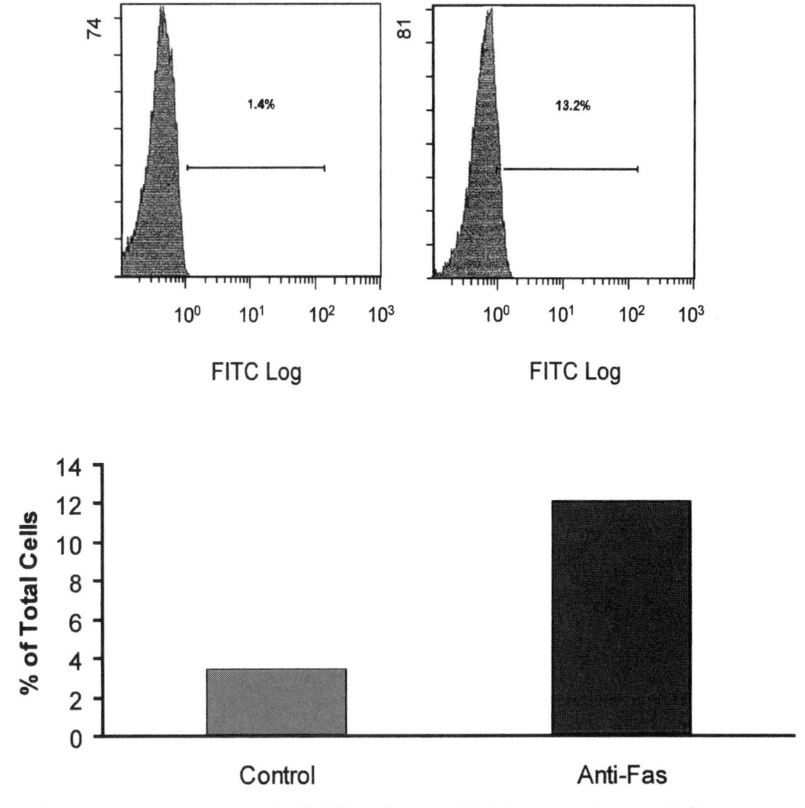

Fig. 3. Apoptosis assays of MG63 cells by TUNEL staining and flow cytometry. Osteosarcoma MG63 cells were incubated with anti-Fas antibody, and then stained with the TUNEL stain as described in **Subheading 3.6.1.** Stained cells were analyzed by flow cytometry. Bars at the bottom graph show the average number of untreated control (gray bar) and anti-Fas-treated (black bar) TUNEL-stained cells. The top diagram shows the number of cells per fluorescence for each experimental group.

1. Follow the instructions in **Subheading 3.6.1., steps 1–11** for preparation of cells, fixing, dehydration, rehydration, and TUNEL labeling.
2. After termination of TdT labeling, transfer 20,000 cells to a glass slide using the Cytospin centrifugation method and incubate slides for 1 min at room temperature. Mount glass slides using Fluorescein-FragEL mounting media (*see* **Note 10**).
3. Analyze the slides by fluorescence microscopy at ×20 magnification. Mounted samples can be stored in the dark.

3.6.3. Apoptosis by Annexin-V Staining

The annexin V-biotin apoptosis detection kit will stain cell membranes, whereas propidium iodide will stain nuclei.

1. Follow the instructions of the annexin V-biotin apoptosis detection kit. Perform all experiments at least twice.
2. Release the cells from the plates with TE and wash twice with 10 mL of cold PBS. Five hundred thousand cells are transferred to a microfuge tube and centrifuged at 1000*g* for 5 min.
3. Resuspend the cells in 0.5 mL of cold binding buffer and add 1.25 μL of annexin V-biotin conjugate, then incubate in the dark for 15 min at room temperature.
4. Centrifuge the cell suspension at 1000*g* for 5 min to remove the supernatant. Resuspend the cells in 0.5 mL of cold binding buffer. Add 15 μL of fluorescein-streptavidin conjugate in 15 μg/mL of binding buffer and add 10 μL of propidium iodide. Place the tubes on ice in the dark.
5. Analyze the cells immediately by flow cytometry (*see* **Note 11**). Untreated cell populations bind little annexin V and show a small percentage of apoptotic cells. Early apopotic cells bind annexin V (green fluorescence) but exclude propidium iodide. Late apoptotic or necrotic cells bind both annexin V and propidium iodide (red cell interior).
6. As an alternative, cells can be transferred to glass slides and examined by fluorescence microscopy.

3.7. Lectin Staining of Cell Surfaces

Biotin-linked lectins are used to detect specific cell-surface carbohydrate structures ***(15)***. At the time of lectin staining in 96-well plates, cells should be confluent to ensure that the same number of cells is present in each well, independent of growth factor treatment (*see* **Note 12**). Controls lack lectins (*see* **Note 13**). Use at least eight samples per experimental group; thus, six lectins per 96-well plate can be used. Because there is variation between cell batches, the experiments should be repeated with another cell batch. Useful lectins are listed in **Fig. 4** (*see also* Chapter 19). Biotinylated lectins are useful for enzyme-linked lectin assay (ELLA) measured with a plate reader (*see* **Fig. 4**). Alternatively, fluorescein-labeled lectins can be used to stain cells for analysis of cells by flow cytometry.

1. Seed the cells at 2500 cells/well of 96-well plates (*see* **Note 14**) in 100 μL of medium. Cells are then grown to complete confluency in several days.
2. Cells are treated with and without growth factors, cytokines, or other agents at the concentrations and conditions previously described. Negative control assays do not contain added lectin (*see* **Note 13**). Positive controls are cell lines known to express specific carbohydrate epitopes.
3. Wash the cells with 200 μL of PBS containing 1 m*M* of $CaCl_2$ and 1 m*M* of $MgCl_2$, and fix with 100 μL of 0.1 *M* cacodylate buffer (pH 7.2) containing 2% paraformaldehyde and 2.5% glutaraldehyde for 15 min.
4. Wash fixed cells three times with 200 μL of PBS containing 0.2% gelatin (PBS-G) and dehydrate in 100 μL of 70–90% graded ethanol solutions, followed by three washes of the same PBS-G.

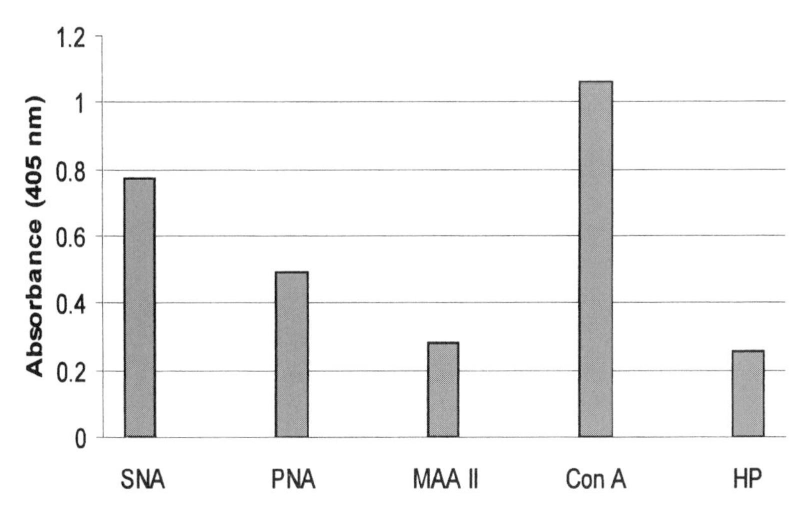

Fig. 4. Lectin binding to osteoblasts in response to tumor necrosis factor-α. Osteoblasts were treated with biotinylated lectins as described in **Subheading 3.7.** All lectins stained the cells significantly above background (no lectins). SNA, *Sambucus nigra* lectin, staining sialyl α2-6Gal(NAc); PNA, Peanut lectin, staining Gal β1-3GalNAc; MAAII, *Maackia amurensis* lectin staining sialyl α2-3Gal; ConA, *Concanavalin* A, staining *N*-glycan core structures; HP, *Helix pomatia*, staining GalNAc.

5. Add biotinylated lectins to the wells at a concentration of 1 μg/mL for 30 min.
6. Wash the cells with 200 μL of PBS-G and add 100 μL of 1:1000 diluted streptavidin–ALP conjugate. Incubate the cells for 1 h at room temperature and then wash with PBS-G.
7. Add 100 μL of ALP substrate (1 mg/mL of pNPP in 0.5 m*M* of $MgCl_2$, pH 9.5) to the cells, followed by incubation for 2 h in the dark. Plates are read at 405 nm with a microplate reader. The absorbance values are statistically evaluated with the *t*-test.

3.8. GT Assays of Cell Homogenates

All enzyme assays should be performed at least twice ***(15,17,19,20)***. Control assays lack exogenously-added acceptor substrate. High radioactivity in controls may indicate degradation of nucleotide sugar donor substrate to radioactive monosaccharide or degradation of radioactive enzyme products and substrates. Inhibitors of degradation may be useful in such cases, e.g., 10 m*M* of adenosine 5′-monophosphate or adenosine triphosphate to inhibit nucleotide sugar degradation or glycosidase inhibitors. It should be verified that the initial linear enzyme rates are measured for all enzyme reactions, and that the enzyme concentration is not saturating. **Table 1** lists several enzymes that are important for GP biosynthesis in bone cells. One hundred million cells are suitable for

Table 1
Assays for Selected Glycosyltransferases Using Bone Cell Homogenates

Enzyme	Substrate	Type of assay
ppGalNAc-T	0.5 m*M* TVTPTPTPTG	AG1x8, Sep-Pak
Core 1 β3-Gal-T	2 m*M* GalNAcα-Bn	AG1x8, Sep-Pak HPLC (C18, 10% AN)
Core 2 β6-GlcNAc-T	2 m*M* Galβ1-3GalNAcα-Bn	AG1x8, Sep-Pak, HPLC (C18, 10% AN)
β4-Gal-T	2 m*M* GlcNAcβ-Bn	AG1x8, Sep-Pak, HPLC (C18, 10% AN)
GlcNAc-T I	0.5 m*M* Manα6 (Manα3) Manβ-octyl	AG1x8, Sep-Pak HPLC (C18, 20% AN)
GlcNAc-T II	0.5 m*M* Manα6 (GlcNAcβ2Manα3) Manβ-octyl	AG1x8, Sep-Pak HPLC (C18, 20% AN)
GlcNAc-T III-V	0.5 m*M* GlcNAcβ2 Manα6 (GlcNAcβ2Manα3) Manβ-octyl	AG1x8, Sep-Pak HPLC (C18, 20% AN)
Core 1 α3-sialyl-T	2 m*M* Galβ1-3GalNAcα-Bn	HPLC (NH2, 80% AN, 20% K-Phosphate)
α3/6-sialyl-T	2 m*M* Galβ1-4GlcNAc	HPLC (NH2, 80% AN, 20% K-Phosphate)

AN, acetonitrile in the mobile phase; Bn, benzyl; C18, C18 column; NH2, amine column; ppGalNAc-T, polypeptide GalNAc-transferase; -T, -transferase.

initial studies of GTs, whereas 10 million cells may be sufficient to assay two or three specific enzymes.

Excess nucleotide sugar donor substrates are effectively removed by AG1x8 anion exchange, but a high background owing to nucleotide sugar degradation may give misleading results. These assays should therefore always be confirmed by separation of enzyme product on HPLC. Sialyltransferase (ST) products are acidic and should not be passed through AG1x8, but can be separated by HPLC using an amine column (*see* Chapter 11) or by high-voltage electrophoresis ***(14,19,20)***. Hydrophobic aglycone groups in acceptor substrates facilitate the separation of enzyme products by hydrophobic chromatography. Standards for chromatography can be purchased from suppliers of glycobiology products or can be prepared enzymatically using the appropriate acceptor and donor substrates, and the assays described in the following sections and as previously reported ***(14,19,20)***. The biosynthetic pathways constructed with these

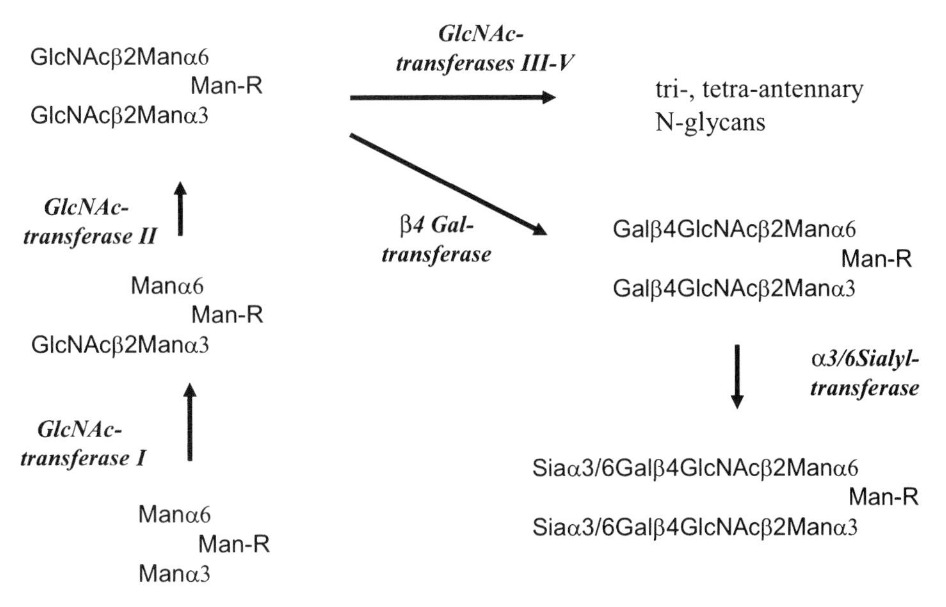

Fig. 5. *N*-glycosylation pathways of osteoblastic cells. Primary osteoblasts have GlcNAc-transferase I and II activities that synthesize bi-antennary *N*-glycan structures. GlcNAc-transferase III, IV, and V activities are below detectable levels. Bi-antennary *N*-glycans can be extended and modified by β4-Gal-transferase and sialyltransferases. Sia, sialic acid.

assays (*see* **Figs. 5** and **6**) should be related to lectin staining and growth and apoptosis for each cell type.

3.8.1. Preparation of Cell Homogenates

1. Rinse cells on plates with 1 mL of TE, and then treat with 3 mL of TE until cells detach (<5 min). The ideal cell number of 100 million cells for enzyme studies can usually be obtained from four to six 75-cm^2 plates.
2. Add 8 mL of medium to each plate to neutralize trypsin, centrifuge the cells at 2000*g* for 2 min, and wash twice with 5 mL of PBS. The cells do not require sterile conditions.
3. Add 5 vol of 0.25 *M* sucrose and homogenize the cells with 15 strokes in a glass hand homogenizer with a Teflon pestle. The protein content is determined by the Bio-Rad protein assays and should be 2 to 10 mg of protein/mL. Store homogenates in small aliquots indefinitely at –80°C.

3.8.2. GT Donor Substrates

Nonradioactive and radioactive nucleotide sugar solutions are mixed to achieve a 10-m*M* final concentration with a specific activity of 1–4 µCi/µmol

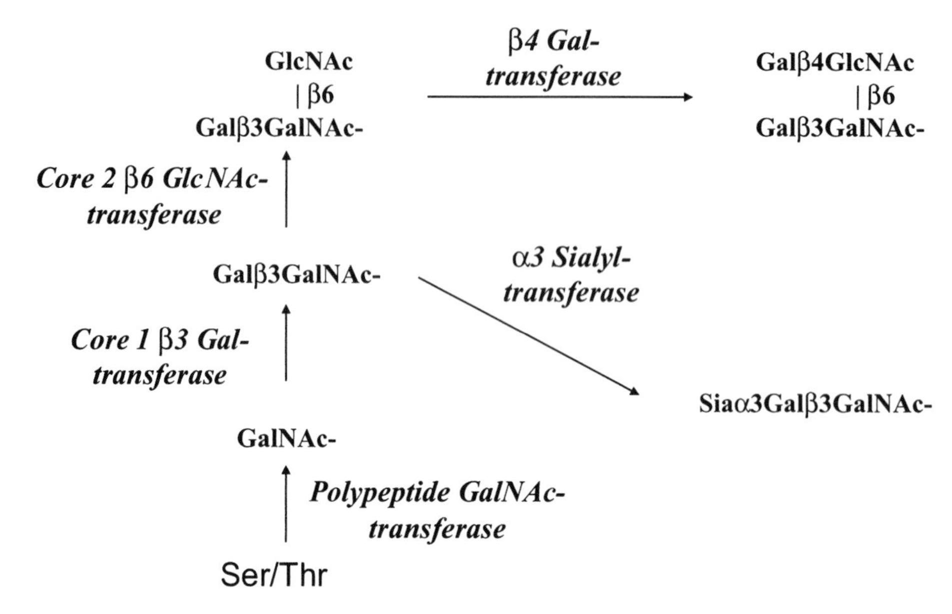

Fig. 6. *O*-glycosylation pathways of osteoblastic cells. Primary osteoblasts can synthesize mucin type *O*-glycans with core1 structure, Gal β1-3GalNAc, and the branched core2 structure, GlcNAc β1-6 (Gal β1-3) GalNAc, which can be further modified by β4-Gal-transferase. Sialyltransferases probably act on both types of core structures.

(*see* **Note 15**). The concentrations are confirmed in 1:1000 aqueous dilutions by measuring the UV absorbance at the wavelength appropriate for the nucleotide (*see* **ref. *21***; UDP, 262 nm [$\varepsilon = 10{,}000$]; CMP, 271 nm [$\varepsilon = 9100$]). The radioactivity of an aliquot is then measured by scintillation counting to calculate the specific radioactivity as dpm/nmol. GDP-containing donor substrates absorb as described in Chapter 3.

3.8.3. GT Acceptor Substrates

Most acceptor substrates can be synthesized in peptide synthesis facilities (polypeptide GalNAc-transferase substrate) or can be purchased from Sigma or from companies selling carbohydrate derivatives. Oligosaccharide substrates can also be synthesized from simpler precursors using GTs.

3.8.3.1. Enzymatic Synthesis of GlcNAc-Transferase II Substrate

1. Synthesize GlcNAc-transferase II substrate Manα6 (GlcNAcβ2Manα3) Manβ-octyl from Manα6 (Manα3)Manβ-octyl under the conditions of GlcNAc-T I assays but using nonradioactive UDP-GlcNAc. Pool 100 small-scale assays (1 h incubation time). The enzyme source can be bone cells, but recombinant GlcNAc-transferase I would be more efficient *(22)*.

2. After the incubation, pass the enzyme products through AG1x8 and purify on HPLC using a C18 column and acetonitrile:water mixtures (20:80). Identify product peaks by running a small-scale assay first (using radioactive nucleotide sugar). Collect fractions of 2 mL (2 min) and pool the fractions that show an absorbance peak (at 195 nm) at the elution time of the radioactive product.
3. Concentrate the pooled fractions by flash evaporation and lyophilization. The purity and concentration is ascertained by HPLC, as compared with standards (*see* **Note 16**).
4. The identity as well as the purity of the enzymatically synthesized compound is determined by matrix-assisted laser desorption ionization-mass spectrometry and proton-NMR spectroscopy, compared with standards. The β1-2 linkage in the product has characteristic NMR parameters ***(22,23)***. Repeat HPLC separations if the product is contaminated by substrate.

3.8.3.2. Enzymatic Synthesis of GlcNAc-Transferase III–V Substrate

GlcNAc-transferase III, IV, and V utilize the same substrate, GlcNAcβ2 Manα6 (GlcNAcβ2Manα3) Manβ-octyl (2,2octyl). This compound can be synthesized from Manα6 (Manα3)Manβ-octyl under the conditions of GlcNAc-transferase I and II assays for bone cells by using nonradioactive UDP-GlcNAc and incubating for an extended period of time, which will allow the action of GlcNAc-transferases I and II to yield 2,2octyl. Bone cells have active GlcNAc-transferases I and II but no significant activity of GlcNAc-transferases III–V. Thus, the enzyme product will be a mixture of Manα6 (Manα3)Manβ-octyl, Manα6 (GlcNAcβ2Manα3) Manβ-octyl, and GlcNAcβ2Manα6 (GlcNAcβ2Manα3) Manβ-octyl. At least 200 times small-scale assays may be necessary to yield 50–100 nmol of 2,2octyl necessary to identify and use as a substrate. This would require about 200 million cells, which can be obtained from a cancer cell culture that has high GlcNAc-transferase I and II and lacks GlcNAc-transferases III–V activities (*see* **Note 17**).

1. Scale up a GlcNAc-transferase I assay 200 times, using nonradioactive UDP-GlcNAc and homogenate from bone cancer cells or other sources of GlcNAc-transferases I and II which lack GlcNAc-transferases III–V, and incubate at least 24 h.
2. Isolate 2,2octyl product by HPLC as described in **Subheading 3.8.3.1.**
3. Identify and analyze 2,2octyl by HPLC, mass spectrometry, and proton-NMR as previously described.

3.8.4. Glycosyltransferase Assays

Optimal conditions for assaying bone-cell enzymes should resemble those published for other cells or tissues ***(14,15,19,20,22,23)***. Most important is to choose the right acceptor substrate. Separation methods for enzyme products are listed in **Table 1**. Assays can be complemented with studies of messenger ribonucleic acid expression levels (*see* **Note 18**).

1. Add substrates, buffers, detergents, degradation inhibitors, and 10 μL of enzyme source to a total volume of 40 μL. If the volume is exceeded, dry solutions in a tabletop vacuum centrifuge (but not detergents or proteins). Always prepare cocktails containing ingredients that are common to all assays in order to keep the concentrations comparable in all assays. Do not add nucleotides and $MnCl_2$ to the same cocktail, as they will cause a precipitate to form. Leave all assay tubes on ice. Act quickly after the addition of the enzyme. Prepare two assays without substrate to determine the background radioactivity (endogenous assays).
2. Incubate assay mixtures for a fixed time period (usually 30–60 min) in a 37°C waterbath.
3. Stop the reactions by adding 600 μL of cold water. At this stage, the tubes can be frozen and stored.

3.8.4.1. AG1x8 Assays for Neutral Acceptor Substrates and Products

1. Pass the assay mixture through 0.4 mL of AG1x8, Cl-form in a plugged Pasteur pipet. Wash the column twice with 600 μL of water and collect all eluates into a scintillation vial. Save the eluates of additional assays or a portion of the assays for HPLC analysis of products. Add 5 mL of scintillation fluid to eluates of each assay and count radioactivity for at least 3 min.
2. Express enzyme activity as nmol product per hour incubation per mg protein per assay.

3.8.4.2. Sep-Pak Assays for Hydrophobic Acceptor Substrates and Products (*see also* Chapters 12 and 16)

1. For neutral enzyme products possessing a hydrophobic group (e.g., benzyl, octyl, etc.), the assay mixtures can be purified on a C18 Sep-Pak column and equilibrated in water. Assay mixtures are applied slowly (2 mL/min) and washed slowly with 5 mL of water. Collect 1 mL of fractions (fractions A–E).
2. Elute the product (and substrate) with 3 mL of methanol. Collect 1 mL of fractions (F–H) and count fractions E–H. Most of the enzyme product is in fractions F and G.
3. Wash the column with 6 mL of methanol and 6 mL of water, and use for the next assay (*see* **Note 19**).

3.8.4.3. HPLC Separations of Enzyme Products

1. For HPLC analyses of assay mixtures, concentrate AG1x8 eluates or fractions F and G from Sep-Pak assays by lyophilization, flash evaporation, or in a tabletop vacuum centrifuge. Add 120 μL of water to dissolve residue (or solvent if necessary to dissolve product). Filter the mixtures through a small syringe filter.
2. Equilibrate and prepare the HPLC system in order to elute the enzyme product well-separated from free radioactive sugar and other possible radioactive components. The product usually elutes at 20–40 min separated from free sugars. Benzyl, nitrophenyl, and octyl derivatives can be separated using a C18 column at 1 mL flow/min and a low percentage (0–20%) of acetonitrile in the mobile phase.

If the hydrophobic group in the substrate is small (e.g., methyl) or absent, an amine column can be used with a high percentage of acetonitrile (70–90%). Run standards first.

3. Inject an aliquot of the concentrated AG1x8 eluate or 80 μL of Sep-Pak fractions F and G into an equilibrated HPLC. Measure the absorbance at 195 nm and collect 2-min fractions.
4. Count all fractions until 20 min past elution time of the product in one assay. In duplicate assays, count only the fractions expected to have radioactivity.

3.8.4.4. HPLC Separations of Acidic Enzyme Products

ST products using small molecule acceptor substrates can be separated from free sialic acid and CMP-sialic acid by ion exchange, electrophoresis, or gel-filtration chromatography. The method described here relies on an HPLC separation.

1. Sialylated standard compounds for the acceptor substrates indicated in **Table 1** must be prepared by enzymatic synthesis, as described in **Subheading 3.8.3.** The STs necessary for the synthesis of sialylated standards can be recombinant, pure enzymes. Rat colon mucosal homogenate is also rich in STs.
2. Purify the product by combined gel filtration (e.g., Bio-Gel P4) and HPLC using an amine column and acetonitrile/15 m*M* of K-phosphate (pH 5.4; 80:20), or the conditions described in Chapter 11.

4. Notes

1. 0.1% Triton X-100 can be used instead of NP-40.
2. Mouse IgM anti-Fas antibody from TNB Laboratories (St. John's, Newfoundland, CAN) can also be used, and is thought to be more effective than IgG antibody.
3. Cells from young and healthy individuals may grow faster. However, it is very difficult to obtain samples from young or normal individuals at surgery, and samples from autopsy are not recommended owing to possible postmortem tissue degradation.
4. The outgrowth method results in the pure osteoblast cultures. However, growth is slower than with cells obtained by collagenase digestion.
5. Ascorbic acid may be a useful addition, especially when the production of bone matrix is studied.
6. Antibody to osteocalcin is expensive but very specific. Other tests are for osteonectin utilizing antibody LF-37, or for biglycan with anti-biglycan antibody LF-15.
7. The supernatant can be transferred into Quartz cuvets and absorbance is read at 405 nm with a UV spectrometer.
8. The conditions given here are useful for the bone cell lines mentioned in this protocol but if similar cell types are used, these conditions are not necessarily ideal and need to be reestablished.
9. The degree of apoptosis can also be assessed by using caspase assays or other commercial apoptosis kits. Before induction of apoptosis, cells could be modified

with respect to their cell-surface carbohydrates with GT inhibitors or glycosidases to study the role of glycosylation in apoptosis.

10. Cells can also be placed directly on a slide without the Cytospin method.
11. Cells are dehydrated and can be stored at 4°C for several weeks.
12. The growth conditions on 96-well plates may not be exactly the same as those in a 75-cm^2 flask. However, it has been observed that TNF and TGF have an effect on cells under these conditions. The changes in glycosylation observed in confluent cells are expected to be owing to a direct effect of growth factors, not to changes in growth.
13. To confirm lectin binding and specificity, lectins can be used under similar conditions to stain purified GPs, blotted onto nitrocellulose membranes. Negative control assays can be carried out with the addition of a high concentration of free carbohydrates, which compete with the lectin ligands. This should reduce lectin binding. Inhibitors of glycosylation (e.g., tunicamycin for *N*-glycans, GalNAcα-benzyl for *O*-glycans) added to the growth medium can be used to prevent the synthesis of carbohydrate epitopes on cell-surface GPs. In this case, the turnover time of existing GPs must be considered. A more expensive method to produce cells lacking specific carbohydrates is the use of glycosidases on whole-cell suspensions. Both glycosidase and inhibitor treatments are not usually completely efficient, but the change in specific lectin binding would confirm their actions.
14. More cells may be required for slowly growing cells.
15. The nucleotide sugar solution can be less concentrated for pure enzymes, and a higher specific radioactivity may be required to detect relatively inactive enzymes.
16. Hexoses such as Man do not absorb at 195 nm and require other methods for detection. Thus GlcNAc-transferase I substrate with 3 Man residues and an octyl group does not absorb, but the amide-carbonyl group of GlcNAc absorbs at this wavelength, and GlcNAc-containing products can be detected.
17. A possible cell line for this purpose is prostate cancer cells from bone metastasis PC-3, obtained from ATTC (CRL-1435). It would also be convenient to use purified or large amounts of recombinant GlcNAc-transferases I and II for the enzymatic synthesis of 2,2octyl (*see* Chapter 13). This would eliminate *N*-acetylglucosaminidase action that could degrade GlcNAc-transferase products.
18. GTs are often found as different members of transferase gene families. Reverse transcriptase polymerase chain reaction is an additional method to examine the glycodynamics of bone cells (*see* Chapters 11 and 13).
19. Columns are labeled "for one time use" and have a limited lifespan, but can probably function well after using them 10 times.

Acknowledgments

This work was supported by Materials and Manufacturing Ontario and a Research Scientist Award from The Arthritis Society to I. Brockhausen.

References

1. Gehron Robey, P., Fedarko, N. S., Hefferan, T. E., et al. (1993) Structure and molecular regulation of bone matrix proteins. *J. Bone Miner. Res.* **8,** S483–S487.
2. Midura, R. J. and Hascall, V. C. (1996) Bone sialoprotein—a mucin in disguise? *Glycobiology* **6,** 677–681.
3. de Pollak, C., Arnaud, E., Renier, D., and Marie, P. J. (1997) Age-related changes in bone formation, osteoblastic cell proliferation and differentiation during postnatal osteogenesis in human calvaria. *J. Cell. Biochem.* **128,** 128-–139.
4. Malaval, L., Modrowski, D., Gupta, A. K., and Aubin, J. E. (1994) Cellular expression of bone-related proteins during in vitro osteogenesis in rat bone marrow stromal cell cultures. *J. Cell. Physiol.* **158,** 555–572.
5. Mundy, G. R. (1996) Regulation of bone formation by bone morphogenic proteins and other growth factors. *Clin. Orthop. Relat. Res.* **323,** 24–28.
6. Brockhausen, I., Schutzbach, J., and Kuhns, W. (1998) Glycoproteins and their relationship to human disease. *Acta Anat.* **161,** 36–78.
7. Brockhausen, I. and Kuhns, W. (1997) Glycoproteins and human disease. Medical Intelligence Unit, CRC Press and Mosby-Year Book, Chapman & Hall NY, Springer-Verlag, Heidelberg, Germany.
8. Varki, A. (1993) Biological roles of oligosaccharides: all of the theories are correct. *Glycobiology* **3,** 97–130.
9. Hanasaki, K., Varki, A., Stamenkovic, I., and Bevilacqua, M. P. (1994) Cytokine-induced beta-galactoside alpha-2,6-sialyltransferase in human endothelial cells mediates alpha 2,6-sialylation of adhesion molecules and CD22 ligands. *J. Biol. Chem.* **269,** 10,637–10,643.
10. Keppler, O. T., Peter, M. E., Hinderlich, S., et al. (1999) Differential sialylation of cell surface glycoconjugates in a human B lymphoma cell line regulates susceptibility for CD95 (APO-1/Fas)-mediated apoptosis and for infection by a lymphotropic virus. *Glycobiology* **9,** 557–569.
11. Koya, D., Dennis, J. W., Warren, C. E., et al. (1999) Overexpression of core 2 N-acetylglycosaminyltransferase enhances cytokine actions and induces hypertrophic myocardium in transgenic mice. *FASEB J.* **13,** 2329–2337.
12. Hiraishi, K., Suzuki, K., Hakomori, S., and Adachi, M. (1993) Le y antigen expression is correlated with apoptosis (programmed cell death). *Glycobiology* **3,** 381–390.
13. Leist, M. and Wendel, A. (1995) Tunicamycin potently inhibits tumor necrosis factor-induced hepatocyte apoptosis. *Eur. J. Pharmacol.* **292,** 201–204.
14. Brockhausen, I., Lehotay, M., Yang, J., et al. (2002) Glycoprotein biosynthesis in porcine aortic endothelial cells and changes in the apoptotic cell population. *Glycobiology* **12,** 33–45.
15. Yang, X., Lehotay, M., Anastassiades, T., Harrison, M., and Brockhausen, I. (2004) The effect of TNFalpha on glycosylation pathways in bovine synoviocytes. *Biochem. Cell Biology* **82,** 559–568.

16. Delmotte, P., Degroote, S., Lafitte, J. J., Lamblin, G., Perini, J. P., and Roussel, P. (2002) Tumor Necrosis Factor alpha increases the expression of glycosyltransferases and sulfotransferases responsible for the biosynthesis of sialylated and sulfated Lewis x epitopes in the human bronchial mucosa. *J. Biol. Chem.* **277,** 424–431.
17. Brockhausen, I., Yang, X., and Harrison, M. (2005) Glycodynamics of human osteoblastic cells. *Glycobiology* (Abstract 223) **15,** 1242.
18. Kassem, M., Risteli, L., Mosekilde, L., Melsen, F., and Eriksen, E. F. (1991) Formation of osteoblast-like cells from human mononuclear bone marrow cultures. *APMIS* **99,** 269–274.
19. Brockhausen, I. (2000) O-linked chain glycosyltransferases. Mucin methods and protocols. *Methods Mol. Biol.* **125,** 273–293.
20. Brockhausen, I., Yang, J., Lehotay, M., Ogata, S., and Itzkowitz, S. (2001) Pathways of mucin O-glycosylation in normal and malignant rat colonic epithelial cells reveal a mechanism for cancer-associated Sialyl-Tn antigen expression. *Biol. Chemistry* **382,** 219–232.
21. National Academy of Sciences—National Research Council, Washington, DC (1972) *Specifications and Criteria for Biochemical Compounds, 3rd ed.*
22. Möller, G., Reck, F., Paulsen, H., et al. (1992) Control of glycoprotein synthesis: Properties and substrate specificity of UDP-GlcNAc: Man α3R β2-N-acetylglucosaminyl-transferase I using synthetic substrates. *Glycoconj. J.* **9,** 180–190.
23. Paulsen, H., Meinjohanns, E., Reck, F., and Brockhausen, I. (1993) Building units of oligosaccharides CVIII. Synthesis of modified oligosaccharides of N-glycoproteins for substrate specificity studies of N-acetylglucosaminyltransferase III to V. *Liebigs Ann. Chemie* 737–750.

15

Micromethods for the Characterization of Lipid A-Core and O-Antigen Lipopolysaccharide

Cristina L. Marolda, Piya Lahiry, Enrique Vinés, Soledad Saldías, and Miguel A. Valvano

Summary

Methods for rapid and simple analysis of lipopolysaccharide (LPS) from bacterial whole-cell lysates or membrane preparations have contributed to advancing our knowledge of the genetics of the LPS biogenesis. LPS, a major constituent of the outer membranes in Gram-negative bacteria, has a complex mechanism of synthesis and assembly that requires the coordinated participation of many genes and gene products. This chapter describes a collection of methods routinely used in our laboratory for the characterization of LPS in *Escherichia coli* and other bacteria.

Key Words: Silver staining; SDS-PAGE; tricine; 2-*keto*-3-deoxyoctulosonic acid; purpald; thiobarbituric acid; Pro-Q Emerald; Western blot.

1. Introduction

Lipopolysaccharide (LPS) is a major constituent of the outer membranes in Gram-negative bacteria. LPS consists of several regions (*see* **Fig. 1**) including lipid A, core oligosaccharide (OS), and in some microorganisms, *O*-specific polysaccharide (or O-antigen) that is made of repeating OS subunits *(1,2)*. The lipid A is embedded in the lipid bilayer of the outer membrane and consists of a β-1,6-linked glucosamine disaccharide, which becomes phosphorylated and acylated with a variable number of fatty and hydroxyfatty acid chains *(2)*. The core OS can be subdivided into inner and outer core domains. The outer core usually consists of hexoses and hexosamines, whereas the inner core, depending on the particular species, is composed of one to three residues of 3-deoxy-D-*manno*-octulosonic acid, and two or three residues of L-*glycero*-D-*manno*-heptose *(3)*. LPS plays an important role in maintaining the structural integrity

From: *Methods in Molecular Biology, vol. 347: Glycobiology Protocols*
Edited by: I. Brockhausen © Humana Press Inc., Totowa, NJ

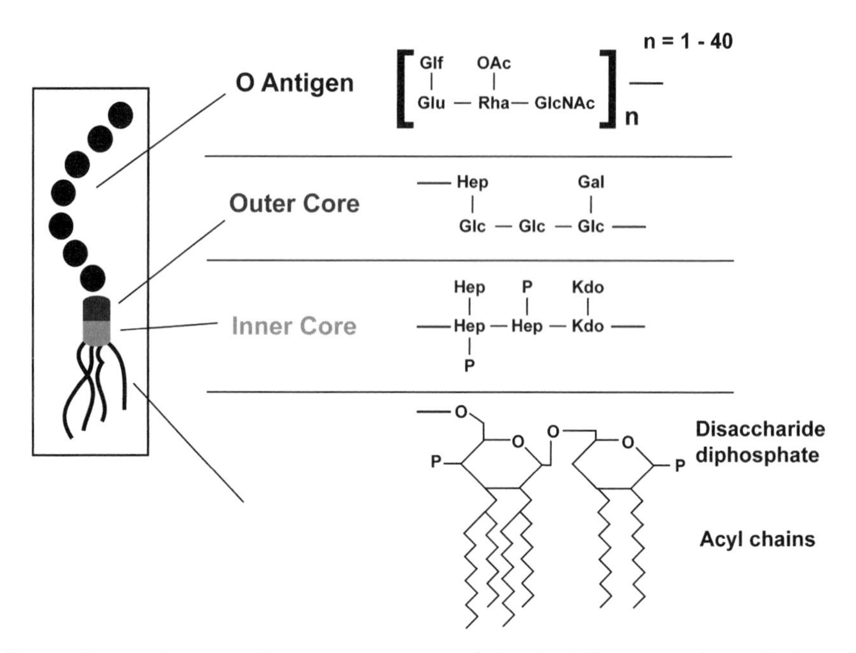

Fig. 1. Lipopolysaccharide (LPS) structure of the O16 O-antigen from *Escherichia coli* K-12. Scheme of the structure of the O16 LPS molecules indicating its various regions: O Antigen, Outer Core, Inner Core, and Lipid A regions. Glf, galactofuranose; Glc, glucose; Rha, rhamnose; GlcNAc, *N*-acetylglucosamine; OAc, *O*-acetyl; Hep, heptose; Gal, galactose; KDO, 2-*keto*-3-deoxyoctulosonic acid; P, phosphate.

of the bacterial outer membrane by interacting with outer membrane proteins as well as divalent cations ***(4)***. Phosphate groups covalently attached to heptose residues in the inner core participate in these ionic interactions, which provide a barrier preventing the passage of hydrophobic substances such as detergents, dyes, and antibiotics across the outer membrane ***(5,6)***.

The biosynthesis of LPS is a complex process involving various steps that occur at the plasma membrane, followed by the translocation of LPS molecules to the bacterial cell surface (*see* **ref.** ***2*** and Chapter 16). The core OS is assembled on preformed lipid A by sequential glycosyl transfer of monosaccharides, while the O-antigen is assembled on undecaprenol-phosphate (Und-P), a polyisoprenoid lipid to which O-antigen is linked via a phosphodiester bond. These pathways eventually converge by the ligation of the O-antigen onto the outer core domain of the lipid A-core OS acceptor, with the concomitant release of Und-P ***(1,2)***. Und-P is also required as a lipid intermediate for the biosynthesis of other cell-surface structures, including peptidoglycan and the enterobacterial common antigen.

Mutations in the core biosynthesis genes (*waa*, formerly *rfa*; for a discussion of the LPS genes nomenclature, *see* **ref.** ***7***) lead to “rough” mutants with an incomplete core, which lacks the site for the attachment of *O*-polysaccharides. *Escherichia coli* mutants lacking heptose in the LPS display a more dramatic phenotype known as "deep rough." This phenotype is characterized by hypersensitivity to novobiocin, detergents, and bile salts, as well as defects in F plasmid conjugation and generalized transduction by the bacteriophage P1 (for a review *see* **ref. *3***). Mutations in *wb** (formerly *rfb*) genes, which are involved in the synthesis of the *O*-polysaccharide, give rise to rough mutants that have a complete core structure. Typically, *wb** gene clusters encode nucleotide sugar synthases (for biosynthesis of the nucleotide sugar precursors specific to O-antigens), and glycosyltransferases (for the sequential and specific addition of sugars that make the *O*-repeating unit). Additional genes encoding functions involved in the assembly of the *O*-polysaccharide are also present in these clusters, such as *wzy* (O-antigen polymerase) and *wzx* (putative O-antigen flippase) in some systems, and *wzm* (membrane component of ABC transporter) and *wzt* (adenosine triphosphate-binding component of ABC transporter) in others ***(1,2)***. In systems containing *wzx* and *wzy*, the average size distribution of the *O*-polysaccharide chain is modulated by the product of *wzz*, a gene usually located in the proximity of *wb** clusters. An *O*-ligase activity encoded by a gene located within the core OS cluster, *waaL*, is required for the transfer of the *O*-polysaccharide onto lipid A-core OS.

The detailed characterization of LPS requires large-scale isolation procedures and chemical methods, which can range from a compositional analysis to sophisticated approaches involving various types of spectroscopy and nuclear magnetic resonance. However, routine and rapid analysis of LPS can also be accomplished using simple procedures in crude lysates of whole bacterial cells or membrane preparations ***(8,9)***. This chapter presents a set of rapid methods used in our laboratory to analyze lipid A-core OS and O-antigen LPS in *E. coli* K-12, a model bacterium. Classically, silver staining has been used to detect LPS molecules separated by electrophoresis in polyacrylamide gels ***(9)***. Also presented here is a comparison of the silver-staining method with two other methods. The methods provided here make possible the characterization of the LPS phenotype of mutants with defects in the biosynthesis of O-antigen and core OS, and can be applied to other Gram-negative bacteria.

2. Materials

2.1. Strains and Growth Medium

2.1.1. Strains

1. The following *E. coli* K-12 strains are used (abbreviated genotypes in parenthesis): W3110 (*F*$^-$ λ^- IN(*rrnD-rrnE*) ***(10)***; CLM37 (W3110Δ*wecA*) ***(11)***; EVV10 (W3110Δ*wzy*::Kn; Vines and Valvano, unpublished); EVV16 (W3110Δ*wzz*::Kn);

χ705 (*gmhA*+) ***(10)***; χX711 (*F⁻leu proAB Str*R T3R *arg*) ***(12)***; D21e7 (*proA trp, his, lac, rps*L *ampA tsx*-81 *rfa*) ***(12)***; CS2051 (CmR,TcR, Δ*rfa*GPMN) ***(12)***; and D31m4 (*proA trp his lac rps*L *ampA lps*) ***(12)***.

2. Transformation of these strains with pMF19 allows for the expression of O16 antigen in *E. coli* K-12 ***(13)***, provided that the core OS is completely synthesized. Plasmid pMF19 encodes the rhamnosyltransferase WbbL whose gene is mutated in *E. coli* K-12 ***(14,15)***, thus allowing the formation of a complete O16-specific polysaccharide in *E. coli* K-12.

2.1.2. Media

1. All strains are grown in Luria-Bertani (LB) Agar media (Difco).
2. When required, antibiotics are added as follows: Kanamycin (Kn, Roche) is prepared as a 20 mg/mL stock solution in deionized water and the final concentration in the plates is 40 μg/mL; chloramphenicol (Cm, Sigma Aldrich) is prepared as a 30 mg/mL stock solution in 100% ethanol, and the final concentration in the plates is 30 μg/mL; tetracycline (Tc, Sigma) is prepared as a 20 mg/mL stock solution in 50% ethanol and the plates contain 20 μg/mL. All antibiotic stock solutions are stored in dark bottles at 4°C.

2.2. LPS Preparation

1. Phosphate-buffered saline (PBS), pH 7.2.
2. Lysis buffer: 2% (w/v) sodium dodecyl sulphate (SDS), 4% β-mercaptoethanol (2-ME), and 0.5 *M* of Tris-HCl (pH 6.8); store at 4°C.
3. Proteinase K (Roche) is prepared in a 20-mg/mL stock solution and stored at 4°C.
4. 90% phenol: 90% phenol, 0.1% 2-ME, and 0.2% 8-hydroxyquinoline (Fisher Scientific). Store at 4°C (this solution will crystallize at this temperature and should be warmed prior to use).
5. Ethyl ether saturated with 10 m*M* of Tris-HCl and 1 m*M* of ethylenediaminetetraacetic acid (EDTA), pH 8.0.
6. DNase I (Sigma).

2.3. 2-Keto-3-Deoxyoctulosonic Acid Determination

2.3.1. Thiobarbituric Acid Assay

1. 2-*Keto*-3-deoxyoctulosonic (KDO) standard: Prepare a stock solution of 1 mg/mL of KDO (Sigma) in deionized water and store it in small aliquots at –20°C. For the standard curve, prepare 50-μL aliquots with the following final concentrations of KDO: 0, 4, 8, 16, and 20 μg/mL. Prepare these standards fresh each time.
2. 0.5 *M* of sulphuric acid (BDH).
3. 0.1 *M* of periodic acid (Sigma). Store at room temperature.
4. 0.2 *M* of sodium arsenite (Sigma); dissolve the powder in 0.5 *M* of HCl and store at room temperature.

5. 0.6% (w/v) thiobarbituric acid (TBA; Sigma). Dissolve the powder in warm (55°C) deionized water water until the solution clears. This solution should be prepared fresh.
6. *n*-Butanol saturated with 0.5 *M* of HCl. Store this solution at 4°C.

2.3.2. Purpald Assay

1. All solutions for this assay should be prepared fresh.
2. KDO standard: Prepare samples for the standard curve in 50-µL aliquots with the following concentrations of KDO: 0, 4, 8, 16, 20, 40, and 60 µg/mL.
3. 32 m*M* and 64 m*M* of $NaIO_4$ solutions (Sigma).
4. Purpald reagent: 136 m*M* of 4-amino-3-hydrazino-5-mercapto-1,2,4-triazole (Sigma) dissolved in 2 *N* of NaOH.
5. 2-Propanol.

2.4. SDS-Polyacrylamide Gel Electrophoresis

2.4.1. Tris-Glycine-SDS-Polyacrylamide Gel Electrophoresis

1. Acrylonitrite-butadiene acrylate (ABA) (44:0.8): 88 g of acrylamide (Roche) and 1.6 g of *bis*-acrylamide (Bio-Rad). Dissolve acrylamide and *bis*-acrylamide in approx 130 mL of warm deionized water and raise the total volume to 200 mL with deionized water after the powder is completely dissolved (*see* **Note 1**).
2. Running buffer: 25 m*M* of Tris-HCl, 0.2 m*M* of glycine, and 0.1% SDS.
3. Ammonium persulfate (APS, Roche): Make a 10% (w/v) solution in deionized water and store it in small aliquots at –20°C.
4. Resolving gel (12.5%): 5.68 mL of ABA (44:80), 0.35 mL of 10% SDS, 4.37 mL of 1.5 *M* Tris-HCl (pH 8.0), 4.2 g of urea, 4.2 mL of deionized water, 0.035 mL of 10% APS, and 0.025 mL of *N,N,N,N′*-tetramethyl ethylenediamine (TEMED) (*see* **Note 2**).
5. Stacking gel: 0.5 mL of ABA (44:0.8), 0.05 mL of 10% SDS, 2.5 mL of 0.25 *M* Tris-HCl (pH 6.8), 1.83 mL of deionized water, 0.015 mL of 10% APS, and 0.02 mL of TEMED.
6. 3X Loading dye: 0.187 *M* of Tris-HCl (pH 6.8), 6% SDS, 30% glycerol, 0.03% bromophenol blue, and 15% 2-ME.

2.4.2. Tricine-SDS-Polyacrylamide Gel Electrophoresis

1. ABA 46.5:3: 23.25 g of acrylamide and 1.5 g of *bis*-acrylamide to 50 mL with deionized water. Store at 4°C (if crystals appear, the stock solution can be warmed to 50°C; *see* **Note 1**).
2. ABA 48:0.5: 24 g of acrylamide and 0.75 g of *bis*-acrylamide to 50 mL with deionized water. Store at 4°C (if crystals appear, the stock solution can be warmed to 50°C).
3. Gel buffer: 0.3% SDS and 3.0 *M* of Tris-HCl (pH 8.45). Store at 4°C.
4. Anode buffer: 0.2 *M* of Tris-HCl (pH 8.9). A 5X concentrated solution can be made and diluted to 1X before use.

5. Catode buffer: 0.1 *M* of tricine (Sigma), 0.1% SDS, and 0.1 *M* of Tris-HCl (pH 8.25). A 5X solution can also be made and diluted to 1X before use.
6. Resolving gel (14%): 2.84 mL of ABA 46.5:3, 3.32 mL of gel buffer, 1.04 mL of glycerol, 2.78 mL of deionized water, 0.02 mL of 10% APS, and 0.012 mL of TEMED. Allow to polymerize for 30 min at room temperature (*see* **Note 2**).
7. Stacking gel: 0.25 mL of ABA 48:1.5, 0.775 mL of gel buffer, 2.1 mL of deionized water, 0.03 mL of 10% APS, and 0.007 mL of TEMED. Allow to polymerize for 20 min at room temperature.

2.5. LPS Detection

2.5.1. Silver Stain

1. Fixing solution: 60% methanol and 10% acetic acid.
2. 7.5% Acetic acid.
3. 0.2% (w/v) Periodic acid in deionized water. This solution should be prepared fresh.
4. 19.4% (w/v) Silver nitrate: 1.6 g of $AgNO_3$ in 8 mL of deionized water. This solution should be prepared fresh.
5. 0.36% Sodium hydroxide: 3.6 g of NaOH in 1000 mL of deionized water. Store at room temperature.
6. Staining solution: this solution is unstable and it is recommended that it be prepared during the last wash prior to staining (*see* **Subheading 3.**). The ingredients for the staining solution should be added in the following order: 42 mL of 0.36% NaOH, 2.8 mL of NH_4OH concentrated solution (Fisher), 8 mL of 19.4% $AgNO_3$ (added dropwise and swirled until the solution becomes clear; *see* **Note 3**), and 148 mL of deionized water (*see* **Note 4**).
7. Developing solution: 0.05% citric acid, 10% methanol, and 0.019% formaldehyde.

2.5.2. Zinc Sulphate

1. 10 m*M* of zinc sulphate.
2. 0.2 *M* of imidazole (Sigma).

2.5.3. Pro-Q® Emerald 300 Lipopolysaccharide Gel Stain Kit (Molecular Probes P-20495)

1. Fixing solution (60% methanol and 10% acetic acid).
2. *N,N*-dimethylformamide (DMF).
3. 3% Acetic acid.
4. Components A, B, C, and D are provided with the kit (Invitrogen-Molecular Probes, Burlington, Ontario, Canada, cat. no. P20495).

2.5.4. Western Blotting

1. Nitrocellulose membrane (Protran, Perkin Elmer) or polyvinylidene difluorite membrane (Roche).
2. Transfer buffer (25 m*M* of Tris-HCl, 190 m*M* of glycine, and 20% methanol).
3. Tris-buffered saline (TBS): 50 m*M* of Tris-HCl (pH 7.6) and 150 m*M* of NaCl.
4. Western blocking reagent (Roche).

5. Primary antibody: O16 polyclonal rabbit antibody (The Gastroenteric Disease Center, Wiley Laboratory, University Park, PA). This can be replaced with any other appropriate monoclonal or polyclonal antibody.
6. Secondary antibody: IRDye800CW affinity purified anti-rabbit IgG antibodies (Rockland Immunochemicals). This antibody is for fluorescence detection with an Odyssey infrared imaging system (LI-COR Biosciences). Other secondary antibodies can be used according to the preferred detection method.

3. Methods

3.1. LPS Preparation

The procedure for LPS preparation described here is based on the previous micro methods *(8,9)*. Some modifications have been introduced to completely remove the protein content from the samples and improve the reproducibility and quality of the preparations.

1. Grow the bacteria in LB agar plates containing the appropriate antibiotics, but bacterial cells can also be grown in liquid media. The authors' experience shows that larger amounts of LPS are obtained from bacteria grown in solid media.
2. Collect bacteria from the plate with a sterile Dacron or cotton swab and suspend them in 2 mL of 1X PBS (pH 7.2). The amount of material to be collected will depend on the amount of bacteria present in the plates.
3. Measure the optical density of the suspension at 600-nm wavelength and adjust the density to 2.0.
4. Transfer 1.5 mL of the normalized suspension to a microcentrifuge tube, centrifuge in a microcentrifuge at 10,000*g* for 1 min, and discard the supernatant. If more LPS is required, increase the volume or adjust the turbidity to 3.0 or 4.0 in **step 3**.
5. Resuspend the bacterial pellet in 150 µL of lysis buffer and boil for 10 min. When high concentrations of bacterial cells are used, the suspension will sometimes become viscous. If this happens, add DNase I (Sigma) at a final concentration of 100 µg/mL and incubate at 37°C for 30 min.
6. Add 10 µL of proteinase K (stock at 20 mg/mL), vortex, and incubate at 60°C for 60 min. In this step the samples can be incubated longer, which is sometimes beneficial when dealing with some microorganisms (such as *Burkholderia* or *Ralstonia sp*).
7. Add 150 µL of prewarmed (70°C) 90% phenol solution and incubate at 70°C for 15 min (vortex every 5 min). With this step the proteins that were digested by proteinase K and residual undigested proteins are extracted. It is not recommended to extend this step beyond 15 min.
8. Incubate for 10 min on ice. This facilitates the separation of the aqueous and phenolic phases.
9. Centrifuge the samples in a microcentrifuge at maximum speed (10,000*g*) for 10 min and transfer approx 80–100 µL of the clear aqueous phase (top phase) to

a clean tube, making sure that the white interface containing proteins is not transferred. The presence of proteins can interfere with the detection procedures.

10. Add 10 vol (usually 500 μL) of ethyl ether saturated with Tris-EDTA to the clear phase and mix vigorously. The ether will remove any traces of phenol present in the samples (*see* **Note 5**).
11. Centrifuge samples in a microcentrifuge at maximum speed (10,000*g*) for 1 min and discard the ether phase (top phase) by aspiration.
12. Keep the samples at –20°C. Unless there is a need for KDO determinations, it is better to add the loading dye at this stage, mix, and keep the samples in the freezer. Under these conditions samples can safely be stored for several months.

3.2. KDO Determination

KDO is a unique and conserved sugar component of the LPS molecules. Therefore, the determination of the KDO can be used to estimate the LPS concentration in the samples. The TBA method ***(16)*** takes advantage of the fact that, following acid hydrolysis at 100°C and periodate oxidation, KDO yields formylpyruvic acid. This compound reacts with TBA, resulting in a chromogen with maximal absorption at 550 nm. This assay has some limitations owing to the fact that in some cases KDO residues are substituted with other groups that prevent reaction with TBA, and in others certain bacteria contain only one KDO substituted at the C-4 or C-5 position, which is not susceptible to periodic acid digestion.

The Purpald method relies on the principle that vicinal glycol groups of the sugar residues such as KDO and heptose in LPS are subjected to periodate oxidation at room temperature, yielding quantitative formaldehyde measurable by purpald reagent ***(17)***. This assay has sensitivity comparable to that of the TBA assay but circumvents the acid hydrolysis and boiling steps, enabling measurements in a large number of samples at the same time using microtiter plates.

3.2.1. Thiobarbituric Acid

1. Start with 50 μL of samples (unknown and controls for the standard curve). Add 50 μL of 0.5 *M* H_2SO_4, vortex, and boil for 8 min. Boil only the unknowns as there is no need to boil the controls (*see* **Note 6**).
2. Cool the samples at room temperature for 10 min.
3. Add 50 μL of 0.1 *M* periodic acid, vortex, and incubate at room temperature for 10 min.
4. Add 200 μL of 0.2 *M* sodium arsenite in 0.5 *M* of HCl, mix by vortexing, and then add 800 μL of 0.6% freshly prepared TBA. Vortex again and boil for 10 min. During this step the pink color will start to appear. The intensity of the color increases with time.
5. Cool the samples at room temperature for 30–40 min and then split them into two 600-μL portions using clean microfuge tubes.

6. Add 750 μL of *n*-butanol equilibrated with 0.5 *M* of HCl to each tube, vortex, and centrifuge for 4 min. After centrifugation, a pink precipitate may appear that should be discarded.
7. Recover and combine the organic phases (pink) into a spectrophotometer cuvet and read the absorbance at 552 nm and 509 nm. Readings should be taken as soon as possible as the color is unstable.
8. To calculate the concentration of KDO, the values obtained from the absorbance at 552 nm subtracted by the absorbance at 509 nm for each dilution of the standard are plotted against the concentration of the standard in μg/mL. This should generate a linear standard curve. The concentration of the unknowns (LPS samples) is calculated by extrapolating the absorbance at 552 nm subtracted by the absorbance at 509 nm using the standard curve.

3.2.2. Purpald

1. Dilute the samples to be tested in deionized water. The dilution will depend primarily on the amount of LPS present in the samples; the authors' lab generally tries 1/500, 1/1000, or even 1/5000.
2. Place 50 μL of samples and standard controls in microtiter plates (prepare standards and samples in duplicates).
3. Add 50 μL of 32 m*M* $NaIO_4$ to each well, mix, and incubate at room temperature for 25 min.
4. Add 50 μL of purpald reagent, mix, and incubate at room temperature for 20 min. A yellow color appears that clears very quickly and is replaced by a purple color that intensifies with time.
5. Add 50 μL of 64 m*M* $NaIO_4$ to stop the reaction.
6. Add 20 μL of 2-propanol to clear the solution of the foam that is formed in the previous step.
7. Read absorbance at 550 nm.

3.3. SDS-Polyacrylamide Gel Electrophoresis

There are various systems to separate the LPS samples for analysis. This chapter presents the two systems used in our laboratory. The Tris-Glycine system provides better resolution of the O-antigen region (upper portion of the gel), whereas the Tricine system provides better resolution of the core region (lower portion). As an example of results with these two buffer systems, compare **Fig. 2A** with **Fig. 2B**, and **Fig. 3A** with **Fig. 3B**.

3.3.1. Tris-Glycine-SDS-Polyacrylamide Gel Electrophoresis

1. Prepare the 12.5% acrylamide solution for resolving gel and degas it for 5 min before adding the APS and TEMED. Once TEMED has been added, pour it immediately on the casting gel unit as it will start polymerizing right away. Very carefully add a layer of isopropanol, isobutanol, Tris-Glycine running buffer, or 0.1% SDS to avoid any contact with the air. Let polymerize for 20–30 min.

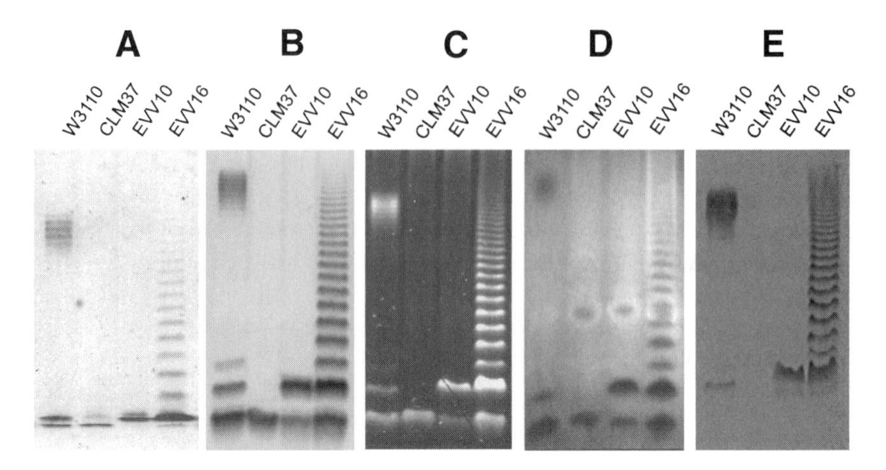

Fig. 2. Analysis of O-antigen in *Escherichia coli* K-12 strains. All strains used to prepare lipopolysaccharide (LPS) shown in **panels A–E** were transformed with pMF19 to enable the expression of O16-specific polysaccharide ***(14)***. Strain W3110 is the parental *E. coli* K-12 expressing O16 antigen with a typical bimodal distribution of O-antigen polysaccharides, as indicated by the cluster of bands shown in the region that corresponds to polysaccharides with 14–24 O-antigen subunits. CLM37 has a mutation in *wecA*, the gene encoding the initiating enzyme for the synthesis of the O-antigen subunit ***(1,18)***. EVV10 has a deleted *wzy* gene, which encodes the O-antigen polymerase ***(10)***. Consequently, this strain only forms bands corresponding to the O-antigen subunit attached the lipid A-core oligosaccharide, which are clearly visible in panels B–E. Strain EVV16 has a mutation in *wzz*, which encodes a protein involved in the regulation of the preferential length of the *O*-polysaccharide chains ***(10)***. Therefore, this mutant forms *O*-polysaccharides with a uniform distribution (monomodal distribution). Samples in panel A were separated by sodium dodecyl sulfate-polyacrylamide gel electrophoresis (SDS-PAGE) on a 12.5% polyacrylamide gel with the Tris-Glycine system. Samples in panels B–E were separated by SDS-PAGE on a 14% polyacrylamide gel with the Tricine system. Gels in panels A and B were silver-stained. Gel in panel C was stained with the Pro-Q Emerald 300 Lipopolysaccharide Kit. Gel in panel D was stained with the zinc sulphate method. Panel E is a scan of a Western blot reacted with an O16 polyclonal rabbit antibody where the specific bands were detected by fluorescence with an Odyssey infrared imaging system (LI-COR Biosciences).

2. Rinse the resolving gel with running buffer and layer it with the stacking solution that has only been degassed for 2 min prior to the addition of TEMED. Place the comb carefully and let polymerize for 20–30 min.
3. Assemble the gel apparatus and add the running buffer to the lower and upper chambers.
4. Adjust the LPS samples to the same concentration of KDO (load 4–10 ng/lane) in 15-μL aliquots and add sufficient loading dye to make it 1X.

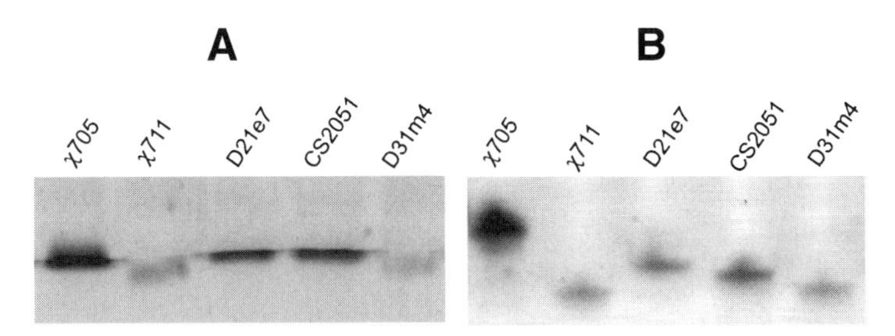

Fig. 3. Analysis of lipid A-core oligosaccharide (OS) in *Escherichia coli* K-12 strains. Strain χ705 is a derivative of W3110 and forms a complete lipid A-core OS. Strain χ711 is a derivative of χ705 with a deletion in the *gmhA* gene that encodes an enzyme required for the synthesis of adenosine 5′-diphosphate-heptose ***(3,12)***. This strain forms a lipid A-core OS devoid of heptose residues, as identified by a fast-migrating band in the gel. The other strains have mutation is various glycosyltransferases involved in the assembly of the outer core OS, resulting in truncated lipid A-core OS forms that are detected as bands of intermediate migration between parental and heptoseless core types. Samples in **panel A** were separated by sodium dodecyl sulfate-polyacrylamide gel electrophoresis (SDS-PAGE) on a 12.5% polyacrylamide gel with the Tris-Glycine system. Samples in **panel B** were separated by SDS-PAGE on a 14% polyacrylamide gel with the Tricine system.

5. Load equal volumes of all the samples in a total volume no greater than 15 μL and add loading dye to make it 1X. Overloading will result in crooked and fuzzy bands. Run the electrophoresis at 20 mAmp until the dye has reached the bottom of the gel.
6. Place the gel into the solution appropriate for the next step after the run is finished.

3.3.2. Tricine-SDS-Polyacrylamide Gel Electrophoresis

1. Prepare the 14% ABA 46.5:3 solution for the resolving gel and degas for 5 min before adding the APS and the TEMED. Once the TEMED has been added, pour immediately as it will start polymerizing right away. Carefully add a layer of isopropanol, isobutanol, catode running buffer, or 0.1% SDS to avoid any contact with the air. Let polymerize for 20–30 min.
2. Rinse the resolving gel with catode running buffer and layer it with the stacking solution that has only been degassed for 2 min. Carefully place the comb and let polymerize for 20–30 min.
3. Assemble the gel apparatus and add the catode buffer to the upper chamber and the anode buffer to the lower chamber.
4. Adjust the LPS samples to the same amount of KDO (we usually use 4–10 ng/lane). Load equal volumes for all the samples in a total volume no greater than 15 μL and add loading dye to make it 1X. Overloading will result in crooked and fuzzy bands.

5. Load the samples into the wells and run them at 50 V until the dye has reached the resolving gel (30–40 min), then switch to 130 V and run the samples for 20–30 min after the dye has left the gel.

3.4. LPS Detection

3.4.1. Silver Stain

The silver-stain protocol is very sensitive to metals present in the water, so it is recommended that milli-Q deionized water be used throughout the procedure. All the containers used for this procedure should be made out of glass that has been rinsed with 70% ethanol. It is better to always use the same dish, slowly discarding the liquid after every wash. It is also very important to wear gloves at all times and to avoid touching the gel (even with gloved hands). Metal or plastic painter's spatulas are recommended to prevent the gel from sliding out of the dish while doing the changes of the different solutions (*see* **Notes 7** and **8**). All the solutions should be prepared fresh before use.

1. Soak the gel overnight (or longer) in 200 mL of fresh fixing solution. The gel may turn white during the initial stages of fixing but it should clear up by the next day.
2. Wash the gel with 200 mL of 7.5% acetic acid for 30 min.
3. Discard the acetic acid, add 200 mL of 0.2% periodic acid, and rock for 30 min.
4. Wash the gel with deionized water, rocking for an hour and changing the water every 15 min. These washes should be done with ample water, typically 500 mL or more for each wash. Prepare the staining solution during the last wash.
5. Add the staining solution and rock for 15 min.
6. Wash the gel with deionized water for 45 min, changing the water every 15 min.
7. Add the developer and rock until brown or yellow bands start to appear (5–15 min).
8. Stop developing with several changes of deionized water (three or four quick changes and then rocking three times for 10 min). Note that despite the several washes with deionized water the bands will continue to develop, getting darker as times passes.
9. Scan or take a picture as soon as possible. If this cannot be done, continue rocking until the data can be recorded. Leaving the gel standing will result in the gel sticking to the bottom of the glass dish and a layer of silver nitrate covering the surface of the gel (*see* **Note 9**). An example of the results obtained with silver-stained gels is shown in **Fig. 2A,B**.

3.4.2. Pro-Q Emerald 300 (Molecular Probes)

This procedure is preferred if money is not a problem. As the developing is based in fluorescence, the gel must be very clean and devoid of dust; filtering the resolving solution is recommended before adding the APS and TEMED. All the steps in this procedure can be done in plastic containers and with rocking at all times.

1. Fix overnight with the same fixing solution as for the silver stain.
2. Wash twice with 300 mL of 3% acetic acid for 20 min each time.
3. Incubate with 25 mL of oxidizing solution for 30 min.
4. Discard the liquid and wash three times with 3% acetic acid for 20 min each time.
5. Incubate for 90–120 min with the stain solution in the dark. Do not stain overnight.
6. Wash twice for 20 min in the dark with 3% acetic acid.
7. Use a 300-nm transilluminator with six 15-W bulbs to detect the bands. The gel can be photographed with a camera with the appropriate filters. An example of the results obtained with this method is shown in **Fig. 2C**.

3.4.3. Zinc Sulphate

This protocol is less conventional but is quicker than the previous ones, although the resolution is not as good as with the previous procedures.

1. After running the gel, soak it in 800 mL of boiling deionized water and continue boiling for 15 min. Discard the water and repeat this step two more times. Make sure that the water is warmed before changing it. These washes will remove the SDS from the gel.
2. Transfer the gel into a glass dish containing 300 mL of 10 m*M* zinc sulfphate and rock for 15 min.
3. Wash the gel quickly three times with deionized water and then soak it in 200 mL of 0.2 *M* imidazole. Rock until the gel turns opaque and white bands appear for approx 1–3 min.
4. Stop the developing of the bands by rinsing the gel three times for 1 min with deionized water.
5. The gel can be photographed to record the data. An example of the result obtained with this method is shown in **Fig. 2D**.

3.4.4. Western Blot

1. After running the samples on either a Tris-Glycine- or a Tricine-SDS-polyacrylamide gel electrophoresis (SDS-PAGE), soak the gel for a couple of minutes in transfer buffer.
2. Transfer the bands onto nitrocellulose or polyvinylidene difluorite membrane for 60–70 min at 250 mAmp.
3. Wash the membrane quickly with 100 mL of TBS.
4. Block with 30 mL of 5% Western blocking reagent or with 5% skim milk at room temperature for 90 min (it can also be blocked at 4°C overnight).
5. Wash three times with TBS for 15 min each time.
6. Incubate the membrane with the primary antibody (O16 polyclonal rabbit antibody) diluted in TBS with rocking at 4°C overnight. If the primary antibody is a monoclonal antibody, the membrane can be incubated at room temperature for 1–2 h.
7. Wash three times with TBS for 20 min each time.

8. Incubate with the secondary antibody (IRDye800CW affinity-purified anti-rabbit IgG antibodies) at room temperature for 60 min.
9. Wash three times with TBS again for 10 min each time.
10. Detect the specific bands by fluorescence with an Odyssey infrared imaging system (LI-COR Biosciences). The bands can also be detected with chemilluminescence using a secondary antibody conjugated with horseradish peroxidase. **Figure 2E** shows an example of the results obtained with this method. Please note that the lowest band detected by the O16-specific antiserum corresponds to lipid A-core OS plus one O-antigen subunit.

4. Notes

1. **Caution:** Gloves and masks should be worn when handling acrylamide and *bis*-acrylamide, as they are neurotoxic substances.
2. The volume of resolving gel is good for casting two small gels of 8 × 10 cm. The amounts of each component can be upscaled as needed.
3. If the solution does not clear, add more ammonium hydroxide.
4. **Caution:** Ammonium hydroxide vapors from the concentrated NH_4OH solution are very toxic. Work under an appropriate fume hood.
5. **Caution:** This step should be performed in an appropriate fume hood because ethyl ether is extremely volatile.
6. The boiling time can be increased up to 15–20 min to obtain maximal hydrolysis of KDO. This is especially useful for LPS samples prepared from nonenteric bacteria (*see* http://www.cmdr.ubc.ca/bobh/methods/KDOASSAY.html).
7. The silver stain will not work properly at room temperatures below 18°C. If this is the case, perform the procedures by placing the dish on top of a 30°C water bath to keep it warm.
8. Samples from strains deficient in lipid A-core OS show a reduced intensity in the staining than that of samples from strains with a complete core. More sample should be loaded onto the gels in these cases (*see* **Fig. 3A**,**B**).
9. The stained gel can be dried onto filter paper (Whatmann No. 1) by placing the gel on the paper, carefully removing air bubbles, and covering it with clear wrap or cellophane paper on a Bio-Rad gel dryer with vacuum at 80°C for 60–90 min.

Acknowledgments

The authors would like to thank all other present and past members of the Valvano laboratory who have contributed over the years to continually improve the detection of O-antigen and lipid A-core OS. Our research was supported by grants from the Natural Sciences and Engineering Research Council of Canada, the Canadian Institutes of Health Research, The Canadian Bacterial Diseases Network, and the Canadian Cystic Fibrosis Foundation. Dr. Valvano holds a Canada Research Chair in Infectious Diseases and Microbial Pathogenesis.

References

1. Valvano, M. A. (2003) Export of O-specific lipopolysaccharide. *Front. Biosci.* **8,** 452–s471.
2. Raetz, C. R. H. and Whitfield, C. (2002) Lipopolysaccharide endotoxins. *Annu. Rev. Biochem.* **71,** 635–700.
3. Valvano, M. A., Messner, P., and Kosma, P. (2002) Novel pathways for biosynthesis of nucleotide-activated glycero-manno-heptose precursors of bacterial glycoproteins and cell surface polysaccharides. *Microbiology* **148,** 1979–1989.
4. Hancock, R. E. W., Karunaratne, D. N., and Bernegger-Egli, C. (1994) Molecular organization and structural role of outer membrane macromolecules. In *Bacterial Cell Wall, Vol. 27* (Ghuysen, J. M. and Hackenbeck, R., eds.), Elsevier Science, Amsterdam, NY, pp. 263–279.
5. Nikaido, H. (1994) Prevention of drug access to bacterial targets: permeability barriers and active efflux. *Science* **264,** 382–388.
6. Nikaido, H. and Vaara, M. (1985) Molecular basis of bacterial outer membrane permeability. *Microbiol. Rev.* **49,** 1–32.
7. Reeves, P. R., Hobbs, M., Valvano, M. A., et al. (1996) Bacterial polysaccharide synthesis and gene nomenclature. *Trends Microbiol.* **4,** 495–503.
8. Hitchcock, P. J. and Brown, T. M. (1983) Morphological heterogeneity among *Salmonella* lipopolysaccharide chemotypes in silver-stained polyacrylamide gels. *J. Bacteriol.* **154,** 269–277.
9. Tsai, C. M. and Frasch, C. E. (1982) A sensitive silver stain for detecting lipopolysaccharides in polyacrylamide gels. *Anal. Biochem.* **119,** 115–119.
10. Vinés, E., Marolda, C. L., Balachandran, A., and Valvano, M. A. (2005) Defective O-antigen polymerization in *tolA* and *pal* mutants of *Escherichia coli* in response to extracytoplasmic stress. *J. Bacteriol.* **187,** 3359–3368.
11. Linton, D., Dorrell, N., Hitchen, P. G., et al. (2005) Functional analysis of the *Campylobacter jejuni N*-linked protein glycosylation pathway. *Mol. Microbiol.* **55,** 1695–1703.
12. Brooke, J. S. and Valvano, M. A. (1996) Biosynthesis of inner core lipopolysaccharide in enteric bacteria identification and characterization of a conserved phosphoheptose isomerase. *J. Biol. Chem.* **271,** 3608–3614.
13. Marolda, C. L., Vicarioli, J., and Valvano, M. A. (2004) Wzx proteins involved in O-antigen biosynthesis function in association with the first sugar of the O-specific lipopolysaccharide subunit. *Microbiology* **150,** 4095–4105.
14. Feldman, M. F., Marolda, C. L., Monteiro, M. A., Perry, M. B., Parodi, A. J., and Valvano, M. A. (1999) The activity of a putative polyisoprenol-linked sugar translocase(Wzx) involved in *Escherichia coli* O-antigen assembly is independent of the chemical structure of the O repeat. *J. Biol. Chem.* **274,** 35,129–35,138.
15. Liu, D. and Reeves, P. R. (1994) *Escherichia coli* K12 regains its O-antigen. *Microbiology* **140,** 49–57.

16. Osborn, M. J. (1963) Studies in the Gram-negative cell wall. I. Evidence for the role of 2-keto-3-deoxyoctonoate in the lipopolysaccharide of *Salmonella typhimurium*. *Proc. Natl. Acad. Sci. USA* **50,** 499–506.
17. Lee, C. H. and Tsai, C. M. (1999) Quantification of bacterial lipopolysaccharides by the purpald assay: measuring formaldehyde generated from 2-keto-3-deoxyoctonate and heptose at the inner core by periodate oxidation. *Anal. Biochem.* **267,** 161–168.
18. Amer, A. O. and Valvano, M. A. (2002) Conserved aspartic acids are essential for the enzymic activity of the WecA protein initiating the biosynthesis of O-specific lipopolysaccharide and enterobacterial common antigen in *Escherichia coli*. *Microbiology* **148,** 571–582.

16

Assay for a Galactosyltransferase Involved in the Assembly of the O7-Antigen Repeat Unit of *Escherichia coli*

Inka Brockhausen, Walter A. Szarek, John G. Riley, and Jason Z. Vlahakis

Summary

Gram-negative bacteria have lipopolysaccharides terminating in repeating oligosaccharides which comprise the O-antigen. The glycosyltransferases (GTs) assembling the O-chain utilize lipid-linked acceptor substrates and nucleotide sugar donor substrates. The natural undecaprenol-linked acceptor substrates are not readily available, precluding the characterization of O-chain GTs. This chapter describes an assay for a galactosyltransferase (GalT) involved in the synthesis of the O7 antigen of *Escherichia coli* VW187. The glycolipid GlcNAcα-pyrophosphate bound to a phenoxyundecyl moiety, which resembles the natural substrate, was synthesized and employed in GT assays using bacterial membranes as the enzyme source. The assay is simple and allows optimization and characterization of the enzyme reaction. Similar protocols can be used to assay other GTs in the O-chain biosynthesis pathways.

Key Words: Lipopolysaccharide; *Escherichia coli*; galactosyltransferase; lipid-linked substrate intermediates; chemical synthesis.

1. Introduction

Gram-negative bacteria have lipopolysaccharides (LPS) in the outer membrane, with the carbohydrate exposed to the environment. Components of LPS have a number of serious effects in the host, and can elicit inflammatory responses *(1)*. The O-chain is the outermost component of LPS and is defined as an antigen. For example, the antigenic O-chain of the LPS of *Escherichia coli* strain VW187 (O7:K1) consists of the repeat unit sequence VioNAcβ1-2

From: *Methods in Molecular Biology, vol. 347: Glycobiology Protocols*
Edited by: I. Brockhausen © Humana Press Inc., Totowa, NJ

[Rhaα1-3] Manα1-4 Galβ1-3 GlcNAcα1-3, which is repeated many times in the complete O-chain *(2)*. The synthesis of this oligosaccharide (OS) is thought to occur at the cytosolic face of the plasma membrane. During O-chain synthesis, sugars are added sequentially to a lipid intermediate based on undecaprenol-pyrophosphate to form the repeat unit. As is true for the mammalian glycolipid synthesis, the repeat unit is growing at the nonreducing end of the OS. Subsequently the repeat units are linked to form the polymeric OS, which is then attached to the core OS linked to lipid A to form LPS (*see* **refs. *3–5*** and Chapter 15).

The LPS protects bacteria from the immune system of the host and is an important virulence factor, although the outermost O-chain is not essential for survival in laboratory culture. Thus rough forms of bacteria which lack O-chain and smooth forms which contain O-chain both grow in culture, but strains isolated from infected hosts contain the O-chain.

In the past it has been very difficult to assay the enzymes involved in O-chain synthesis, since adequate amounts of natural substrate are not readily available. This chapter describes a facile assay for a galactosyltransferase (GalT) involved in O-chain synthesis that relies on the synthesis of a lipid-linked acceptor substrate *(6)*. The first glycosyltransferase (GT) reaction in the O7-chain pathway is the addition of *N*-acetylglucosamine-phosphate (GlcNAc-phosphate) to undecaprenol-phosphate. The second GT in the pathway, uridine 5′-diphosphate (UDP)-Gal:GlcNAc-Rβ1,3-GalT, utilizes the lipid-linked substrate GlcNAcα1-pyrophosphoryl-undecaprenol (GlcNAc-PP-Und) in vivo to which galactose (Gal) is added in β1-3 linkage *(7)*.

The assay described here is based on the use of synthetic GlcNAcα-pyrophosphate bound to a phenoxyundecyl moiety (GlcNAcα1-O-PO_2-O-PO_2-O-$[CH_2]_{11}$-O-Phenyl)$^{2-}$(GlcNAc-PP-PhU) as a substrate for *E. coli* β1,3-GalT; *see* **Fig. 1**). The method is highly specific and sensitive, and can be developed into assays for other GTs that utilize undecaprenol-phosphates and -pyrophosphates. The acceptor substrate binds to hydrophobic chromatography columns and can easily be detected by absorbance measurements. The assay allows characterization of the reaction and monitoring of enzyme activity during purification.

2. Materials

2.1. Preparation of Bacterial Enzyme

2.1.1. Bacterial Growth

1. Luria-Bertani (LB) broth: 0.5 g of yeast extract/mL (DIFCO, Detroit, MI), 1 g of tryptone/mL, and 1 g of NaCl adjusted to pH 7.0 with NaOH (Sigma-Aldrich, Oakville, Ontario).

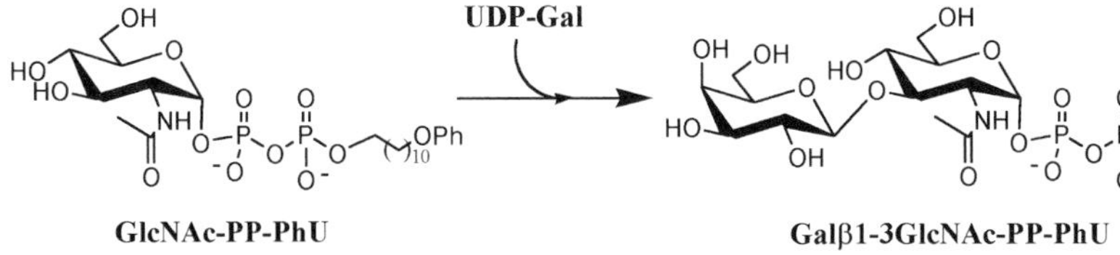

Fig. 1. Reaction of *Escherichia coli* β1,3-galactosyltransferase using synthetic GlcNAcα-pyrophosphate-phenoxyundecyl acceptor substrate.

2. Shaker water bath at 37°C, autoclave, 0.22-μm filters, sterile tubes, flasks, and pipets.

2.1.2. Preparation of Bacterial Membranes for Enzyme Assays

1. Centrifuge (Centronics M-1200 with an IEC MultiRF Rotor, Boehringer-Mannheim, Laval, Quebec, Canada).
2. Sonicator (Sonic Dismembrator Model 100, Fisher Scientific, Toronto, ON, Canada).
3. Sonication buffer: 10 m*M* of Tris/acetate (pH 8.5), 50 m*M* of sucrose, and 1.2 m*M* of ethylenediaminetetraacetic acid.
4. Protein dye to determine protein content of homogenates (Bio-Rad, Mississauga, ON, Canada) and 4 mg/mL of bovine serum albumin (BSA) as a protein standard.

2.2. Chemical Synthesis of Acceptor Substrate GlcNAc-PP-PhU

1. ^{1}H, ^{13}C, and ^{31}P nuclear magnetic resonance (NMR) spectroscopy: Bruker Avance 300-, 400-, 500- and 600-MHz spectrometers.
2. Mass spectrometry: MDS Sciex QSTAR XL spectrometer (Applied Biosystems, Foster City, CA).
3. Thin-layer chromatography plates: glass- or aluminum-backed, Silica Gel 60 F_{254} (Silicycle, Quebec City, CAN). Reactions are monitored by charring of the plates after spraying with either 5% H_2SO_4 in ethanol or phosphomolybdic acid (PMA) in ethanol (Aldrich, Milwaukee, WI).
4. Chromatography: silica gel, high-performance liquid chromatography (HPLC) system.
5. Chemicals and reagents (Aldrich): sodium hydride, sodium chloride, 2-acetamido-2-deoxy-D-glucopyranose, dry *N,N*-dimethylformamide, benzyl bromide, diethyl ether, ethyl acetate, sodium sulfate, dichloromethane, acetic anhydride, pyridine, palladium/charcoal (Pd/C), methanol, H_2 gas, Celite, 1*H*-tetrazole, sodium sulfite, petroleum ether, *m*-chloroperoxybenzoic acid, *bis*(benzyloxy)(diisopropylamino)phosphine, IR-120 resin, and Sephadex G15.

2.3. GalT Assay

1. 10 m*M* of 1000–10,000 cpm/nmol donor substrate UDP-[^{3}H]Gal (*see also* Chapter 14).
2. 5-m*M* Solution of synthetic acceptor substrate GlcNAc-PP-PhU in methanol.

3. 0.5 *M* of 2-*N*-morpholino-ethanesulfonate buffer (pH 7.0) and 0.1 *M* of $MnCl_2$.
4. C18 Sep-Pak columns (Whatman, Clifton, NJ) regenerated with 4 mL of methanol, followed by 6 mL of water (*see* Chapter 14).
5. Ready Safe scintillation fluid and scintillation counter LS 6500 (Beckman, Mississauga, Canada) or equivalent.

2.4. Purification of Enzyme Product

1. HPLC system: two pumps, injector, controller, ultraviolet monitor (Waters, Milford, CA), recorder, fraction collector, C18 column (Phenomenex, Torrance, CA), and acetonitrile–water mixtures as the mobile phase (*see also* Chapter 14).
2. Rotary evaporator or tabletop vacuum centrifuge.

2.5. Analysis of Enzyme Product

1. Jack bean and bovine testicular β-galactosidase and coffee bean α-galactosidase (Sigma); 0.2 *M* of Na_2HPO_4/0.1 *M* of citric acid buffer, pH 4.3; 0.25 *M* of glycine buffer, pH 10.0 (Sigma); Galα-*p*-nitrophenyl and Galβ-*p*-nitrophenyl (Sigma); and ultraviolet spectrophotometer.
2. D_2O (Aldrich) and ^1H NMR spectrometer Avance (Bruker).
3. Matrix-assisted laser desorption ionization (MALDI) mass spectrometer Voyager DE-STR MALDI-TOF (Applied Biosystems).

3. Methods

3.1. Preparation of Bacterial Enzyme

3.1.1. Bacterial Growth

1. Transfer scrapings from frozen glycerol stock of bacteria (kept at –80°C) into a sterile tube of 12 mL of LB broth (with loose cap) and gently aerate (shake at 150 rpm) overnight in a water bath. The appearance of the suspension should be cloudy.
2. Transfer a 10-mL portion of this overnight culture into a sterile 500-mL flask with 250 mL of LB broth; cover the mixture with a cotton plug and aluminum foil, and shake at 250 rpm at 37°C. Incubate the culture for 2.5 h until the solution is cloudy and the absorbance at 600 nm is 0.8. Harvest the bacteria by centrifugation at 4900*g* for 10 min (IEC MultiRF Rotor). For washing, resuspend the bacteria in phosphate-buffered saline (PBS) and centrifuge the mixture.
3. Suspend the bacterial pellet in 10 mL of PBS/10% glycerol and store as 1-mL aliquots in 2-mL Eppendorf tubes at –20°C (short term) or at –80°C (long term) for enzyme assays.

3.1.2. Preparation of Bacterial Membranes for Enzyme Assays

1. Thaw frozen bacterial cells in PBS/10% glycerol and centrifuge at 4900*g* for 10 min, and wash with 1 mL of PBS (pH 7.2).

2. Suspend bacteria in 0.5 mL of sonication buffer. Rupture and homogenize cells in Eppendorf tubes with the sonicator at setting 3 for two pulses of 15 s with a 2-min interval to allow for cooling on ice.
3. Determine protein concentrations with the Bio-Rad (Bradford) protein assay using BSA as a standard.

3.2. Chemical Synthesis of Acceptor Substrates (see *Scheme 1*)

The key step in the synthesis of the GlcNAc-PP-PhU acceptor substrate **A** involves the coupling of the GlcNAc-phosphate head group **E** and the phospholipid tail **H** followed by appropriate deprotection. The relevant experimental procedures are given in this section, and details of the characterization are given in **ref. *6*** (*see* **Note 1**).

3.2.1. Synthesis of 3,4,6-Tri-O-Acetyl-2-Acetamido-2-Deoxy-α-D-Glucopyranosyl Dihydrogen Phosphate (E) (see *Scheme 2*)

1. Selective benzylation of the hydroxyl group at the anomeric position is accomplished by adding sodium hydride (1.23 g, 51 mmol) to a solution of 2-acetamido-2-deoxy-D-glucopyranose (GlcNAc) (**B**) (8.84 g, 40 mmol) in dry *N,N*-dimethylformamide (20 mL). Stir the mixture until the evolution of gas has subsided and add benzyl bromide (13.7 g, 81 mmol). Stir the resulting solution for 6 h and quench by the addition of 100 mL of water. Extract the mixture sequentially with 2 × 100 mL of ether and 2 × 100 mL of ethyl acetate. Dry the combined extracts over sodium sulfate and concentrate under reduced pressure to give a residue which is fractionated by flash chromatography on silica gel using dichloromethane/methanol (85/15 [v/v]) as eluent to give the intermediate benzyl 2-acetamido-2-deoxy-α-D-glucopyranoside as a white solid (7.90 g, 25.38 mmol, 63%). A sample (506 mg, 1.63 mmol) of this product is treated with a 1/1 (v/v) solution of acetic anhydride/pyridine (10 mL). Stir the solution overnight at room temperature and add 10 mL of ice water. Extract the aqueous layer with ethyl acetate (3 × 15 mL), dry the extracts over sodium sulfate, and concentrate to give a residue which, by flash chromatography on silica gel using ethyl acetate as the eluent, affords benzyl 2-acetamido-3,4,6-tri-O-acetyl-2-deoxy-α-D-glucopyranoside (**C**) as a white solid (545 mg, 1.25 mmol, 77%).
2. Remove the benzyl group in **C** (obtained in a separate preparation) by stirring a mixture of **C** (9.98 g, 22.81 mmol) and 10% Pd/C (1.50 g) in 200 mL of methanol overnight under an atmosphere of H_2. Filter the reaction mixture through Celite and concentrate the filtrate. To a solution of a portion (508 mg, 1.46 mmol) of the residue (2-acetamido-3,4,6-tri-O-acetyl-2-deoxy-D-glucopyranose) and 1*H*-tetrazole (832 mg, 11.88 mmol) in 20 mL of dry dichloromethane at –20°C, add *bis* (benzyloxy)(diisopropylamino)phosphine (1.84 g, 5.33 mmol) dropwise over a period of 2 min. Stir the mixture for 5 h at room temperature and then treat at –40°C with 77% 3-chloroperoxybenzoic acid (2.60 g, 11.60 mmol); stir the mixture at 0°C for 30 min, and then overnight at room temperature. Add 100 mL

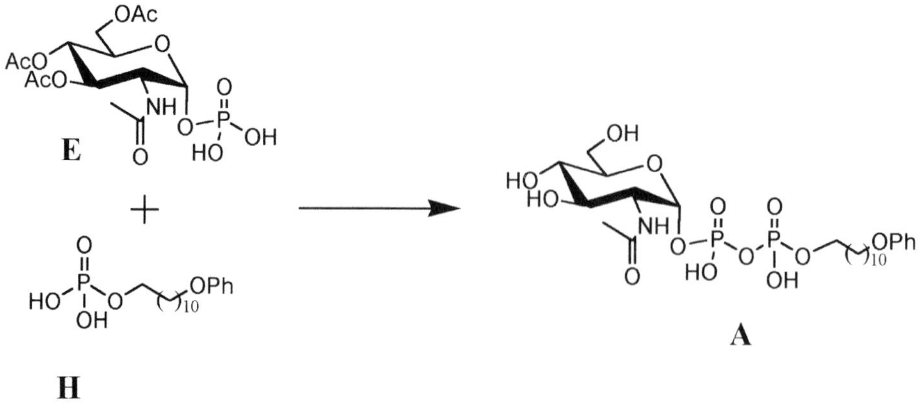

Scheme 1

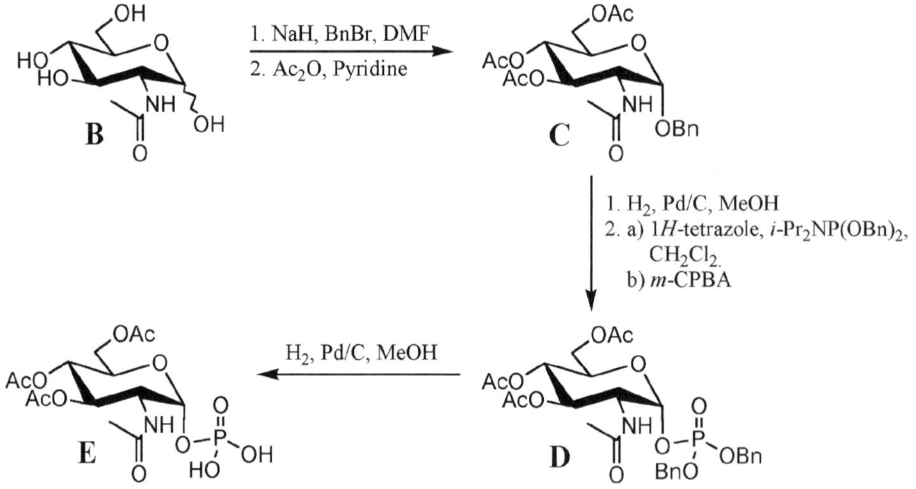

Scheme 2

of ethyl acetate and wash the mixture sequentially with saturated sodium sulfite (3 × 50 mL), water (2 × 50 mL), and brine (2 × 50 mL). Dry the organic layer over sodium sulfate and concentrate. Purify the residue by flash chromatography on silica gel using ethyl acetate/petroleum ether (3/1 [v/v]) as eluent to give the dibenzyl phosphate **D** as a colorless oil (617 mg, 1.02 mmol, 70%).

3. A mixture of **D** (251 mg, 0.41 mmol) and 10% Pd/C (0.30 g) in 15 mL of methanol is stirred overnight under an atmosphere of H_2. Filter the reaction mixture through Celite and concentrate to give 2-acetamido-3,4,6-tri-*O*-acetyl-2-deoxy-α-D-glucopyranosyl dihydrogen phosphate (**E**) as a white solid (170 mg, 0.40 mmol, 98%).

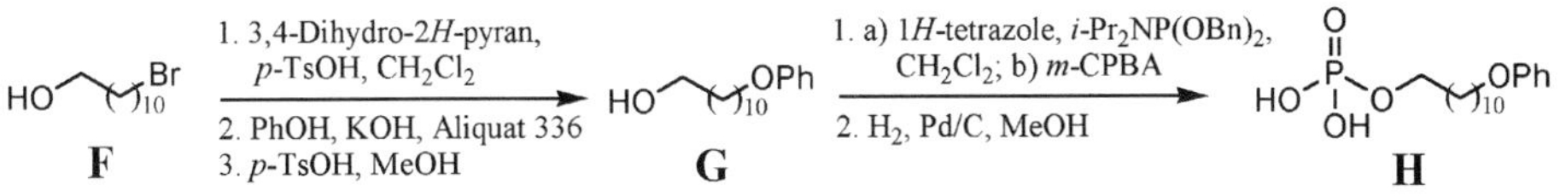

HO–$(CH_2)_{10}$–CH₂Br **F** → 1. 3,4-Dihydro-2*H*-pyran, *p*-TsOH, CH_2Cl_2; 2. PhOH, KOH, Aliquat 336; 3. *p*-TsOH, MeOH → HO–$(CH_2)_{10}$–CH₂OPh **G** → 1. a) 1*H*-tetrazole, *i*-$Pr_2NP(OBn)_2$, CH_2Cl_2; b) *m*-CPBA; 2. H_2, Pd/C, MeOH → HO–P(O)(OH)–O–$(CH_2)_{10}$–CH₂OPh **H**

Scheme 3

*3.2.2. Synthesis of 11-Phenoxyundecyl Dihydrogen Phosphate (**H**) (see **Scheme 3**)*

1. A solution of 11-bromo-1-undecanol **F** (7.54 g, 30 mmol) and 3,4-dihydro-2*H*-pyran (12.60 g, 150 mmol) in 140 mL of dry dichloromethane is treated with *p*-toluenesulfonic acid monohydrate (0.06 g, 0.32 mmol) at 0°C with stirring for 10 min, and then at room temperature for 16 h. The mixture is then washed sequentially with a saturated aqueous solution of 210 mL of sodium hydrogencarbonate and brine (2 × 50 mL). The organic phase is dried over magnesium sulfate and concentrated to give crude 2-[(11-bromoundecyl)oxy]tetrahydropyran (10.0 g) as a colorless oil.
2. Stir a mixture of phenol (0.94 g, 10 mmol), potassium hydroxide (0.70 g, 12.50 mmol), and Aliquat 336 (81 mg, 0.20 mmol) at 85°C for 5 min, and then add a sample (3.35 g, 10 mmol) of the crude 2-[(11-bromoundecyl) oxy]tetrahydropyran. Stir the mixture for 6 h, add 75 mL of diethyl ether at room temperature, and pass the mixture through Florosil. Concentration of the effluent provides a residue which is purified by flash chromatography on silica gel using ethyl acetate/hexanes (7/1 [v/v]) as eluent to give the intermediate 2-[(11-phenoxyundecyl)oxy]tetrahydropyran as a colorless liquid (2.75 g, 7.89 mmol, 79%). The tetrahydropyranyl protecting group is removed by treatment of this product with 100 mg of *p*-toluenesulfonic acid in 40 mL of methanol. Stir the mixture for 1.5 h and then concentrate. A solution of the residue in 50 mL of chloroform is washed with an aqueous saturated solution of sodium hydrogencarbonate (2 × 50 mL), dried over magnesium sulfate, and concentrated to afford 11-phenoxy-1-undecanol (**G**) as a white solid (1.88 g, 7.11 mmol, 90%).
3. Add *bis*(benzyloxy)(diisopropylamino)phosphine (1.22 g, 3.53 mmol) dropwise over a period of 2 min to a solution of **G** (508 mg, 1.92 mmol) and 1*H*-tetrazole (271 mg, 3.87 mmol) in 20 mL of dry dichloromethane at –30°C. The mixture is then stirred at room temperature for 15 h. Cool the mixture to –40°C and treat with 77% *m*-chloroperoxybenzoic acid (2.16 g, 9.6 mmol), and stir at 0°C for 1 h and then at room temperature for 1 h. Add 100 mL of ethyl acetate and wash the organic phase sequentially with an aqueous solution of saturated sodium sulfite (3 × 50 mL), water (2 × 50 mL), and brine (2 × 50 mL). Dry the organic layer over sodium sulfate and concentrate. Purification of the residue by flash chromatography on silica gel using ethyl acetate/petroleum ether (2/3 [v/v]) as eluent gives the intermediate dibenzyl 11-phenoxyundecyl phosphate as a white

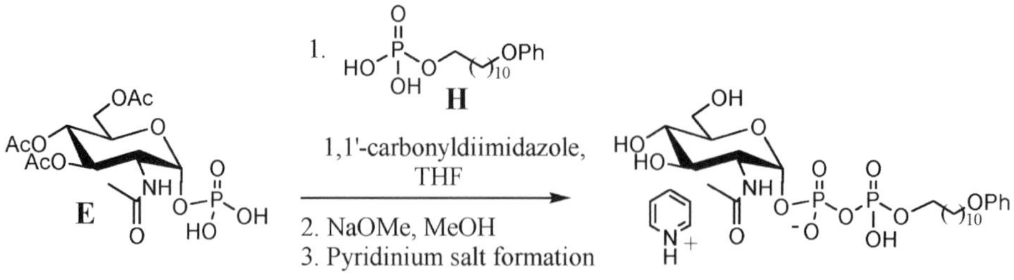

1. HO–P(=O)(OH)–O–(CH2)10–OPh **H**

1,1'-carbonyldiimidazole, THF

2. NaOMe, MeOH

3. Pyridinium salt formation

Scheme 4

solid (751 mg, 1.43 mmol, 74%). A portion (517 mg, 1.02 mmol) of this product together with 10% Pd/C (0.20 g) in 15 mL of methanol is stirred overnight under an atmosphere of H_2. The mixture is filtered through Celite and concentrated to give 11-phenoxyundecyl dihydrogen phosphate (**H**) as a white solid (170 mg, 0.49 mmol, 48%).

*3.2.3. Synthesis of the Pyridinium Salt of P¹-2-Acetamido-2-Deoxy-α-D-Glucopyranosyl P²-11-Phenoxyundecyl Hydrogen Diphosphate (A) (see **Scheme 4**)*

1. Both sugar head group **E** (250 mg, 0.59 mmol) and phospholipid tail **H** (223 mg, 0.65 mmol) are dried separately by coevaporation (three times) with a solution of three drops of diisopropylamine in 3 mL of toluene. Compound **H** is dried under high vacuum for 30 min and dissolved in 15 mL of dry tetrahydrofuran. Add to this solution 1,1′-carbonyldiimidazole (417 mg, 2.53 mmol) and stir the reaction mixture at room temperature for 1.5 h. Add 212 μL of methanol and stir the solution for another hour. Concentrate the mixture and dry the residue under high vacuum for 1 h. A solution of **E** in 15 mL of dry tetrahydrofuran is then added to dry **H**, the mixture is stirred at room temperature for 48 h, and then concentrated. Purification by a Sephadex G-15 size-exclusion column using 0.1% ammonium hydrogencarbonate in methanol as eluent gives the intermediate ammonium P¹-2-acetamido-3,4,6-tri-*O*-acetyl-2-deoxy-α-D-glucopyranosyl P²-11-phenoxyundecyl hydrogen diphosphate as a white solid (365 mg, 0.47 mmol, 80%).
2. The ammonium salt is converted into the pyridinium salt as follows: a solution of the ammonium salt (25 mg, 32 μmol) prepared earlier in sodium methoxide in methanol (33 m*M*, 4 mL) is stirred at room temperature for 20 min. An excess of IR-120 resin in the pyridinium form is then added and the suspension is stirred for 40 min. The mixture is filtered and the resin is washed with methanol. The filtrate and washings are combined and concentrated to give the pyridinium salt of P¹-2-acetamido-2-deoxy-α-D-glucopyranosyl P²-11-phenoxyundecyl hydrogen diphosphate (**A**, GlcNAc-PP-PhU) as a white solid (20 mg, 24 μmol, 74%).

3.3. GalT Assay

The GalT assay conditions described here can be modified to determine the optimal conditions of assay and to characterize the enzyme reaction with respect to linearity with time and protein concentration, and dependence on metal ions (*see* **Fig. 2**), detergents (*see* **Fig. 3**), and possible cofactors *(7)*. The assays can be also be adjusted to assay other enzyme-requiring lipid-linked acceptor substrates (*see* **Note 2**).

1. The standard assay mixtures are carried out in a 40-μL total volume, containing 0.5 m*M* of pyridinium salt of GlcNAc-PP-PhU (**A**) added in methanol solution, 5 m*M* of $MnCl_2$, 75 m*M* of 2-*N*-morpholino-ethanesulfonate buffer (pH 7.0), 0.5 m*M* of UDP-[^{3}H]Gal, and 20 μL of enzyme homogenate (with a total of 20 μL sonication buffer present). All assays are carried out in duplicate. As a background control, in each experiment two assays lack acceptor substrate (*see* **Note 3**).
2. Mixtures are incubated for 10 min (*see* **Note 4**). The reactions are stopped by the addition of 0.7 mL of ice-cold water and freezing.
3. The incubation mixture is then applied to a regenerated C18 Sep-Pak column at a flow rate not more than 2 mL/min. Columns are washed with 5 mL of water to elute the hydrophilic radioactive components (UDP-[^{3}H]Gal and [^{3}H]Gal). The hydrophobic product (and substrate) is eluted with 5 mL of methanol. Fractions of 1 mL are collected (A–J) into scintillation vials. The radioactive product is present in the first three methanol fractions (F–H).
4. Radioactivity is then determined in these fractions, as well as in the one before (E) and the one after (I). 5 mL of scintillation fluid is added to fractions E–I, and radioactivity is determined by scintillation counting.
5. After use, the C18 Sep-Pak columns are regenerated by washing them with 4 mL of methanol and then with 6 mL of water at a flow rate of 2–3 mL/min, and can be used again at least four times. If stored for longer time periods, regenerate again just before use. The columns should be tested for their effectiveness after repeated use.

3.4. Purification of Enzyme Product

To isolate enzyme product as a radioactive standard for further structural analysis or as a nonradioactive substrate for the assay of other O-chain GTs, small-scale assays are scaled up, aiming at the production of at least 100 nmol of enzyme product (*see* **Note 5**).

1. Fractions eluted with methanol from the Sep-Pak columns that contain enzyme product are concentrated by rotary evaporation or using a tabletop vacuum centrifuge.
2. An HPLC system is adjusted to achieve an elution of the GlcNAc-PP-PhU acceptor substrate at about 40 min. This interval should be sufficient to separate enzyme product and acceptor substrate. Run standard compounds to verify separation.

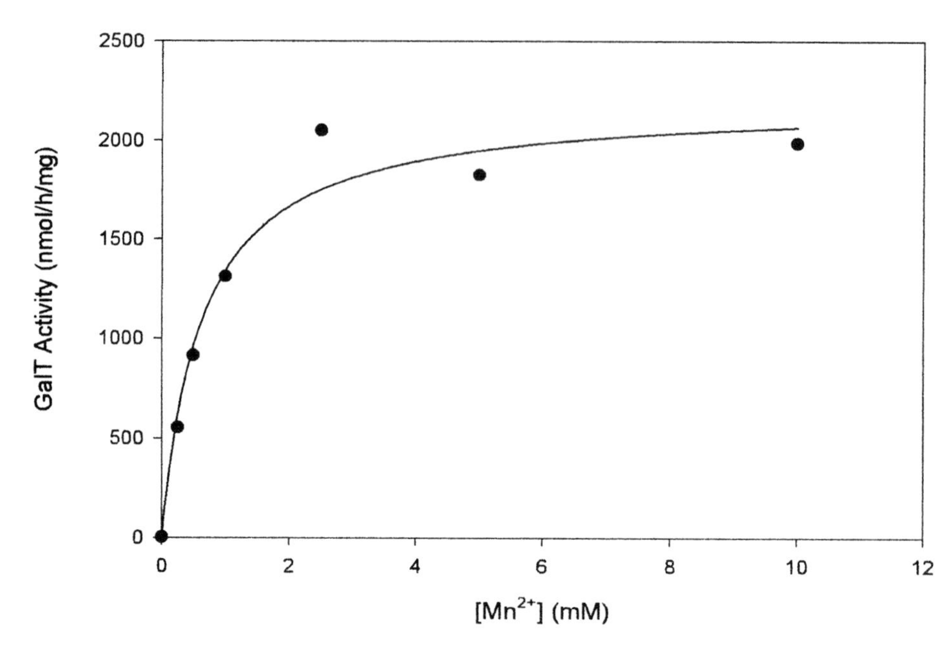

Fig. 2. Dependence of Gal-transferase activity on $MnCl_2$ concentration in the assay mixture.

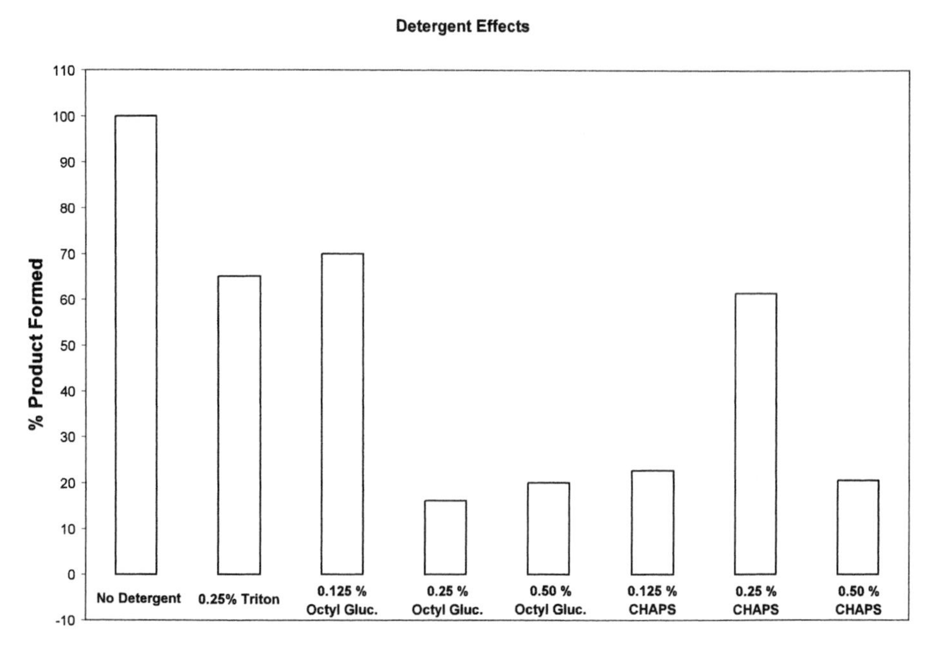

Fig. 3. Dependence of Gal-transferase activity on detergents in the assay. Octyl Gluc, octyl glucoside.

3. Residues are taken up in methanol and a small aliquot is injected into an HPLC system, using a C18 column and acetonitrile: water (11:89) at a flow rate of 1 mL/min. To monitor elutions, the absorbance at 195 nm is measured; fractions of 2 mL are collected, and the radioactivity is measured by scintillation counting (*see* **Fig. 4**).
4. After determining the elution time of the enzyme product, the entire large-scale assay preparation eluted from Sep-Pak columns is separated by HPLC. Collected fractions containing the enzyme product are combined and evaporated, and taken up in a small amount of methanol.

3.5. Analysis of Enzyme Product

1. The absorbance and radioactivity patterns of purified enzyme product eluted by HPLC are sensitive indicators of purity. The molecular weight and purity of the enzyme product is determined by MALDI mass spectrometry, run in the negative ion mode.
2. Using a flash evaporator and a vacuum pump, enzyme product is exchanged three times with 99.99% deuterium oxide and analyzed by ^{1}H NMR spectroscopy. This will indicate the linkage of the newly-added sugar.
3. For further linkage analysis, radioactive enzyme product is analyzed for its susceptibility to cleavage by galactosidases. The specificity of galactosidases helps to identify the linkage point and anomeric configuration of the newly-added sugar in the enzyme product. Incubate an aliquot (2000 cpm) of enzyme product at 37°C for 60 min with 0.01–1 U (μmol/min) of α- or β-galactosidase in 100 μL containing citric acid/sodium phosphate buffer and 0.01% BSA at pH 4.3. After the incubation, add 800 μL of water and pass the mixture through a C18 Sep-Pak column. Undigested product should elute with methanol as described in **Subheading 3.3.** Digested product elutes as free [^{3}H]galactose when eluted with water, and as non-radioactive substrate when eluted with methanol. A comparison of controls lacking the enzyme and the digested samples will indicate the susceptibility of the product to cleavage by the galactosidase. The positive control substrates are Galα-*p*-nitrophenyl or Galβ-*p*-nitrophenyl. After the incubation with control substrates the solution is made alkaline with glycine buffer, and *p*-nitrophenolate ion is detected by measuring the absorbance at 400 nm.

4. Notes

1. The acceptor substrate GlcNAc-PP-PhU *(A)* has amphipathic properties similar to those of the endogenous substrate GlcNAc-PP-Und, is soluble in methanol–water mixtures, and can be added directly to the reaction mixtures. The compound is stable at –20°C in methanol for at least 6 mo. The pyridinium salt of GlcNAc-PP-PhU can bind to C18 Sep-Pak cartridges and can be eluted with methanol and subsequently analyzed by HPLC.
2. Pyrophosphate-containing lipids having a phenoxyundecyl group may also be acceptor substrates for other GTs that catalyze the subsequent steps of O7 chain synthesis, and for GTs of other strains of Gram-negative bacteria. For example,

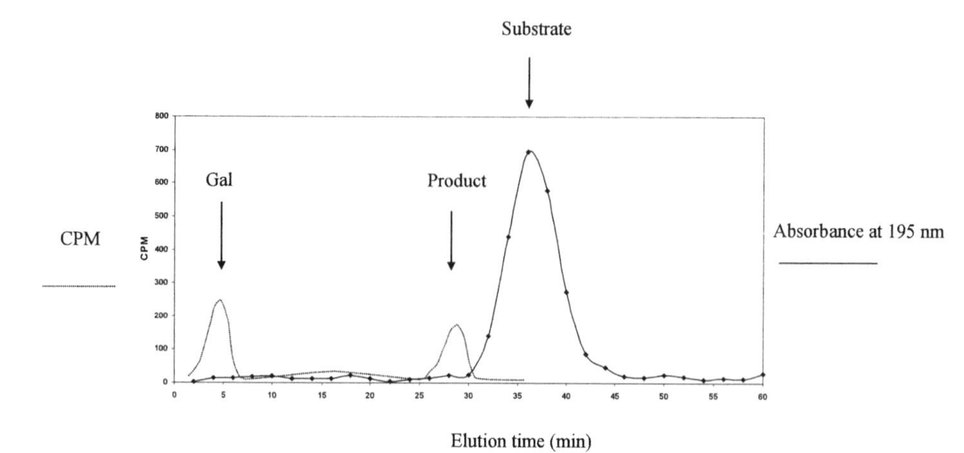

Fig. 4. High-performance liquid chromatography (HPLC) separation of enzyme product, GlcNAc-PP-PhU substrate, and free galactose. After the Gal-transferase assay, the enzyme product is purified by C18 Sep-Pak, which removes UDP-Gal, and then injected into HPLC using a C18 column with acetonitrile:water (12:88) at 1 mL/min. The graph shows the absorbance at 195 nm of the substrate GlcNAc-PP-PhU, and the radioactivity in 2-mL fractions of the assay mixture. The arrows indicate the elution of [^{3}H]Gal (Gal) of GlcNAc-PP-PhU (Substrate) and the enzyme product (Product).

the Gal-GlcNAc-PP-PhU can be a substrate for the next enzyme in the sequence of O7 chain synthesis, a mannosyltransferase.

3. The topology of the reaction catalyzed by membrane-bound enzymes is important. Detergents may make the enzyme inaccessible to the substrate. It is therefore important to carefully choose the right detergent. Sonication renders the Gal-transferase active without detergent (*see* **Fig. 3**).
4. A longer incubation time may yield more product, but with crude enzyme preparations there is often significant breakdown of both substrate and product.
5. A product yield of less than 100 nmol still allows identification by galactosidase digestions, mass spectrometry, and HPLC. To obtain a proof of anomeric linkage in the enzyme product by one- and two-dimensional NMR methods, at least 200 nmol of highly pure product is required. The pyrophosphate linkage in the product can be cleaved by nucleotide pyrophosphatase and the phosphate can be removed by phosphatase, leaving a stable disaccharide. This facilitates the purification of the oligosaccharide by HPLC and identification of the sugar linkage *(7)*.

Acknowledgments

This work was supported by a grant from the Natural Sciences and Engineering Research Council (to Drs. Brockhausen and Szarek). Dr. Brockhausen has been supported by a Research Scientist Award from The Arthritis Society.

References

1. Raetz, C. R. and Whitfield, C. (2002) Lipopolysaccharide endotoxins. *Annu. Rev. Biochem.* **71,** 635–700.
2. L'vov, V. L., Shashkov, A. S., Dmitriev, B. A., and Kochetkov, N. K. (1984) Structural studies of the O-specific side chain of the lipopolysaccharide from *Escherichia coli* O:7. *Carbohydr. Res.* **126,** 249–259.
3. Valvano, M. A. (2003) Export of O-specific lipopolysaccharide. *Front. Biosci.* **8,** s452–s471.
4. Alexander, D. C. and Valvano, M. A. (1994) Role of the rfe gene in the biosynthesis of the *Escherichia coli* O7-specific lipopolysaccharide and other O-specific polysaccharides containing *N*-acetylglucosamine. *J. Bacteriol.* **176,** 7079–7084.
5. Samuel, G. and Reeves, P. (2003) Biosynthesis of O-antigens: genes and pathways involved in nucleotide sugar precursor synthesis and O-antigen assembly. *Carbohydr. Res.* **338,** 2503–2519.
6. Montoya-Peleaz, P. J., Riley, J. G., Szarek, W. A., Valvano, M. A., Schutzbach, J. S., and Brockhausen, I. (2005) Identification of a UDP-Gal: GlcNAc-R galactosyltransferase activity in *Escherichia coli* VW187. *Bioorg. Med. Chem. Lett.* **15,** 1205–1211.
7. Riley, J., Menggad, M., Montoya-Peleaz, P. J., et al. (2005) The wbbD gene of *E. coli* strain VW187 (O7:K1) encodes a UDP-Gal: GlcNAcα-pyrophosphate-R β1,3-galactosyltransferase involved in the biosynthesis of O7-specific lipopolysaccharide. *Glycobiology* **15,** 605–613.

17

High-Throughput Quantitation of Metabolically Labeled Anionic Glycoconjugates by Scintillation Proximity Assay Utilizing Binding to Cationic Dyes

Karen J. Rees-Milton and Tassos P. Anastassiades

Summary

Rapid, quantitative methods suited to a large number of samples are required for studies into the determination of disease etiology and in the evaluation of drugs and biological agents. This chapter describes an assay for anionic glycoconjugates (GCs), including glycosaminoglycans, which are major gene products of chondrocytes appearing in the extracellular matrix. The assay utilizes the electrostatic interaction between negatively charged sulfate and carboxyl groups of anionic GCs synthesized and secreted by chondrocytes with the cationic dye Alcian blue, immobilized to scintillant-coated 96-well plates. Metabolic labeling with D-[1, 6-^{3}H (N)]-glucosamine allows all anionic GCs, including cartilage-specific and hyperglycosylated variants of fibronectin, to be quantitated. If $Na_2{}^{35}SO_4$ is used for the metabolic labeling instead, only glycosaminoglycans and proteoglycans will be quantitated. The samples are counted using a multi-detector instrument for scintillation proximity assays, such as the Wallac 1450 Microbeta® Trilux, designed for detection of samples in 96-well plates and, as such, can be a high-throughput system. The bound anionic GCs can be visualized by sodium dodecyl sulfate-polyacrylamide gel electrophoresis after quantitation by elution with denaturing buffers. The method can be modified to include predigestion of the sample with a specific lyase, e.g., chondroitinase ABC or testicular hyaluronidase. To separate polyanions from other digested material after ethanol precipitation, the sample can be assayed as described in this chapter for a particular subtype of anionic GC. This assay addresses the need for high-throughput applications in arthritis and other medical and biological problems.

Key Words: Anionic glycoconjugates; proteoglycans; glycosaminoglycans; fibronectin; metabolic radiolabeling; cell culture; Alcian blue; high-throughput screening; immobilization; scintillation proximity assay; flashplates.

From: *Methods in Molecular Biology, vol. 347: Glycobiology Protocols*
Edited by: I. Brockhausen © Humana Press Inc., Totowa, NJ

1. Introduction

Articular cartilage contains anionic glycoconjugates (GCs), including the proteoglycans (PGs). Anionic GCs function primarily as structural and signaling molecules. Cultures of articular chondrocytes provide an in vitro model system for the investigation of the synthesis of anionic GCs and their regulation by endogenous and exogenous factors. If the growth medium is supplemented with D-[1, 6-^{3}H (N)]-Glucosamine hydrochloride, all anionic GCs will be radiolabeled, including hyperglycosylated and cartilage-specific variants of fibronectin *(1,2)*. The anionic glycoproteins secreted into the growth medium by bovine articular chondrocytes grown in monolayer culture and precipitated with Toluidine blue are shown in **Fig. 1**. The anionic glycoproteins can be visualized with "Stains-All" staining with silver nitrate counterstaining *(3)* of gels (*see* **Fig. 1**) after precipitation of the anionic GCs with the cationic dye, Toluidine blue *(1)*. "Stains-All" is a cationic carbocyanine dye that will stain sialoglycoproteins and phosphoproteins blue and all other proteins red/pink *(3)*. In addition to the PGs, a highly glycosylated variant of fibronectin (hgFN), is synthesized and secreted by bovine articular chondrocytes in monolayer culture in the presence of transforming growth factor-β1 (5 ng/mL) and stains blue (*see* **Fig. 1A**). This molecule, hgFN, is now considered to be a marker for differentiated chondrocytes *(2)*. If the medium is instead supplemented with $Na_2{}^{35}SO_4$, the only anionic GCs prominently radiolabeled are the glycosaminoglycans (GAGs) of the PGs. Such studies will be useful in diseases of cartilage tissue and in the evaluation of drugs and biological agents. These investigations require rapid, quantitative methods suited to a large number of samples. Methods to quantitate anionic GCs that make use of the negatively charged sulfate and carboxyl groups utilize cationic dyes, such as Alcian blue, in order to form complexes with the anionic GCs *(4–6)*.

High-throughput scintillation proximity assays (SPAs) can also involve the immobilization of radiolabeled molecules to polystyrene 96-well microtiter plates. Such a high-throughput determination of anionic GCs has been hindered because very hydrophilic, negatively charged anionic GCs do not readily bind to hydrophobic polystyrene surfaces. Attempts at enhancing adsorption of anionic GCs to polystyrene surfaces have included ultraviolet or irradiation activation of plastic surfaces, removal of GAGs by chondroitinase ABC digestion to enhance adsorption of hydrophobic parts of the core protein to the plastic, and rendering the polystyrene surface cationic by binding of cationic molecules, such as protamine sulfate, poly-L-lysine, and spermine *(7,8)*.

This chapter describes a high-throughput assay for anionic GCs, including GAGs, in an in vitro cell culture system that involves scintillation counting of radiolabeled anionic GCs to Alcian blue-coated polystyrene plates *(9)*. The

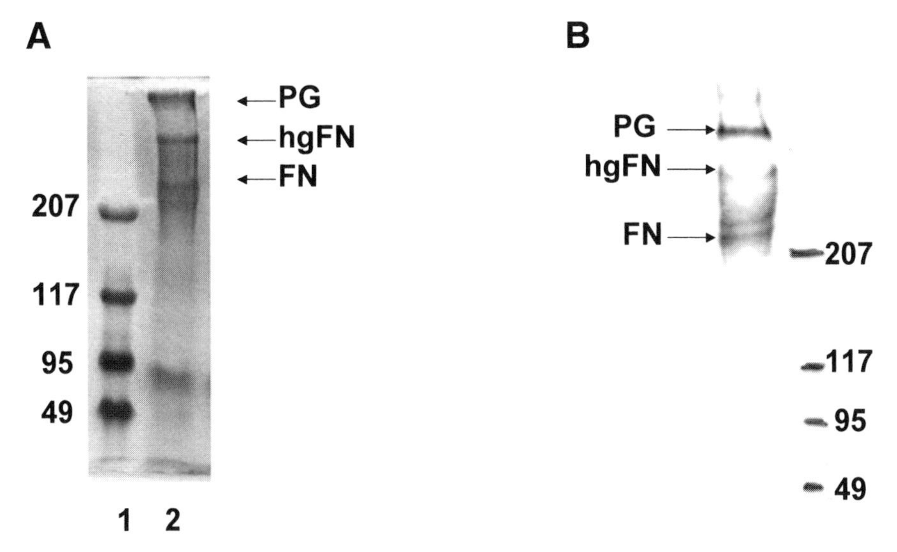

Fig. 1. **(A)** The anionic glycoproteins secreted into the growth medium by bovine chondrocytes grown in monolayer culture **(lane 2)** can be visualized by "Stains-All" staining with silver nitrate counterstaining of sodium dodecyl sulfate-polyacrylamide gel electrophoresis (SDS-PAGE) of Toluidine blue-precipitated anionic glycoconjugates. **(B)** Anionic glycoproteins bound to Alcian blue-coated flashplates eluted with 4 *M* of GuHCl can be visualized by autoradiography of SDS-PAGE. The position of the molecular weight markers (207, 117, 95, and 49 kDa, and **lane 1**) are indicated. The anionic glycoproteins secreted are indicated (PG, proteoglycan; hgFN, highly glycosylated fibronectin; FN, fibronectin).

assay utilizes the electrostatic interaction between negatively-charged sulfate and carboxyl groups of anionic GCs with the cationic dye Alcian blue, immobilized to scintillant-coated 96-well plates. The samples are counted using a multidetector instrument with scintillation proximity capability, such as the Wallac 1450 Microbeta® Trilux, which is designed for liquid scintillation detection of samples in 96-well plates and, as such, can be a high-throughput system. The linearity of the assay with increasing sample is demonstrated by Rees-Milton and Anastassiades *(9)*. The method can be modified to include a predigestion of the sample with a specific lyase, e.g., chondroitinase ABC or testicular hyaluronidase. To separate polyanions from other digested anionic material after ethanol precipitation, the sample can be assayed for a particular subtype of anionic GC as described in this chapter. The bound anionic glycoproteins can be eluted with increasing concentrations of guanidine hydrochloride (GuHCl; *see* **Fig. 2**), and the anionic glycoproteins eluted with 4 *M* of GuHCl can be visualized by sodium dodecyl sulfate-polyacrylamide gel electrophoresis

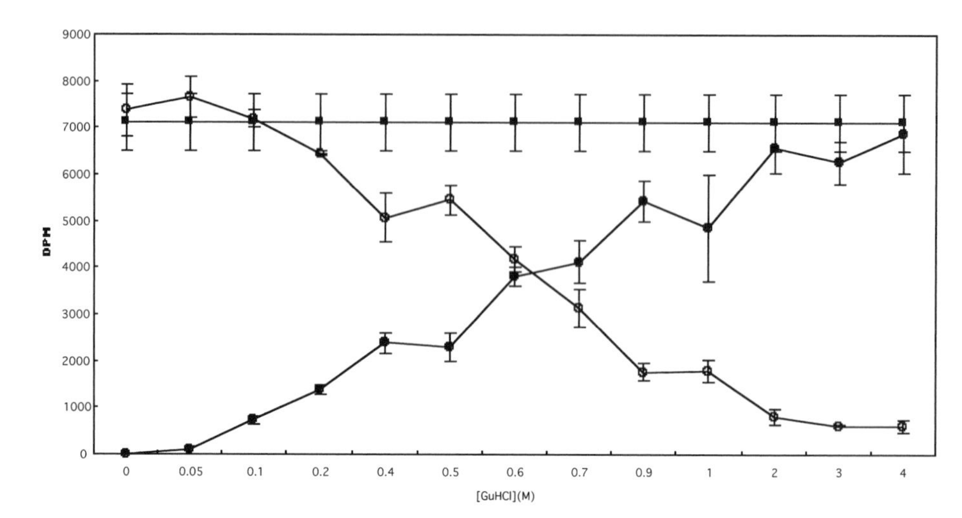

Fig. 2. Radioactivity (DPM) released with increasing concentrations of GuHCl from the Alcian blue-coated flashplates (●). The DPM still bound to the flashplate at each concentration of GuHCl (■) and total DPM initially bound to the flashplate (○) are shown.

(SDS-PAGE) and autoradiography (*see* **Fig. 1B**). This chapter describes the quantitation of anionic GCs in cell-culture medium, but the assay could be used for the quantitation of anionic GCs in serum or synovial fluid.

The anticipated utilization of this assay will include drug development for damaged connective tissues (as in arthritis and neurobiology), in problems of differentiation, and in cancer.

2. Materials

See **Notes 1** and **2**.

2.1. Cell Culture and Radiolabeling

1. Dulbecco's Modified Eagle's Medium (DMEM)-F12 with 10% fetal bovine serum (FBS; Gibco/Invitrogen, Burlington, Ontario, Canada).
2. Solution of 0.25% trypsin and 1 m*M* of ethylenediaminetetraacetic acid (EDTA; Gibco).
3. Store 10,000 U/mL of penicillin G sodium/10,000 μg/mL of streptomycin in 0.85% saline, and 250 μg/mL of Fungiozone amphotericin B (Gibco) in 1-mL aliquots at –20°C and use to supplement media at 10 U/mL, 10 μg/mL, and 0.25 μg/mL, respectively.
4. Store 1 mCi/mL of D-[1, 5-^{3}H (N)]-GlcN.HCl (glucosamine hydrochloride) in 90% ethanol and 1 mCi/mL of $Na_2{}^{35}SO_4$ in distilled water at 4°C (Perkin Elmer

Canada Inc., Woodbridge, Ontario, Canada). Dilute the $Na_2{}^{35}SO_4$ with distilled water to give a working solution of 2 µCi/µL and store at 4°C.

5. Add proteinase inhibitors to collected cell-culture media and store at –20°C. Add 0.25 *M* of EDTA stock solution (pH 8.0, stored at room temperature) to collected media to give a final concentration of 11.5 m*M*. Dissolve *N*-ethylmaleimide (NEM) and phenylmethyl sulfonyl fluoride (PMSF) in isopropanol just before use at 50 mg/mL and 0.1 *M*, respectively, and add to collected growth media to give a final concentration of 1.3 mg/mL and 1 m*M*, respectively. Dissolve benzamidine hydrochloride hydrate in distilled water just prior to use at 0.5 *M* and add to growth medium to give a final concentration of 10 m*M*.
6. Phosphate-buffered saline (PBS), pH 7.4, is made as a 10X stock (80 g of NaCl, 2 g of KCl, 14.4 g of Na_2HPO_4, and 2.4 g of KH_2PO_4) and store at 4°C. PBS is used at 1X concentration.
7. Isolate bovine articular chondrocytes from the articular cartilage of bovine ankle joints ***(1)***.

2.2. Alcian Blue Coating of Flashplates

1. Flashplates (white 96-well polystyrene microplates with plastic scintillator-coated wells) should be stored in the dark (Perkin Elmer).
2. Alcian blue coating solution: 1% w/v Alcian blue (Sigma-Aldrich, Oakville, Ontario, Canada), 3% v/v acetic acid. The Alcian blue is dissolved by stirring overnight. The solution is then filtered using Whatman number 1 filter paper and stored at room temperature in the dark for several months.
3. Washing solution: 1X PBS, pH 7.4.
4. Plastic sealing tape (Perkin Elmer).

2.3. Immobilization of Anionic GCs to Alcian Blue-Coated Flashplates and Quantitation Using the SPA

1. Multi-detector scintillation counter, such as the Wallac 1450 Microbeta Trilux scintillation counter.
2. Washing solution: 1X PBS, pH 7.4.

2.4. Elution of Anionic GCs From Alcian Blue-Coated Flashplates

1. Elution buffer (SDS-PAGE denaturing buffer): 6 *M* of urea, 1.6% w/v of SDS, 8% v/v of glycerol, 0.04% w/v of bromophenol blue, 4% v/v of β-mercaptoethanol, and 0.05 *M* of Tris-HCl, pH 6.8.

2.5. Visualization of Eluted Anionic GCs: Tris-Acetate Gradient SDS-PAGE and Autoradiography

The Tris-acetate gradient gel is based on the Invitrogen system for the resolution of high-molecular-weight proteins (Gibco).

1. Running buffer: 50 *m*M of Tris-HCl, 50 m*M* of tricine, and 0.1% w/v of SDS, pH 8.2.

2. Stacking gel: 4% w/v of acrylamide, 0.1% w/v of *bis*-acrylamide, and 0.125 *M* of Tris-HCl (pH adjusted to 7.0 with acetic acid).
3. Separating gel: 4–8% w/v of acrylamide, 0.1–0.2% w/v of *bis*-acrylamide, and 0.375 *M* of Tris-HCl (pH adjusted to 7.0 with acetic acid). The gel is made by mixing a 4% and an 8% acrylamide gel solution (each of 0.375 *M* Tris-HCl, pH 7.0) using a gradient maker.
4. Ammonium persulfate (APS): Prepare a 10% w/v solution in distilled water and use within 2 h.
5. *N,N,N′,N′*-tetramethyl ethylenediamine (TEMED; Bio-Rad Laboratories Inc., Mississauga, ON, Canada).
6. Prestained high range SDS-PAGE standards (Bio-Rad).
7. Fixative: 30% v/v of ethanol and 10% v/v of acetic acid.
8. Amplify fluorographic reagent (Amersham Biosciences, Baie d'Urfe, Quebec, Canada).
9. Whatman thick chromatography paper.
10. Kodak X-OMAT AR film (XAB).

3. Methods

3.1. Cell Culture and Radiolabeling

1. The bovine chondrocytes are passaged at confluency with 2 mL per 75-cm^2 cell-culture flask of trypsin/EDTA to provide new maintenance cultures in 75-cm^2 flasks, and experimental cultures in six-well plates (35-mm wells). The chondrocytes are seeded at 5×10^4 cells/well in a total volume of 4 mL of DMEM-F12, 10% FBS.
2. When the cells in the six-well plates reach confluency (4 d), replace the growth medium with 4 mL of fresh medium containing 5 μCi/mL of either ^{3}H-GlcN or $Na_2{}^{35}SO_4$ (*see* **Notes 3** and **4**).
3. Collect the growth medium after 4 d and supplement with proteinase inhibitors. Store at –20°C.

3.2. Alcian Blue Coating of Flashplates

1. This protocol is based on the method of Laglace et al. ***(10)*** with some modifications.
2. Add 200 μL of Alcian blue coating solution to each well of a Perkin-Elmer flashplate (white 96-well polystyrene microplate with plastic scintillator-coated wells).
3. Cover the plate with plastic sealing tape and incubate at 37°C in the dark for 60 min, and lightly agitate on a rocking platform.
4. Pipet off the Alcian blue coating solution and air-dry the plates at room temperature overnight in the dark.
5. Rinse the plate wells twice with 300 μL/well of washing solution.
6. Air-dry the plates at room temperature overnight in the dark.
7. Plates can be stored for several weeks in the dark at room temperature.

3.3. Immobilization of Anionic GCs to Alcian Blue-Coated Flashplates and Quantitation Using the Wallac 1450 Microbeta Trilux Scintillation Counter

See **Note 5**.

3.3.1. Immobilization of Anionic GCs to Alcian Blue-Coated Flashplates

1. Add up to 100 μL of radiolabeled cell-culture medium to the Alcian blue-coated Flashplate wells (*see* **Note 6**).
2. Make up the total volume added to each well to 100 μL with fresh DMEM-F12, 10% FBS.
3. Incubate the plates at 4°C overnight in the dark.
4. Remove the medium and wash the wells twice with 300 μL of washing solution.

3.3.2. Quantitation Using the SPAs

1. Count the flashplates using a Wallac 1450 Microbeta Trilux scintillation counter.
2. Normalization and counting protocols should be set up as described in the New England Nuclear technical information (available at http://las.perkinelmer.com/content/ApplicationNotes/1450-1401-01.pdf). The protocols should use only the upper photomultiplier tube; the selected label should be "other," and the window selected for counting tritium and ^{35}S, 175-360 and 175-650, respectively.

3.4. Elution of Anionic GCs From Alcian Blue-Coated Flashplates

1. Add 200 μL of elution buffer to the Flashplate well.
2. Gently agitate plate, using an orbital shaker, at room temperature for 40 min.
3. Transfer the elution buffer to a clean microtube.

3.5. Visualization of Eluted Anionic GCs: Tris-Acetate Gradient (4–8%) SDS-PAGE and Autoradiography

1. These instructions assume the use of a Bio-Rad Mini-Protean three-gel system. The plates should first be scrubbed clean with distilled water, then with 100% (v/v) ethanol.
2. Prepare a 0.75-mm-thick 4–8% separating gel from 2 mL each of the 4% and 8% separating gel solutions: 4 μL of TEMED and 4 μL of 10% APS are added to each solution, and the solutions are transferred one to each chamber of a gradient maker. The gradient gel is formed by mixing, using a magnetic stirrer, of the two solutions as the 4% solution flows to 8%, and then between the two plates.
3. Pour the stacking gel after 20 min (after first adding 50 μL of TEMED and 50 μL of 10% APS). Insert the comb.
4. Remove the comb and rinse the wells with water after 60 min using a blunt-ended Hamilton syringe. The gel can be run or stored at this point. To store the gel, wrap with paper towels soaked in distilled water and then plastic wrap. The gels

can be stored at 4°C for up to 3 d. After storage, the gels should be warmed to room temperature before running.

5. Add running buffer to the inside and outside chambers to run the gel.
6. Gel samples should be incubated at 95°C for 10 min before loading 20 μL.
7. The Bio-Rad markers should be heated at 40°C for 1 min.
8. Run the gel at a constant current of 20 mA until the dye front just reaches the bottom of the gel.
9. Transfer the gel to a container with 50–100 mL of fixative after electrophoresis and gently agitate the gel on a rotary shaker at room temperature for 30 min.
10. Remove the fixative and replace with 50–100 mL of Amplify. Agitate gently for 30 min.
11. Rinse the gel for 10 s in distilled water before transferring to thick Whatman paper, covering with plastic wrap, and placing in a Model 543 gel dryer.
12. Dry the gel at 80°C for 60 min.
13. Expose the gel to X-ray film at –70°C for 1–2 wk before developing.

4. Notes

1. Unless otherwise stated, all reagents can be purchased from Fisher Scientific Co., ON, Canada.
2. Unless otherwise stated, all solutions should be prepared in water that has a resistivity of 18.2 mΩ-cm.
3. GAGs and PGs synthesized and secreted into the medium by cultured cells can be radiolabeled with $Na_2{}^{35}SO_4$, and later quantitated by immobilization to Alcian blue-coated flashplates.
4. If the growth medium is supplemented instead with D-[1,6-^{3}H(N)]-glucosamine hydrochloride, any newly-synthesized glycan will be radiolabeled. Any anionic GC, including GAGs and PGs, synthesized and secreted by cells in culture will be tritium-labeled and can later be quantitated by immobilization to Alcian blue-coated flashplates.
5. Unincorporated $Na_2{}^{35}SO_4$ and ^{3}H-glucosamine will not be immobilized and hence not quantitated. Only label incorporated into polyanions will be immobilized to polycation, Alcian blue, coated plates.
6. Add 1, 5, 10, 50, or 100 μL of radiolabeled medium to each Alcian blue-coated well. Each volume should be repeated in triplicate.

Acknowledgment

This research was supported by the National Science and Engineering Research Council of Canada (STP 246 039 01).

References

1. Rees-Milton, K. J., Terry, D., and Anastassiades, T. P. (2004) Hyperglycosylation of fibronectin by TGF-beta1-stimulated chondrocytes. *Biochem. Biophys. Res. Commun.* **317(3),** 844–850.

2. Steffy, M. A., Miura, N., Todhunter, R. J., et al. (2004) The potential limitations of cartilage-specific $(V+C)^-$ fibronectin and cartilage oligomeric matrix protein as osteoarthritis biomarkers in canine synovial fluid. *Osteoarthr. Cartil.* **12,** 818–825.
3. Goldberg, M. A. and Warner, K. J. (1997) The staining of acidic proteins on polyacrylamide gels: Enhanced sensitivity and stability of "Stains-All" staining in combination with silver nitrate. *Anal. Biochem.* **251,** 227–233.
4. Hronowski, L. and Anastassiades, T. P. (1979) Quantitation and interaction of glycosaminoglycans with Alcian blue in dimethyl sulfoxide solutions. *Anal. Biochem.* **93,** 60–72.
5. Hronowski, L. and Anastassiades, T. P. (1988) Detection and quantitation of proteoglycans extracted from cell culture medium and cultured cartilage slices. *Anal. Biochem.* **174,** 501–511.
6. Hronowski, L. and Anastassiades, T. P. (1980) Characterization of glycosaminoglycan-alcian blue complexes by elution from cellulose acetate utilizing different $MgCl_2$ concentrations. *Anal. Biochem.* **107,** 393–405.
7. Karamanos, N. K., Aletras, C. A., Antononopoulos, C. A., Hjerpe, A., and Tsiganos, C. P. (1990) Chondroitin proteoglycans from squid skin. Isolation, characterization and immunological studies. *Eur. J. Biochem.* **192,** 33–38.
8. Vynios, D. H., Vamacas, S. S., Kalapaxis, D. L., and Tsiganos, C. P. (1998) Aggrecan immobilization onto polystyrene plates through electrostatic interactions with spermine. *Anal. Biochem.* **260,** 64–70.
9. Rees-Milton, K. J. and Anastassiades, T. P. (2003) Measurement of metabolically-labeled anionic glycoconjugates in media from bovine articular chondrocyte cultures by immobilization to polystyrene plates treated with solid scintillant and coated with Alcian blue. *Anal. Biochem.* **315,** 273–276.
10. Lagace, J., Arsenault, E. A., and Cohen, E. A. (1994) Alcian blue-treated polystyrene plates for use in an ELISA to measure antibodies against synthetic peptides. *J. Immunol. Methods* **175,** 131–135.

18

Quantitative Analysis of Mucins in Mucosal Secretions Using Indirect Enzyme-Linked Immunosorbent Assay

Pablo Argüeso and Ilene K. Gipson

Summary

Mucins are extremely large and highly *O*-glycosylated glycoproteins synthesized and secreted by all wet-surfaced epithelia. Within the mucosal secretion they protect the underlying epithelium by forming a selective diffusion barrier against harmful substances and microorganisms, and act as lubricants to minimize shear stress. Variation in the character and quantity of secreted mucins is important to maintain the normal function of the epithelia (e.g., reproductive tract) but it may also reflect disease states. Understanding of the role of mucins has been limited by the lack of specific methods to detect and quantitate mucins in biological samples. Continuous progress in the development of specific antibodies against different mucin gene products has allowed the establishment of immunological techniques to perform these analyses. This chapter describes two protocols to quantify individual mucins from small-volume samples using indirect enzyme-linked immunosorbent assay (ELISA). These protocols allow a rapid, reproducible, and sensitive assay for these large molecules.

Key Words: Enzyme-linked immunosorbent assay (ELISA); mucin; *O*-glycosylation; mucosal secretion; quantitation.

1. Introduction

Mucins are a heterogeneous group of extraordinarily large glycoproteins that constitute the major structural components of the mucus gel that covers all wet-surfaced epithelia of the body *(1)*. Based on molecular characterization, mucins are defined by the presence of central tandem repeats of amino acids with their extensive *O*-glycosylation, which comprises up to 80% of the mass of the mature molecule *(1)*. Mucins participate in the maintenance of a wet-surfaced

From: *Methods in Molecular Biology, vol. 347: Glycobiology Protocols*
Edited by: I. Brockhausen © Humana Press Inc., Totowa, NJ

phenotype and in the protection of the underlying epithelium. Epithelia producing and secreting mucins often express several of the mucin genes, giving rise to a heterogeneous mucus secretion *(2)*. Several methods are available to detect and analyze soluble mucins in mucosal secretions. Some of them take advantage of the high content of O-linked carbohydrates on mucins for their analysis. These include solution assays (e.g., colorimetric assays for sialic acid; *see* **ref. *3***); membrane-based methods (e.g., dot-blotting and periodate-Schiff staining; *see* **ref. *4***); and lectin-based methods (*see* **ref. *5*** and Chapter 19). However, since these methods are based on the recognition of common carbohydrate moieties on mucins, they neither differentiate between members of the mucin family nor allow mucin quantitation owing to the varying glycosylation of mucins in the central tandem repeat region. Continuous progress in the development of antibodies of defined specificity against individual mucin–protein cores has allowed a more precise detection and quantitation of the different types of mucins in mucosal secretions. Enzyme-linked immunosorbent assay (ELISA) is a powerful method that allows quantitation of antigens in biological samples using antibodies *(6)*. With indirect ELISA, the mucin is passively adsorbed to a microtiter plate and detected using an unlabeled primary antibody followed by a labeled secondary antibody. This assay is very flexible, since the labeled secondary antibody recognizes all antibodies produced by a given species and can be used with a wide variety of primary antibodies against individual mucins. This method is also highly sensitive, can handle large numbers of samples that may then be analyzed rapidly, and is useful when only small-volume samples are available. Although ELISA has proven to be a valuable method, interpretation of data should be performed carefully, especially when analyzing and quantitating mucins, owing to their high carbohydrate content and to the "sticky" nature of these molecules. Several considerations should be taken into account when quantifying mucins by ELISA:

1. Antibodies with a defined specificity toward a mucin protein epitope should be employed. In those cases in which antibody recognition is affected by the glycosylation status of the mucin, chemical or enzymatic deglycosylation may be required to increase the success of the quantitation.
2. For quantitative purposes, the antibodies employed must recognize epitopes outside the tandem repeat region of the mucin.
3. It is essential to ensure that the signals obtained are within the linear range for the technique.
4. Whenever possible, the technique should be performed with a mucin standard.
5. To validate the assay, results of mucin quantitation should be confirmed using an additional technique based on a different principle (e.g., Western blot or semiquantitative real-time reverse transcription-polymerase chain reaction [RT-PCR]).

This chapter describes two protocols developed in our laboratory to quantify soluble mucins using indirect ELISA. The first protocol was developed to quantify *MUC5B* mucin in human cervical secretions using a mucin standard *(7)*. For absolute quantitation, a source of pure *MUC5B* protein would have been desirable for the generation of a standard curve. However, pure mucin standards are rarely (if ever) available. Because of the lack of pure *MUC5B*, a mixture of cervical mucins including *MUC5B* was purified from midcycle samples to serve as a cervical mucin standard in our protocol. The second protocol was developed for the analysis and relative quantification of mucin from small samples (human tears) using serial dilution curves *(8)*. Both protocols are designed to compare the relative amount of a specific mucin between different samples, but not to compare amounts of different mucins within a single sample (in this case, pure standards for each mucin are required). These techniques do not require prior mucin purification from the sample to be analyzed, and have the potential of being applied to the analysis of soluble mucins from other mucosal secretions.

2. Materials

1. Extraction buffer: 0.1 *M* of NH_4HCO_3, 0.5 *M* of NaCl, 5 m*M* of ethylenediaminetetraacetic acid, 2 m*M* of *N*-ethylmaleimide, and 0.02% of NaN_3. This solution should be stored at 4°C.
2. Protease inhibitor cocktail tablets (Complete Mini; Roche Applied Science, Indianapolis, IN).
3. Bicinchoninic acid (BCA) or MicroBCA™ Protein Assay Kit (Pierce, Rockford, IL).
4. Gel filtration column: Sepharose CL-4B (Agarose Bead Technologies, Tampa, FL).
5. Blue Dextran (fresh weight 2,000,000; Sigma Chemical, St. Louis, MO).
6. RNase-A and DNase I (Roche).
7. High-quality cesium chloride.
8. Polyallomer ultracentrifuge tubes.
9. Dialysis tubes (molecular weight cutoff 12,000–14,000).
10. 16- × 150-mm Disposable borosilicate tubes.
11. Liquified phenol 90% w/w (Fisher Scientific, Pittsburgh, PA).
12. Sulfuric acid, concentrated ACS (Fisher).
13. Galactose (1 mg/mL stock in distilled water [dH_2O]; Sigma).
14. 96-Well flat-bottom microtiter plates manufactured from polystyrene with high binding-certified surface chemistry (Corning Inc., Corning, NY).
15. Adhesive sealing films for microplates (SealPlate™; Excel Scientific, Wrightwood, CA).
16. Nonsterile solution basins (Labcor, Frederick, MD).
17. Multichannel 40- to 200-μL micropipet.
18. Coating buffer: 50 m*M* of carbonate/bicarbonate, pH 9.6 (buffer capsules available from Sigma).

19. Blocking buffer (*see* **Note 1**): phosphate-buffered saline (PBS) containing 3% (w/v) gelatin from cold-water fish skin (Sigma).
20. Washing buffers (*see* **Note 2**): PBS (10 m*M* of sodium phosphate [pH 7.4] and 150 m*M* of NaCl) with or without 0.05% (v/v) NP-40 detergent.
21. Primary poly- or monoclonal antibodies recognizing nonglycosylated regions of the mucin (*see* **Note 3**).
22. Antibody control: species- and isotype-matched, nonspecific immunoglobulin.
23. Peroxidase-conjugated secondary antibody (Sigma).
24. Substrate for peroxidase: Tetramethylbenzidine liquid substrate system (Sigma).
25. Neuraminidase, recombinant from *Arthrobacter ureafaciens* (Glyko, Rockville, MD).
26. Microtiter plate spectrophotometer and software (Softmax Pro; Molecular Devices, Sunnyvale, CA).

3. Methods

3.1. Quantitation of Mucin by ELISA Using a Mucin Standard

In this method, a purified mucin preparation from the same source as that to be assayed is used as a mucin standard for determining relative amounts of mucin in individual crude mucus samples. This method has been used to determine the relative amounts of *MUC5B* mucin in human cervical secretions *(7)*.

3.1.1. Purification of Mucins for Standard

1. Solubilize mucin secretions from several individual samples in an equal volume of ice-cold extraction buffer (*see* **Heading 2.**) containing protease inhibitors. The samples should be maintained at 4°C during the purification protocol.
2. Pool samples (for cervical mucin standard, we have used as little as 12 mL of cervical mucus at a protein concentration of 3.6 mg/mL as starting material) and homogenize in a Dounce homogenizer to break up mucus strands.
3. Centrifuge at 48,000*g* at 4°C for 45 min.
4. Dialyze the clear supernatant against dH_2O overnight with two changes to fresh water during the next day; dialyze for 2–3 h after each change. Lyophilize on a freeze-drier and reconstitute in PBS.
5. Centrifuge at 16,000*g* at 4°C for 20 min and recover the supernatant. Determine protein concentration by using the Pierce BCA Protein Assay.
6. Pass the clear supernatant over a 3- × 80-cm Sepharose CL-4B size exclusion column that has been calibrated with 1% Blue Dextran to determine the void volume peak. Monitor absorbance at 280 nm.
7. Elute the column with PBS (0.8 mL/min) and collect the void volume. Dialyze against dH_2O and lyophilize.
8. Reconstitute the mucin preparation in PBS containing 1 m*M* of $MgSO_4$, 0.02% NaN_3, and protease inhibitors. Determine the protein concentration using the Pierce BCA Protein Assay. Digest with RNase A and DNase I (1 mg nuclease/100 mg protein) at room temperature for 3 h (*see* **Note 4**) and dialyze overnight against dH_2O.

9. Separate mucins by cesium chloride density centrifugation (isopycnic density gradient centrifugation). Add cesium chloride powder until the concentration reaches 0.54 g/mL. Measure the density of the sample prior to loading in polyallomer ultracentrifuge tubes by weighing a known volume. Centrifuge at 164,000*g* at 4°C for 72 h in a swinging bucket rotor (e.g., TH641 rotor; Sorvall), stop the centrifuge without brake, remove tubes in vertical position, and gently collect fractions (~1 mL each) by piercing the tube, starting at the top. Analyze the fractions for density.
10. Dialyze fractions with a density of 1.3–1.5 g/mL against dH_2O and determine the protein content of each fraction using the Pierce BCA Protein Assay, and the hexose content using the phenol-sulfuric acid assay ***(9)***. For the phenol-sulfuric acid assay:
    a. Prepare a series of galactose standards in dH_2O (ranging from 25 to 1000 μg/mL).
    b. Mix 50 μL of each standard with 550 μL of dH_2O in borosilicate tubes.
    c. Mix 50 μL of each cesium chloride fraction with 550 μL of dH_2O in separate tubes.
    d. Add 50 μL of liquified phenol reagent to each tube and vortex.
    e. Immediately add 1.5 mL of concentrated H_2SO_4 and vortex carefully (*see* **Note 5**).
    f. Read absorbance at 490 nm against blank (600 μL of dH_2O + 50 μL of phenol + 1.5 mL of H_2SO_4).
    g. Determine the hexose concentration of the cesium chloride fractions.
11. Pool the fractions with a hexose content at least twice the protein content. Determine the final protein concentration using the Pierce BCA Protein Assay. This pooled preparation of purified mucins can be aliquoted and frozen at –80°C for use as mucin standard.

3.1.2. ELISA (see ***Note 6****)*

1. Coat wells in the microtiter plate with 100 μL of sample (for concentration, *see* **Subheadings 3.1.3.** and **3.1.4.**) diluted in coating buffer (*see* **Heading 2.** and **Note 7**).
2. Cover the plate with an adhesive sealing film. Incubate the plate at 4°C overnight.
3. Remove the sealing film. Shake off the antigen solution and wash the plate with PBS (fill and empty wells three times with a squirt bottle or plate washer).
4. Add 300 μL of blocking buffer (*see* **Heading 2.** and **Note 8**).
5. Incubate at room temperature for 2 h on shaker. During the incubation, cover the plate with a sealing film.
6. Shake off the blocking buffer and wash the plate three times with PBS.
7. Add 100 μL/well of primary antibody (for concentration, *see* **Subheading 3.1.3.**) in 1% (w/v) fish skin gelatin in PBS. As a negative control to determine background, a species- and isotype-matched, nonspecific immunoglobulin can be used.
8. Incubate at room temperature for 1 h on shaker. During the incubation, cover the plate with a sealing film.
9. Take out substrate to equilibrate at room temperature.
10. Shake off the primary antibody and wash the plate three times with PBS containing 0.05% (v/v) NP-40 and once with PBS.

11. Add 100 µL of peroxidase-conjugated secondary antibody at 1:10,000 dilution (*see* **Note 9**) in 1% (w/v) fish skin gelatin in PBS (*see* **Note 10**).
12. Incubate at room temperature for 1 h on shaker. During the incubation, cover the plate with a sealing film.
13. Shake off the secondary antibody and wash the plate twice with PBS containing 0.05% (v/v) NP-40 and twice with PBS.
14. Add 100 µL of substrate (tetramethylbenzidine).
15. Incubate the plate at room temperature for 30 min.
16. Stop the reaction by adding 50 µL of 0.5 *M* H_2SO_4 to each well.
17. Read absorbance in the spectrophotometer at 450 nm.
18. Subtract the mean absorbance value of the negative-control (background) wells from all wells.
19. Plot protein concentration against absorbance units.

3.1.3. Titration of Antigen (Crude Mucus Sample) and Antibody

1. Pool approx 80 µg total protein of crude mucus from several subjects.
2. Add 100 µL of coating buffer to each well in the microtiter plate using a multichannel micropipet.
3. Add 100 µL of crude mucus samples (containing approx 10 µg of total protein in the coating buffer) to the wells in column 1.
4. Serially dilute the samples (twofold dilution) across columns 1–11 in the microtiter plate (*see* **Note 11**). Reserve the last well (column 12) for the blank (coating buffer alone) or negative control.
5. Proceed with **Subheading 3.1.2., steps 2–6**.
6. Add 100 µL of primary antibody (~1:100 dilution) to wells in row A and serially dilute the antibody (twofold dilution) in 1% (w/v) fish skin gelatin in PBS down rows A–G of the plate in the opposite direction of the antigen, obtaining a chessboard titration of antigen–antibody. Reserve row H for the blank or negative control.
7. Proceed with **Subheading 3.1.2., steps 8–18**.
8. Plot the entire range of dilutions against absorbance units. Determine the optimal crude mucus concentration to coat the wells and the optimal antibody dilution. The optimal amounts produce midlinear responses in a linear regression analysis.

3.1.4. Titration of Mucin Standard

1. Serially dilute the mucin standard (twofold dilution, starting at approx 10 ng of total protein) in coating buffer across the microtiter plate in the upper row. Reserve the last well for the blank or negative control.
2. Proceed with **Subheading 3.1.2., steps 2–19**. Use the primary antibody dilution selected from **Subheading 3.1.3., step 8**.
3. Determine the range of mucin standard protein concentration that produces a linear response using a linear regression analysis. Prepare a series of standards that produce a linear response, aliquot them, and freeze in amber glass vials at –80°C.

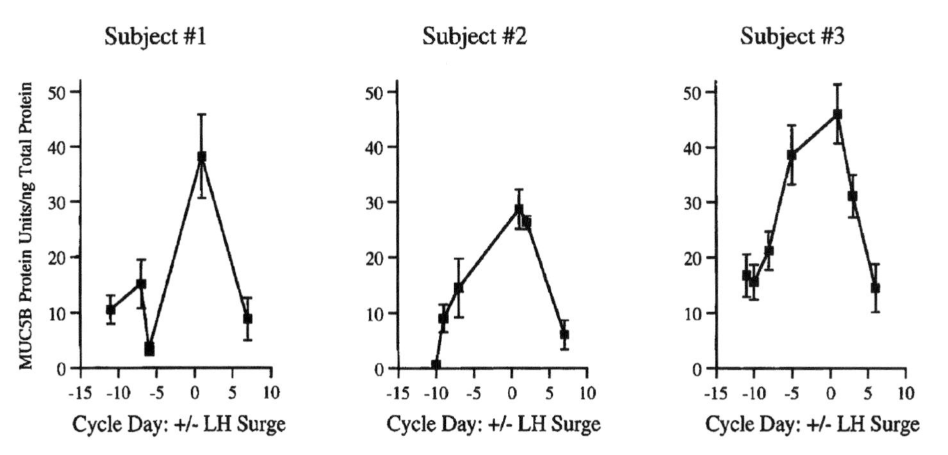

Fig. 1. Change in *MUC5B* mucin content in three subjects through the menstrual cycle, as determined by enzyme-linked immunosorbent assay. The values are expressed as *MUC5B* units per nanogram of total cervical mucus protein, and they reflect the amount of *MUC5B* present relative to that present in picogram amounts of cervical mucin standard. The quantity of *MUC5B* is plotted against the cycle day relative to luteinizing hormone surge. LH, luteinizing hormone. (Reproduced with permission from **ref.** *7*.)

3.1.5. Quantification of Mucin in Individual Samples

1. Coat individual crude mucus samples in triplicates diluted in 100 µL of coating buffer. The optimal crude mucus protein concentration should be established in **Subheading 3.1.3., step 8**. A series of standards that produce linear responses should be run in the same microtiter plate. For the negative control, add coating buffer alone.
2. Proceed with **Subheading 3.1.2., steps 2–18**.
3. Plot the absorbance values of the standard against the protein concentration and perform a linear regression analysis.
4. Interpolate the average absorbance values obtained for each individual sample against the standard curve to obtain the concentration of mucin in the sample. Because the purified mucin standard contains a mixture of mucins, absolute values of mucin concentration cannot be obtained. The amount of mucin in the sample is converted to mucin units to be used in relative quantitation (*see* **Fig. 1**). A mucin unit is defined as the picogram amount of mucin standard corresponding to the optical density reading obtained per sample (*see* **Note 12**).

3.2. Quantitation of Mucin by ELISA Using Serial Dilution Curves

This method allows the relative quantitation of individual mucins in mucosal secretions when limited amounts of starting material are available and it is not possible to generate a purified mucin standard. This method has been used to determine the relative levels of *MUC5AC* mucin in human tears *(8)*.

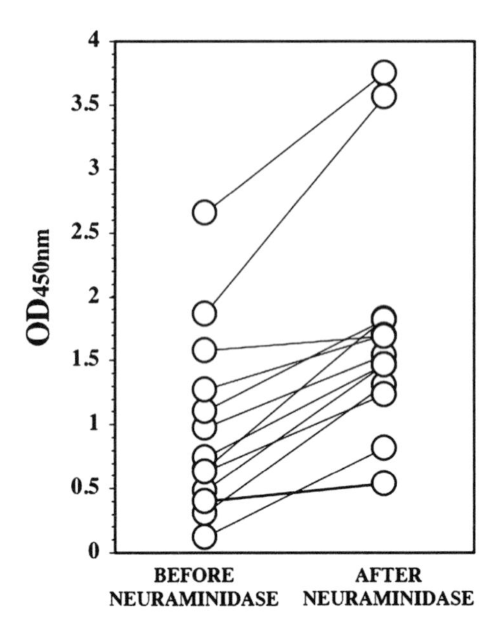

Fig. 2. As determined by enzyme-linked immunosorbent assay, the release of neuraminic acid from glycoconjugates present in tear samples efficiently enhances recognition of the *MUC5AC* protein backbone by a chicken polyclonal antibody designated 791. (Reproduced with permission from **ref.** ***8***.)

3.2.1. Preparation of Samples

1. Centrifuge mucin secretions at 16,000*g* at 4°C for 30 min.
2. Recover the clear supernatant and determine the amount of total protein on a small aliquot using the MicroBCA Protein Assay Kit (*see* **Notes 13** and **14**).
3. If antibody recognition depends on the glycosylation status of the mucin, chemical or enzymatic deglycosylation may be required to enhance mucin recognition (*see* **Fig. 2**). For deglycosylation, digest individual mucin samples (~10 µg total protein) with 2.5 mU of neuraminidase from *Arthrobacter ureafaciens* in 50 m*M* of sodium phosphate (pH 5.5) at 37°C for 2 h.

3.2.2. Titration of Antigen (Crude Mucus Sample) and Antibody

1. Proceed with the protocol described in **Subheading 3.1.3., steps 1–7**.
2. Plot the entire range of dilutions against absorbance units. Determine the optimal range of crude mucus dilutions to coat the wells and the optimal antibody dilution.

*3.2.3. Quantification of Mucin in Individual Samples (*see ***Note 15****)*

1. Add 100 µL of coating buffer to each well in the microtiter plate using a multichannel micropipet.

2. Add 100 μL of samples (for starting concentration in the serial dilution, *see* **Subheading 3.2.2., step 2**) diluted in coating buffer to the wells in column 1.
3. Serially dilute samples (twofold dilution) across columns 1–11 in the microtiter plate. Reserve the last well (column 12) for the blank or negative control.
4. Proceed with **Subheading 3.1.2., steps 2–18**. Use the primary antibody dilution selected from **Subheading 3.2.2., step 2**.
5. Plot data relating the protein concentration of all serial dilutions of the sample (as a $\log_{10}$ twofold series) to the optical density (*see* **Fig. 3**). Determine the range of linear response for each sample (*see* **Note 16**). Perform a linear regression analysis with points in the center of the graph that are nearly linear. To compare relative amounts of mucin from different samples, absorbance values from a single protein concentration point within the midlinear range of the regression line should be used.

4. Notes

1. A common problem when utilizing antibodies against mucins is the nonspecific adsorption of the antibody to the microtiter well even after the addition of the blocking reagent. The optimal blocking substance for each antibody should be determined empirically. Common blocking reagents used with antimucin antibodies include normal horse serum, bovine serum albumin fraction V, and fish skin gelatin with or without nonionic detergents.
2. The optimal washing buffer should be determined empirically. Commonly used washing buffers typically contain PBS to maintain isotonicity and may include detergents such as 0.05% NP-40 or 0.05% Tween-20.
3. The most common sources of mucin antigen to prepare antibodies are purified mucins, deglycosylated mucins, fresh tissue and cultured cells, synthetic peptides, and recombinant proteins ***(11)***. Antibodies made toward synthetic peptides or recombinant proteins to be used in quantitation should mimic unique nontandem repeat regions of the specific mucin as identified using the Protein–Protein Basic Local Alignment Search Tool (BLASTP; available at http://www.ncbi.nlm.nih.gov/BLAST).
4. Invert samples gently every 15 min.
5. The rapid mixing of sulfuric acid is essential to the success of the assay.
6. When analyzing mucosal secretions collected at different time-points from a single individual, care must be taken to run all samples in the same ELISA experiment.
7. Most proteins adsorb to plastic surfaces as a result of hydrophobic interactions. Strong interactions with the plastic are usually achieved by optimizing the coating buffer (i.e., modifying pH or the ionic strength, and adding detergents to denature the protein), the temperature, and the time of adsorption. Different conditions should be investigated where problems are encountered.
8. Pipet the solution with a multichannel micropipet using a 55-mL solution basin as a reservoir for the buffer.
9. The conjugated secondary antibody must be pretitrated to give optimal color development (1:10,000 is a standard dilution).

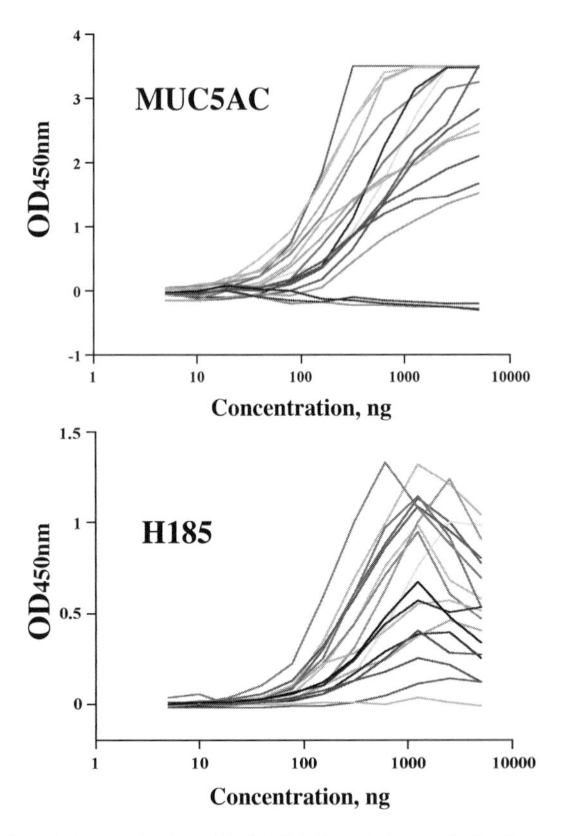

Fig. 3. Example of the analysis of *MUC5AC* and the carbohydrate epitope H185 on *MUC16* ***(10)*** in individual tear samples using enzyme-linked immunosorbent assay. Plots relate the protein concentration of the serial dilutions to the optical density. *MUC5AC* mucin is present in 15 of the tear secretions analyzed and absent in three samples. H185 is detected in 16 tear secretion samples and absent in two samples. The linear range of detection for both antibodies can be established between approx 40 and 800 ng of total protein. At high protein concentrations, the graph shows a plateau region in some samples where the absorbance values are similar; increasing the concentration of antigen either has no effect or a negative (hook) effect in the binding of the antibody. This area represents a condition in which all antigen is saturated with antibody. For relative quantitation, absorbance units are compared after coating the plates with 500 ng of total protein.

10. Secretions must be tested and found to be negative for endogenous peroxidase activity by incubation with peroxidase substrate alone. If positive, other enzyme-conjugated secondary antibodies can be used (e.g., alkaline phosphatase).
11. Put tips into column 1 and mix contents by pipeting up and down eight times. Collect 100 μL and transfer to column 2, mix, and transfer to column 3, and so on to column 11. After the last mixing, discard the 100 μL left in the tips.

12. So-called add-back experiments can be performed to verify the accuracy of the quantitation of mucin in crude mucus samples. Known amounts of purified mucin are added to previously quantitated amounts of crude mucus samples and analyzed by ELISA. The value obtained from the add-back experiment should approximate the additive value obtained from the purified mucin plus the crude mucus sample.
13. Retain the pellet and solubilize in 20 μL of 0.1% sodium dodecyl sulfate. Perform protein assay on a 5-μL aliquot of the pellet. If the amount of protein in the pellet is sufficient for ELISA, the assay could also be performed on this material in order to determine the amount of insoluble mucin.
14. If the samples are not analyzed immediately they should be stored at –80°C.
15. The assay should be performed at different times and analyzed by two different investigators to assure reproducibility and reliability of the ELISA. In this method, owing to the lack of a standard, it is necessary to perform the analyses with batches of antibody that have the same binding efficiency.
16. The plotted data generally produces sigmoidal curves. There is a region in the curve that contains three to five points in the center that are nearly linear.

Acknowledgments

This research was supported by NIH/NEI Grants no. R01EY014847 to Dr. Argüeso and R01EY03306 to Dr. Gipson.

References

1. Gendler, S. J. and Spicer, A. P. (1995) Epithelial mucin genes. *Annu. Rev. Physiol.* **57,** 607–634.
2. Moniaux, N., Escande, F., Porchet, N., Aubert, J. P., and Batra, S. K. (2001) Structural organization and classification of the human mucin genes. *Front. Biosci.* **6,** D1192–D1206.
3. Pearce, E. I. and Major, G. N. (1978) The colorimetric analysis of sialic acid in human saliva and bovine salivary mucin. *J. Dent. Res.* **57,** 995–1002.
4. Goso, Y. and Hotta, K. (1994) Dot blot analysis of rat gastric mucin using histochemical staining methods. *Anal. Biochem.* **223,** 274–279.
5. Jackson, A., Kemp, P., Giddings, J., and Sugar, R. (2002) Development and validation of a lectin-based assay for the quantitation of rat respiratory mucin. *Novartis Found. Symp.* **248,** 94–105; discussion 106–112, 277–282.
6. Crowther, J. R. (1995) ELISA. Theory and practice. *Methods Mol. Biol.* **42,** 1–218.
7. Gipson, I. K., Moccia, R., Spurr-Michaud, S., et al. (2001) The amount of MUC5B mucin in cervical mucus peaks at midcycle. *J. Clin. Endocrinol. Metab.* **86,** 594–600.
8. Argueso, P., Balaram, M., Spurr-Michaud, S., Keutmann, H. T., Dana, M. R., and Gipson, I. K. (2002) Decreased levels of the goblet cell mucin MUC5AC in tears of patients with Sjogren syndrome. *Invest. Ophthalmol. Vis. Sci.* **43,** 1004–1011.

9. Dubois, M., Gilles, K. A., Hamilton, K. K., Rebers, P. A., and Smith, F. (1956) Colorimetric method for determination of sugars and related substances. *Anal. Chem.* **28,** 350–356.
10. Argueso, P., Spurr-Michaud, S., Russo, C. L., Tisdale, A., and Gipson, I. K. (2003) MUC16 mucin is expressed by the human ocular surface epithelia and carries the H185 carbohydrate epitope. *Invest. Ophthalmol. Vis. Sci.* **44,** 2487–2495.
11. Real, F. X., de Bolos, C., and Oosterwijk, E. (2000) Polyclonal and monoclonal techniques. *Methods Mol. Biol.* **125,** 353–368.

19

Molecular Dissection of the Mouse Zona Pellucida

An Electron-Microscopic Perspective Utilizing High-Resolution Colloidal-Gold Labeling Methods

Frederick W. K. Kan

Summary

The zona pellucida (ZP) is an extracellular coat that encloses growing oocytes, ovulated eggs, and preimplantation embryos in mammals. The ZP contains receptors that mediate initial interactions between the sperm and egg and the relative species-specificity during gamete interaction. It also prevents polyspermy and protects the developing embryo prior to implantation. The current model of the organization of mouse ZP depicts this extracellular matrix as an extensive three-dimensional array of long interconnected filaments of a structural repeat. Using high-resolution colloidal-gold labeling methods in combination with specific lectins and monoclonal antibodies against three major ZP glycoproteins, we have characterized the outer and inner mouse ZP during folliculogenesis and shown the modifications of the ZP after ovulation and fertilization. Our immunogold labeling results also indicate the involvement of a vesicular aggregate, a specialized subcellular compartment in the oocyte, in the synthesis and secretion of ZP glycoproteins.

Key Words: Zona pellucida (ZP); ZP glycoproteins; immunocytochemistry; oocyte; folliculogenesis; structure and assembly of ZP.

1. Introduction

The zona pellucida (ZP) is a glycoprotein (GP) matrix of uniform thickness that surrounds mammalian oocytes. In the mouse (which is the most well-studied animal model), the ZP is primarily composed of three sulfated GPs (ZP1, ZP2, and ZP3) that are highly glycosylated ***(1–3)***. ZP3 has been shown to act as the primary ligand that binds to the plasma membrane of acrosome-intact sperm ***(4,5)***, whereas ZP2 serves as the secondary ligand that

From: *Methods in Molecular Biology, vol. 347: Glycobiology Protocols*
Edited by: I. Brockhausen © Humana Press Inc., Totowa, NJ

preferentially binds to the inner acrosomal membrane exposed after acrosome reaction ***(6)***. ZP1 is believed to provide the structural integrity of the zona matrix by interconnecting filaments of ZP2/ZP3 heterodimeric repeats through intermolecular disulfide bonds ***(7)***. Because the ZP GPs play an important role during sperm–egg interactions in the fertilization process, there has been extensive investigation into the molecular basis of sperm binding to the ZP resulting in mammalian fertilization ***(8,9)***. The information regarding the pathway of secretion and assembly of the three major ZP GPs is very limited. Monoclonal antibodies (MAbs) prepared against individual ZP1, ZP2, and ZP3 GPs ***(10–12)*** and lectins that recognize specific sugar residues, when used in combination with different sizes of colloidal-gold (CG) particles in suspension, can provide routine protocols for studying the stoichiometric disposition of the three GPs in the zona matrix. We have used electron microscopy in conjunction with high-resolution CG labeling methods to reveal the distribution of ZP1, ZP2, and ZP3 GPs in the mouse zona matrix during folliculogenesis ***(13)***, and the detection of glycosidic residues in the hamster ZP after ovulation ***(14)*** and fertilization ***(15)***. When using MAbs against individual ZP GPs and lectins that recognize specific sugar residues (*see* **Table 1**), results obtained with high-resolution CG labeling methods also indicate that the vesicular aggregate, a specialized complex of membranous vesicles in the ooplasm, is likely to serve as an intermediary in the synthesis and secretion of ZP GPs ***(13,16)***.

2. Materials

2.1. Animals

1. Sexually mature female CD1 mice (Charles River Laboratories, St-Constant, PQ, Canada).
2. Sexually mature female golden hamsters (*Mesocricetus auratus*; Charles River Laboratories).

2.2. MAbs and Secondary Antibodies

1. Two sets of MAbs are used. The first set consists of MAbs M-1.4 (rat IgG), IE-3 (rat IgG), and IE-10 (rat IgG), which are specific for ZP1, ZP2, and ZP3, respectively. The second set consists of anti-ZP1, anti-ZP2, and anti-ZP3 antibodies that are directly bound to CG particles with a diameter of 10 or 15 nm, 5 or 15 nm, and 5 or 10 nm, respectively, as per the instructions of the manufacturer of the CG (BBInternational, Cardiff, UK). These antibodies were generously provided by Dr. Jurrien Dean (Laboratory of Cellular and Developmental Biology, National Institute of Diabetes and Digestive and Kidney Diseases, National Institutes of Health, Bethesda, MD).
2. Rabbit anti-rat IgG polyclonal antibody (Sigma, St. Louis, MO).

Table 1

Lectins Used for the Localization of Oligosaccharide Content of Hamster Ovarian Zona Pellucida (ZP) During Folliculogenesis or for the Localization of Glycoconjugate Content in Hamster ZP Following Ovulation and Fertilization

Taxonomic name of the source	Abbreviation	Carbohydrate binding specificity	Inhibiting sugar
Aleuria aurantia	AAA	Complex type *N*-linked oligosaccharides with an α 1,6 fucosyl residue at the innermost GlcNAc. Terminal α 1,2-linked Fuc residues	L-Fucose
Arachis hypogea	PNA	Terminal Galβ1,3 GalNAc disaccharides	D-Galactose
Datura stramonium	DSA	Bi, tri, and tetraantennary sugar chains with at least one *N*-acetyllactosamine	*N*-acetyllactosamine
Dolichos biflorus	DBA	Terminal α-linked GalNAc	D-GalNAc
Galanthus nivalis	GNA	Terminal Man α 1,3 in high mannose *N*-linked Oligosaccharides	Methyl-α-D-manno-pyranoside
Helix pomatia	HPA	Terminal α- and β-linked GalNAc	D-GalNAc
Limax flavus	LFA	Neu5Ac α 2,3/6 Gal, Neu5Ac α 2,3/6 GalNAc	Neu5Ac
Lotus tetragonolobus	LTA	Type 2 chains containing Fuc residues on C2 of the Gal of Galβ1,4GlcNAc	L-Fucose
Maackia amurensis	MAA	Neu5Ac α 2,3/6 Galβ 1,4GlcNAc	*N*-acetylneuraminyllactose
Ricinus communis	RCA I	Terminal Galβ1,4GlcNAc disaccharides	D-Galactose
Sambucus nigra	SNA	Neu5Ac α 2,6Gal/GalNAc	*N*-acetylneuraminyllactose
Triticum vulgaris	WGA	Terminal non-reducing GlcNAc or Neu5Ac	D-GlcNAc and/or Neu5Ac
Ulex Uropaeus	UEA I	Terminal α-L-Fuc	L-Fucose

2.3. CG-Conjugated Lectins, Unconjugated Lectins, Simple Sugars, and Secondary Antibodies

1. The lectins used in our studies are listed in **Table 1**.
2. CG-conjugated lectins (*Dolichos biflorus* agglutinin [DBA] and *Datura stramonium* agglutinin [DSA]) and unconjugated lectins (*Helix Pomatia* agglutinin [HPA], *Limax flavus* agglutinin [LFA], *Lotus Tetragonolobus* agglutinin [LTA], *Ricinus Communis* agglutinin I [RCA I], *Ulex Uropaeus* agglutinin I [UEA I], and wheat germ agglutinin [WGA]) from EY Laboratories, Inc., San Mateo, CA.
3. D-Galactose (D-Gal), L-fucose (L-Fuc), *N*-acetylglucosamine (GlcNAc), *N*-acetyl-D-galactosamine (GalNAc), *N*-acetylneuraminic acid (Neu5Ac), *N*-acetylneuraminyllactose, methyl-α-D-mannopyranoside (Methyl-α-Man), Methyl-α-galactopyranoside (Methyl-α-Gal), and *N*-acetyllactosamine (EY Laboratories).
4. Digoxigenin (DIG)-labeled lectins (*Aleuria aurantia* [AAA], *Maackia amurensis* agglutinin [MAA], *Sambucus nigra* agglutinin [SNA], *Galanthus nivalis* agglutinin [GNA], and peanut agglutinin [PNA]) and mouse monoclonal IgG anti-DIG antibodies (Roche Diagnostics, Laval, Quebec, Canada).
5. Goat anti-mouse IgG + IgM-gold complex (15 nm) (Cedarlane Laboratories, Ltd., Hornby, Ontario, Canada).
6. Ovomucoid (Ovo; trypsin inhibitor) and fetuin (Sigma).

2.4. Collection and Fixation of Ovarian Tissue, Cumulus Masses, and Fertilized Eggs for Immunocytochemistry

1. Polyethylene glycol (PEG; MW 20,000), polyvinylpyrrolidone (PVP; MW 10,000), tetrachloroauric acid ($HAuCl_4 \cdot 2H_2O$), pregnant mare serum (PMSG), human chorionic gonadotropin (hCG), neuraminidase (type V, from *Clostridium perfringens*), bovine serum albumin (BSA), and galactose oxidase (GO; from *Dactylium dendroides*; Sigma).
2. Protein A (Invitrogen Canada, Inc., Burlington, ON, Canada).
3. 2.5% Glutaraldehyde (CANEMCO, St-Laurent, PQ, Canada) is prepared in 0.1 *M* of sodium cacodylate buffer (CB; CANEMCO), pH 7.4.
4. 1% Osmium tetroxide (CANEMCO) is prepared in double-distilled water.
5. 200- and 300-mesh nickel grids and Lowicryl K4M embedding kit (CANEMCO).
6. LR White™ embedding kit (Electron Microscopy Sciences, Fort Washington, PA).

3. Methods

3.1. Preparation of CG (15 nm)

1. CG solutions with a mean particle diameter of 15 nm are prepared following the sodium method of Frens *(17)*.
2. Prepare a 1% solution of tetrachloroauric acid (gold chloride solution) by dissolving 0.5 g of $HAuCl_4 \cdot 2H_2O$ in 50 mL of double-distilled water (*see* **Note 1**).
3. To prepare a volume of 100 mL of 15-nm CG complex, add 1 mL of 1% gold chloride solution to a 500-mL Erlenmeyer flask (use of a siliconized flask is preferred). Add 99 mL of double-distilled water to the flask and place the flask on

top of a hot plate with the temperature setting adjusted to the maximum. Place a magnetic stirring bar into the flask to agitate the solution while the latter is being heated on the hot plate.

4. The following is a critical step: as soon as the solution is brought to a boil, rapidly add 2.5 mL of a freshly prepared 1% (w/v) aqueous sodium citrate solution into the flask using a 10-mL pipet.
5. Allow the solution to boil for 10 min at a maximum temperature setting of 100°C. During this time, the solution will turn into a red wine color indicating the formation of gold particles in dispersion in the solution.
6. After boiling for 10 min, reduce the heat to a setting of 3 (on a hot plate with a temperature setting ranging from 0 to 5) and let the boiling continue for another 10 min. At the end, remove the flask from the hot plate and let it cool in the sink with running water or let it cool on ice. A freshly prepared 15-nm CG solution should have a pH between 6.0 and 6.3 (*see* **Note 2**).

3.2. Preparation of Protein A-Gold Complex (15 nm) for Immunocytochemistry

1. Add 40 mL of CG solution (15 nm) to a 50-mL Erlenmeyer flask.
2. Adjust the pH of the CG solution to 5.6 using 0.1 *N* of HCl.
3. Dissolve 4 mg of protein A in 1 mL of double-distilled water (use 1 mg for every 10 mL of gold solution) and then add the protein A solution rapidly into the Erlenmeyer flask, and mix the two solutions well with a magnetic stirring bar.
4. Withdraw 1–2 mL of protein A-complex from the flask and transfer it to an Eppendorf tube, and test to find out if a well-conjugated protein A–gold complex is obtained (*see* **Note 3**).
5. Before storage, add 0.4 mL of 1% (w/v) PEG in double-distilled water to the protein A–gold complex (use 0.1 mL of PEG per 10 mL of protein A–gold complex).
6. A freshly prepared protein A–gold complex can be kept at 4°C for at least 2 mo.

3.3. Preparation of Ovomucoid–Gold Complex (15 nm) for WGA Labeling

1. Adjust the pH of the CG solution to 5.5 with 0.1 *N* of HCl.
2. To prepare an ovomucoid (Ovo)–gold complex, dissolve 1.2 mg of Ovo (trypsin inhibitor) in 8 mL of double-distilled water in a polypropylene tube. Increase the amount of Ovo in proportion to the final volume of the desired CG solution (i.e., 240 µg of Ovo/40 mL gold solution). Add 1.6 mL of the Ovo solution to 40 mL of 15-nm CG solution after the pH has been adjusted to 5.5.
3. Mix the Ovo and CG solutions together by shaking gently until a dark purplish color is observed. The appearance of a dark purplish color indicates a well-conjugated Ovo–CG complex (*see* **Note 4**).
4. Once a well-conjugated Ovo–CG complex is prepared and before storing the Ovo–CG complex for further use, add 0.1 mL of 1% PEG to every 10 mL of Ovo–CG complex and mix the PEG and Ovo–CG complex together gently. Then centrifuge the Ovo–CG complex at 600*g* at room temperature to get rid of

large aggregates. Collect the supernatant and centrifuge the supernatant at 56,000*g* at 4°C for 30 min in an ultracentrifuge. At the end of centrifugation, discard the supernatant and resuspend the pellet in 1 mL of phosphate-buffered saline (PBS) containing 0.02% PEG and 0.05% PVP per 10 mL of starting volume of Ovo–CG complex.

5. The final Ovo–CG complex can be stored at 4°C for further use. Normally, a freshly-prepared Ovo–CG complex can be used for two months or more until a precipitate is formed, indicating that a new complex should be made and used.

3.4. Preparation of Fetuin–CG Complex (15 nm) for LFA Labeling

1. Adjust the pH of the CG solution to 6.0 with 0.1 *N* of HCl.
2. To prepare a fetuin–CG complex, use 35 μg of fetuin for every 10 mL of CG solution. For a volume of 40 mL of gold solution, dissolve 1 mg of fetuin in 1 mL of double-distilled water. Then transfer 140 μL of the fetuin solution to 40 mL of the gold solution after the pH of has been adjusted to 6.0.
3. Mix the fetuin and CG solutions together by shaking gently until a dark purplish color is observed.
4. Repeat **Subheading 3.3., step 4** before storage. A freshly prepared fetuin–gold complex can be stored at 4°C for at least 2 mo.

3.5. Preparation of Lectin–CG Complex (15 nm)

1. For preparation of *Helix Pomatia* agglutinin (HPA)–, *Lotus Tetragonolobus* (LTA)–, *Ricinus Communis* agglutinin I (RCA I)–, and *Ulex Uropaeus* I (UEA I)–CG complexes, adjust the pH of the 15-nm CG solutions to 7.4, 6.3, 8.0, and 6.3, respectively.
2. Dissolve 0.13 mg of HPA, 260 μg of LTA, 0.3 mg of RCA I, and 0.6 mg of UEA I, respectively, in 2 mL of double-distilled water. Add 2 mL of each of the lectin solutions, respectively, to 40 mL of 15-nm CG solution, adjusted previously to the corresponding pH mentioned in **step 1**.
3. Gently mix each of the lectin–gold complexes prepared in **step 2**. Note that each of the four lectin–gold complexes should retain a red wine color upon testing with 10% NaCl; otherwise, repeat the preparation using the correct amount of lectin and proper pH.
4. Once a well-conjugated lectin–gold complex is prepared, add 0.4 mL of PEG to the lectin–gold complex and mix well.
5. Centrifuge the lectin–gold complex at 600*g* at room temperature to get rid of large aggregates. Collect the supernatant and then centrifuge the supernatant at 56,000*g* at 4°C for 30 min in an ultracentrifuge. At the end of the centrifugation, discard the supernatant and resuspend the pellet in 1 mL of PBS containing 0.02% PEG per 10 mL of starting volume of lectin–gold complex.
6. A fresh lectin–gold complex can be kept at 4°C for 1–2 mo until a precipitate is formed.

3.6. Preparation of Animals for the Collection of Ovarian Tissue

1. Sexually mature female CD1 mice (3–4 wk old) and golden hamsters (7–8 wk old) are used.
2. The animals are sacrificed by cervical dislocation. The skin covering the abdomen region of the animals is wetted with 70% ethanol and then the abdomen is incised with a pair of surgical scissors. The ovary (which is normally embedded in a thick fat pad) can be located by following the uterus that leads proximally to the oviduct and ovary. The ovaries are excised and then washed briefly in PBS, pH 7.4.
3. The tissue is fixed at room temperature for 2 h by immersion in 2.5% glutaraldehyde (GA) in 0.1 *M* of CB, pH 7.4. After fixation, the ovarian tissue is washed three times in 0.1 *M* of CB. (The tissue can be left overnight in the same buffer at 4°C before further processing.)
4. For postembedding labeling, the ovarian tissue samples are trimmed into small cubes, dehydrated in a series of graded ethanol, and then embedded in Lowicryl K4M (samples from CD1 mice) for immunolocalization of ZP1, ZP2, and ZP3 GPs, or embedded in LR White (samples from golden hamsters) for detection of lectin-binding glycoconjugates according to routine procedures ***(18)***.

3.7. Preparation of Animals for the Collection of Superovulated Oocytes

1. Sexually mature golden hamsters are cycled for at least two consecutive weeks to acertain their regularity (*see* **Note 5**). The animals are injected intraperitoneally with 25 IU of PMSG the day before the expected estrus, and with 25 IU of hCG 48 h later.
2. The animals are sacrificed by cervical dislocation 17 h after injection of hCG. The oviducts are excised with the ovary and a small proximal portion of the uterus is attached. Tissue samples are placed in PBS in an eight-well porcelain dish.
3. The ampulla region of the oviduct is identified under a dissecting microscope (*see* **Note 6**). Once the ampulla region is located, the wall of the ampulla is torn with a pair of fine steel tweezers. The cumulus mass oozes out from the torn opening of the ampullary wall.
4. The cumulus masses are transferred with a tapered-end glass pipet into a well containing fresh PBS for a brief wash. They are then transferred to another well containing 2.5% (v/v) GA in 0.1 *M* of CB (pH 7.4), and fixed by immersion in the fixative at room temperature for 1 h. After fixation, the cumulus masses are washed three times in 0.1 *M* of CB.
5. For postembedding labeling, the cumulus masses are dehydrated and then embedded in LR White according to routine procedures ***(18)***.

3.8. Preparation of Animals for the Collection of Fertilized Eggs

1. Female hamsters are cycled for 2 to 3 consecutive weeks to ascertain their regularity.

2. For mating, female hamsters are placed with males in the evening before the estrus stage.
3. For collection of fertilized eggs, females are sacrificed by cervical dislocation on day 1 after the animals were mated.
4. Repeat **Subheading 3.7., step 3** for collection of fertilized oocytes.
5. Fertilized eggs are fixed at room temperature for 1 h by immersion in 2.5% (v/v) GA in 0.1 *M* of CB.
6. Fertilized eggs are washed three times with 0.1 *M* of CB after fixation and then embedded in a 3% (w/v) gelatin solution crosslinked with 1% (v/v) GA (*see* **Note 7**).
7. For post-embedding labeling, the hardened gelatin gels containing the fertilized eggs are trimmed, dehydrated, and then embedded in LR White following routine procedures *(18)*.

3.9. Preparation of Ultra-Thin Sections of Ovarian Tissue and Cumulus Masses for Immunocytochemistry

1. One-μm-thick sections of Lowicryl K4M-embedded ovaries (CD1 mice) and LR White-embedded specimens (golden hamster ovaries and cumulus masses) are prepared on an ultramicrotome.
2. In these sections, areas of interest are first selected on a light microscope.
3. Once the areas of interest have been selected from the 1-μm-thick sections, the block face of the Lowicryl-embedded mouse ovaries and the LR White-embedded hamster ovaries and cumulus masses is trimmed to include the areas selected on the light microscope.
4. Prepare ultra-thin sections (~80–100 nm) on an ultramicrotome with a diamond knife.
5. Mount sections on formvar-coated nickel grids for subsequent immunolabeling (*see* **Note 8**).

3.10. Immunolocalization of ZP1, ZP2, and ZP3 GPs in Ovarian Sections

1. Thin sections of Lowicryl-embedded ovarian tissue samples (with the sections on the formvar-coated nickel grids facing down) are incubated on a drop of 0.01 *M* of PBS containing 1% bovine serum albumin (PBS–BSA) for 10 min.
2. The formvar-coated nickel grid carrying the tissue sections is removed from the PBS–BSA by holding the edge of the grid firmly with a pair of fine-steel tweezers.
3. Excess PBS–BSA on the surface of the sections is removed by tapping the hand gently on the lab bench while holding the grid firmly by the tweezers.
4. The grid is then transferred onto a drop of the monoclonal anti-ZP1, anti-ZP2, or anti-ZP3 antibody diluted in 0.01 *M* of PBS and incubated, with the sections facing down, at room temperature for 1 h.
5. After labeling, the grid is washed three times for 5 min each in 0.01 *M* of PBS. This washing is done by floating the grid with sections facing down, successively, on wells of a 9-well porcelain dish containing the washing buffer.

6. After washing, the sections are incubated for 1 h on a drop of rabbit anti-rat IgG polyclonal antibody diluted in 0.01 *M* of PBS.
7. Wash again as in **step 5** and transfer the grid onto a drop of PBS–BSA for 10 min.
8. Incubate the grid on a drop of 15-nm protein A–gold complex at room temperature for 30 min.
9. At the end of the incubation, the grid is washed with 0.01 *M* of PBS and then double-distilled water. This is done by a jet-wash with PBS followed by another jet-wash with double-distilled water.
10. After washing, the sections are allowed to air-dry on a filter paper.
11. All labeled sections are counter-stained with uranyl acetate and lead citrate for electron microscope examination. An example of immunolocalization of ZP1 GP in the ZP of an ovarian oocyte is shown in **Fig. 1**.

3.11. Immunolabeling, Double Labeling, and Triple Immunolocalization Using Direct Monoclonal Antibody–CG Conjugates

1. Sections of Lowicryl K4M-embedded ovaries are incubated with 1% ovalbumin in 0.01 *M* of PBS at room temperature for 10 min.
2. Repeat **Subheading 3.10., steps 2** and **3**.
3. The sections are labeled at room temperature for 1 h with either a single antibody-gold conjugate solution or with a combination of two or three different antibody-gold conjugates mixed in equal volumes for 1 h prior to use.
4. Repeat **Subheading 3.10., step 5**.
5. Incubate the sections for 30 min with protein A-gold complex following washing.
6. For washing, repeat **Subheading 3.10, step 9**.
7. All labeled sections are counter-stained with uranyl acetate and lead citrate before they are examined on an electron microscope. An example of triple immunolocalization of ZP GPs in the ZP of an ovarian oocyte is shown in **Fig. 2**.

3.12. Immunocytochemical Controls for Localization of ZP1, ZP2, and ZP3 GPs

To establish the specificity of the immunolabeling, the following immunocytochemical controls are carried out by:

1. Substituting the primary or secondary antibody with the corresponding buffer.
2. Substituting the antibody–gold conjugates with the protein A–gold complex alone.
3. Substituting the primary antibody with an antibody that recognizes a known GP expressed exclusively in another tissue (e.g., the oviduct) and not in the ovary.

3.13. Detection of Lectin-Binding Glycoconjugates in Follicular Oocytes and Superovulated Oocytes

3.13.1. One-Step Method for Labeling With HPA, UEA I, LTA, and DBA

1. Incubate thin sections of LR White embedded ovaries or cumulus masses on a drop of 0.01 *M* of PBS (pH 7.4) in 0.5% BSA at room temperature for 5–10 min.

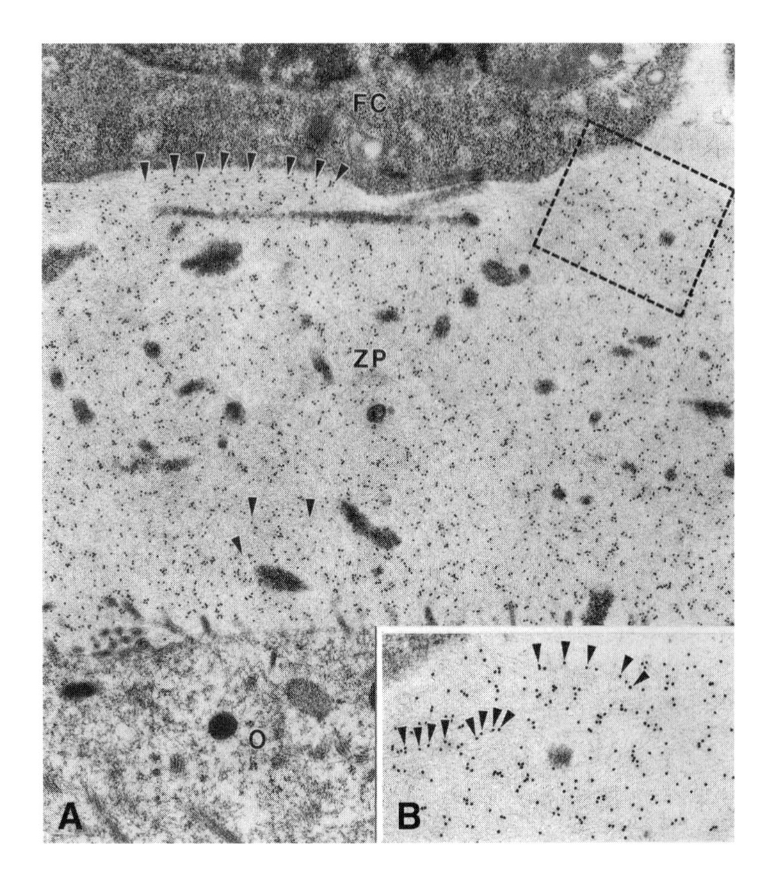

Fig. 1. Electron micrograph of a multi-laminar secondary follicle labeled with anti-zona pellucida (ZP)3 antibody followed by incubation with rabbit anti-rat IgG polyclonal antibody, and then labeling with protein A–gold complex. **(A)** The ZP shows labeling by gold particles of filaments located in the outer region of the zona matrix (arrowheads). Follicular cells are devoid of any labeling by anti-ZP3 antibody. Original magnification: ×21,000. **(B)** Higher magnification of the framed box (A) showing the gold particles (arrowheads) directly superimposed on the linear arrays of filaments. Original magnification: ×58,000 *(13)*.

2. Repeat **Subheading 3.10, steps 2** and **3**.
3. The sections are then transferred onto a drop of lectin–gold complex as in **Subheading 3.10, step 4** and labeled with lectin–gold complex at room temperature for 1 h.
4. Repeat **Subheading 3.10., steps 9–11**. An example of detection of HPA-binding glycoconjugates in the ZP of ovarian oocytes and oviductal oocytes is shown in **Fig. 3**.

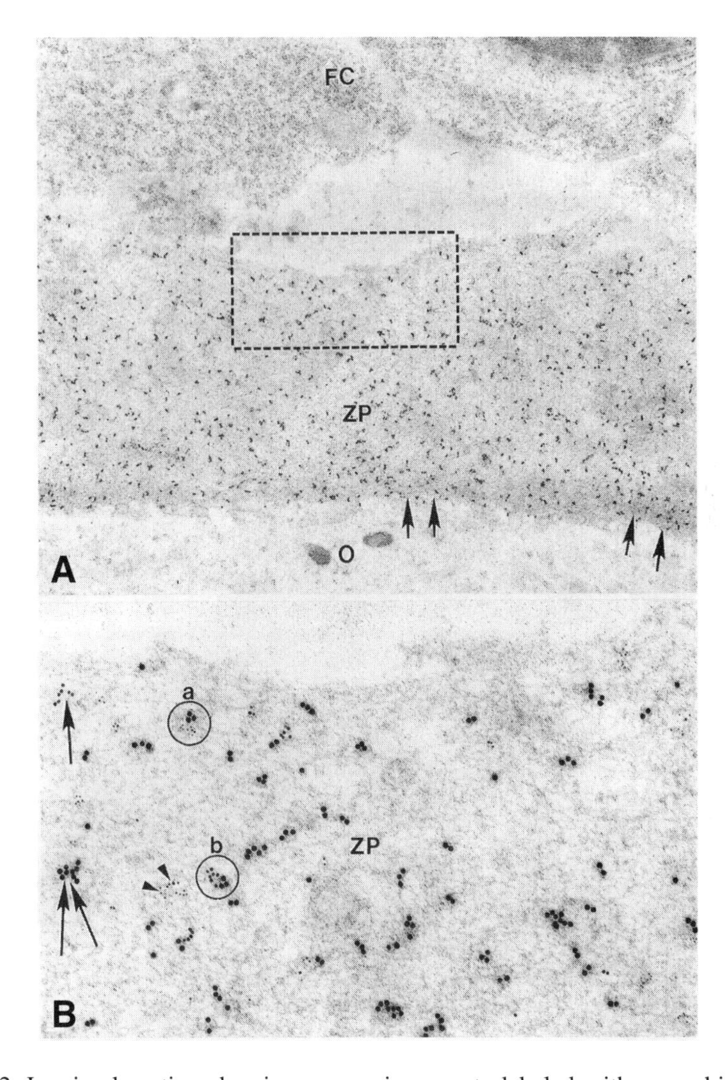

Fig. 2. Lowicryl section showing an ovarian oocyte labeled with a combination of anti-zona pellucida (ZP)1, anti-ZP2, and anti-ZP3 antibodies tagged directly to 15-, 10-, and 5-nm gold particles, respectively. **(A)** At low magnification, labeling by the three antibodies occupies the entire ZP, with a relatively higher concentration of gold labeling detected in the inner layer of the ZP (arrows). Follicular cells are not labeled. Original magnification: ×21,000. **(B)** A high-magnification view of the area in the framed box reveals the distribution of gold labeling by the various antibody–gold conjugates in the outer layer of the ZP. Labeling by gold particles is often seen as clusters of 10-nm gold particles (arrow), 15-nm gold particles (double arrows), or 5-nm gold particles (arrowheads). Association between two particle sizes such as 15- and 5-nm (circle a) or 15- and 10-nm (circle b) is often seen. Original magnification: ×90,000 ***(13)***.

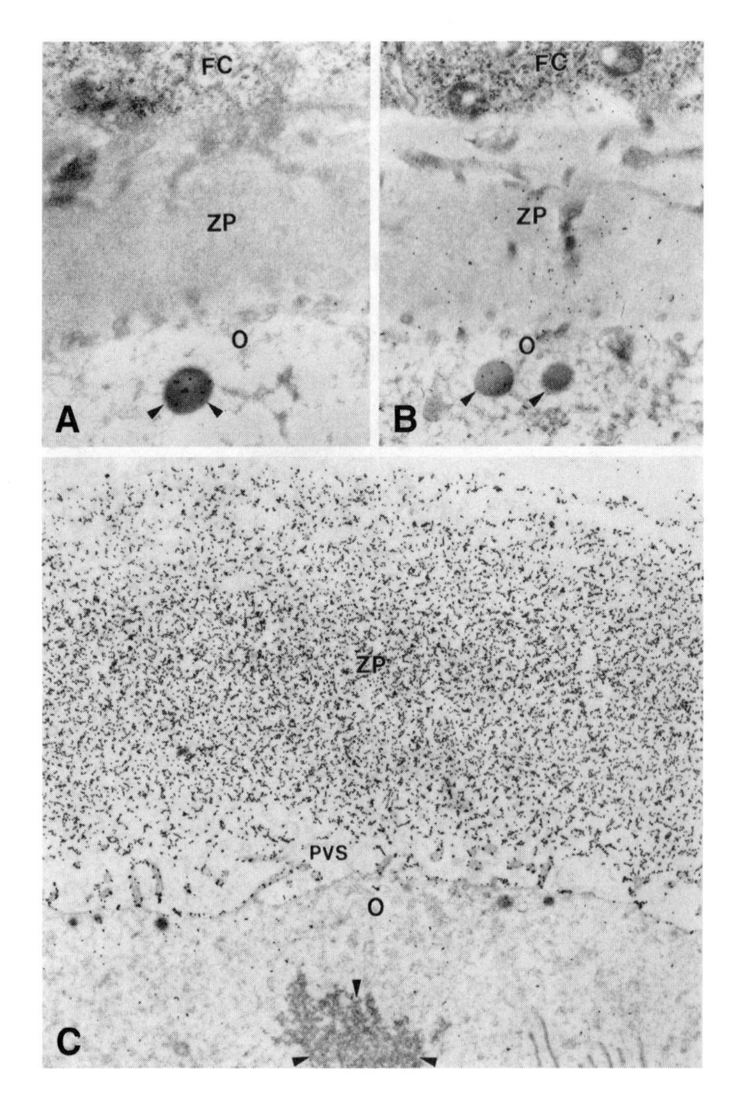

Fig. 3. HPA–gold labeling. **(A)** The well-developed zona pellucida (ZP) of a preantral ovarian follicle is virtually devoid of any labeling. In the ooplasm (O), however, gold particles are shown to be associated with a cortical granule (arrowheads). **(B)** A multilaminar secondary follicle in a tissue section digested with neuraminidase prior to labeling with HPA displays a moderate intensity of the labeling over the entire ZP. Note that one cortical granule is moderately labeled whereas the other one is devoid of any labeling by gold particles (arrowheads). **(C)** In the ovulated oocyte, however, numerous gold particles are found uniformly distributed throughout the ZP. The vesicular aggregate of the ovulated oocyte (arrowheads) is only labeled with a few gold particles. FC, follicular cells; PVS, perivitelline space. Original magnifications: A, ×23,000; B, ×17,500; C, ×16,000 ***(14)***.

3.13.2. Two-Step Method for LFA and WGA Labeling

1. Repeat **Subheading 3.13.1., steps 1** and **2**.
2. The grid is then transferred onto a drop of LFA (100 μg/mL in *tris*-buffered saline) or WGA (125 μg/mL in PBS) solution, respectively, for incubation at room temperature for 1 h.
3. For washing, repeat **Subheading 3.10., step 5**.
4. After washing, the sections are floated on a drop of Ovo–gold complex (for WGA) or fetuin-gold complex (for LFA) for 30 min.
5. Repeat **Subheading 3.10., steps 9–11** before examining the section on an electron microscope.

3.13.3. Three-Step Method for Labeling With AAA, GNA, MAA, PNA, and SNA

1. Repeat **Subheading 3.13.1., steps 1** and **2**.
2. The grid is then transferred to a drop of one of the DIG-labeled lectins (20 μg/mL) for a 1-h incubation at room temperature.
3. For washing, repeat **Subheading 3.10., step 5**.
4. After washing, the sections are floated on a drop of unlabeled anti-DIG antibody (2 μg/mL in PBS) for 1 h.
5. After repeating **step 3** for washing, the grid is transferred onto a drop of goat anti-mouse IgG-CG conjugate (10 nm) diluted 1:10 in PBS-BSA, and the sections are labeled with the antibody–gold conjugate at room temperature for 1 h.
6. After labeling, repeat **Subheading 3.10., steps 9–11** before examining the section on an electron microscope.

3.13.4. Cytochemical Controls for Lectin–Gold Labeling

To establish the specificity of the lectin–gold labeling, the following controls are carried out by:

1. Substituting the conjugated or unconjugated lectin with the corresponding buffer.
2. Preincubating the lectin with the corresponding hapten–sugar inhibitor (*see* **Table 1**) used at a range of concentrations from 0.1 to 1 *M*: D-Gal (for RCA I), methyl-α-man (for GNA), D-GalNAc (for HPA and DBA), D-GlcNAc (for WGA), *N*-acetyllactosamine (for DSA), L-Fuc (for AAA, UEA-I, and LTA), Neu5Ac (for LFA), and *N*-acetylneuraminyllactose (for MAA and SNA).
3. Preincubate the grids with neuraminidase for LFA, WGA, MAA, and SNA (*see* **Note 9**), and GO for PNA (*see* **Note 10**).

4. Notes

1. A 1% gold solution can be kept at room temperature in a brown glass bottle for several months for repeated use. A fresh solution should be prepared when a precipitate starts forming on the bottom of the glass container.
2. A freshly prepared 15-nm CG solution can be kept at 4°C for several months for further use. Before using the gold solution to prepare CG–lectin complexes,

centrifuge the gold solution at 600*g* for 30 min to get rid of undesirable large gold aggregates.

3. A well-conjugated protein A–gold complex should have the color of red wine. To test if a good complex has been formed, transfer 0.5–1 mL of the complex to a 1.5-mL Eppendorf tube, add three to four drops of 10% aqueous NaCl solution, and mix gently. The retention of a red wine color indicates a well-conjugated complex. The appearance of a violet color indicates the presence of free, unconjugated gold particles and that not enough protein A–gold has been added.
4. A good Ovo–CG complex should retain the dark purplish color. Appearance of a black color indicates that either an insufficient amount of Ovo has been used for conjugation, or the pH of the gold solution has not been properly adjusted. To verify if a good Ovo–CG complex has been formed, transfer 0.5–1.0 mL of the complex to a 1.5-mL Eppendorf tube, add 0.1 mL of 10% aqueous NaCl solution, and mix gently. Occurrence of precipitation after addition of 10% NaCl indicates failure of or incomplete conjugation of Ovo to gold particles, and preparation of the Ovo–CG must be repeated.
5. Female hamsters are tested for the typical vaginal discharge that is usually found in abundance at the metestrus stage of the estrous cycle. We usually observe the estrous cycle of female hamsters for 2 to 3 consecutive weeks to ascertain their regularity.
6. Note that the oviduct of the golden hamster is highly convoluted, but the ampulla can be distinguished from the other regions by its enlargement as a bulge having a diameter larger than the rest of the oviduct. Under the dissecting microscope, the cumulus mass containing oviductal oocytes can sometimes be seen floating freely in the lumen of the ampulla.
7. To facilitate the subsequent preparation of ultra-thin sections for lectin labeling, we normally place several fertilized eggs together in a 3% (w/v) gelatin solution. A small volume of the gelatin solution (3–5 μL) containing the fertilized eggs is transferred into a porcelain well containing approx 1 mL of 1% GA in 0.1 *M* of PBS, which crosslinks the gelatin to form a gel. Hence, a group of 5–10 eggs can be embedded in the same gelatin gel, making it easier to locate the eggs during the sectioning procedures.
8. Nickel grids are coated with a thin film of a 2% solution of formvar. The formvar film acts as a support for the tissue sections placed on the nickel grids. Ideally, the formvar film is further coated with a very thin carbon layer in carbon coating equipment. The carbon coat provides the tissue sections with a firm adherence so that the sections will not be displaced during the subsequent labeling and washing procedure.
9. WGA has an affinity for both GlcNAc and Neu5Ac residues, whereas LFA shows an affinity only for Neu5Ac residues. Treatment of tissue sections with neuraminidase prior to WGA–Ovo gold or LFA–fetuin gold labeling is performed to establish the presence of sialic acid. Grids carrying tissue sections are treated with 1 U/mL of neuraminidase type V (from *Clostridium perfringens*) in acetate buffer (pH 5.0) at 37°C for 3 h to remove Neu5Ac residues.

10. For GO treatment, grids carrying the sections are incubated with 50 U/mL of GO in 0.01 *M* of PBS (pH 7.4) at 37°C for 24 h in a moist chamber.

Acknowledgments

The author would like to thank Changnian Shi for technical assistance. This work was supported by grant MOP-69034 from the Canadian Institutes for Health Research (CIHR).

References

1. Shimizu, S., Tsuji, M., and Dean, J. (1983) In vitro biosynthesis of three sulfated glycoproteins of murine *zonae pellucidae* by oocytes grown in follicle culture. *J. Biol. Chem.* **258,** 5858–5863.
2. Wassarman, P. M. (1988) Zona pellucida glycoproteins. *Annu. Rev. Biochem.* **57,** 415–442.
3. Dunbar, B. S., Avery, S., Lee, V., et al. (1994) The mammalian zona pellucida: its biochemistry, immunochemistry, molecular biology and developmental expression. *Reprod. Fertil. Dev.* **6,** 59–76.
4. Bleil, J. D. and Wassarman, P. M. (1980) Mammalian sperm-egg interaction: identification of a glycoprotein in mouse egg zonae pellucidae possessing receptor activity for sperm. *Cell* **20,** 873–882.
5. Bleil, J. D. and Wassarman, P. M. (1988) Galactose at the nonreducing terminus of O-linked oligosaccharides of mouse egg zona glycoprotein ZP3 is essential for the glycoprotein's sperm receptor activity. *Proc. Natl. Acad. Sci. USA* **85,** 6778–6782.
6. Bleil, J. D., Greve, J. M., and Wassarman, P. M. (1988) Identification of a secondary sperm receptor in the mouse egg zona pellucida: role in maintenance of binding of acrosome-reacted sperm to eggs. *Dev. Biol.* **128,** 376–385.
7. Greve, J. M. and Wassarman, P. M. (1988) Mouse egg extracellular coat is a matrix of interconnected filaments possessing a structural repeat. *J. Mol. Biol.* **181,** 253–264.
8. Rankin, T. and Dean, J. (2000) The zona pellucida: using molecular genetics to study the mammalian egg coat. *Rev. Reprod.* **5,** 114–121.
9. Hoodbhoy, T. and Dean, J. (2004) Insights into the molecular basis of sperm-egg recognition in mammals. *Reproduction* **127,** 417–422.
10. East, I. J. and Dean, J. (1984) Monoclonal antibodies as probes of the distribution of ZP-2, the major sulfated glycoprotein of the murine zona pellucida. *J. Cell Biol.* **98,** 795–800.
11. East, J. D., Gulyas, B. J., and Dean, J. (1985) Monoclonal antibodies to murine zona pellucida protein with sperm receptor activity: effects on fertilization and early development. *Dev. Biol.* **109,** 268–273.
12. Rankin, T. L., Tong, Z.-B., Castle, P. E., et al. (1998) Human ZP3 restores fertility in Zp3 null mice without affecting order-specific sperm binding. *Development* **125,** 2415–2424.
13. El-Mestrah, M., Castle, P. E., Borossa, G., and Kan, F. W. K. (2002) Subcellular distribution of ZP1, ZP2, and ZP3 glycoproteins during folliculogenesis and

demonstration of their topographical disposition within the zona matrix of mouse ovarian oocytes. *Biol. Reprod.* **66,** 866–876.

14. El-Mestrah, M. and Kan, F. W. K. (2001) Distribution of lectin-binding glycosidic residues in the hamster follicular oocytes and their modifications in the zona pellucida after ovulaton. *Mol. Reprod. Dev.* **60,** 517–534.
15. El-Mestrah, M. and Kan, F. W. K. (2002) Variation in modifications of sugar residues in hamster zona pellucida after *in vivo* fertilization and *in vitro* egg activation. *Reproduction* **123,** 671–682.
16. Avilé, M., El-Mestrah, M., Jaber, L., Castells, M. T., Ballesta, J., and Kan, F. W. K. (2000) Cytochemical demonstration of modification of carbohydrates in the mouse zona pellucida during folliculogenesis. *Histochem. Cell Biol.* **113,** 207–219.
17. Frens, G. (1973) Controlled nucleation for the regulation of the particle size in monodispersed gold suspension. *Nature Phys. Sci.* **241,** 20–22.
18. Bendayan, M., Nanci, A., and Kan, F. W. K. (1987) Effect of tissue processing on colloidal gold cytochemistry. *J. Histochem. Cytochem.* **35,** 983–996.

20

Soluble Adamantyl Glycosphingolipid Analogs as Probes of Glycosphingolipid Function

Clifford Lingwood, Skanda Sadacharan, Maan Abul-Milh, Murugespillai Mylvaganum, and Marcus Peter

Summary

Despite the extensive structural characterization of glycosphingolipids (GSLs), their functions in cell physiology and pathobiology remain elusive. This is largely owing to the fact that they are difficult to handle, being insoluble in aqueous media, and that no one gene alone determines their synthesis. The heterogeneity of the lipid moiety provides a further confounding factor. GSLs are central components within lipid rafts, which are major foci for transmembrane signaling and interactions between eukaryotic cells and microbial pathogens. GSL receptor function often requires the lipid moiety, and lipid-free sugar analogs are ineffective inhibitors. In order to overcome some of these problems, we have synthesized adamantyl GSL analogs which, in part, mimic GSL membrane receptor function in solution. These compounds are made by replacing the endogenous fatty acid with an adamantan frame. This rigid hydrophobic structure surprisingly increases the water solubility of the conjugate and retains receptor function. These GSL mimics provide probes to study GSL receptor function within cells. They compete with native GSLs for ligand binding and are taken up by cells to potentially alter GSL-mediated interaction. We are focused on two derivatives, adamantyl globotriaosyl ceramide and adamantyl sulfogalactosyl ceramide, and have used these analogs to probe GSL function in microbial pathology and hsp70 function. This chapter describes the syntheses and uses of these mimics.

Key Words: Human immunodeficiency virus (HIV); verotoxin; hsp70; microbial pathogenesis.

1. Introduction

Glycosphingolipids (GSLs) are unique cell-surface molecules in that they comprise a hydrophilic sugar molecule which can be recognized by a variety of extracellular ligands, conjugated to a hydrophobic ceramide lipid moiety within the

From: *Methods in Molecular Biology, vol. 347: Glycobiology Protocols*
Edited by: I. Brockhausen © Humana Press Inc., Totowa, NJ

membrane, which can be organized into a variety of cholesterol-enriched microdomains. These domains play an important role in subsequent cell-surface and intracellular trafficking. Such domains are central in many signal transduction pathways *(1)*, and provide a portal for pathogenic microorganism–host cell interaction *(2)*. Conjugation of a hydrophilic and hydrophobic structure can easily be appreciated as a potential dynamic union in that essentially repulsive thermodynamic forces are covalently brought together (*see* **Notes 1** and **2**). The unusual receptor biology of cell membrane GSLs, in terms of their frequent dependence not only on the carbohydrate structure but on the lipid moiety as well *(3)*, while perhaps reflecting an extreme situation, may provide a model of other recognition processes in which particular adjacent hydrophobic structures can have a significant impact on ligand binding to a hydrophilic epitope. In this regard, we have shown that binding of protein ligands to sulfotyrosine-lipid conjugates is remarkably dependent on the lipid structure *(4)*.

1.1. Importance of the Lipid Moiety in GSL Receptor Function

Our studies on GSL receptor biology have been centered essentially on two GSLs, globotriaosyl ceramide (galα1-4gal β1-4 glucosyl ceramide, Gb_3) in relation to its function as a receptor for the *Escherichia coli*-derived verotoxin (VT) family of AB_5 subunit toxins *(5)*, and the receptor role of sulfatide (3′ sulfogalactosyl ceramide, SGC) and its binding to the N-terminal adenosine triphosophatase (ATPase) domain of the heat-shock protein (hsp)70 chaperone family *(6)*. In both of these systems, it has been shown that the lipid moiety can dramatically affect receptor function and that the lipid-free sugar is virtually devoid of ligand-binding ability *(7–9)*. In both cases, the recognition epitopes are found exclusively within the carbohydrate moiety and changes within these structures can completely ablate binding *(10,11)*. Therefore, the problem does not lie in the fact that the lipid moiety is part of the recognition epitope. The effect is more subtle and concerns the presentation of the carbohydrate for binding, both in terms of solid-phase recognition *(12)* and the more complex situation in the cell membrane; that is, the cholesterol-enriched microdomain structure within the plasma membrane *(13)*. These properties greatly complicate the analysis of GSL receptor biology. The insolubility of GSLs in aqueous media provides a great barrier to understanding GSL receptors. The biology of GSLs is largely inferred from their binding ligands.

We have proposed an interface area between the hydrophobic and hydrophilic components of glycolipids as they are situated in the membrane bilayer leaflet *(14,15)*. This interface forms a hydrogen bond network that is influenced by the cholesterol and phospholipid content of the GSL microenvironment. This network serves to restrict the conformation of the anomeric carbohydrate linkage *(16)* and the solvation of the membrane-apposed sugar residues. The restriction of solvation in dense lipid raft structures might generate a "less hydrated" domain in which displacement of water molecules for ligand/carbohydrate binding may not be as necessary than in the otherwise more solvent exposed non-raft GSL environment. Indeed, lipid rafts have been directly visualized in living cells by examining the non-uniform distribution of a fluorescent probe, which detects rafts as less hydrated domains in the generally more hydrated bulk phospholipid plasma membrane ***(17)***. While the multivalency of such aggregated GSLs within a lipid raft will certainly play a significant role in augmenting binding of multivalent ligands, these effects are also seen for monovalent binding ligands, supporting the idea of additional thermodynamic properties involved.

This effect of the lipid moiety on GSL recognition may have surprising and far-reaching implications. We have recently demonstrated structural mimicry between SGC and tyrosine phosphate recognition ***(4)***. The sulfogalactose moiety of SGC could be substituted with tyrosine sulfate and retain ligand binding. Such ligands include cSrcSH2 domains. Binding to tyrosine sulfate-containing lipids has been found to be significantly affected by the lipid composition. We propose that this effect related to the ability of SH2 domains to selectively bind phosphotyrosine residues in the context of a specific peptide sequence (microenvironment) and that these effects are related to the solvation of the receptor (phosphotyrosine or sulfogalactose).

1.2. Soluble GSL Mimics

1.2.1. Problems With Lipid-Free Analogs

The central importance of the lipid moiety in GSL receptor function and the essential exclusion of such glycoconjugates from an aqueous environment would seem to be an insurmountable hurdle in generating receptor-active soluble GSL analogs. For VT/Gb_3 binding, a high degree of multivalency of lipid-free carbohydrate sequences has been successful in the generation of soluble receptor analog inhibitors of increased efficacy ***(18,19)***. Nevertheless, many dimeric, trimeric, and even tetrameric Gb_3-sugar constructs are not recognized by VT ***(9)***. Despite the best efforts of synthetic chemistry to tailor-make soluble carbohydrate dendrimers to fit the known crystal structure of VT, the random orientation of the carbohydrate coupled within an acrylamide-based polymer has proven to be a more inhibitory receptor analog ***(20)***, indicating the considerable gap in development of structure-based glycolipid-carbohydrate soluble mimics in terms of matching the binding affinity of the membrane-presented Gb_3 glycolipid. The tailor-made dendrimers (a pentamer of trisaccharide dimers) have proved most effective by crosslinking opposing B subunit pentamers in a "sandwich," rather than crosslinking the two binding sites within each B subunit monomer as designed ***(18)***.

Our early studies had shown that fatty acid acyl chain length, unsaturation *(12)*, and hydroxylation *(21)* could have a profound effect on VT/Gb_3 binding. We considered that in order to make an effective inhibitor of VT binding to membrane Gb_3, this effect must be taken into account. Studies using the lipid-free Gb_3 oligosaccharide (OS) identified the primary binding site within the B subunit pentamer as site 2 *(22,23)*, which is essentially not bound by the membrane-located Gb_3 glycolipid *(24)*, illustrating the danger in using lipid-free OSs to identify the molecular basis of GSL receptor function.

1.2.2. Adamantyl GSLs

Our approach has been to develop monomeric glycolipid analogs which retain receptor function and yet are water-soluble (*see* **Notes 2** and **3**). This approach involves the removal of the fatty acid of the parent GSL and substituting a rigid hydrocarbon frame such as adamantan on norbornane. These hydrophobic frames, though in themselves more nonpolar than the fatty acid they replace, generate glycoconjugates that preferentially partition into aqueous media *(25,26)*. The alkyl chain of the sphingosine may wrap the hydrophobic adamantane frame to intramolecularly reduce exposure of hydrophobic residues in an aqueous medium *(14)*. Such a configuration may be below the maximum hydrophobic size necessary to disturb water organization *(27)*, which could explain the solubility of adamantyl GSLs.

Thus far, we have made adamantyl globotriaosyl ceramide (AdaGb_3) as an inhibitor of VT/Gb_3 binding *(25)*, adamantyl sulfogalactosyl ceramide (AdaSGC) as an inhibitor of hsp70/SGC binding (*see* **ref. *26*** and **Fig. 1**), and adamantyl galactosyl ceramide (AdaGalCer) and adamantyl lactosyl ceramide (AdaLacCer; largely as control) glycolipid analogs. Most recently, we have shown AdaGb_3 is an inhibitor of HIV infection in vitro *(27a)*. In each case, the selective partitioning into aqueous media is observed (as expected, less dramatic for AdaGalCer) and receptor binding function is retained within the aqueous medium (see **ref. *9*** and **Notes 1–6**).

2. Materials

2.1. Solvents

1. Chloroform ($CHCl_3$), methanol (MeOH), dichloromethane (CH_2Cl_2, DCM), hydrochloric acid (HCl), acetonitrile (CH_3CN), dimethyl formamide (DMF), acetic anhydride $(COCl)_2$, and diethyl ether (Et_2O; Caledon, Georgetown, Ontario, Canada, or Aldrich, Milwaukee, WI). DMF and CH_2Cl_2 are dried over a 1:1 mixture of 3 and 4 Å molecular seives.
2. Ethanol (EtOH; Commercial Alcohols Inc., Brampton, Ontario, Canada).

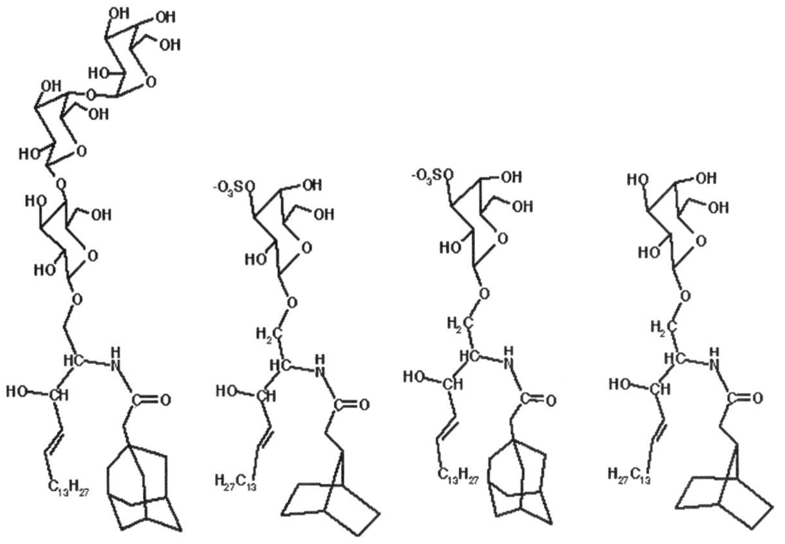

adamantylGb$_3$ norbornylSGC adamantylSGC norbornylGC

Fig. 1. Structure of water-soluble glycosphingolipid receptor mimics. The fatty acid of the parent glycosphingolipid has been replaced with a rigid hydrophobic frame, α-adamantane or norbornane.

2.2. Reagents

1. Sodium chloride (NaCl), acetic acid (CH_3COOH), and potassium hydroxide (KOH; Fisher Scientific Company, Fair Lawn, NJ).
2. Activated charcoal (BDH Chemicals, Toronto, Ontario, Canada).
3. Ammonia (NH_3), phosphoric acid (H_3PO_4), and triethylamine (Et_3N; Sigma, St. Louis, MO).
4. 1-Ethyl-3-(3-dimethylamino-propyl)carbodiimide, CH_3COONa, 2,2-azino-*bis* (3-ethylbenz-thiazoline-6-sulfonic acid, and polyisobutylmethacrylate (Sigma).
5. Adamantane acetic acid ($AdaCH_2COOH$) and norbornane acetic acid (Aldrich), and 1-Hydroxy-7-azabenzotriazole (HOAT; Fluka, Oakville, Ontario, Canada).
6. Evergreen microtiter plates (DiaMed Lab Supplies Inc., Mississauga, Ontario, Canada).
7. Silica-gel 60 (40–63 microns) and aluminum-backed nanosilica thin-layer chromatography (TLC) plates (Caledon).
8. Reverse-phase C18 cartridges (Waters, Mississauga, Ontario, Canada).

2.3. Glycolipids (see *Note 3*)

1. Gb_3 is purified from human kidney by chloroform/methanol extraction, saponification, and silicic acid chromatography *(28)*.

2. SGC is purified from bovine brain extract (Sigma, cat. no. B1014, which contains 10% SGC, 30% GalCer, 30% sphingomyelin, and some gangliosides) as for Gb_3 in **item 1** with an additional diethylamino ethanol cellulose ion exchange step ***(29)***. Purified Gb_3 and SGC are commercially available from Sigma.
3. Purified GalCer and LacCer (Sigma).

3. Methods

3.1. Synthesis of AdaGb$_3$ (see Notes 1 and 2)

1. Purified Gb_3 is deacylated in 1 *M* of NaOH in MeOH at 70°C for 72 h under N_2 (60–70% conversion). The reaction mixture is desalted by applying to a Sep-Pak cartridge, washing with water, and eluting with MeOH. LysoGb_3 is separated from Gb_3 by silica gel chromatography. All subsequent manipulations are carried out under a nitrogen atmosphere.
2. 5 µL of DMF is added to a solution of 50 µL of $COCl_2$ in 1 mL of DCM. AdaCH_2 COOH (1 mL of a 0.2-m*M* solution in DCM) is then slowly added over 30 min. After stirring at room temperature for 2 h, excess $COCl_2$ and solvent are removed under a stream of N_2 and residual AdaCH_2COCl is dissolved in 1 mL of DCM.
3. Suspend 0.5 mg of LysoGb_3 in 0.5 mL of DCM and 15 µL of pyridine, and then two 5-µL aliquots of the AdaCH_2COCl solution are added at 30-min intervals. The reaction mixture is dried under N_2 and monitored by TLC (C:M:H_2O, 80:20:2).
4. The product is purified on a 0.5- × 2-cm mini silica column (yield = 0.35 mg, 70%). AdaGalCer, norbornyl GalCer, and AdaLacCer are similarly made from the deacylated parent GSL.

3.2. Synthesis of AdaSGC (see Notes 3–5)

1. Deacylated SGC (lysoSGC) is made by hydrolysis of SGC with 0.3 *M* of KOH in aqueous methanol (9 MeOH:1 H_2O) at 80°C for 8 d ***(30)***. The reaction is neutralized to pH 7.0–8.0 with concentrated HCl, flash evaporated to a minimum volume, and applied to a C-18 Sep-Pak cartridge. The column is washed with water and SGC and lysoSGC eluted with 0.2 *M* of NH_3 in methanol. This procedure yields approx 65% conversion to lysoSGC. To avoid complications during the synthesis of adaSGC, the lysoSGC can be purified (>99%) by silica-gel-column chromatography from SGC.
2. Adamantane acetic acid (or 2-norbornane acetic acid; 1.3 mg, approx 2 equiv), HOAT (30 µL of 0.2 *M* solution in 5:1 CH_3CN/Et_3N, 6 µmol), and solid 1-ethyl-3-(3-dimethylamino-propyl)carbodiimide (2–3 mg, 10–15 µmol) are added in the given order to a solution of lysoSGC (2 mg, 4 µmol) in 2 mL of 5:1 CH_3-CN/Et_3N and stirred at 60°C for 3 h.
3. The progress of the reaction is monitored by TLC ($CHCl_3$:MeOH:0.88% KCl; 65:25:4), LysoSGC migrates below SGC and AdaSGC migrates similarly to SGC. Upon completion, the reaction mixture is dried under a stream of N_2.
4. The crude product is dissolved in $CHCl_3$:MeOH (98:2) and loaded on to a silica column (0.5 × 10 cm of silica gel in $CHCl_3$:MeOH; 98:2). HOAT and excess adamantane acetic acid are eluted with CH_3COOH:$CHCl_3$ (4:1). Product is eluted

with $CHCl_3$:MeOH:H_2O; 80:20:2 (6- × 4-mL fractions are collected). The estimated yield by TLC should be 70%.

The purity of adamantyl GSLs is verified by mass spectrometry (electrospray ionization mass spectrometry or matrix-assisted laser desorption ionization mass spectrometry).

4. Notes

1. *A raft mimic for VT1 and human immunodeficiency virus (HIV) gp120?* AdaGb$_3$ provides a receptor mimic which more closely mirrors the membrane presentation of Gb$_3$ glycolipid ***(25)***, in part by reflecting Gb$_3$ within lipid rafts but in a soluble format ***(31)***. While Gb$_3$ requires cholesterol, AdaGb$_3$ could form VT1-binding lipid "rafts" on a sucrose gradient without cholesterol ***(31)***. The rigid adamantan frame coupled to Gb$_3$ may mimic to some degree the intercalation of cholesterol within rafts to generate a more rigid lipid domain.

   Gb$_3$ has been implicated in the mechanism of HIV infection, particularly the process of viral/host fusion ***(32,33)***. The HIV gp120 binds to several GSLs ***(34,35)***, but Gb$_3$ is a poor receptor ***(36)***. Coupling Gb$_3$ to bovine serum albumin markedly enhances gp120/Gb$_3$ binding ***(36)***. The slow (hours) sigmoidal binding and insertion of gp120 into a Gb$_3$ air/water interface monolayer became more rapid and exponential in the presence of cholesterol ***(31)***, suggesting that raft structures are important. Gp120 insertion into pure AdaGb$_3$ monolayers was exponential and saturated within 10 min, indicating again that AdaGb$_3$ can behave as a mimic of Gb$_3$ rafts. Although AdaSGC efficiently competes with SGC for gp120 binding ***(13)***, gp120 binding to SGC was not increased by cholesterol and AdaSGC was not more effectively bound by gp120 than SGC ***(31)***. AdaGb$_3$ was found to form a more rigid monolayer than Gb$_3$ with a larger molecular area ***(31)***. In contrast, AdaSGC shows a smaller molecular area and is less rigid that the parental SGC (*see* **Fig. 2**). Therefore, although the same substitution has been made in these two soluble GSL analogs, their relative physical behavior is different; only AdaGb$_3$ behaves as a raft mimic and is a candidate inhibitor for HIV. Our current studies are verifying this potential.
2. *Inhibition of VT binding.* Unlike the lipid-free Gb$_3$ OS, AdaGb$_3$ retains high-affinity VT1 binding ***(25)***. The multivalent binding of the VT B subunit to Gb$_3$ is complex. We have shown apparent cooperative binding kinetics at defined Gb$_3$ concentrations, at least for VT1 ***(37)***. This may explain our observation that incubation of VT2 with AdaGb$_3$ was found to promote, rather than inhibit, the cytopathology of this toxin and in the mouse ***(38)***, although protection against VT1 and VT2 cytotoxicity in vitro was observed. This might suggest that receptor analog binding in one or more of the pentameric binding sites could, in fact, promote membrane Gb$_3$ binding in the remaining sites. This serves as a caution in the development of bioactive receptor analog inhibitors. In excess, AdaGb$_3$ clearly competes for VT1/Gb$_3$ binding by receptor enzyme-linked immunosorbent assay and in vitro cell cytotoxicity assay (*see* **Fig. 3**). VT2/Gb$_3$ binding is distinct from VT1 ***(39)***, and in hindsight, our selection of VT2 to test the efficacy of AdaGb$_3$

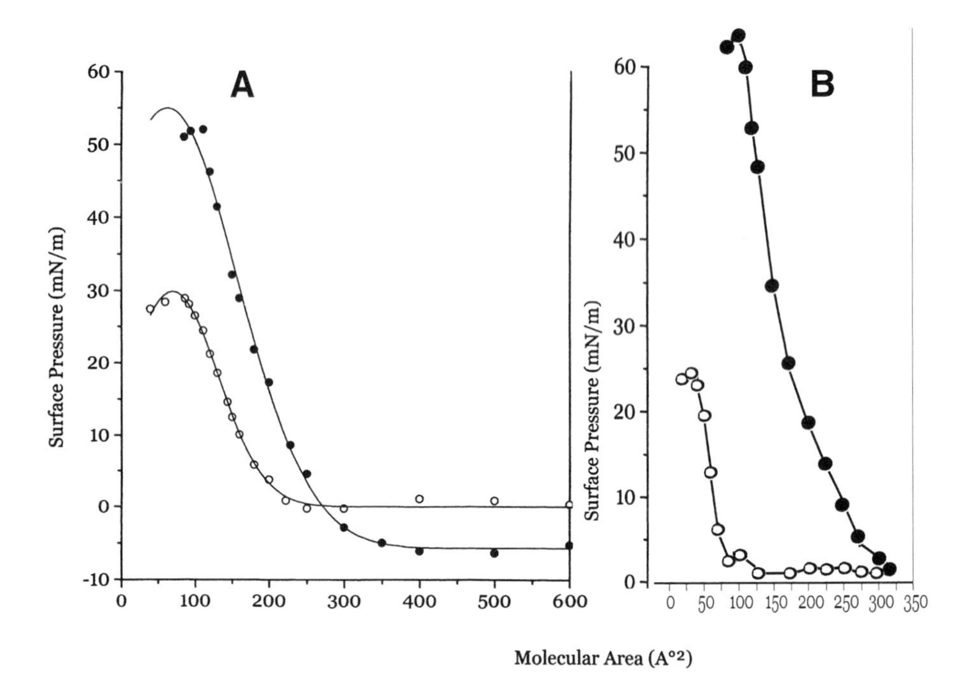

Fig. 2. Surface pressure profiles of monolayers of **(A)** galα1-4gal β1-4 glucosyl ceramide (Gb_3) and **(B)** 3′ sulfogalactosyl ceramide (SGC) and their adamantyl derivatives. Monolayers of the native glycosphingolipid (GSL) and its adamantyl analog were prepared in a Langmuir trough as described ***(31)***. The effect of compressing the monolayer (reducing the molecular area) on surface pressure was determined. (A) Gb_3 (○), adamantyl Gb_3 (●); (B) adamantyl SGC (○), SGC (●). Adamantyl Gb_3 is more rigid and has a larger molecular area than Gb_3, whereas adamantyl SGC is less rigid and has a smaller molecular area than SGC. SGC is more rigid than Gb_3 but has a similar molecular volume.

in vivo, owing to its lower LD50 in mice ***(38)***, was problematic. Internalization of VT1 can occur through both clathrin and caveolae-mediated mechanisms ***(40)***. VT2 internalization is in part distinct (in progress). The lipid raft dependency of VT1 and VT2 internalization is different, and susceptibility to $AdaGb_3$ protection may also be distinct.

The rigidity of the adamantan frame is necessary to maintain the receptor function of $AdaGb_3$. Substitution of the sphingoid amine with a tertiary butylacetamido function compared with an adamantyl acetamide showed that only the adamantyl conjugate retained VT1 binding ***(9)***. $AdaGb_3$ was used to determine the efficacy of amino substitution within the galabiose of Gb_3. These studies showed amino substitution of the 2 or 6 hydroxyls of the terminal galactose-retained binding, but substitution at the terminal 4OH or subterminal 6OH prevented VT1 and VT2 recognition ***(9)***.

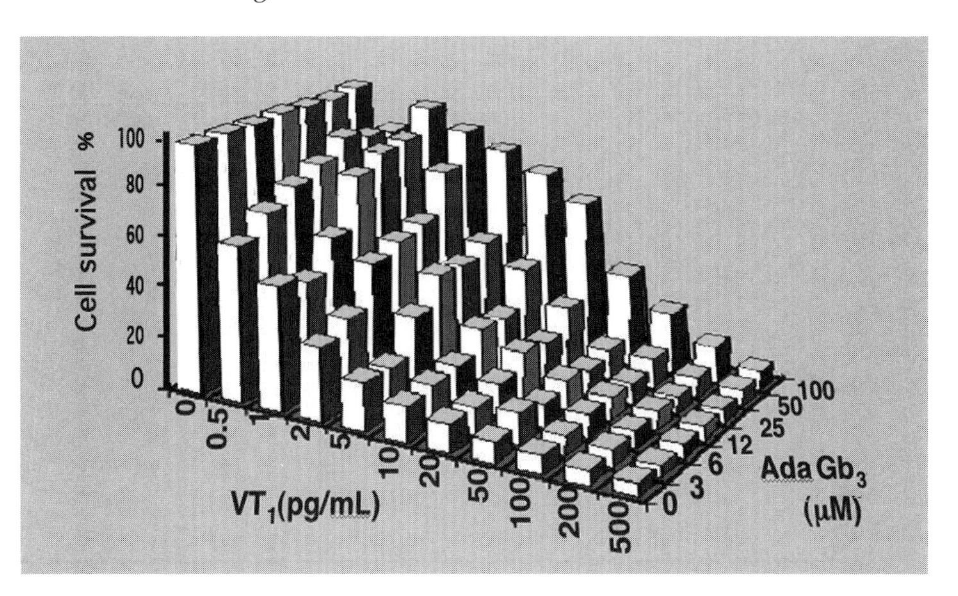

Fig. 3. Adamantyl galα1-4gal β1-4 glucosyl ceramide (Gb_3) inhibition of verotoxin (VT)1 cytotoxicity. Increasing doses of VT1 were preincubated with increasing concentrations of adamantyl Gb_3 for 1 h at room temperature prior to the addition to the microtiter plate of vero cell cultures. Cell survival after 72-h culture was measured by crystal violet staining of residual cells.

3. *Inhibition of Hsp70 ATPase.* SGC binding is a conserved property of the hsp70 family of protein chaperones ***(8,41,42)***. The SGC binding site is contained within the N-terminal ATPase domain ***(43)***. Site-specific mutagenesis-identified arginine 342 is a particularly important amino acid residue for SGC binding ***(43)***. This arginine forms a sandwich with arginine 272 in the hydrophobic and electronic coordination of the bound adenine ring within the ATPase substrate binding site ***(45)***. Since the SGC binding site is adjacent or even partially overlapping with the ATPase cleft, SGC binding might affect the ATPase activity of hsp70. Using AdaSGC, we demonstrated that this was indeed the case. AdaSGC was an effective competitor of hsp70 binding to native SGC ***(11)***, and proved to be a noncompetitive inhibitor of hsp70 ATPase activity ***(26)***. ATP hydrolysis was necessary for inhibition, suggesting that the SGC bound preferentially to the adenosine 5′ diphosphate-bound form of the enzyme. Inhibition was dose-dependent between 100 and 300 μ*M*. AdaSGC was effective in the presence or absence of hsp40, indicating that cochaperone binding was not affected. These studies suggest that interaction with SGC could be an unappreciated mechanism of regulation of hsp70 chaperone function, and our current studies are investigating the potential use of AdaSGC to modify hsp70 chaperone function within cells.

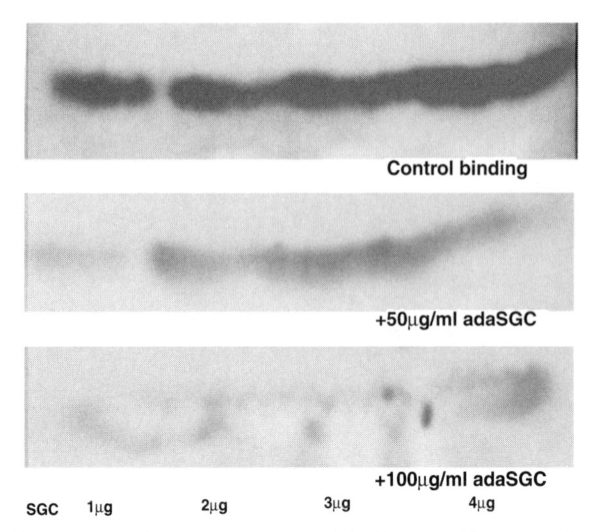

Fig. 4. Inhibition of *Helicobacter pylori* binding to 3′ sulfogalactosyl ceramide (SGC) in vitro. Log phase *H. pylori* were monitored for binding to increasing concentrations of SGC as indicated by thin-layer chromatography overlay as previously ***(49)***, after preincubation with adamantyl SGC for 30 min.

4. *Inhibition of bacterial binding.* Our studies have also implicated hsp70s on the surface of pathogenic bacteria, particularly following stress, as adhesins mediating bacterial/SGC binding ***(41,45,46)***. We have found that pretreatment of such an organism *(Helicobacter pylori)* with AdaSGC reduced their subsequent binding to the parental SGC (*see* **Fig. 4**), indicating that soluble GSL analogs could be used to inhibit host cell GSL recognition by pathogenic microorganisms.
5. *Inhibition of virus binding.* While sialic acid containing glycoconjugates provides the major receptor for the influenza virus on host cells ***(47)***, binding of the influenza hemagglutinin to SGC has also been reported ***(48)***. We therefore investigated the potential of AdaSGC to inhibit influenza host–cell binding. Initial screening was carried out by human rbc hemagglutination using a crude hemagglutinin preparation. The results (*see* **Fig. 5**) showed that norbornyl SGC (0.8–1.5 μ*M*), AdaSGC (2.5–1.3 μ*M*), and SGC coupled to 1,3adamantandiacetic acid (1.5–2.4 μ*M*), inhibited flu virus-mediated agglutination of human red blood cells (*see* **Fig. 5**). However, despite the inhibition of hemagglutination, these analogs proved ineffective to prevent viral/cell infection in vitro (not shown). These studies therefore imply that sulfated GSLs on red blood cells make a significant contribution to the hemagglutination activity of the influenza virus, but sialated glycoconjugates probably provide the major means by which cellular infection is achieved.
6. *Future potential.* The generation of soluble GSL mimics that bind in a manner similar to their membrane-located parental species offers several avenues for

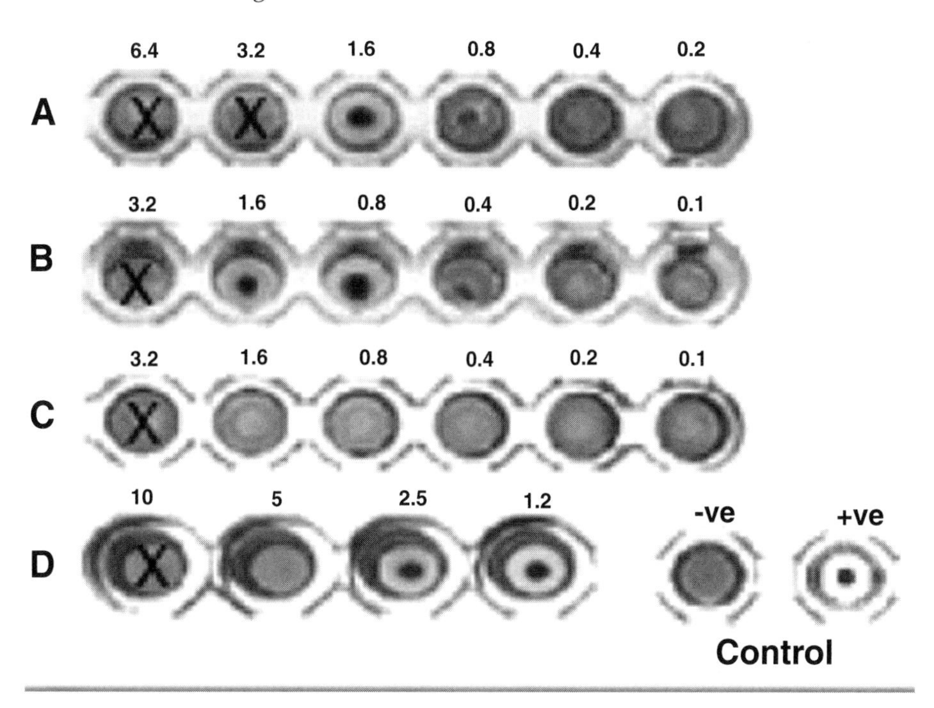

Fig. 5. Inhibition of influenza virus-mediated hemagglutination by soluble 3′ sulfogalactosyl ceramide (SGC) analogs. Crude viral hemagglutinin was preincubated with increasing concentrations of **(A)** norbornyl SGC, **(B)** adamantyl SGC, **(C)** adamantyl GalCer, or **(D)** acetoadamantyl SGC as indicated, and added to a washed suspension of human red blood cells in citrate buffer and incubated at room temperature for 30 min. Rbc agglutination results in a uniform carpet of aggregated cells, whereas unagglutinated rbc accumulate in the center of the round-bottomed wells. Wells marked X, hemolysed cells.

future research. These compounds provide a new window to examine the role of GSL receptor function in microbial pathogenesis and approaches to prophylaxis. Optimization of the carbohydrate moiety for ligand binding could then be followed by multimeric presentation, as has been found effective for many carbohydrate receptors. AdaGSLs offer an opportunity to define the basis of protein/GSL binding using standard physical procedures—nuclear magnetic resonance and crystallography, largely unfeasible for native GSLs. These compounds provide potential soluble substrates/inhibitors for GSL anabolic and catabolic enzymes, and as such could provide new ways to specifically modulate cellular GSL metabolic pathways. The linkage between SGC and tyrosine phosphate recognition that we have shown *(4)* implies that AdaSGC and its analogs have potential in regulating cellular signal transduction pathways.

References

1. Golub, T., Wacha, S., and Caroni, P. (2004) Spatial and temporal control of signaling through lipid rafts. *Curr. Opin. Neurobiol.* **14,** 542–550.
2. van der Goot, F. G. and Harder, T. (2001) Raft membrane domains: from a liquid-ordered membrane phase to a site of pathogen attack. *Semin. Immunol.* **13,** 89–97.
3. Lingwood, C. A. (1996) Aglycone modulation of glycolipid receptor function. *Glycoconj. J.* **13,** 495–503.
4. Lingwood, C., Mylvaganam, M., Minhas, F., Binnington, B., Branch, D., and Pomes, R. (2005) The sulfogalactose moiety of sulfoglycosphingolipids serves as a mimic of tyrosine phosphate in many recognition processes: prediction and demonstration of SH2 domain/sulfogalactose binding. *J. Biol. Chem.* **280,** 12,542–12,547.
5. Lingwood, C. A. (2003) Shiga Toxin Receptor Glycolipid Binding: Pathology and Utility. *Methods in Molecular Medicine, vol. 73, Shigatoxin-Producing Escherichia Coli: Methods and Protocols* (Philpott, D. and Ebel, F., eds.), Humana Press, Totowa, NJ, pp. 165–186.
6. Lingwood, C. A. (1999) Glycolipid receptors for verotoxin and Helicobacter pylori: role in pathology. *Biochim. Biophys. Acta.* **1455,** 375–386.
7. Boyd, B., Zhiuyan, Z., Magnusson, G., and Lingwood, C. A. (1994) Lipid modulation of glycolipid receptor function: Presentation of galactose α1-4 galactose disaccharide for Verotoxin binding in natural and synthetic glycolipids. *Eur. J. Biochem.* **223,** 873–878.
8. Mamelak, D. and Lingwood, C. (1997) Expression and sulfogalactolipid binding of the recombinant testis-specific cognate heat shock protein 70. *Glycoconj. J.* **14,** 715–722.
9. Mylvaganam, M., Binnington, B., Hansen, H., Magnusson, G., and Lingwood, C. (2002) Interaction of verotoxin 1 B subunit with globotriaosyl ceramide analogues: Aminosubstituted (aminodeoxy) adamantylGb_3Cer provides insight into the nature of the Gb_3Cer binding sites. *Biochem. J.* **368,** 769–776.
10. Nyholm, P. G., Magnusson, G., Zheng, Z., Norel, R., Binnington-Boyd, B., and Lingwood, C. A. (1996) Two distinct binding sites for globotriaosyl ceramide on verotoxins: molecular modelling and confirmation by analogue studies and a new glycolipid receptor for all verotoxins. *Chem. Biol.* **3,** 263–275.
11. Mamelak, D., Mylvaganam, M., Tanahashi, E., et al. (2001) The aglycone of sulfogalactolipids can alter the sulfate ester substitution position required for hsc70 recognition. *Carbohydr. Res.* **335,** 91–100.
12. Kiarash, A., Boyd, B., and Lingwood, C. A. (1994) Glycosphingolipid receptor function is modified by fatty acid content: Verotoxin 1 and Verotoxin 2c preferentially recognize different globotriaosyl ceramide fatty acid homologues. *J. Biol. Chem.* **269,** 11,138–11,146.
13. Nutikka, A. and Lingwood, C. (2004) Generation of receptor active, globotriaosyl ceramide/cholesterol lipid 'rafts' in vitro: a new assay to define factors affecting glycosphingolipid receptor activity. *Glycoconj. J.* **20,** 22–38.
14. Mylvaganam, M. and Lingwood, C. A. (2003) In *Carbohydrate-based Drug Discovery, Vol. 2* (Wong, C.-H., ed.), Wiley-VCH, Weinheim, Germany, pp. 761–780.

15. Lingwood, C. A. and Mylvaganam, M. (2003) In *Methods in Enzymology, vol. 363* (Lee, Y. C. and Lee, R. T., eds.), Academic Press, San Diego, pp. 264–284.
16. Strömberg, N., Nyholm, P.-G., Pascher, I., and Normark, S. (1991) Saccharide orientation at the cell surface affects glycolipid receptor function. *Proc. Natl. Acad. Sci. USA* **88,** 9340–9344.
17. Gaus, K., Gratton, E., Kable, E. P., et al. (2003) Visualizing lipid structure and raft domains in living cells with two-photon microscopy. *Proc. Natl. Acad. Sci. USA* **100,** 15,554–15,559.
18. Kitov, P. I., Sadowska, J. M., Mulvey, G., et al. (2000) Shiga-like toxins are neutralized by tailored multivalent carbohydrate ligands. *Nature* **403,** 669–672.
19. Nishikawa, K., Matsuoka, K., Kita, E., et al. (2002) A therapeutic agent with oriented carbohydrates for treatment of infections by Shiga toxin-producing Escherichia coli O157:H7. *Proc. Natl. Acad. Sci. USA* **99,** 7669–7674.
20. Watanabe, M., Matsuoka, K., Kita, E., et al. (2004) Oral therapeutic agents with highly clustered globotriose for treatment of Shiga toxigenic Escherichia coli infections. *J. Infect. Dis.* **189,** 360–368.
21. Binnington, B., Lingwood, D., Nutikka, A., and Lingwood, C. (2002) Effect of globotriaosyl ceramide fatty acid hydroxylation on the binding by Verotoxin 1 and Verotoxin 2. *Neurochem. Res.* **27,** 807–813.
22. Ling, H., Boodhoo, A., Hazes, B., et al. (1998) Structure of the Shiga toxin B-pentamer complexed with an analogue of its receptor Gb_3. *Biochem.* **37,** 1777–1788.
23. Thompson, G., Shimizu, H., Homans, S., and Donohue-Rolfe, A. (2000) Localization of the binding site for the oligosaccharide moiety of Gb_3 on verotoxin 1 using NMR residual dipolar coupling measurements. *Biochem.* **39,** 13,153–13,156.
24. Soltyk, A. M., MacKenzie, C. R., Wolski, V. M., et al. (2002) A mutational analysis of the Globotriaosylceramide binding sites of Verotoxin VT1. *J. Biol. Chem.* **277,** 5351–5359.
25. Mylvaganam, M. and Lingwood, C. (1999) $AdaGb_3$—a monovalent soluble glycolipid mimic which inhibits verotoxin binding to its glycolipid receptor. *Biochem. Biophys. Res. Comm.* **257,** 391–394.
26. Whetstone, D. and Lingwood, C. (2003) 3′Sulfogalactolipid binding specifically inhibits Hsp70 ATPase activity in vitro. *Biochem.* **42,** 1611–1617.
27. Chandler, D. (2002) The two faces of water. *Nature* **417,** 491.
27a. Lund, N., Branch, D. R., Mylvaganam, D., et al. (2006) A novel soluble mimic of the glycolipid globotriaosyl ceramide inhibits HIV infection. *AIDS* **20,** 1–11.
28. Pellizzari, A., Pang, H., and Lingwood, C. A. (1992) Binding of verocytotoxin 1 to its receptor is influenced by differences in receptor fatty acid content. *Biochem.* **31,** 1363–1370.
29. Lingwood, C. A., Murray, R. K., and Schachter, H. (1980) The preparation of rabbit antiserum specific for mammalian testicular sulfogalactoglycerolipid. *J. Immunol.* **124,** 769–774.
30. Koshy, K. and Boggs, J. (1983) Partial synthesis and physical properties of cerebroside sulfate containing palmitic acid or alpha-hydroxy palmitic acid. *Chem. Phys. Lip.* **34,** 41–53.

31. Mahfoud, R., Mylvaganam, M., Lingwood, C. A., and Fantini, J. (2002) A novel soluble analog of the HIV-1 fusion cofactor, globotriaosylceramide(Gb_3), eliminates the cholesterol requirement for high affinity gp120/Gb_3 interaction. *J. Lipid Res.* **43,** 1670–1679.
32. Puri, A., Hug, P., Jernigan, K., Rose, P., and Blumenthal, R. (1999) Role of glycosphingolipids in HIV-1 entry: requirement of globotriosylceramide (Gb3) in CD4/CXCR4-dependent fusion. *Biosci. Rep.* **19,** 317–325.
33. Puri, A., Hug, P., Jernigan, K., et al. (1998) The neutral glycosphingolipid globotriaosylceramide promotes fusion mediated by a CD4-dependent CXCR4-utilizing HIV type 1 envelope glycoprotein. *Proc. Natl. Acad. Sci. USA* **95,** 14,435–14,440.
34. Hammache, D., Yahi, N., Maresca, M., Pieroni, G., and Fantini, J. (1999) Human erythrocyte glycosphingolipids as alternative cofactors for human immunodeficiency virus type 1 (HIV-1) entry: Evidence for CD4-induced interactions between HIV-1 gp120 and reconstituted membrane microdomains of glycosphingolipids (Gb3 and GM3). *J. Virol.* **73,** 5244–5248.
35. Delezay, O., Hammache, D., Fantini, J., and Yahi, N. (1996) SPC3, a V3 loop-derived synthetic peptide inhibitor of HIV-1 infection, binds to cell surface glycosphingolipids. *Biochemistry* **35,** 15,663–15,671.
36. Mylvaganam, M. and Lingwood, C. A. (1999) A convenient oxidation of natural glycosphingolipids to their "ceramide acids" for neoglycoconjugation: Bovine serum albumin-glycoceramide acid conjugates as investigative probes for HIV coat protein gp120-glycosphingolipid interactions. *J. Biol. Chem.* **274,** 20,725–20,732.
37. Peter, M. and Lingwood, C. (2000) Apparent cooperativity in multivalent verotoxin globotriaosyl ceramide binding: Kinetic and saturation binding experiments with radiolabelled verotoxin [^{125}I]-VT1. *Biochim. Biophys. Acta* **1501,** 116–124.
38. Rutjes, N., Binnington, B., Smith, C., Maloney, M., and Lingwood, C. (2002) Differential tissue targeting and pathogenesis of Verotoxins 1 and 2 in the mouse animal model. *Kid. Intl.* **62,** 832–845.
39. Chark, D., Nutikka, A., Trusevych, N., Kuzmina, J., and Lingwood, C. (2004) Differential carbohydrate epitope recognition of globotriaosyl ceramide by verotoxins and monoclonal antibody: Role in human renal glomerular binding. *Eur. J. Biochem.* **271,** 1–13.
40. Khine, A. A., Tam, P., Nutikka, A., and Lingwood, C. A. (2004) Brefeldin A and filipin distinguish two Globotriaosyl ceramide/Verotoxin-1 intracellular trafficking pathways involved in Vero cell cytotoxicity. *Glycobiology* **14,** 701–712.
41. Boulanger, J., Faulds, D., Eddy, E. M., and Lingwood, C. A. (1995) Members of the 70kDa heat shock protein family specifically recognize sulfoglycolipids: role in gamete recognition and mycoplasma related infertility. *J. Cell. Physiol.* **165,** 7–17.
42. Mamelak, D., Mylvaganam, M., Whetstone, H., et al. (2001) Hsp70s contain a specific sulfogalactolipid binding site. Differential aglycone influence on sulfogalactosyl ceramide binding by prokaryotic and eukaryotic hsp70 family members. *Biochemistry* **40,** 3572–3582.
43. Mamelak, D. and Lingwood, C. (2001) The ATPase domain of Hsp70 possesses a unique binding specificity for 3′sulfogalactolipids. *J. Biol. Chem.* **276,** 449–456.

44. Sondermann, H., Scheufler, C., Schneider, C., Höfeld, J., Hartl, F.-U., and Moarefi, I. (2001) Structure of a Bag/Hsc70 complex: convergent functional evolution of Hsp70 nucleotide exchange factors. *Science* **291,** 1553–1557.
45. Huesca, M., Goodwin, A., Bhagwansingh, A., Hoffman, P., and Lingwood, C. A. (1998) Characterization of a stress protein (hsp70) induced by acidic pH, a putative sulfatide binding adhesin, from Helicobacter pylori. *Infect. Immun.* **66,** 4061–4067.
46. Hartmann, E. and Lingwood, C. A. (1997) Brief heat shock induces a long-lasting alteration in the glycolipid receptor binding specificity of Hemophilus influenzae. *Infect. Immun.* **65,** 1729–1733.
47. Skehel, J. J. and Wiley, D. C. (2000) Receptor binding and membrane fusion in virus entry: the influenza hemagglutinin. *Ann. Rev. Biochem.* **69,** 531–569.
48. Suzuki, T., Sometani, A., Yamazaki, Y., et al. (1996) Sulphatide binds to human and animal influenza A viruses, and inhibits the viral infection. *Biochem. J.* **318,** 389–393.
49. Huesca, M., Borgia, S., Hoffman, P., and Lingwood, C. A. (1996) Acidic pH changes receptor binding of Helicobacter pylori: a binary adhesion model in which surface heat-shock (stress) proteins mediate sulfatide recognition in gastric colonization. *Infect. Immun.* **64,** 2643–2648.

21

Diagnosis of Krabbe Disease by Use of a Natural Substrate

John W. Callahan and Marie-Anne Skomorowski

Summary

This chapter describes in detail a practical procedure for the preparation of radiolabeled galactocerebroside and its use in the assay of galactocerebrosidase (GalCase), the enzyme deficient in globoid cell leukodystrophy (Krabbe disease). The reference range for leukocytes and fibroblasts is 0.9–4.4 and 8–36 nmoles substrate hydrolyzed per hour per milligram of protein, respectively. Because of its low abundance this enzyme is difficult to assay in certain situations, such as prenatal diagnosis by chorionic villus sampling. To obviate this a modified assay is used where only the radiolabeled substrate is included in the incubation. This provides a clear separation between affected samples and unaffected controls. The methods detailed here should be reproducible in any laboratory.

Key Words: Krabbe disease; galactocerebrosidase; galactosylceramide; galactose oxidase; borotritiide reduction.

1. Introduction

1.1. Krabbe Globoid Cell Leukodystrophy

Krabbe globoid cell leukodystrophy is largely a disease of early infancy characterized by marked irritability, progressive neurodeterioration with signs of peripheral neuropathy, elevated cerebrospinal fluid protein levels, and with a clinical onset of 4 to 6 mo *(1)*. Older patients (juvenile and adult onset variants) constitute about 10% of all proven cases and display the same neurological features of the infantile form, but with a more protracted course. In all variants the nervous system, particularly the central and the peripheral system myelin (*see* **Fig. 1**), is the exclusive site of clinical and pathological

From: *Methods in Molecular Biology, vol. 347: Glycobiology Protocols*
Edited by: I. Brockhausen © Humana Press Inc., Totowa, NJ

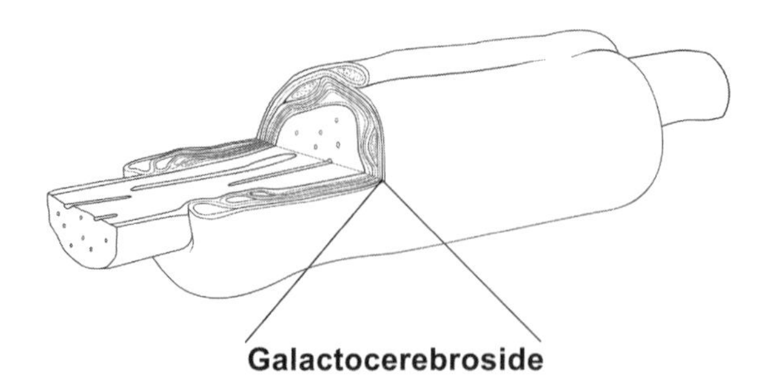

ß-D-galactose

N-acylsphingosine

Galactosylceramide (Galactocerebroside)

Fig. 1. The myelin sheath **(Upper)** is depicted containing one of the two major glycosphingolipids of the myelin, galactocerebroside **(Lower)**.

manifestations. At the terminal stage, the white matter is almost totally devoid of myelin and the oligodendroglia are replaced by severe astrocytic gliosis and the unique abnormal globoid cells. Unlike the other sphingolipidoses, abnormal accumulation of the affected lipid, galactosylceramide (GalC; *see* **Fig. 1**), does not occur in the whole tissue, although the globoid cells appear to be the site of a local accumulation of GalC. Krabbe leukodystrophy is autosomal-recessive, with at least 61 different disease-causing mutations currently known in the GALC gene (Chr14q31) that encodes galactosylceramidase (GalCase; *see* **refs.** ***1*** and ***2***).

1.2. Enzymic Diagnosis

GalCase (galactocerebrosidase [GalCase]; E.C. 3.2.1.46) is a β-galactosidase that catalyzes hydrolysis of GalC and galactosylsphingosine. For diagnosis of affected patients, leukocytes and cultured fibroblasts have been utilized as the enzyme source with good results *(1)*. Although reliable, assays with serum require additional care, as the activity of GalCase is low; therefore, the specific activity of the substrate should be increased to 5–10 times that indicated in the usual procedure. Leukocytes and fibroblasts are equally reliable. Overlap in the activities in leukocytes is usually observed between the normal control and the heterozygote population, and in our opinion the carrier state cannot be reliably determined by enzyme assay alone *(3)*. Several affected fetuses have been diagnosed *in utero* with chorionic villus (CVS) samples, cultured chorionic trophoblasts, and cultured amniotic fluid cells as the enzyme source *(1)*. Although highly-purified GalCase displays activity toward the usual artificial substrates, such as 4-methylumbelliferyl β-D-galactoside, its contribution to the hydrolysis of the artificial substrates in tissue homogenates is minimal owing to its very low abundance in lysosomes. An artificial chromogenic substrate specific for GalCase *(4)* has been described. Recently, a sensitive assay employing a synthetic galactocerebroside (GalC) analog and tandem mass spectrometry verified the possibility of detecting Krabbe disease as part of a newborn screening program using blood spots as the enzyme source *(5)*. However, the optimum substrate for the enzyme to date remains the natural glycosphingolipid, GalC.

To assay the enzyme, tissue extracts are incubated with GalC labeled with tritium in the galactose moiety *(6–8)*. The most common procedure used to introduce tritium into GalC is by oxidation by galactose oxidase (GO; *see* **Reaction 1**) and reduction by [^{3}H] sodium borohydride (*see* **Reactions 2, 3** and **ref.** *7*). The tritium is located on carbon 6 of the galactose moiety as shown:

Reaction 1:
Galactocerebroside + galactose oxidase → 6-dehydro-galactocerebroside +H_2O_2

Reaction 2:
H_2O_2 + catalase/horseradish peroxidase →2 H_2O

Reaction 3:
6-dehydro-galactocerebroside + ^{3}H –$NaBH_4$ → 6-^{3}H-galactocerebroside

The enzymatically liberated [6-^{3}H] galactose is determined after it is separated from the unreacted GalC by solvent partitioning. This chapter provides instructions for the preparation of radioactive GalC, the substrate for GalCase, details of the assay system, and points out assay modifications helpful in decision making.

2. Materials

2.1. Reagents for Preparation of the Radiolabeled Substrate

1. 10 m*M* of sodium phosphate buffer at a final pH of 7.0 ± 0.05.
2. Tetrahydrofuran, redistilled before use or a freshly opened bottle, and diethylether (DE; Fisher Scientific, Toronto, Canada).

3. Chloroform/methanol, 2/1 (v/v) and 4/1 (v/v).
4. Methanol/water/sodium chloride: Combine 50 mg of NaCl, 50 mL of water, and 50 mL of methanol.
5. 1 m*M* of sodium hydroxide.
6. 1 *M* of acetic acid.
7. Sodium borotritiide (crystalline, 5–10 Ci/mmole, 25 mCi; Amersham, Montreal, Canada). It is recommended that the crystalline salt (purple-colored in a sealed ampule) be used. The material supplied gives less than optimal results when dissolved in a weak base.
8. Sodium borohydride (Sigma, St. Louis, MO, cat. no. S9125), crystalline.
9. GalCs, bovine brain (Sigma, cat. no. C4905) in 2:1 chloroform:methanol (2 mg/mL).
10. GO (Sigma, cat. no. G-7907), horseradish peroxidase (Sigma, cat. no. P6140), and catalase (Sigma, cat. no. C9322) in 2-mL sodium phosphate buffer.
11. Silicic acid (Unisil), chromatography grade.
12. 5% Palladium on activated charcoal (Baker Chemicals, Poole, UK).

2.2. Reagents for Enzyme Assay

1. McIlvaine buffer, pH 4.2: Mix 0.1 *M* of citric acid and 0.2 *M* of disodium hydrogen phosphate dihydrate to a final pH of 4.2 ± 0.1. Store at 4°C, where it is stable for several months.
2. Substrate [^{3}H] GalC. Remove from the stock chloroform/methanol solution of tritiated GalC into a glass vial sufficient to provide 80,000 cpm per assay. Add to this GalC to give a final specific activity of 1000 cpm/nmole (assume a molecular weight of 0.8 mg/nmole; this gives 80,000 cpm and 64 μg per assay) and dry under a nitrogen stream.
3. Sodium taurocholate (Sigma, cat. No. S9640). Prepare 10 mg/mL of methanol (this is stable for several months if stored at 4°C).
4. Oleic acid: Dilute from a stock liquid (Sigma, cat. no. O1008) 1 mg/mL of hexane. Alternatively, dissolve 1.218 mg of sodium oleate per mL of methanol. This is stable for several months if stored at 4°C.
5. 100 mg of D-galactose (Sigma, cat. no. G0750) in 100 mL of water.
6. Chloroform/methanol (2/1, v/v).
7. Ideal Solvents upper phase (chloroform/methanol/water; 3/48/47, v/v/v). Ideal Solvents lower phase (chloroform/methanol/water; 86/14/1, v/v/v).

3. Methods

3.1. Enzymatic Oxidation of Substrate

1. Weigh 10 mg of GalC into a large (20-mL) teflon-capped glass tube and add 2 mL of tetrahydrofuran (*see* **Fig. 2**).
2. Dissolve 450 U of GO, 1 mg of horseradish peroxidase, and 1 mg of catalase in 2 mL of sodium phosphate buffer.
3. Add the enzyme solution to the GalC solution and mix well. Close the tube tightly and stir gently overnight at room temperature. The solution may turn cloudy, but this does not affect the reaction.

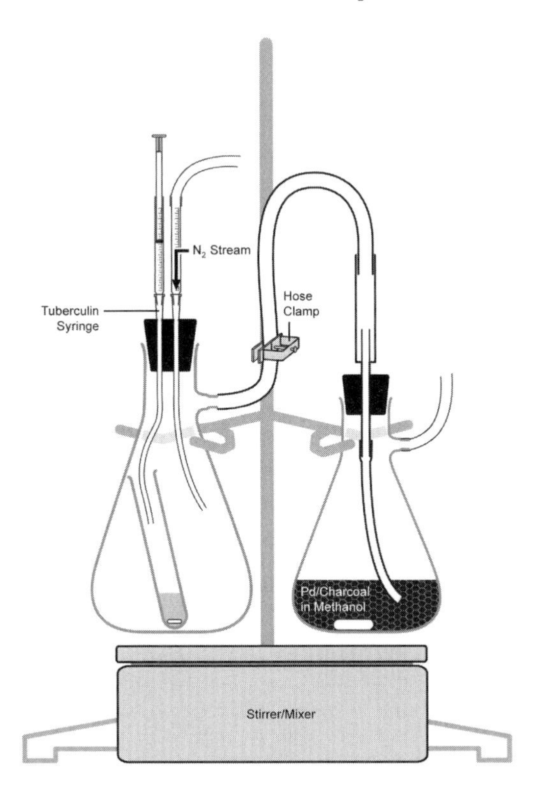

Fig. 2. Once the enzyme reaction is completed and the oxidized galactocerebroside is recovered, the reaction flask with the tuberculin syringes, one for delivery of the dissolved borotritiide and the other for the nitrogen stream, are set up as shown. The hose clamp is tightened to isolate the reaction flask during the tritiation. It is released and a small nitrogen flow is started immediately before the acetic acid is added to destroy the borohydride, thereby stopping the reaction. Any hydrogen gas generated is trapped on the Pd/charcoal suspended in methanol.

4. Stop the reaction with 5 vol of 20 mL of chloroform/methanol, 2/1 (v/v). Mix well and allow the phases to separate (the sample may be centrifuged to separate the phases if needed). The denatured protein occurs at the water–chloroform interface.
5. Remove the infranatant (lower phase) into a separate glass teflon-capped tube (a 5- or 10-mL tube is sufficient) taking care not to include the proteinaceous interphase material or the upper phase, and dry under a nitrogen stream. To speed up evaporation a warm water bath or heat block may be used, but avoid high temperatures.

3.2. Reduction of Oxidized Substrate

1. In the chemical fume hood, set up the radioactive trap (*see* **Fig. 2**) in a 250- or 500-mL side-arm Erlenmeyer flask with a rubber stopper fitted with two 18-gage needle attachments (one for adding the radiolabel and the acetic acid, and one for the nitrogen stream). This flask must be large enough so that the tube containing the oxidized GalC is completely inside the flask (lean it against one side of the flask and below the side-arm; *see* **Fig. 2**).
2. Dissolve the oxidized GalC in 2 mL of tetrahydrofuran, add a stir bar flea, and place the tube inside the trap Erlenmeyer. Cap the vessel with the rubber stopper and the syringe attachments. Attach a rubber tube to the side arm and prepare the trap.
3. The trap is a second side-arm flask (250 mL is sufficient) containing 100 mg of palladium on activated charcoal suspended in methanol. Add a small stir bar and provide continuous but gentle stirring.
4. Attach a rubber tube from the side-arm of the reaction flask to the rubber stopper of the trap. The trap is rubber-stoppered but has an 18-gage needle inserted, to which is attached a small length of polyethylene tubing with one end attached to the end of the needle and the other immersed in the methanol solution (*see* **Fig. 2**).
5. Open the seal on the isotope container inside the hood and add 1 mL of the 1-m*M* sodium hydroxide solution. The purple compound will dissolve immediately to give a clear, colorless solution (*see* **Note 1**).
6. Draw up the solution into a 1-mL syringe, attach it to the needle with the tubing directed into the test tube, and deliver the liquid to the reaction tube through the needle inserted into the tube.
7. Gently purge the system with nitrogen, clamp the side-arm tubing to close the system, and let it stand overnight with gentle stirring at room temperature.

3.3. Stopping the Reaction and Capturing the Excess Radiolabel

1. Dissolve 50 mg of sodium borohydride in 0.1 mL of 1 m*M* NaOH and add it to the reaction flask through the needle head. Incubate for 2 h.
2. Mineral acid destroys the borohydride with evolution of hydrogen and tritium gases, so there must be a small but positive flow of nitrogen gas into the trap. This will direct the evolved gas through the methanol containing the palladium on activated charcoal, whereupon it will bind the generated hydrogen, convert it from the gaseous state to a bound ionic state, and effectively trap the tritium that is not incorporated into the GalC.
3. Open the blocked tube on the trap to allow a gentle flow of nitrogen gas. Fill a tuberculin syringe tube with 1 mL of 1 *M* acetic acid, add it to the top of the reaction vessel again through the needle top, and add the acid to the reaction vessel very slowly to control the evolution of gas. Before the total volume of the acid is added the gas evolution should subside. Stir for 10 min with nitrogen gas flowing.
4. At this point the trap can be disassembled, but set all the glassware used aside in the hood for decontamination. Immediately place into a deep bucket filled with decontamination detergent and warm water.

3.4. Recovering the Labeled Product

1. Add 5 vol of chloroform/methanol, 4/1 (v/v), cap the tube, mix the solutions, and allow the phases to separate.
2. Remove the upper phase waste to a container of celite (an adsorbent for radioactive liquid waste). Wash the lower phase with one fifth vol of methanol:water: sodium chloride, invert the capped tube, allow the phases to separate, and remove the aqueous upper phase to the celite container as before.
3. Repeat **step 2** seven more times.
4. Dry the lower phase under a nitrogen stream, reconstitute in a known volume (10 mL) of 2:1 chloroform:methanol, and count a small (5-µL) aliquot.
5. This procedure typically generates a product in the range of 150 µCi with high specific radioactivity, and one that should last for several years if stored dried in the cold in a closed tube.

3.5. Silicic Acid Chromatography for Final Purification of the Labeled Product

Chromatography on silicic acid (adapted from Vance and Sweeley; *see* **ref.** ***9***) provides an excellent purification step for the initial starting material, for final purification of the labeled product, and for removing the products of radiochemical decay. The details are described in the following steps.

1. Activate a bottle of chromatography-grade silicic acid overnight in a hot (100°C) oven. Weigh 4 g into a glass beaker, slurry in a small volume (20 mL of DE), and pour the slurry into a glass column fitted with a stopcock and a pad of glass wool at the bottom of the column bed.
2. Wash the column with 50 mL of DE.
3. Dissolve the labeled GalC in 0.2–0.5 mL of 2:1 chloroform:methanol and then dilute the sample with pure chloroform to raise the proportions to more than 9:1. Load the sample onto the column and collect the effluent.
4. Elute the column batchwise into a round-bottom flask using reagent-grade solvents in the following order:
   a. 100 mL of chloroform to elute neutral lipids. A small amount of labeled material elutes in this fraction; generally some nonspecific neutral lipid that may be present in the initial enzyme solution, or may arise during the reaction and processing from the breakdown of the parent substrate.
   b. 200 mL of acetone:methanol, 9:1, to elute glycolipids. Labeled GalC elutes in this fraction with a high level of purity.
   c. 100 mL of methanol to elute phospholipids. Some residual GalC is recovered.
5. Use the rotary evaporator to dry the samples, then suspend the residues in a small volume of 2:1 chloroform:methanol (5–10 mL is generally sufficient). Count an aliquot; it is now ready to be used in the assay for GalCase.
6. In a recent typical preparation, 9.2 mg of labeled product was recovered from 10.9 mg of GalC starting material. Its radioactivity was 5.41×10^9 cpm after

column chromatography. In the enzyme assay, a blank value of 1810 cpm from 66,101 cpm of substrate was obtained. This value is too high, so the sample was Folch extracted, brought to complete dryness, and reassayed. The final product contained 9.2 mg, and the specific radioactivity was 0.45 mCi/nmole. The assay blank value was 145 cpm out of a total 70,000 cpm.

7. To prepare a working aliquot, dissolve the total product in 5 mL of 2:1 chloroform:methanol, remove 0.5 mL to a clean glass tube with a teflon-lined cap, and dry the rest under a nitrogen stream. Store the stock standard in the cold (–20°C is preferred).

3.6. Assay Incubation Mixture (see *Note 2*)

1. In a 13- × 100-mm screw-capped glass test tube, dry together 0.03 mL of the substrate, 0.1 mL of sodium taurocholate, and 0.03 mL of the oleic acid solutions under a stream of nitrogen.
2. Suspend the residue in 0.1 mL of the citrate–phosphate buffer and disperse evenly by vortex mixing or a brief sonication.
3. Add the enzyme source (suspend cells in water, freeze–thaw five times, and briefly sonicate) and water to a total volume of 0.2 mL. For leukocytes or fibroblasts, 100–300 μg of protein can be present in the assay tube without risking the non-linear portion of the reaction.
4. Incubate the tubes at 37°C in a shaking water bath for 2 h.
5. Add 0.3 mL of the carrier galactose and 2.5 mL of chloroform/methanol, 2/1 (v/v) at the end of the incubation.
6. Partition the solutes by vortex mixing and centrifuging for 5 min at 1000*g*.
7. Remove the lower phase with a disposable capillary pipet, and wash the upper phase twice with the addition of 1 mL of the pure solvent lower phase.
8. The final upper phase is quantitatively transferred to a scintillation vial.
9. Dry the contents under a stream of nitrogen in a warm heating block.
10. Dissolve the dried residue in 0.5 mL of water.
11. Add 5 mL of universal scintillant and vortex mix. The radioactivity is determined and corrected for the blank tube counts. A liquid scintillation counter (such as the Beckman LS 6500) is calibrated before use with stock-quenched tritium reference and background calibration standards supplied with the instrument.

3.7. Quality Control

1. Check for specificity by assaying a sample from a known affected patient.
2. Check for sample integrity by assaying another lysosomal hydrolase(s), such as β-galactosidase, the enzyme deficient in GM1 gangliosidosis and Morquio disease type B ***(10)***.

3.8. Calculations

1. Units of activity:

$$\text{(CPM of sample} - \text{CPM of blank)} \times \text{Spec radioactivity (nmoles/cpm)} = \text{nmoles substrate hydrolyzed/incubation time (hours)}$$

2. Specific activity: Units of activity per mg of protein, where protein is determined by the Lowry et al. ***(11)*** procedure using bovine serum albumin as the standard.

3.9. Reference Intervals (see Notes 3–5)

1. Leukocytes: 0.9–4.4 nmoles of substrate hydrolyzed/h/mg of protein.
2. Fibroblasts: 8–36 nmoles of substrate hydrolyzed/h/mg of protein.

4. Notes

1. The high level of specific radioactivity achieved during the labeling procedure can result in radiochemical decay. This is detected as an increase in the enzyme assay blank value as a percent of the total counts used in each assay. For small amounts of substrate, if the blank values rise from the expected values of less than 400 cpm, the aliquot of the substrate solution required for the assay should be subjected to a Folch extraction before use. Alternatively, the column chromatographic procedure can be used for cleanup of a larger amount.
2. Linearity limits of assay: the assay conditions described are well within the linearity capability of the assay. Where the activity is in the affected range it is recommended that the assay be repeated with twice the usual enzyme sample volume, whereupon the activity values continue to be well below the usual twofold increment seen in normal control samples. Alternatively, the "no substrate" assay can be performed as described in **Note 4**.
3. When values for the activity are in the 0–5% range of normal values, the patient is likely to be affected with the disease. This assay is specific for diagnosis of the enzyme deficiency in leukocytes, fibroblasts, and other enzyme sources (CVS and amniocytes), but is not always reliable to detect the carrier state. Carrier detection may be possible in specific families, but this should be checked on an individual basis. The reasons for this are partly owing to the low activity in leukocytes, and to the inability to reliably predict the impact the diverse mutations (both polymorphisms and disease-causing) have on the life cycle of the enzyme (folding in the endoplasmic reticulum, post-translational processing, lysosomal localization, and maturation). The specific active site residues of the enzyme are not defined.
4. In the course of kinetic studies, the low activity obtained in the full assay with leukocytes as the enzyme source can be circumvented with limited amounts of substrate in the incubation. For example, an assay containing 80,000 cpm of radiolabeled substrate (about 0.14 n*M* of GalC), but with no cold carrier added, will generate more than 10,000 cpm of product per hour per mg of protein, depending on the enzyme source, with a blank value of 200–300 cpm. Patients with Krabbe disease generate less than 100 cpm over the blank at all protein concentrations. This is called the "no substrate" assay. It provides a way of diagnosing the disease when activity is usually low (as in leukocytes), when there is a minimum of enzyme source available as sometimes occurs in the case of CVS samples if the serum is used as the enzyme source, or when attempting to discriminate a low carrier value from an affected level.

5. Several affected fetuses have been diagnosed with direct CVS sampling, cultured trophoblasts, and cultured amniotic fluid cells as the enzyme source. Affected patients with high GalCase activity have not been reported. Enzymic activities of known carriers (specifically where the major common mutation has been identified) do not give an intermediate value between normal controls and affected patients with a high level of consistency. Therefore carrier detection by enzyme assay is considered to be unreliable.

Acknowledgment

The authors would like to thank Gulanaar Hassam for her excellent technical assistance.

References

1. Wenger, D. A., Suzuki, K., Suzuki, Y., and Suzuki, K. (2001) Galactosylceramide lipidosis: Globoid cell leukodystrophy (Krabbe disease). In *The Metabolic and Molecular Basis of Inherited Disease, 8th ed., Vol. III* (Scriver, C. R., Beaudet, A. L., Sly, W. S., and Valle, D., eds.), McGraw-Hill, New York, pp. 3669–3694.
2. Wenger, D. A., Rafi, M. A., and Luzi, P. (1997) Molecular genetics of Krabbe disease (Globoid cell leukodystrophy): diagnostic and clinical implications. *Human Mutation* **10,** 268–279.
3. Randell, E., Connolly-Wilson, M., Duff, A., Skomorowski, M.-A., and Callahan, J. W. (2000) Evaluation of the accuracy of enzymatically determined carrier status for Krabbe disease by DNA-based testing. *Clin. Biochem.* **33,** 217–220.
4. Gal, A. E., Brady, R. O., Pentchev, P. G., et al. (1977) A practical chromogenic procedure for the diagnosis of Krabbe's Disease. *Clin. Chim. Acta* **77,** 53–59.
5. Li, Y., Scott, C. R., Chamoles, N. A., et al. (2004) Direct multiplex assay of lysosomal enzymes in dried blood spots for newborn screening. *Clin. Chem.* **50,** 1785–1796.
6. Harzer, K., Knoblich, R., Rolfs, A., Bauer, P., and Eggers, J. (2002) Residual galactosylsphingosine (psychosine) beta-galactosidase activities and associated GALC mutations in late and very late onset Krabbe disease. *Clin. Chim. Acta* **317,** 77–84.
7. Radin, N. S. and Arora, R. C. (1971) A simplified assay method for galactosylceramide beta-galactosidase. *J. Lipid Res.* **12,** 256–257.
8. Radin, N. S., Hof, L., Bradley, R. M., and Brady, R. O. (1969) Lactosylceramide galactosidase: comparison with other sphingolipid hydrolases in developing rat brain. *Brain Res.* **14,** 497–502.
9. Vance, D. E. and Sweeley, C. C. (1967) Quantitative determination of the neutral glycosylceramides in human blood. *J. Lipid Res.* **8,** 621–630.
10. Lowden, J. A., Callahan, J. W., Gravel, R. A., Skomorowski, M. A., Becker, L., and Groves, J. (1981) Type 2 GM1-gangliosidosis with long survival and neuronal ceroid lipofuscinosis. *Neurology* **31,** 719–724.
11. Lowry, O. H., Rosebrough, J. J., Farr, A. L., and Randall, R. J. (1951) Protein measurement with the Folin phenol reagent. *J. Biol. Chem.* **193,** 265–275.

22

In Vitro Assays of the Functions of Calnexin and Calreticulin, Lectin Chaperones of the Endoplasmic Reticulum

Breanna S. Ireland, Monika Niggemann, and David B. Williams

Summary

Calnexin and calreticulin are molecular chaperones of the endoplasmic reticulum (ER) whose folding-promoting functions are directed predominantly toward aspargine-linked glycoproteins. This is a consequence of calnexin and calreticulin being lectins with specificity for the early oligosaccharide (OS)-processing intermediate, $Glc_1Man_9GlcNAc_2$. In addition, they interact with non-native conformers of glycoprotein polypeptide chains to prevent aggregation and recruit the thiol oxidoreductase ERp57 to catalyze glycoprotein disulfide formation/isomerization. In vitro assays of these functions have contributed greatly to our understanding of how calnexin and calreticulin promote glycoprotein folding. This chapter describes the isolation of $Glc_1Man_9GlcNAc_2$ OS, as well as the assay used to measure OS binding. Furthermore, details are provided of assays that detect ERp57 binding by calnexin and calreticulin, as well as the abilities of these chaperones to suppress the aggregation of non-native protein substrates.

Key Words: Aggregation; calnexin; calreticulin; endoplasmic reticulum; ERp57; glycoprotein folding; lectin; molecular chaperone; oligosaccharide.

1. Introduction

Asparagine (Asn)-linked glycoproteins (GPs) fold within the endoplasmic reticulum (ER) with the assistance of a variety of molecular chaperones and folding enzymes. Prominent among these are the type I membrane chaperone calnexin (Cnx) and its soluble homolog calreticulin (Crt). Cnx and Crt interact transiently with most, if not all, newly-synthesized GPs, enhancing the efficiency of folding and subunit assembly ***(1–5)***. They also participate in ER quality control, ensuring that misfolded or incompletely folded GPs are not

From: *Methods in Molecular Biology, vol. 347: Glycobiology Protocols*
Edited by: I. Brockhausen © Humana Press Inc., Totowa, NJ

exported along the secretory pathway ***(2,3,6,7)***. The specificity of these chaperones for GPs is because of the fact that they are lectins that bind to the early Asn-linked processing intermediate $Glc_1Man_9GlcNAc_2$ ***(8–12)***. The single terminal glucose unit of this "monoglucosylated" oligosaccharide (OS) is crucial for recognition by the lectin sites of Cnx and Crt. Cnx and Crt also bind to ERp57, a thiol oxidoreductase similar to protein disulfide isomerase, which is capable of catalyzing disulfide bond formation and isomerization ***(13,14)***.

Cnx and Crt exert their chaperone and quality-control functions through an elegant interplay of functional sites. Their lectin site binds to a nascent GP once $Glc_1Man_9GlcNAc_2$ OSs are formed by the action of ER glucosidases I and II ***(9,15)***. In addition, a polypeptide binding site on Cnx and Crt recognizes non-native conformers of the GP polypeptide chain ***(10,16,17)***. This dual mode of interaction substantially enhances Cnx/Crt recognition of GP substrates when compared to non-lectin chaperones such as the Hsp70 chaperone, BiP ***(18)***. Folding is promoted in two ways. Nonproductive aggregation reactions are suppressed by the polypeptide-based interactions ***(16,17)***, and disulfide bond formation/isomerization is catalyzed by chaperone-bound ERp57 ***(14,19)***. How GP dissociation from Cnx/Crt occurs is less well-characterized but likely occurs when the peptide binding site shifts to a low-affinity state, possibly upon adenosine triphosphate (ATP) release or hydrolysis ***(16,17)***, combined with removal of the terminal glucose residue by glucosidase II ***(15)***. If the GP rapidly acquires a native state, no rebinding to the chaperone occurs. However, if folding/assembly is delayed, rebinding can occur through peptide site recognition of the non-native polypeptide chain and through reattachment of the terminal glucose residue by uridine 5′-diphosphate-glucose:GP glucosyltransferase. The latter enzyme only acts on non-native GP substrates ***(20)***. Prolonged cycles of chaperone interaction result in the ER retention associated with quality control.

Mammalian Cnx consists of a 461-residue ER luminal segment, a 22-residue transmembrane domain, and a 90-residue cytoplasmic tail ***(21)***. The X-ray crystal structure of the luminal segment of Cnx has revealed a globular domain that resembles legume lectins, and an extended-arm domain that is composed of two tandemly repeated sequence motifs ***(22)***. Mammalian Crt is a 400-residue soluble protein that shares approx 39% sequence identity with the luminal segment of Cnx ***(23)***. It also possesses an arm domain which, in nuclear magnetic resonance studies, has been shown to be similar to but shorter than that of Cnx ***(24)***. It is presumed that Crt possesses a globular lectin domain based on the identical lectin specificities of Cnx and Crt ***(12)***. Nuclear magnetic resonance studies ***(25)*** as well as binding experiments with chaperone deletion mutants ***(26)*** have revealed that ERp57 binds to the tip of the arm domains of both Cnx and Crt.

Much of our progress in understanding the functions of Cnx and Crt has been based on in vitro assays that have permitted an assessment of their OS binding specificities ***(12,26,27)***, their abilities to bind ERp57 ***(26,27)***, and their capacities to prevent protein aggregation ***(16–18,26,27)***. This chapter describes such assays using bacterially expressed recombinant Cnx and Crt. In addition, the procedure for isolating radiolabeled $Glc_1Man_9GlcNAc_2$ OS is described.

2. Materials

2.1. Purification of $Glc_1Man_9GlcNAc_2$ OS

1. D-[2-^{3}H]Mannose, specific activity 15 Ci/mmol (Amersham Pharmacia Biotech, Oakville, ON, Canada, cat. no. TRK.364).
2. *Saccharomyces cerevisiae alg8* strain (*MATα, ade2-101, gal$^-$-, mal$^-$-, alg8-1*), obtained from Dr. Mark Lehrman, University of Texas SW Medical Center, Dallas, TX.
3. YPAD medium: 1% yeast extract, 2% peptone, 2% glucose, and 40 µg/mL of adenine sulfate.
4. Low-glucose YPAD medium: 1% yeast extract, 2% peptone, 0.1% glucose, and 40 µg/mL of adenine sulfate.
5. Chloroform:methanol, 2:1.
6. Chloroform:methanol:H_2O, 10:10:3.
7. Methanol.
8. Tetrahydrofuran (80% in H_2O).
9. 11.8 *M* HCl.

2.2. OS Binding Assay

1. Purified glutathione-*S*-transferase (GST), GST-Cnx, and GST-Crt.
2. Glutathione-agarose beads (Sigma-Aldrich, Oakville, ON, Canada, cat. no. G-4510).
3. Buffer A: 10 m*M* of *N*-2-hydroxyethylpiperazine-*N*′-ethanesulfonic acid (HEPES; pH 7.5), 150 m*M* of NaCl, and 5 m*M* of $CaCl_2$.
4. 20 m*M* of reduced glutathione in buffer A.
5. 1 m*M* of ethylene glycol *bis*-2-aminoethyl ether-*N,N′,N″,N′*-tetraactic acid (EGTA; pH 7.5).

2.3. ERp57 Binding Assay

1. Purified GST, GST-Cnx, and GST-Crt.
2. Glutathione-agarose beads (Sigma, cat. no. G-4510).
3. Purified human ERp57.
4. Rabbit anti-ERp57 antiserum (StressGen Bioreagents, Ann Arbor, MI, cat. no. SPA-585E).
5. *N*-hydroxysuccinimidyl (NHS)-[^{14}C]acetate (specific activity 15,000 cpm/nmol).
6. Buffer A: 10 m*M* of HEPES (pH 7.5), 150 m*M* of NaCl, and 5 m*M* of $CaCl_2$.
7. Factor Xa (Amersham Biosciences).
8. Human ERp57 complementary deoxyribonucleic acid (cDNA) in a pET vector, obtained from Dr. David Thomas, McGill University.

2.4. Aggregation Suppression Assay

1. Purified GST, GST-Cnx, and GST-Crt.
2. Jack bean α-mannosidase (Sigma, cat. no. M-7257), citrate synthase (Sigma, cat. no. C-3260), and malate dehydrogenase (Sigma, cat. no. 410-13).
3. 6 *M* of guanidinium hydrochloride.
4. Buffer A: 10 m*M* of HEPES (pH 7.5), 150 m*M* of NaCl, and 5 m*M* of $CaCl_2$.

3. Methods

All of the following assays make use of Cnx and Crt purified as GST fusion proteins from *Escherichia coli*. This is convenient in binding assays since the GST-fusion proteins can be readily immobilized on glutathione-agarose beads. To avoid complications associated with the use of membrane proteins, only the ER luminal segment of Cnx is fused to GST. pGEX plasmids (Amersham Biosciences) containing GST fused via its C-terminus to cDNA encoding rabbit Crt or the ER luminal segment of dog CNX (both lacking their N-terminal signal sequences) can be obtained from the authors. Alternative sources can be found in **refs.** ***23,28–30***.

Details of GST-fusion protein expression and purification will not be given here, as they are common methods that can be obtained from Amersham Biosciences. In brief, expression is induced in protease-deficient BL21 *E. coli* cells (Amersham) using 0.5 m*M* of isopropyl-β-D-thiogalactopyranoside (IPTG) at 30°C for 3 h. The majority of the fusion protein is soluble within the bacterial cytosol. Standard purification using glutathione-agarose affinity chromatography results in a yield of 5–20 mg of protein/L of bacterial culture with a purity of 80–90%. Further purification can be achieved by Mono Q anion exchange fast protein liquid chromatography (1.5- × 10-cm column, Amersham) using a linear gradient of 50 m*M* to 1 *M* of NaCl in 20 m*M* of HEPES (pH 7.2) and 5 m*M* of $CaCl_2$. With a flow rate of 1 mL/min, the first 38% of the gradient occurs in 10 min, with a slow increase to 45% over the next 35 min and a final increase to 100% over 5 min. The fusion proteins elute between 38% and 45% and have a purity in excess of 90%. Following dialysis against buffer A, protein concentration is determined by the Bio-Rad assay using a bovine serum albumin (BSA) standard.

3.1. Purification of $Glc_1Man_9GlcNAc_2$ OS

The *alg8* strain of *S. cerevisiae* lacks a glucosyltransferase activity involved in dolichol-OS biogenesis and produces $Glc_1Man_9GlcNAc_2$-P-P-dolichol rather than the normal $Glc_3Man_9GlcNAc_2$-P-P-dolichol precursor *(31)*. In the following protocol (originally developed by Dr. M. Lehrman), $Glc_1[^3H]Man_9GlcNAc_2$-P-P-dolichol is isolated from *alg8* yeast radiolabeled with $[^3H]$mannose and then the OS is released by hydrolysis of the pyrophosphate linkage. This OS can then be used in binding experiments with Cnx or Crt.

Unless mentioned otherwise, all centrifugation steps are performed in a Sorvall H4000 swinging bucket rotor at room temperature.

1. Prepare a starter culture of *alg8* yeast by growing to stationary phase in 2 mL of YPAD at 30°C. Innoculate 100 mL of YPAD medium with 10 µL of a starter culture and grow at 30°C to mid-log phase (A_{600} of 0.7–0.9). This normally takes approx 16 h.
2. Transfer the yeast culture to two 50-mL plastic centrifuge tubes and centrifuge at 1800*g* for 5 min. Remove the supernatant, resuspend the pelleted yeast in 4 mL of low-glucose YPAD, and transfer to a 15- × 125-mm round-bottom glass centrifuge tube with a teflon-lined screw cap. Centrifuge again at 1800*g* for 5 min and remove the supernatant.
3. Resuspend the yeast in 400 µL of low-glucose YPAD containing 1 mCi/mL [^{3}H]mannose. Incubate at 30°C for 10 min. Following the incubation, centrifuge at 1800*g* for 3 min, remove the supernatant as completely as possible, and discard as radioactive waste.
4. Resuspend the yeast pellet in 4 mL of chloroform:methanol (2:1). Sonicate in a sonicator water bath for 2 min and then centrifuge at 2600*g* for 5 min. Using a Pasteur pipet, remove the supernatant and discard as radioactive waste. Repeat this chloroform:methanol 2:1 extraction two more times, and then dry the pellet under a stream of air for approx 3 min in a 50°C water bath.
5. Resuspend the dried pellet in 4 mL of H_2O, sonicate for approx 2 min, centrifuge at 2600*g* for 5 min, and discard the supernatant. Repeat this aqueous extraction three to five times.
6. Following the last aqueous extraction, resuspend the pellet in 500 µL of methanol, centrifuge at 2600*g* for 5 min, discard the supernatant, and dry under a stream of air as in **step 4**.
7. Add 1 mL of chloroform:methanol:water (10:10:3), sonicate for 2 min, and centrifuge at 2600*g* for 5 min. *Save the supernatant, which contains the dolichol-P-P-OS*. Repeat the extraction two more times and pool the supernatants. Count 1% of the sample by liquid scintillation counting. Approximately 100,000–500,000 cpm should be detected at this stage. Dry down the pooled supernatants in the 50°C water bath under an air stream for approx 35 min. A small amount of cloudy aqueous solution will remain.
8. Suspend in 5 mL of 80% tetrahydrofuran in H_2O. With the lid of the glass centrifuge tube loosened, heat in a water bath at 50°C for 30 min. To hydrolyze the radiolabeled OS from dolichol, add 40 µL of concentrated HCl (11.8 *M*) to give a final HCl concentration of 0.1 *M*. Mix and heat at 50°C for 60 min. Evaporate until dry under an air stream at 50°C (~45 min).
9. Resuspend in 1 mL of H_2O, vortex, and add 2.5 mL of chloroform:methanol (2:1). Vortex well, then centrifuge at 2600*g* for 5 min. The radiolabeled OS will be in the top aqueous layer. Quantify radioactivity by liquid scintillation counting. A typical final yield is approx 100,000 cpm. This preparation may be used directly in OS binding assays.

3.2. OS Binding Assay

This assay has been used for a variety of purposes, including assessment of the OS binding specificities of Cnx and Crt *(12)*, localizing their lectin sites through the use of deletion mutants *(26)*, and identifying specific residues involved in binding OS *(27)*. By determining the specific activity of the $Glc_1[^3H]Man_9GlcNAc_2$ OS, the assay may also be used to determine dissociation constants for the chaperones.

1. In all subsequent steps, agarose beads are sedimented in microcentrifuge tubes by centrifugation for 30 s at top speed in a tabletop microcentrifuge.
2. Wash glutathione-agarose beads (Sigma) twice in buffer A by suspension in buffer and centrifugation. Resuspend in buffer A, distribute the required number of aliquots corresponding to 50 μL of packed beads into microcentrifuge tubes, centrifuge, and remove the supernatant.
3. To immobilize fusion proteins on the beads, incubate in triplicate 0.1 nmol of GST (mol wt = 26,000; 2.6 μg) and either GST-Cnx (mol wt = 78,510; 7.85 μg) or GST-Crt (mol wt = 72,620; 7.26 μg) with 50 μL of washed glutathione-agarose beads in a final volume of 100 μL (adjusted with buffer A) for 30 min at room temperature on a rocker (*see* **Notes 1** and **2**). Wash the beads three times in buffer A to remove any unbound fusion protein. Remove the supernatant as completely as possible. Set aside one of the triplicate samples for later elution and assessment of protein amount (Bio-Rad assay) or integrity (sodium dodecyl sulfate-polyacrylamide gel electrophoresis) as desired.
4. Resuspend the immobilized fusion proteins in 100 μL of buffer A containing 2000 cpm of $Glc_1[^3H]Man_9GlcNAc_2$ OS. Incubate at 4°C for 30 min with occasional gentle mixing.
5. Centrifuge the beads and remove the supernatant completely (*see* **Note 3**). Wash beads once *very rapidly* with 100 μL of buffer A to minimize losses owing to the weak OS-chaperone association (K_d ~ 1–2 μ*M*; *see* **ref. *30***). Remove supernatant completely.
6. Resuspend beads in 200 μL of 1 m*M* EGTA (pH 7.5) and elute the bound OS by heating at 50°C for 5 min. Count radioactivity by liquid scintillation.
7. Results are expressed as the average binding of duplicate GST-Cnx or GST-Crt samples (~300–400 cpm) less the binding observed for the duplicate GST negative control (50–100 cpm). When assessing binding to various mutant forms of GST-Cnx/Crt, the binding is reported as the average percentage binding compared to wild-type GST-Cnx/Crt control, which is set to 100%.

3.3. ERp57 Binding Assay

We have used two different forms of this assay to localize the ERp57 binding site on Cnx and Crt *(26)* and to assess the integrity of the arm domain in various chaperone mutants *(27)*. The more quantitative assay measures $[^{14}C]$acetyl-ERp57 binding to immobilized GST-Cnx/Crt with quantification by liquid scintillation counting. The second, more qualitative, assay assesses ERp57 binding to immobilized GST-Cnx/Crt by immunoblotting.

1. Sources of ERp57: In brief, human ERp57 cDNA in a pET vector is used to transform BL21 *E. coli* cells, and expression is induced with 0.5 m*M* of IPTG at 30°C for 3 h. After sonication and centrifugation, the bacterial extract is applied to a Mono-Q anion exchange column and developed with a 0.05- to 0.5-*M* NaCl gradient. ERp57 that elutes at approx 0.3 *M* of NaCl is dialyzed and concentrated using a Centricon-10 concentrator (Amicon). Alternatively, a pGEX-3X plasmid-encoding GST fused to the mature N-terminus of mouse ERp57 can be created. It is expressed and purified as previously described for GST-Cnx/Crt (*see* **Heading 3.**), but the GST is removed by cleavage with Factor Xa (20 U in 7 mL at 4°C for 48 h). Cleaved GST is removed by passage through glutathione-agarose.
2. ERp57 (30 μ*M*) is radiolabeled by incubation at room temperature for 30 min in 100 μL of 0.1 *M* $NaHCO_3$, pH 8.0, with a 100-fold molar excess of NHS [^{14}C]acetate (specific activity 15,000 cpm/nmol). Unreacted NHS-[^{14}C]acetate is removed by rapid gel filtration using a Nap-25 column (Amersham) equilibrated in buffer A.
3. The immobilization of GST and either GST-Cnx or GST-Crt on glutathione-agarose beads is identical to that of the OS binding assay (*see* **Subheading 3.2., steps 1–3**). Duplicate samples of immobilized fusion proteins are incubated at room temperature for 30 min in 100 μL of buffer A containing 0.1% Nonidet P-40, 100 μg/mL of BSA, and 1.5 μ*M* of [^{14}C]acetyl-ERp57. Centrifuge and remove the supernatant completely.
4. Wash the agarose beads twice rapidly with 200–500 μL of buffer A, removing the supernatants completely. Elution of bound [^{14}C]acetyl-ERp57 is carried out by resuspending the beads in 200 μL of buffer A containing 10 m*M* of reduced glutathione and incubating at room temperature for 15 min. Following centrifugation, the supernatant is counted by liquid scintillation. Specific binding is calculated as cpm bound to GST-Cnx or GST-Crt minus cpm bound to GST.
5. The immunoblotting version of the assay is carried out as above with the exception that 0.1 μ*M* of unlabeled ERp57 replaces 1.5 μ*M* of [^{14}C]acetyl-ERp57. Eluted samples are analyzed by 10% sodium dodecyl sulfate-polyacrylamide gel electrophoresis under reducing conditions, and ERp57 is detected by immunoblotting with anti-ERp57 antiserum and visualized with the enhanced chemiluminescence system (Amersham; *see* **Note 4**).

3.4. Aggregation Suppression Assay

As molecular chaperones, Cnx and Crt possess the ability to suppress the aggregation of denatured protein substrates (both glycosylated and nonglycosylated) in vitro ***(16,17,32)***. In the case of nonglycosylated substrates such as citrate synthase (CS) and malate dehydrogenase (MDH), this aggregation suppression is mediated through a polypeptide binding site located primarily within the globular domains of Cnx and Crt ***(26)***. For GP substrates such as α-mannosidase or IgY that contain $Glc_1Man_9GlcNAc_2$ OSs, a major contribution to aggregation suppression is made through the lectin sites of Cnx and Crt as well ***(18)***. Consequently, the choice of substrate can report on the polypeptide or lectin functions of Cnx and Crt ***(27)***.

3.4.1. Aggregation Suppression of Nonglycosylated Substrates

CS and MDH are mitochondrial proteins that aggregate at 43–45°C without a requirement for prior chemical denaturation. They have been used extensively as substrates to study the aggregation suppression of chaperones of the Hsp60 ***(33)***, Hsp70 ***(18)***, Hsp90 ***(34)***, and small heat-shock protein families ***(35)***.

1. CS or MDH are desalted by Nap25 gel filtration in buffer A.
2. Either substrate is diluted to 1 μM in 500 μL of buffer A in the absence or presence of GST-Cnx or GST-Crt. Several substrate/chaperone molar ratios are used in duplicate, ranging from 1:0.5 to 1:8. GST is used as a negative control in place of the chaperone.
3. The solution is heated to 45°C in a heating block, and aggregation is monitored at 5- to 10-min intervals over 60 min by measuring light scattering (apparent absorbance) at 360 nm. By staggering start times, about 10 samples may be analyzed at once. Periodic mixing is advised to keep aggregates in suspension. Alternatively, a thermally controlled spectrophotometer equipped with an automatic cell changer may be used.
4. The degree of aggregation suppression is expressed relative to substrate alone. Because aggregation suppression by Cnx and Crt is enhanced in the presence of ATP, 3 m*M* of ATP may be added to incubation mixtures ***(16,17)***. This assay has been used to demonstrate the polypeptide-based aggregation suppression functions of Cnx/Crt ***(16,17)***, examine the role of other ligands and cofactors in this function ***(16,17)***, compare aggregation suppression capability to that of the Hsp70 chaperone, BiP ***(18)***, and determine the domain where the polypeptide binding site resides ***(26)***.

3.4.2. Aggregation Suppression of Monoglucosylated GP Substrates

Both jack bean α-mannosidase and chicken IgY possess $Glc_1Man_9GlcNAc_2$ OSs ***(36,37)*** and have been used as substrates to monitor aggregation suppression by Cnx or Crt ***(18)***. Because α-mannosidase is commercially available, the method for this GP will be described.

1. Jack bean α-mannosidase is desalted by Nap25 gel filtration in buffer A and lyophilized. It is dissolved in 6 *M* of guanidinium hydrochloride at a concentration of 24 μ*M* and denatured at room temperature for 60 min.
2. Add 7.5 μL of denatured α-mannosidase to the bottom of a 10–15°C cuvet and dilute very rapidly with 492.5 μL of buffer A (precooled to 10–15°C) containing 3 m*M* of ATP and various amounts of GST-Cnx, GST-Crt, or GST (negative

control). The final concentration of denatured α-mannosidase is 0.36 μ*M*, and molar ratios of substrate/chaperone typically used are 1:0.5 to 1:2 (*see* **Note 5**).

3. Samples are prepared in duplicate and aggregation is monitored by measuring light scattering at 360 nm in 30 s intervals over 5 min. This assay has been used to demonstrate the ability of Cnx and Crt to suppress the aggregation of monoglucosylated GPs, assess the relative roles of lectin and polypeptide sites in this function, compare the GP aggregation suppression capability of Cnx to that of the Hsp70 chaperone, BiP, and characterize potential lectin-deficient mutants of Cnx/Crt ***(18,27)***.

4. Notes

1. When pipeting agarose beads with a micropipet, cut off the end of the micropipet tip to allow easy entry of the beads.
2. The reader may find it more convenient to combine the triplicate samples for the immobilization step, i.e., use 150 μL of glutathione-agarose beads and three times the amount of fusion protein. After washing away unbound protein, divide into three equal aliquots.
3. A long, gel-loading micropipet tip works well to completely remove supernatants, as its narrow internal diameter does not permit the entry of beads.
4. With the immunoblotting version of the ERp57 binding assay, it is common to observe weak binding of ERp57 to the GST negative control. This can be minimized by increasing the volume of buffer A used to wash the beads following incubation with ERp57. The Triton and BSA concentrations may also be increased in the binding assay. It is also crucial to use new glutathione-agarose beads rather than recycled beads.
5. Because of the extreme aggregation propensity of chemically-denatured α-mannosidase, the aggregation assay is initiated rapidly, conducted at low temperature, and is of short duration. If the initial mixing of chaperone and α-mannosidase is not completed instantly, extensive aggregation will occur prior to the onset of aggregation measurements.

References

1. Vassilakos, A., Cohen-Doyle, M. F., Peterson, P. A., Jackson, M. R., and Williams, D. B. (1996) The molecular chaperone calnexin facilitates folding and assembly of class I histocompatibility molecules. *EMBO J.* **15,** 1495–1506.
2. Danilczyk, U. G., Cohen-Doyle, M. F., and Williams, D. B. (2000) Functional relationship between calreticulin, calnexin, and the endoplasmic reticulum luminal domain of calnexin. *J. Biol. Chem.* **275,** 13,089–13,097.
3. Gao, B., Adhikari, R., Howarth, M., et al. (2002) Assembly and antigen-presenting function of MHC class I molecules in cells lacking the ER chaperone calreticulin. *Immunity* **16,** 99–109.
4. Parodi, A. J. (2000) Role of N-oligosaccharide endoplasmic reticulum processing reactions in glycoprotein folding and degradation. *Biochem. J.* **348,** 1–13.

5. Leach, M. R. and Williams, D. B. (2003) Calnexin and calreticulin, molecular chaperones of the endoplasmic reticulum. In *Calreticulin, 2nd ed.* (Eggleton, P. and Michalak, M., eds.), Landes Bioscience, Georgetown, TX, pp. 49–62.
6. Jackson, M. R., Cohen-Doyle, M. F., Peterson, P. A., and Williams, D. B. (1994) Regulation of MHC class I transport by the molecular chaperone, calnexin (p88, IP90). *Science* **263,** 384–387.
7. Rajagopalan, S. and Brenner, M. B. (1994) Calnexin retains unassembled major histocompatibility complex class I free heavy chains in the endoplasmic reticulum. *J. Exp. Med.* **180,** 407–412.
8. Ou, W. J., Cameron, P. H., Thomas, D. Y., and Bergeron, J. J. (1993) Association of folding intermediates of glycoproteins with calnexin during protein maturation. *Nature* **364,** 771–776.
9. Hammond, C., Braakman, I., and Helenius, A. (1994) Role of N-linked oligosaccharide recognition, glucose trimming, and calnexin in glycoprotein folding and quality control. *Proc. Natl. Acad. Sci. USA* **91,** 913–917.
10. Ware, F. E., Vassilakos, A., Peterson, P. A., Jackson, M. R., Lehrman, M. A., and Williams, D. B. (1995) The molecular chaperone calnexin binds Glc1Man9 GlcNAc2 oligosaccharide as an initial step in recognizing unfolded glycoproteins. *J. Biol. Chem.* **270,** 4697–4704.
11. Spiro, R. G., Zhu, Q., Bhoyroo, V., and Soling, H. D. (1996) Definition of the lectin-like properties of the molecular chaperone, calreticulin, and demonstration of its copurification with endomannosidase from rat liver Golgi. *J. Biol. Chem.* **271,** 11,588–11,594.
12. Vassilakos, A., Michalak, M., Lehrman, M. A., and Williams, D. B. (1998) Oligosaccharide binding characteristics of the molecular chaperones calnexin and calreticulin. *Biochemistry* **37,** 3480–3490.
13. Oliver, J. D., Roderick, H. L., Llewellyn, D. H., and High, S. (1999) ERp57 functions as a subunit of specific complexes formed with the ER lectins calreticulin and calnexin. *Mol. Biol. Cell* **10,** 2573–2582.
14. Zapun, A., Darby, N. J., Tessier, D. C., Michalak, M., Bergeron, J. J., and Thomas, D. Y. (1998) Enhanced catalysis of ribonuclease B folding by the interaction of calnexin or calreticulin with ERp57. *J. Biol. Chem.* **273,** 6009–6012.
15. Hebert, D. N., Foellmer, B., and Helenius, A. (1995) Glucose trimming and reglucosylation determine glycoprotein association with calnexin in the endoplasmic reticulum. *Cell* **81,** 425–433.
16. Ihara, Y., Cohen-Doyle, M. F., Saito, Y., and Williams, D. B. (1999) Calnexin discriminates between protein conformational states and functions as a molecular chaperone in vitro. *Mol. Cell* **4,** 331–341.
17. Saito, Y., Ihara, Y., Leach, M. R., Cohen-Doyle, M. F., and Williams, D. B. (1999) Calreticulin functions in vitro as a molecular chaperone for both glycosylated and non-glycosylated proteins. *EMBO J.* **18,** 6718–6729.
18. Stronge, V. S., Saito, Y., Ihara, Y., and Williams, D. B. (2001) Relationship between calnexin and BiP in suppressing aggregation and promoting refolding of protein and glycoprotein substrates. *J. Biol. Chem.* **276,** 39,779–39,787.

19. Molinari, M. and Helenius, A. (1999) Glycoproteins form mixed disulphides with oxidoreductases during folding in living cells. *Nature* **402,** 90–93.
20. Sousa, M. and Parodi, A. J. (1995) The molecular basis for the recognition of misfolded glycoproteins by the UDP-Glc:glycoprotein glucosyltransferase. *EMBO J.* **14,** 4196–4203.
21. Wada, I., Rindress, D., Cameron, P. H., et al. (1991) SSR alpha and associated calnexin are major calcium binding proteins of the endoplasmic reticulum membrane. *J. Biol. Chem.* **266,** 19,599–19,610.
22. Schrag, J. D., Bergeron, J. J., Li, Y., et al. (2001) The structure of calnexin, an ER chaperone involved in quality control of protein folding. *Mol. Cell* **8,** 633–644.
23. Baksh, S. and Michalak, M. (1991) Expression of calreticulin in Escherichia coli and identification of its Ca2+ binding domains. *J. Biol. Chem.* **266,** 21,458–21,465.
24. Ellgaard, L., Riek, R., Herrmann, T., et al. (2001) NMR structure of the calreticulin P-domain. *Proc. Natl. Acad. Sci. USA* **98,** 3133–3138.
25. Frickel, E. M., Riek, R., Jelesarov, I., Helenius, A., Wuthrich, K., and Ellgaard, L. (2002) TROSY-NMR reveals interaction between ERp57 and the tip of the calreticulin P-domain. *Proc. Natl. Acad. Sci. USA* **99,** 1954–1959.
26. Leach, M. R., Cohen-Doyle, M. F., Thomas, D. Y., and Williams, D. B. (2002) Localization of the lectin, ERp57 binding, and polypeptide binding sites of calnexin and calreticulin. *J. Biol. Chem.* **277,** 29,686–29,697.
27. Leach, M. R. and Williams, D. B. (2004) Lectin-deficient calnexin is capable of binding class I histocompatibility molecules in vivo and preventing their degradation. *J. Biol. Chem.* **279,** 9072–9079.
28. Ou, W. J., Bergeron, J. J., Li, Y., Kang, C. Y., and Thomas, D. Y. (1995) Conformational changes induced in the endoplasmic reticulum luminal domain of calnexin by Mg-ATP and Ca2+. *J. Biol. Chem.* **270,** 18,051–18,059.
29. Bouvier, M. and Stafford, W. F. (2000) Probing the three-dimensional structure of human calreticulin. *Biochemistry* **39,** 14,950–14,959.
30. Kapoor, M., Srinivas, H., Kandiah, E., et al. (2003) Interactions of substrate with calreticulin, an endoplasmic reticulum chaperone. *J. Biol. Chem.* **278,** 6194–6200.
31. Runge, K. W. and Robbins, P. W. (1986) A new yeast mutation in the glucosylation steps of the asparagine-linked glycosylation pathway. Formation of a novel asparagine-linked oligosaccharide containing two glucose residues. *J. Biol. Chem.* **261,** 15,582–15,590.
32. Mancino, L., Rizvi, S. M., Lapinski, P. E., and Raghavan, M. (2002) Calreticulin recognizes misfolded HLA-A2 heavy chains. *Proc. Natl. Acad. Sci. USA* **99,** 5931–5936.
33. Buchner, J., Schmidt, M., Fuchs, M., et al. (1991) GroE facilitates refolding of citrate synthase by suppressing aggregation. *Biochemistry* **30,** 1586–1591.
34. Jakob, U., Scheibel, T., Bose, S., Reinstein, J., and Buchner, J. (1996) Assessment of the ATP binding properties of Hsp90. *J. Biol. Chem.* **271,** 10,035–10,041.
35. Lee, G. J., Roseman, A. M., Saibil, H. R., and Vierling, E. (1997) A small heat shock protein stably binds heat-denatured model substrates and can maintain a substrate in a folding-competent state. *EMBO J.* **16,** 659–671.

36. Ohta, M., Hamako, J., Yamamoto, S., et al. (1991) Structures of asparagine-linked oligosaccharides from hen egg-yolk antibody (IgY). Occurrence of unusual glucosylated oligo-mannose type oligosaccharides in a mature glycoprotein. *Glycoconj. J.* **8,** 400–413.
37. Kimura, Y., Hess, D., and Sturm, A. (1999) The N-glycans of jack bean alpha-mannosidase. Structure, topology and function. *Eur. J. Biochem.* **264,** 168–175.

23

Quantitative Measurement of Selectin–Ligand Interactions

Assays to Identify a Sweet Pill in a Library of Carbohydrates

Mark E. Beauharnois, Sriram Neelamegham, and Khushi L. Matta

Summary

Soluble oligosaccharides and glycoproteins can inhibit leukocyte adhesion during a range of vascular ailments including inflammation, thrombosis, and cancer metastasis. The design of such molecules in many cases is based on the structure of naturally occurring selectin ligands. In this case, synthetic selectin–ligand mimetics act as competitive inhibitors of cell adhesion. In an alternate approach, cell-permeable, small-molecule oligosaccharides have been shown to alter the metabolic pathways that lead to the biosynthesis of functional selectin–ligands. The addition of such molecules results in glycoproteins that are defective in their ability to bind selectins. Quantitative in vitro testing of the efficacy of the above inhibition strategies ideally requires the application of assays that mimic the in vivo physiological milieu in terms of the valency of selectin and selectin–ligands, the physiological fluid-flow conditions, and the use of blood cells. Assays that are performed in small volumes are preferable when the quantity of available inhibitor is scarce. Finally, the measurements must account for the rapid on- and off-rates of selectin-mediated binding interactions. This chapter addresses these issues by presenting methods to measure selectin function in enzyme-linked immunosorbent assay and flow cytometry-based static assays, cell-adhesion assays performed under shear flow in cone-plate viscometers, and Biacore surface plasmon resonance measurements of molecular-binding kinetics. Examples are presented where such methods are applied to measure the ability of simple oligosaccharides based on sialyl Lewis-X and complex molecules with the core-2 structure to block selectin function. Such methods may be extended to identify potent selectin antagonists in a library of carbohydrates.

Key Words: Selectin; small molecule antagonist; sialyl Lewis-X; neutrophil; leukocyte; aggregation; adhesion; ELISA; flow cytometry; Biacore; surface plasmon resonance; valency;

From: *Methods in Molecular Biology, vol. 347: Glycobiology Protocols*
Edited by: I. Brockhausen © Humana Press Inc., Totowa, NJ

P-selectin glycoprotein ligand-1 (PSGL-1); glycosylation-dependent cellular adhesion molecule-1 (GlyCAM-1); shear stress.

1. Introduction

Cell-surface carbohydrates mediate important molecular recognition processes including bacterial and viral infection, cell adhesion in inflammation, differentiation, immune response, fertilization, and development. A growing body of literature also supports a crucial role for glycans in the progress of the various steps of cancer (*1,2*). It has been estimated that over half of the world's drug leads are derived directly from natural product pools, many of which have functionally active carbohydrate moieties (*3,4*). Calicheamicin, daunomycin, streptomycin, erythromycin, vancomycin, amphotericin, and staurosporine are notable examples of therapeutically relevant natural glycosylated metabolites (*3*). In certain cases the presence of the carbohydrate moiety in these molecules increases hydophilicity, which improves pharmacokinetic properties; in other cases, glycosidic linkage can be crucial for the molecule's activity. For example, sialylated oligosaccharide glycans linked to insulin have been shown to prolong blood glucose-lowering activity (*5*). Certain galactosyl and lactosyl derivatives have been synthesized to examine their antimetastasis potential (*6*). Synthesis of carbohydrate-linked cisplatin analogs have also been reported (*7*).

The focus of this chapter is on the measurement of selectin interactions with natural and synthetic ligands. Selectins are a family of adhesion molecules with three members: L-, E-, and P-selectin. These are expressed on blood leukocytes (L-selectin), stimulated endothelial cells (E- and P-selectin), and activated blood platelets (P-selectin). Specific calcium-dependent recognition of carbohydrates presented primarily on glycoproteins (and to a lesser extent on glycolipids on various cell types) by these selectins results in the formation of molecular bonds between selectins and their ligands. The tensile strength of these molecular interactions is high, and the binding kinetics (both on- and off-rates) are rapid.

The study of selectin interactions is important, as leukocyte binding to vascular endothelial cells via this adhesion molecule is a critical step during normal immune response, lymphocyte homing, and inflammation. Platelet-tumor cell adhesion also plays an important role during cancer metastasis. Finally, P-selectin plays an important role in selectin-mediated heterotypic platelet-leukocyte aggregate formation in circulation, which is thought to contribute to cardiovascular events. Development of specific strategies to control selectin-binding interaction may ameliorate a range of vascular pathologies. Selectin-binding mechanisms and rates also represent a prototypic protein–carbohydrate interaction that forms the basis for similar studies with other families of adhesion molecules, e.g., galectins and siglecs.

Diverse strategies have been examined to control selectin binding function. In one approach, soluble molecules such as sialyl Lewis-X (sLe^X, *N*-acetylneuraminic acid [NeuAc]α2→3galactose [Gal]β1→4[fucose {Fuc}α1→3] *N*-acetylglucosamine [GlcNAc]), which are designed to mimic the carbohydrate structure of natural selectin ligands, have been demonstrated to act as effective, competitive inhibitors of selectin-mediated cell adhesion ***(8,9)***. The efficacy of carbohydrates with complex core-2 glycans that are identical to the functional carbohydrate portion of the natural selectin-ligands P-selectin glycoprotein ligand-1 (PSGL-1) and glycosylation cell adhesion molecule-1 (GlyCAM-1) have also been tested ***(10)***. Finally, the ability of glycoprotein fragments from PSGL-1 ***(11)*** and non-sLe^X-based heparin sulfates ***(12)*** to block selectin function have been examined.

In a second approach, it has been suggested that acetylated modified monosaccharides or primers introduced into cells can alter the biosynthetic pathways that lead to the formation of selectin ligands (*see* **Fig. 1**). **Figure 1A** illustrates the normal biochemical pathway and glycosyltransferases responsible for the assembly of the selectin-ligand on PSGL-1, which consists of an sLe^X moiety linked to the core-2 trisaccharide structure (GlcNAcβ1→6[Galβ1→3 *N*-acetylgalactosamine {GalNAc}α]). **Figure 1B** shows that modified monosaccharides can be incorporated into glycans. In support of this, it has been shown that such analogs can engage cellular metabolic reactions ***(13–17)***. Modified analogs of galactose, GlcNAc, GalNAc, and Mannose have been synthesized, and studies have been conducted in cellular assays ***(18,19)***. Some of these studies have examined the application of these reagents for chemotherapy, e.g., 4-F-GlcNAc can block ovarian carcinoma cell adhesion to extracellular matrix ***(20)***. More recently, Dimitroff et al. ***(21)*** applied the same molecule to inhibit selectin-ligand biosynthesis. The possibility that 4-F-GlcNAc can truncate *O*-glycan extension leading to the synthesis of defunct selectin-ligands is illustrated in **Fig. 1B**. It should be noted that the Gal arm of the glycan can still be extended by enzymes, although these reactions are not shown in this schematic.

Figure 1C illustrates the concept of primers or alternate substrates that act as "bogus" intermediates ***(22–26)***. Here, primers added to cells are acted upon by glycosyltransferases. Thus, instead of glycosylating the natural protein efficiently, the enzymes now act on the cell-permeable alternate substrates, which are available in large quantities. Consequently, the natural protein is either not glycosylated or underglycosylated. Therefore it does not display the normal structures, and glycoprotein functions can be altered. In a well-known example, GalNAcα1→Obenzyl has been shown to act as an inhibitor of endogenous mucin biosynthesis, since it primes the formation of mucin-linked oligosaccharides ***(26)***. In a collaborative study, it has also been observed that the addition of tetrasaccharide 3-*O*-methylgalactosamine (MeGal)β1-4GlcNAcβ1-6

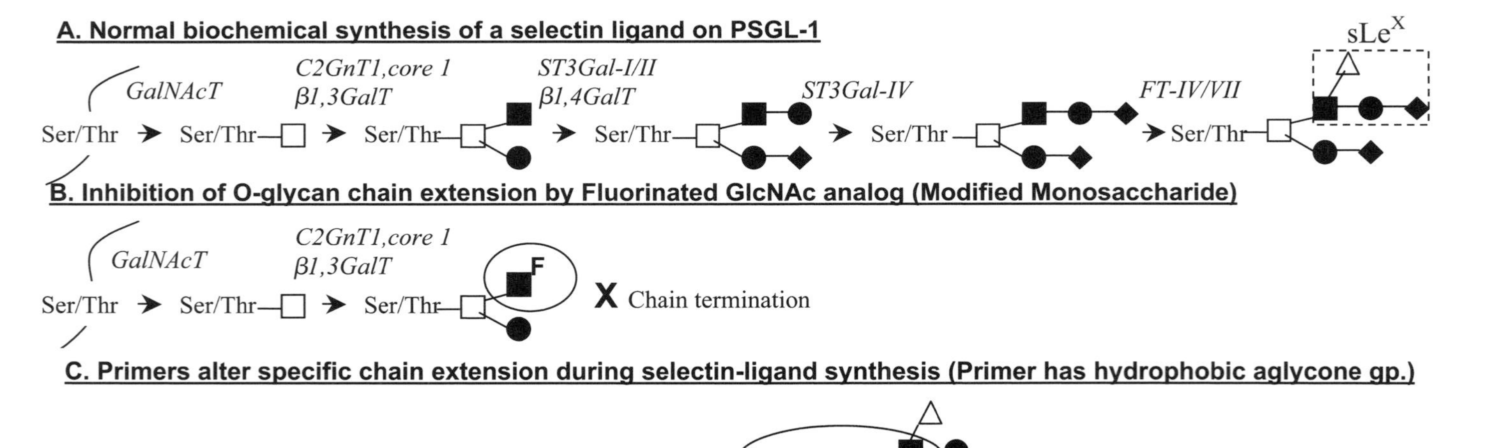
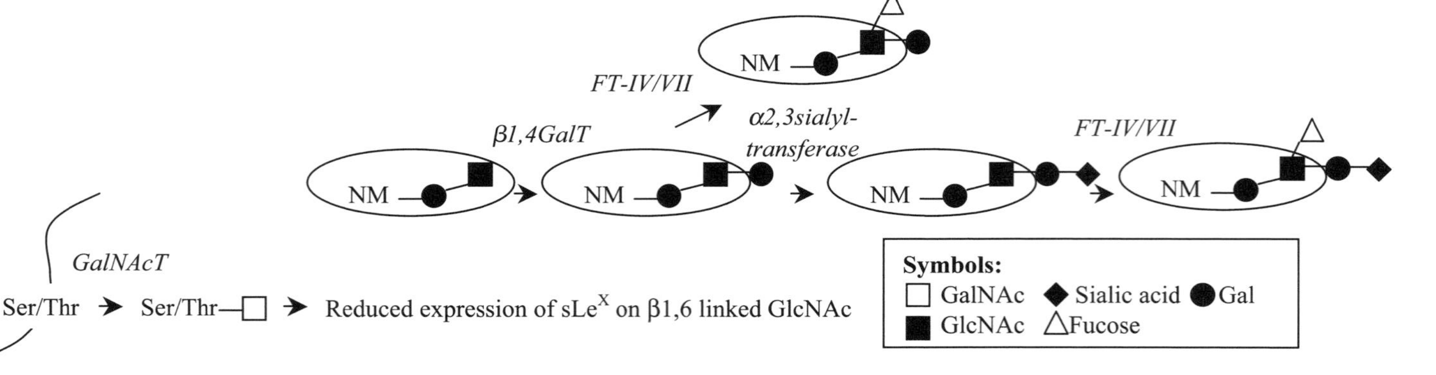

Fig. 1. Strategies for glycosyltransferase inhibition. **(A)** Normal biosynthetic pathway, **(B)** strategies to alter the biochemical pathways leading to the synthesis of selectin-ligands using modified monosaccharide, and **(C)** primer. Inhibitors applied in panels B and C are shown in ovals. Glycosyltransferases participating in each reaction are shown in italics. Sialyl Lewis-X is depicted in the dashed square box in panel A. The primer shown in panel C is fully acetylated GlcNAcβ1,3Galβ-Naphthalenemethanol (*see* **refs.** ***22*** and ***23***). NM, napthalene methanol.

(Galβ1-3)GalNAcα-Obenzyl to gastric cell lines reduces the binding of these cells to the mouse omentum and reduces tumor growth rates ***(25)***. Esko and colleagues ***(22,23)*** have also shown that the acetylated disaccharide GlcNAcβ1-3Gal-naphthalenemethanol is cell-permeable and prevents the formation of functional/adhesive selectin-ligands in U937 lymphoma cells. These cells, when treated with the disaccharide, do not bind E-selectin on stimulated human endothelial cells.

Quantitative assays are required to evaluate selectin inhibition strategies. These must be performed under fluid shear conditions that are relevant to the disease condition of interest because static binding assays may reveal molecular interactions that are not physiologically relevant. Selectins also exhibit a shear threshold feature that can influence the binding kinetics, depending on the magnitude of applied shear stress ***(27)***. The molecular binding rates are also very rapid, so the use of wash steps must be evaluated carefully. It is also best to mimic the physiological valency of selectins and their ligands in the experimental assay, since these can influence the specificity of selectin–ligand recognition ***(28)***. For this reason, isolated blood cells or endothelial cells are better suited for cell adhesion measurements compared with reconstituted systems bearing recombinant proteins or cell lines. These issues are addressed in this chapter by presenting methods to measure selectin function in enzyme-linked immunosorbent assay (ELISA) and flow cytometry-based static assays, cell adhesion assays performed under shear flow in cone-plate viscometers, and Biacore surface plasmon resonance measurements of molecular binding kinetics. The advantages and shortcomings of each approach are discussed.

2. Materials

2.1. Neutrophil Isolation

1. Neutrophil isolation medium (Mono-Poly™ resolving medium, MP Biomedicals, Irvine, CA) and anti-coagulant Heparin (Elkin-Sinn Inc., Cherry Hill, NJ).
2. *N*-2-hydroxyethylpiperazine-*N*′-ethanesulfonic acid (HEPES)-buffered saline: 30 m*M* of HEPES, 110 m*M* of NaCl, 10 m*M* of KCl, 2.1 m*M* of $MgCl_2$, and 10 m*M* of D-glucose (pH 7.2), made in endotoxin-free distilled water.
3. 25% Injectable human serum albumin (HSA; Alpha Therapeutic Corp., Los Angeles, CA).
4. Model Z1-dual Coulter counter (Beckman Coulter, Fullerton, CA).
5. Sorvall Legend swinging bucket centrifuge (Thermo Electron Corporation, Ashville, NC).

2.2. ELISA-Based Selectin Inhibition Assay

1. 384-Well plates (Nalgene Nunc, Rochester, NY).
2. Chimeric P-selectin fusion protein consists of the first 158 amino acids of the mature human protein, including the lectin and epidermal growth factor domains

fused to a 220 amino acid sequence that encodes the mouse IgG2a Fc portion. This construct was produced using the Bac-to-Bac baculovirus expression system (Invitrogen, Carlsbad, CA) according to the manufacturer's instructions. Briefly, recombinant deoxyribonucleic acid encoding the fusion protein was cloned into pFASTBac-1 plasmid and transformed into DH10Bac *Escherichia coli* cells. Recombinant bacmid DNA was purified from the DH10Bac cells and used to transfect SF21 insect cells. After three rounds of amplification, high-titer virus was used to infect SF21 cells for protein production. The cells were grown in EX-CELL 405 serum-free growth media (JRH Biosciences, Lenexa, KS). The growth media containing the recombinant P-selectin IgG was harvested 3–5 d postinfection and stored at 4°C for up to 1 mo or at –20°C for up to 1 yr.

3. Goat anti-human IgG antibody (Zymed Laboratories, South San Francisco, CA) and goat anti-mouse horseradish peroxidase (HRP)-conjugated antibody (Jackson ImmunoResearch Laboratories, West Grove, PA).
4. Recombinant PSGL-1 (19.ek.Fc) containing the 19 N-terminal amino acids of mature PSGL-1 linked to an enterokinase cleavage site that is in turn linked to a human IgG Fc is described elsewhere ***(11)***. The negative control molecule called ek.Fc is identical to 19.ek.Fc, but without the 19 amino acid PSGL-1 fragment. These reagents are stored at –80°C until use.
5. 10X Phosphate-buffered saline (PBS; Invitrogen). 1X PBS is supplemented with 3% (w/v) bovine serum albumin (BSA; Sigma, St. Louis, MO) for the blocking step, and 0.1% (w/v) BSA and 1.5 m*M* of Ca^{2+} for the selectin binding step.
6. o-Phenylenediamine (OPD; Sigma).
7. Synergy HT ELISA plate reader (Bio-Tek Instruments, Winooski, VT).

2.3. Selectin Chimera Static Inhibition Assay

1. L- and P-selectin IgG selectin chimera used consists of the lectin domain, the epidermal growth factor domain, and most of the short consensus repeat domains (2 for L-selectin and 9 for P-selectin) of human selectin fused to a human IgG_1 tail (GlycoTech, Gaithersburg, MD).
2. Goat anti-human $F(ab')_2$ fluorescein isothiocyanate (FITC)-conjugated secondary antibody and goat serum (Jackson).
3. FACSCalibur flow cytometer (Becton Dickinson, Franklin Lakes, NJ).

2.4. Homotypic Neutrophil Inhibition Assay

1. VT550 cone-plate viscometer equipped with a 2° cone (Haake Inc., Paramus, NJ).
2. 1 m*M* stock solutions of formyl peptide (fMLP; Sigma) are made in dimethyl sulfoxide. Store at –80°C for up to 1 yr.
3. 2% Glutaraldehyde solution is made fresh on the day of the experiment by diluting 50% electron microscopy-grade stock solutions (Sigma) in HEPES-buffered saline.
4. FACSCalibur flow cytometer (Becton Dickinson).

2.5. Surface Plasmon Resonance

1. Research-grade CM5 sensor chips (Biacore Inc., Piscataway, NJ).
2. The running buffer is PBS: 10 m*M* of phosphate, 150 m*M* of NaCl (pH 7.0), and 0.005% (v/v) surfactant P-20 supplemented with either 1 m*M* of calcium or 3 m*M* of ethylenediaminetetraacetic acid (EDTA). All other chemicals are from Sigma.
3. Biacore™ 3000 instrument (Biacore).

3. Methods

3.1. Human Neutrophil Isolation

1. Isolated neutrophils are used in some of the binding studies (*see* **Subheadings 3.3.** and **3.4.**). Collect blood from healthy, nonsmoking adult volunteers by venipuncture into a sterile syringe containing 10 U of heparin/mL of blood. Donors certify that they have not taken any medication in the two weeks preceding blood donation. This procedure is approved by the University at Buffalo, Institutional Ethics and Biosafety Review Board.
2. Layer 5 mL of blood on top of 4 mL of Mono-Poly in 15-cc plastic centrifuge tubes. For some donors 5–7% (v/v) H_2O is added to the gradient medium, as this enhances leukocyte separation.
3. Tubes containing layered blood are usually centrifuged at approx 820*g* for 20 min at room temperature. Longer times or higher centrifugation speeds are required for some donors.
4. The polymorphonuclear leukocyte band, which contains more than 90% neutrophils, is removed and transferred to a tube containing 30 mL of HEPES buffer with 0.1% (v/v) human serum albumin (HSA) at 4°C.
5. Centrifuge these neutrophils in HEPES buffer at approx 200*g* at 4°C for 10 min, and resuspend the pellet in 1 mL of ice-cold HEPES buffer with 0.1% (v/v) HSA (without calcium). Measure the cell count using a Coulter counter or hemocytometer. Such separation generally yields approx $1–2 \times 10^6$ cells/mL of blood, and viability is more than 98% as judged using trypan blue exclusion.
6. Store neutrophils for up to 2 h after isolation on ice in calcium-free buffer for experimentation. Add HEPES buffer containing 1.5 m*M* of calcium to the cells 2–10 min prior to each experiment run as described in **Subheadings 3.3., step 1** and **Subheading 3.4., step 2**. Such precautions are necessary in order to prevent undesired neutrophil activation prior to experimentation.

3.2. ELISA-Based Selectin Inhibition Assay (No Fluid Shear in This Assay)

1. Incubate 50 µL of 50 µg/mL of goat anti-human IgG antibody in PBS and incubate at 4°C overnight in the wells of a 384-well plate.
2. Wash the wells once with 100 µL of PBS and block with 50 µL of PBS containing 3% (v/v) BSA at 4°C for 1 h.
3. After another wash step with 100 µL of PBS, add 25 µL of 10 µg/mL fusion protein containing the functional end of PSGL-1, 19.ek.Fc, at room temperature for

2 h in the presence of 0.1% BSA. In some wells, a similar concentration of the control ek.Fc fragment is added.

4. Preincubate a 1:5 dilution of P-selectin IgG cell-culture supernatant with a 1:200 dilution of goat anti-mouse HRP-conjugated secondary Ab in PBS containing 0.1% (v/v) BSA and 1.5 m*M* Ca^{2+} for 15 min in the presence or absence of varying doses of selectin antagonist. Preincubation with HRP-conjugated secondary Ab is necessary in this step in order to enhance multivalency of selectins in this assay (*see* **Note 1**).
5. Following this, add 50 μL of P-selectin IgG complexed with the HRP-conjugated secondary Ab to each well at room temperature for 15 min.
6. The wells are washed three times with 100 μL of PBS. Bound P-selectin IgG is detected by adding 50 μL of OPD, stopping the reaction with 12.5 μL of 4 *M* H_2SO_4 and measuring the absorbance at 590 nm.
7. **Figure 2** presents representative experimental data using the above protocol. As seen in **Fig. 2A**, the binding interaction studied is specific as only wells that had both the immobilized 19.ek.Fc and P-selectin IgG chimera displayed high absorbance at 590 nm. Wells with the control ek.Fc fragment had low signal. **Figure 2B** demonstrates a typical experiment where an sLe^X analog was added at varying dosages to inhibit P-selectin-19.ek.Fc interactions. The measured IC50 (reagent concentration that blocks selectin function by 50%) was 0.55 m*M*.

3.3. Flow Cytometry-Based Selectin Inhibition Assay (No Fluid Shear in This Assay)

1. Incubate 50-μL samples of human L-selectin (4 μg/mL) or P-selectin (3 μg/mL) IgG chimera with 6 μg/mL of goat anti-human $F(ab')_2$ FITC-conjugated secondary antibody in HEPES buffer supplemented with 1.5 m*M* of Ca^{2+} and 1% goat serum at 37°C for 10 min, with or without selectin antagonists present. Preincubation with FITC-conjugated secondary Ab is necessary here in order to enhance the multivalency of selectins in this assay (*see* **Note 1**).
2. Warm isolated neutrophils (*see* **Subheading 3.1.**) to 37°C for 2 min and add at a final concentration of 0.5×10^6 cells/mL.
3. Remove 10-μL samples at fixed time points (30 and 90 s, and 4 min) and dilute in 100 μL of HEPES buffer supplemented with 1.5 m*M* of Ca^{2+} and 1% goat serum.
4. Quantify the amount of L- and P-selectin IgG bound to neutrophils by measuring the fluorescence of the neutrophils using flow cytometry immediately following sample dilution.
5. **Figure 3** shows representative data for the above experiments. In panel A, L-selectin IgG Chimera binding to human neutrophils is shown to be specific, since it can be completely blocked upon addition of either 5 m*M* of EDTA or anti-L-selectin antibody, DREG-56. Selectin-leukocyte binding is also blocked by 60% upon addition of antagonist to leukocyte PSGL-1, thus demonstrating that ligands beside PSGL-1 account for 40% of L-selectin binding to human neutrophils. L-selectin binding to neutrophils is inhibited in a dose-dependent manner

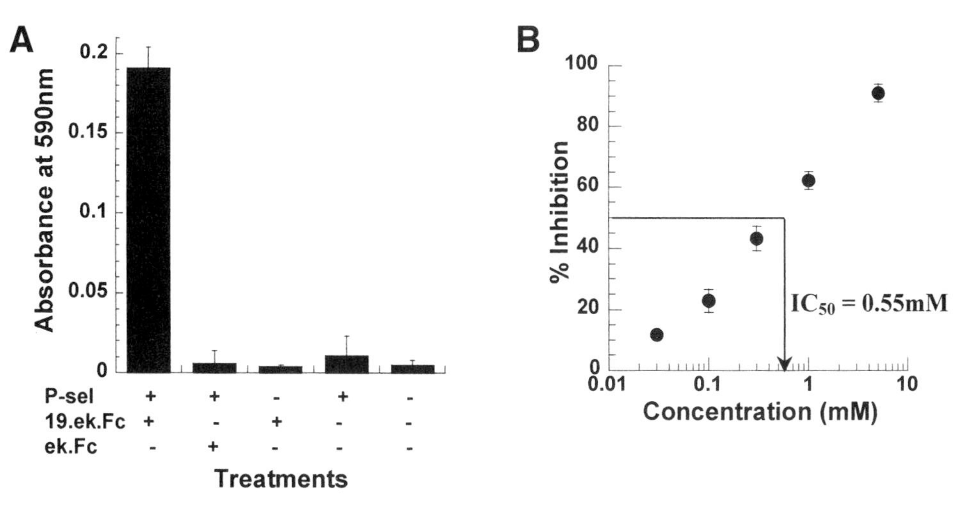

Fig. 2. P-selectin carbohydrate interaction measured using enzyme-linked immunosorbent assay. **(A)** Control experiments where plates were coated with either 19.ek.Fc, ek.Fc, or without any human Fc fusion proteins. P-selectin IgG was added in some wells along with HRP-conjugated Ab, while HRP-conjugated Ab alone was added in the remaining wells. **(B)** Inhibition data in runs where varying doses of TBC1269 (a sialyl Lewis-X analog) is added. IC_{50} measured is approx 0.55 m*M*. Data are mean ± SEM.

upon addition of sLe^X with a methyl glycoside (sLe^X-OMe). The measured IC50 is approx 0.3 m*M*.

3.4. Neutrophil Aggregation Assay (Fluid Shear Applied in This Assay)

1. Isolated human neutrophils (*see* **Subheading 3.1.**) are diluted in HEPES buffer containing 1.5 m*M* of Ca^{2+} and 0.1% HSA, either with or without cell-adhesion blocking reagent. The final cell concentration is 0.5×10^6/mL and the sample volume is 60 µL. This sample is maintained at room temperature for 7 min followed by 3 min at 37°C prior to being placed in the gap between the cone and plate of the viscometer.
2. 10 µL of the sample, the zero second time point, is removed and fixed in 100 µL of ice-cold 2% glutaraldehyde.
3. Add 1 µ*M* of fMLP to stimulate the cells. Fluid shear is immediately applied to the cell suspension by rotation of the cone at a shear rate of 1500/s. fMLP is added in this step to activate β_2-integrins (*see* **Note 2**).
4. Remove 15 µL of the samples from the viscometer at 10 and 40 s poststimulation during the cell-adhesion experiment, and also fix these in 100 µL of cold 2% glutaraldehyde.
5. The neutrophil population in fixed samples is identified using the flow cytometer, based on its characteristic forward- vs side-scatter profile. The number of single

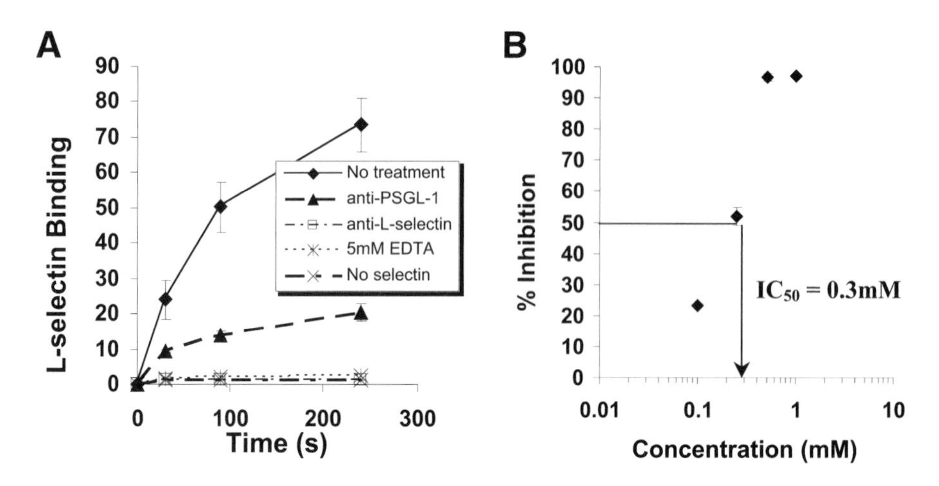

Fig. 3. L-selectin IgG chimera binding to its natural ligands on neutrophils. **(A)** Control experiments that confirm the specificity of the measurements. **(B)** Inhibition data where increasing dosage of sLe^x-OMe progressively blocks selectin binding to leukocytes. The IC_{50} measured was approx 0.3 m*M*. Here, isolated human neutrophils are employed in order to mimic the physiological valency of selectin ligands. Anti L-selectin (DREG-56) and PSGL-1 (KPL-1) antibodies were used at 10 μg/mL. Data are mean ± SEM.

leukocytes and aggregates of various sizes are resolved using the autofluorescence imparted by glutaraldehyde fixation. Samples are typically fixed for 1 h in glutaraldehyde prior to sample reading, as this allows uniform fixation of all cells. The autofluorescence of a neutrophil doublet is twice that of a singlet, a triplet has three times the fluorescence of a singlet, and so on. Note that fluorescence following glutaraldehyde fixation can be read on any channel of a conventional Becton Dickinson flow cytometer. Sample flow cytometry dot plots, histograms, and analysis strategies are presented elsewhere *(29)*.

6. The extent of homotypic adhesion (fraction aggregation) is determined by monitoring the depletion of single neutrophils using the following equation: fraction aggregation = $1 - S/(S + 2D + 3T + 4Q + 5P+)$, where S denotes the number of singlets, D denotes doublets, T denotes triplets, Q denotes quadruplets, and $P+$ denotes pentuplets and larger aggregates. Such data are presented in **Fig. 4A**. Here, neutrophil homotypic aggregation is abrogated by 5 m*M* of EDTA or anti-L-selectin antibody DREG-56.
7. The extent of each selectin inhibitor's ability to block homotypic adhesion is determined as shown in **Fig. 4B.** Statistics in these experiments is accurate owing to the large number of events analyzed in the aggregate sample (*see* **Note 3**).
8. While the above assay is specific for L-selectin function, alternate strategies to measure P- and E-selectin function under fluid shear are described in **Note 4**.

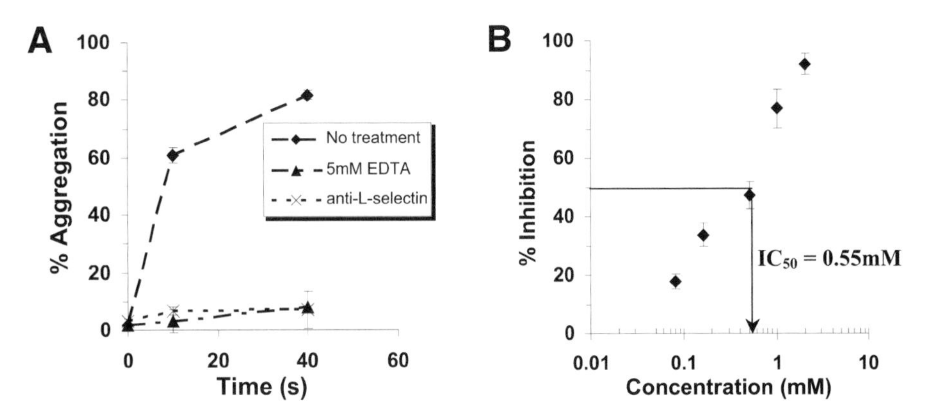

Fig. 4. L-selectin-mediated cell-adhesion assay under fluid shear. L-selectin-mediated homotypic neutrophil aggregation runs were performed under fluid flow either in the presence or absence of inhibitors. **(A)** Control runs confirm the specificity of interactions. Anti L-selectin (DREG-56) was used at 10 µg/mL. **(B)** Addition of sLe^x-OMe blocks neutrophil aggregation. Estimated IC_{50} is 0.55 m*M*. Data are mean ± SEM.

3.5. Surface Plasmon Resonance Assay

1. Immobilize a polyclonal goat anti-mouse Fc specific antibody onto research-grade CM5 sensor chips via amine coupling using the manufacturer's protocol.
2. Inject a threefold dilution of insect cell-culture supernatant containing P-selectin-IgG (*see* **Subheading 2.2.**) at 5 µL/min for 8 min over a single flow cell (the active flow cell).
3. Following noncovalent attachment of P-selectin-IgG to the active flow cell, the flowpath is changed to include a reference flow cell containing only immobilized goat anti-mouse antibody upstream of the active flow cell, and the flow rate is increased to 50 µL/min.
4. Perform four 1-min injections of 1 *M* of NaCl to remove nonspecifically-bound supernatant components from the sensor chip surface.
5. Allow running buffer to flow through the flow cells for 20 min before analyte introduction over the reference and active flow cells at a fixed concentration in running buffer for 1 min.
6. Following analyte wash, two 1-min 20-m*M* HCl injections are performed to regenerate the surface back to unbound goat anti-mouse Fc.
7. The cycle is then repeated, beginning with the loading of fresh P-selectin-IgG for a different dose of the analyte (or different inhibitor). A typical Biacore cycle described above is shown in **Fig. 5** for the sLe^X analog, TBC1269 (*see* **ref. *10*** for additional details). Similar runs can be performed with other small molecules and selectins.

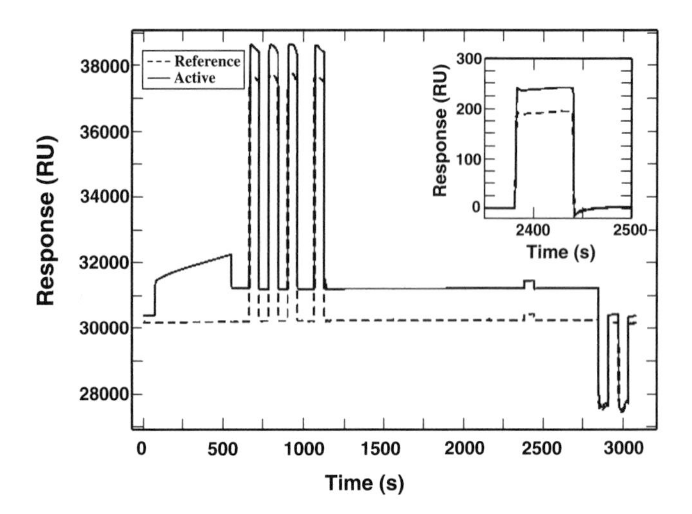

Fig. 5. Surface plasmon resonance measurement of selectin-ligand interaction. Representative experiment cycle where the signal from the active flow cell is depicted by a solid line, while reference flow cell data are in dashed form. Inset presents an expanded view of the net response in the active cell (solid line) and the reference flow cell (dashed line) to 0.14 m*M* of sLe^x analog (TBC1269) injection. (Reproduced with permission from **ref. *10***.)

4. Notes

1. The use of wash steps in selectin-binding assays must be carefully considered, since the time taken for wash steps is typically greater than that required for a single selectin-ligand bond to dissociate. For example, P-selectin PSGL-1 bonds are reported to have a large off-rate of approx 1.4/s and a rapid on-rate of $4.4 \times 10^6/M/s$ ***(30)***. Thus, while the measurement of rapid binding kinetics may be possible for Biacore-based measurements, ELISA (*see* **Subheading 3.2.**) and flow cytometry-based assays (*see* **Subheading 3.3.**) require that either the selectin or its ligand be multivalent in nature. Although not explicitly stated, multivalency is imparted in **Subheadings 3.2.** and **3.3.** by the secondary antibody used for cytometry/ELISA detection.

   The importance of multivalency is illustrated in **Fig. 6**. Here, P-selectin IgG fusion protein binding to human neutrophils is not detectable in the absence of secondary rabbit anti-mouse antibody (Sec. Ab). However, crosslinking the selectin fusion protein enhances selectin multivalency and allows detection.
2. Though functional PSGL-1 is present on unstimulated human neutrophils, formyl peptide (fMLP) is used to stimulate cells in **Subheading 3.4.** in order to activate the β2-integrin subunits on human neutrophils, LFA-1 (CD11a/CD18) and Mac-1 (CD11b/CD18). In these runs, activated integrins replace the aforementioned role of multivalency. They stabilize cell aggregates after formation in the viscometer,

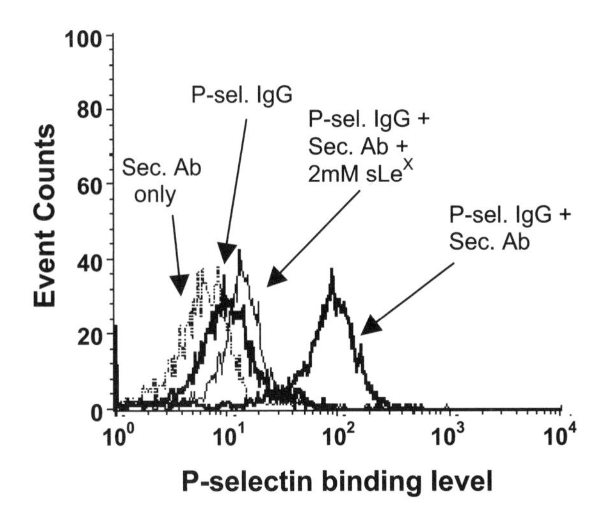

Fig. 6. Multivalency is necessary for flow cytometry detection. P-selectin IgG chimera (*see* **Subheading 2.2.**) was added to 0.5×10^6 neutrophils/mL either in the presence or absence of secondary rabbit anti-mouse $F(ab')_2$ Ab (abbreviated Sec. Ab) at room temperature for 15 min in HEPES buffer containing 1.5 m*M* of Ca^{2+}. Samples were then washed in HEPES buffer and probed with a goat anti-mouse Alexa488-conjugated polyclonal antibody at room temperature for 5 min to quantify P-selectin that was bound to neutrophils. Flow cytometry histogram shows that P-selectin IgG did not remain bound to neutrophils in the absence of Sec. Ab, confirming the need for selectin multivalency for flow cytometry detection. Negative-control runs were performed with Sec. Ab alone in the absence of P-sel. IgG. Multimeric P-selectin IgG binding to neutrophils was blocked by 2 m*M* of sLe^x analog (TBC1269), confirming the specificity of the interaction studied.

and thus enable detection after fixation using flow cytometry. Overall, the homotypic cell aggregation phenomenon is an ensemble effect of both selectins and integrins. In such assays, the counter receptor for L-selectin is primarily PSGL-1 while β2-integrin ligands are diverse, including ICAM-3 ***(31)***. The efficacy of anti-L-selectin antibodies in blocking neutrophil adhesion varies with shear rate, with these reagents being most effective above a wall shear rate of 400/s ***(32)***. Therefore it is best to evaluate selectin function at high shear rates, such as 1500/s used in **Fig. 4**. At shear rates below 400/s, both the contribution of β2-integrins and the shear threshold phenomenon become prominent ***(32)***.

3. The use of viscometer and flow cytometry for measuring cell adhesion under fluid shear has several distinct advantages. First, the volume of the sample or inhibitor used is small (60–100 μL/run) and the runs are rapidly performed. Second, the statistics and reproducibility of such experiments are excellent, because thousands of cell–cell collisions take place within seconds in the viscometer and the flow

cytometry method used for adhesion quantitation typically evaluates several hundred aggregate events in each sample.

4. The experimental system discussed in **Fig. 4** measures L-selectin-mediated homotypic neutrophil adhesion. Note that similar assays can also be developed to study P- and E-selectin-dependent cell adhesion in suspension. Studies of P-selectin-dependent function under shear can be evaluated by measuring the binding of neutrophils that bear PSGL-1 to activated platelets that bear P-selectin *(33)*. E-selectin-binding to neutrophil ligands can also be measured by studying the collision of E-selectin-bearing cells with leukocytes *(34)*.

Acknowledgments

We thank Dr. Peter Vanderslice and Encysive Pharmaceuticals for providing some of the reagents used in the study, Kevin Lindquist for performing the Biacore measurements, and grants from the National Institutes of Health (CA35329, HL63014, HL76211) for financial support.

References

1. Fuster, M. M. and Esko, J. D. (2005) The sweet and sour of cancer: Glycans as novel therapeutic targets. *Nat. Rev. Cancer* **5,** 526–541.
2. Nangia-Makker, P., Conklin, J., Hogan, V., and Raz, A. (2002) Carbohydrate-binding proteins in cancer, and their ligands as therapeutic agents. *Trends Mol. Med.* **8,** 187–192.
3. Yang, J., Hoffmeister, D., Liu, L., Fu, X., and Thorson, J. S. (2004) Natural product glycorandomization. *Bioorg. Med. Chem.* **12,** 1577–1584.
4. Kren, V. and Martinkova, L. (2001) Glycosides in medicine: "The role of glycosidic residue in biological activity." *Curr. Med. Chem.* **8,** 1303–1328.
5. Sato, M., Furuike, T., Sadamoto, R., et al. (2004) Glycoinsulins: dendritic sialyloligosaccharide-displaying insulins showing a prolonged blood-sugar-lowering activity. *J. Am. Chem. Soc.* **126**, 14,013–14,022.
6. Li, Q., Su, B., Li, H., et al. (2003) Synthesis and potential antimetastatic activity of monovalent and divalent beta-D-galactopyranosyl-(1→4)-2-acetamido-2-deoxy-D-glucopyranosides. *Carbohydr. Res.* **338,** 207–217.
7. Chen, Y., Janczuk, A., Chen, X., Wang, J., Ksebati, M., and Wang, P. G. (2002) Expeditious syntheses of two carbohydrate-linked cisplatin analogs. *Carbohydr. Res.* **337,** 1043–1046.
8. Ohnishi, M., Imanishi, N., and Tojo, S. J. (1999) Protective effect of anti-P-selectin monoclonal antibody in lipopolysaccharide-induced lung hemorrhage. *Inflammation* **23,** 461–469.
9. Buerke, M., Weyrich, A. S., Zheng, Z., Gaeta, F. C., Forrest, M. J., and Lefer, A. M. (1994) Sialyl Lewisx-containing oligosaccharide attenuates myocardial reperfusion injury in cats. *J. Clin. Invest.* **93,** 1140–1148.
10. Beauharnois, M. E., Lindquist, K. C., et al. (2005) Affinity and Kinetics of Sialyl Lewis-X and Core-2 Based Oligosaccharides Binding to L- and P-Selectin. *Biochemistry* **44,** 9507–9519.

11. Somers, W. S., Tang, J., Shaw, G. D., and Camphausen, R. T. (2000) Insights into the molecular basis of leukocyte tethering and rolling revealed by structures of P- and E-selectin bound to SLe(X) and PSGL-1. *Cell* **103,** 467–479.
12. Koenig, A., Norgard-Sumnicht, K., Linhardt, R., and Varki, A. (1998) Differential interactions of heparin and heparan sulfate glycosaminoglycans with the selectins. Implications for the use of unfractionated and low molecular weight heparins as therapeutic agents. *J. Clin. Invest.* **101,** 877–889.
13. Woynarowska, B., Dimitroff, C. J., Sharma, M., Matta, K. L., and Bernacki, R. J. (1996) Inhibition of human HT-29 colon carcinoma cell adhesion by a 4-fluoro-glucosamine analogue. *Glycoconj. J.* **13,** 663–674.
14. Yarema, K. J., Goon, S., and Bertozzi, C. R. (2001) Metabolic selection of glycosylation defects in human cells. *Nat. Biotechnol.* **19,** 553–558.
15. Sharma, M., Bernacki, R. J., Paul, B., and Korytnyk, W. (1990) Fluorinated carbohydrates as potential plasma membrane modifiers. Synthesis of 4- and 6-fluoro derivatives of 2-acetamido-2-deoxy-D-hexopyranoses. *Carbohydr. Res.* **198,** 205–221.
16. Paul, B., Bernacki, R. J., and Korytnyk, W. (1980) Synthesis and biological activity of some 1-N-substituted 2-acetamido-2-deoxy-beta-D-glycopyranosylamine derivatives and related analogs. *Carbohydr. Res.* **80,** 99–115.
17. Thomas, R. L., Abbas, S. A., and Matta, K. L. (1988) Synthesis of uridine 5′-(2-acetamido-2,4-dideoxy-4-fluoro-alpha-D-galactopyranosyl) diphosphate and uridine 5′-(2-acetamido-2,6-dideoxy-6-fluoro-alpha-D-glucopyranosyl) diphosphate. *Carbohydr. Res.* **184,** 77–85.
18. Berkin, A., Szarek, W. A., and Kisilevsky, R. (2000) Synthesis of 4-deoxy-4-fluoro analogues of 2-acetamido-2-deoxy-D-glucose and 2-acetamido-2-deoxy-D-galactose and their effects on cellular glycosaminoglycan biosynthesis. *Carbohydr. Res.* **326,** 250–263.
19. Goon, S. and Bertozzi, C. R. (2002) Metabolic substrate engineering as a tool for glycobiology. *J. Carbohydr. Chem.* **21,** 943–977.
20. Woynarowska, B., Skrincosky, D. M., Haag, A., Sharma, M., Matta, K., and Bernacki, R. J. (1994) Inhibition of lectin-mediated ovarian tumor cell adhesion by sugar analogs. *J. Biol. Chem.* **269,** 22,797–22,803.
21. Dimitroff, C. J., Kupper, T. S., and Sackstein, R. (2003) Prevention of leukocyte migration to inflamed skin with a novel fluorosugar modifier of cutaneous lymphocyte-associated antigen. *J. Clin. Invest.* **112,** 1008–1018.
22. Sarkar, A. K., Fritz, T. A., Taylor, W. H., and Esko, J. D. (1995) Disaccharide uptake and priming in animal cells: Inhibition of sialyl Lewisx by acetylated Galb1®4GlcNAcb-O-naphthalenemethanol. *Proc. Natl. Acad. Sci. USA* **92,** 3323–3327.
23. Sarkar, A. K., Rostand, K. S., Jain, R. K., Matta, K. L., and Esko, J. D. (1997) Fucosylation of disaccharide precursors of sialyl Lewisx inhibit selectin-mediated cell adhesion. *J. Biol. Chem.* **272,** 25,608–25,616.
24. Neville, D. C., Field, R. A., and Ferguson, M. A. (1995) Hydrophobic glycosides of N-acetylglucosamine can act as primers for polylactosamine synthesis and can affect glycolipid synthesis in vivo. *Biochem. J.* **307(Pt. 3),** 791–797.

25. Okamura, A., Yazawa, S., Nishimura, T., et al. (2000) A new method for assaying adhesion of cancer cells to the greater omentum and its application for evaluating anti-adhesion activities of chemically synthesized oligosaccharides. *Clin. Exp. Metastasis* **18,** 37–43.
26. Kuan, S. F., Byrd, J. C., Basbaum, C. and Kim, Y. S. (1989) Inhibition of mucin glycosylation by aryl-N-acetyl-alpha-galactosaminides in human colon cancer cells. *J. Biol. Chem.* **264,** 19,271–19,277.
27. Neelamegham, S. (2004) Transport features, reaction kinetics and receptor biomechanics controlling selectin and integrin mediated cell adhesion. *Cell Commun. Adhes.* **11,** 35–50.
28. Galustian, C., Childs, R. A., Yuen, C. T., et al. (1997) Valency dependent patterns of binding of human L-selectin toward sialyl and sulfated oligosaccharides of Le(a) and Le(x) types: relevance to anti-adhesion therapeutics. *Biochemistry* **36,** 5260–5266.
29. Neelamegham, S. and Matta, K. L. (2002) Liposomes containing ligands. Binding specificity to selectins. *Methods Mol. Biol.* **199,** 175–191.
30. Mehta, P., Cummings, R. D., and Mcever, R. P. (1998) Affinity and kinetic analysis of P-selectin binding to P-selectin glycoprotein ligand-1. *J. Biol. Chem.* **273,** 32,506–32,513.
31. Neelamegham, S., Taylor, A. D., Shankaran, H., Smith, C. W., and Simon, S. I. (2000) Shear and time dependent changes in Mac-1, LFA-1, and ICAM-3 binding regulate neutrophil homotypic adhesion. *J. Immunol.* **164,** 3798–3805.
32. Taylor, A. D., Neelamegham, S., Hellums, J. D., Smith, C. W., and Simon, S. I. (1996) Molecular dynamics of the transition from L-selectin- to beta 2-integrin-dependent neutrophil adhesion under defined hydrodynamic shear. *Biophys. J.* **71,** 3488–3500.
33. Xiao, Z., Goldsmith, H. L., McIntosh, F. A., Shankavan, M., and Neelamegham, S. (2006) Biomechanics of P-selectin PSUL-1 bonds: shear threshold and integrin independent cell adhesion. *Biophys. J.* **90,** 2221–2234.
34. McDonough, D. B., Mcintosh, F. A., Spanos, C., Neelamegham, S., Goldsmith, H. L., and Simon, S. I. (2004) Cooperativity between selectins and beta2-integrins define neutrophil capture and stable adhesion in shear flow. *Ann. Biomed. Eng.* **32,** 1179–1192.

24

Binding and Inhibition Assays for Siglecs

Nadine Bock and Sørge Kelm

Summary

Protein–carbohydrate interactions are believed to be important in many biological processes involving cellular communication. Although many of the experimental approaches for studying protein–carbohydrate interactions are similar to those commonly employed in protein–protein interactions, several important aspects have to be taken into account in binding assays with lectin-like receptors such as Siglecs, sialic acid-recognizing immunoglobulin-like lectins. This chapter describes experimental approaches using solid-phase assays with sialylated targets and complexed Siglec Fc-chimeras for binding studies or competitive inhibition assays. Such assays are useful to investigate various aspects of Siglec interactions with their carbohydrate-binding partners, such as the specificity and the molecular basis of sialic acid recognition.

Key Words: Lectins; sialic acids; Siglecs; solid-phase binding assay; target glycoproteins; inhibitor.

1. Introduction

Protein–carbohydrate interactions play diverse roles in recognition phenomena, e.g., between mammalian oligosaccharides and pathogenic agents such as viruses, bacteria, and toxins. Carbohydrate-binding proteins (lectins) are widespread among plants, prokaryotes, and eukaryotes ***(1–3)***. They mediate specific biological functions such as protein trafficking and cell–cell interactions, e.g., in the immune system. Lectins recognize specific structural features of one or more sugars in cell-surface glycans, and in most cases they bind to a range of structurally related glycans. Therefore, different glycoproteins (GPs) or glycolipids can serve as binding partners (ligands, counter receptors) for the same lectin as long as they carry appropriate carbohydrate structures (ligand determinants). The biosynthesis of glycan structures involves several enzymes.

From: *Methods in Molecular Biology, vol. 347: Glycobiology Protocols*
Edited by: I. Brockhausen © Humana Press Inc., Totowa, NJ

Furthermore, some of the glycosyltransferases are competing for the same acceptor substrates. Consequently, the analysis of gene expression does not provide conclusive information about the presence of lectin-binding partners.

In principle, the methods suitable for the analysis of protein–carbohydrate interactions depend on the type of question addressed. In most cases, these methods fall into three categories: determination of the presence of lectin binding partners, the characterization of the lectin binding specificities for defined glycan or glycoconjugate (GC) structures, and the analysis of the molecular interaction between protein and carbohydrate in the binding site. To determine the presence of lectin binding sites on cells, in most cases histochemical methods (lectin histochemistry) and Western blots (lectin blots) provide useful qualitative and partially quantitative information on the type of cells or GC binding the lectin. Enzyme-linked immunosorbent assay-type methods such as those described in this chapter can be used for more quantitative questions, but do not allow discriminating the molecules in a mixture that is binding. To characterize the glycan/GC specificity of lectins and to analyze the molecular interaction as well as screening for inhibitory molecules, assays based on the immobilized carbohydrates or lectins as described in this chapter have been very useful.

Sialic acids (Sia) are a family of acidic monosaccharides based on the structure of neuraminic acid. They are commonly found in all animals of the deuterostome lineage, some proteostomes, and microorganisms ***(4–7)***. A variety of modifications led to a total of more than 40 Sia found in nature. Together with the different types of glycosidic linkages between Sia and the underlying glycans, Sias are major contributors to the variability of GCs. Because they are found mainly at the terminal positions of glycans, they often play key roles in lectin-mediated processes ***(4–7)***. The Sia-binding immunoglobulin-like lectins (Siglecs; *see* **refs. *8–10***) have drawn increasing interest, because they all bind specifically to sialylated GCs with different preferences for the types of Sia and the underlying glycan structures ***(11)***. These binding preferences are likely related to their biological functions, such as cell–cell interactions and signaling functions in the haemopoietic, immune, and nervous systems ***(12,13)***. Examples are the inhibitory effect of Siglecs on the activation of B-cell ***(14)*** and T-cell ***(15)*** receptors. However, little is known about the mechanism of how Sia recognition is linked to cellular signalling events.

To investigate the biological functions of Siglecs, it is necessary to characterize the interaction between these lectins and their ligands. Several binding assays have been used for this purpose. The target glycans to be investigated for Siglec binding can be presented on cells or artificial particles, or on surfaces of microtiter plates. In both cases, the levels of Siglecs bound are determined using either direct or indirect labeling techniques. In principle two setups are possible, direct binding assays or competition assays.

Several direct binding assays have been developed to investigate the specificities of Siglecs. Examples are enzymatically modified cell surfaces ***(16)***, immobilized glycolipids ***(17)***, and glycoarrays based on carbohydrate structures covalently linked to microtiter plates ***(18)***. These assays have the advantage of allowing a direct comparison of the carbohydrate specificities of several Siglecs, in particular the glycoarray provided by the Glycomics consortium (http://www.functionalglycomics.org/static/consortium/), which is a very useful platform for testing the specificities of Siglecs and other lectins.

Alternatively, a suitable target GC molecule can be immobilized to determine Siglec binding, and the compounds to be tested are used as competitive inhibitors ***(14,19,20)***. The concentrations required for 50% inhibition (IC_{50}) are determined from titration experiments. In this case, only one target structure has to be immobilized and different types of competing compounds can be compared without any further derivatization or processing. Because the titration can be done at almost any concentration level, low affinity binders can be tested on the same platform as compounds with very high affinities. The contributions of each functional group in the Sia residue and the influence of Sia modifications can be investigated using synthetic Sia derivatives in such competition assays. This has been useful to elucidate the molecular mechanisms mediating the binding specificities of different Siglecs as well as for the development of specific Siglec inhibitors. Furthermore, competition assays can be used to investigate the presence of soluble Siglec binding partners in samples from biological sources. This chapter focuses on the methodology of such hapten inhibition assays.

For routine screenings, the assays generally have been adapted to microtiter plate formats. The first competition assays with small-molecular-weight Sia derivatives have applied sialylated target cells ***(19,21)***. More recently, assays based on derivatized artificial surfaces have been developed and are described in this chapter. The primary advantages are the replacement of radioactive labels, better reproducibility, and more general applicability. Because of the different glycan specificities of Siglecs, suitable target surfaces can be developed using natural or synthetic GCs, whereas suitable target cells are not available for each Siglec. In addition, the adaptation to the 384-well format reduces the amounts of reagents and samples needed. A comparison of several targets used to investigate the binding specificities of Siglecs is given in **Table 1**.

Relatively low affinities at single binding sites are characteristic of most protein–carbohydrate interactions. Multivalent interactions generally provide stable, high-avidity binding between lectins and their ligands. This "cluster effect" has been described in detail for the hepatic asialoGP (ASGP) receptor ***(22)***. To provide a multivalent presentation of Siglec binding sites, Fc-chimeras have commonly been used. Such "immunoadhesins" are fusion proteins

Table 1
Compatibility of Siglec-Fc-Chimeras With Different Binding Assay Protocols

Target for Siglec binding	Surface-bound targets with enzyme-labeled anti-Fc antibodies and fluorogenic substrate FDP			Cell suspensions as targets with radioiodinated anti-Fc antibodies	
	Bovine fetuin	Bovine IgM	GlycoWell (Neu5Ac)	Human RBCs	Murine AG8 cells
Siglec-1 (sialoadhesin)	No	No	No	Yes	No
Siglec-2 (CD22)					
human (Neu5Gc/Ac)	Yes	Yes	No	Yes	Yes
murine (Neu5Gc)	No	Yes	No	No	Yes
Siglec-3 (CD33)	n.t.	n.t.	n.t.	n.t.	n.t.
Siglec-4 (MAG)	Yes	No	Yes	Yes	No
Siglec-5	n.t.	n.t.	n.t.	Yes	No
Siglec-6	n.t.	n.t.	n.t.	n.t.	n.t.
Siglec-7	No	No	Yes	No	No
Siglec-8	n.t.	n.t.	n.t.	n.t.	n.t.
Siglec-9	No	No	Yes	No	No
Siglec-10	n.t.	n.t.	n.t.	n.t.	n.t.
Siglec-11	n.t.	n.t.	n.t.	n.t.	n.t.

n.t., combination not tested.

combining the lectin-binding region of the adhesion molecule with the Fc part of an immunoglobulin. Advantages are the relatively easy production of soluble proteins from tissue culture supernatant and the possibility of simply obtaining multivalent complexes with anti-Fc antibodies ***(21)***.

For the quantification of bound Siglecs, several suitably labeled anti-Fc antibodies are commercially available. These could either be antibodies providing a direct fluorescent label or enzymes such as alkaline phosphatase (AP). Fluorescein diphosphate (FDP) is probably the most sensitive substrate available for AP providing high sensitivity. If the initial enzymatic velocity is determined from measuring the kinetics of fluorescein production, this assay provides reliable data over at least two orders of magnitude in AP activity and is only limited by the level of nonspecific binding (NSB) of the Fc-chimera antibody complexes to the microtiter plates.

This chapter provides several protocols for the following procedures: desialylation of GPs, truncation of the glycerol side chain of Sia, and acetone precipitation of GPs, binding, and inhibition assays. The aim of this chapter is to provide an assay template (*see* **Fig. 1**) which can easily be adapted to many different purposes in any biochemical laboratory. Obviously, in this type of assay, different tags and target molecules can be used for the detection of Sia-dependent binding by Siglecs as well as other proteins. Examples for experimental data from such assays are presented in **Figs. 2–4**, demonstrating modifications and adaptations of sialylated target surfaces for different Siglec-Fc chimeras and the use of competition experiments.

2. Materials

2.1. Assay Buffers

1. Immobilization buffer: 50 m*M* of $NaHCO_3$, pH 9.5. Best immobilization is achieved at this alkaline pH (*see* **Note 1**).
2. *N*-2-hydroxyethylpiperazine-*N′*-ethanesulfonic acid (HEPES)-buffered saline (HBS): 10 m*M* of HEPES (pH 7.4) and 150 m*M* of NaCl. For convenience, a tenfold concentrated stock solution (10X HBS) is prepared, autoclaved, and diluted when needed (*see* **Note 2**).
3. HBS with 0.05% Tween-20 (HBS-T). This buffer is kept at 4°C because all washing steps are performed on ice to reduce microbial growth.
4. AP buffer: 50 m*M* of Tris-HCl (pH 8.5) and 10 m*M* of $MgCl_2$. This buffer must be set up fresh every day. For convenience, prepare this buffer from 1 *M* of Tris-HCl (pH 8.5) and 1 *M* of $MgCl_2$ stock solutions.

2.2. Substrate and Inhibitor Solutions

1. FDP stock solution: Dissolve FDP (MoBiTec, Goettingen, Germany) to a 10-m*M* stock solution with 10m*M* of Tris-HCl, pH 7.5. Store in aliquots at –20°C. For smaller assays, 1-m*M* aliquots might be useful. Protect from direct light (*see* **Note 3**).

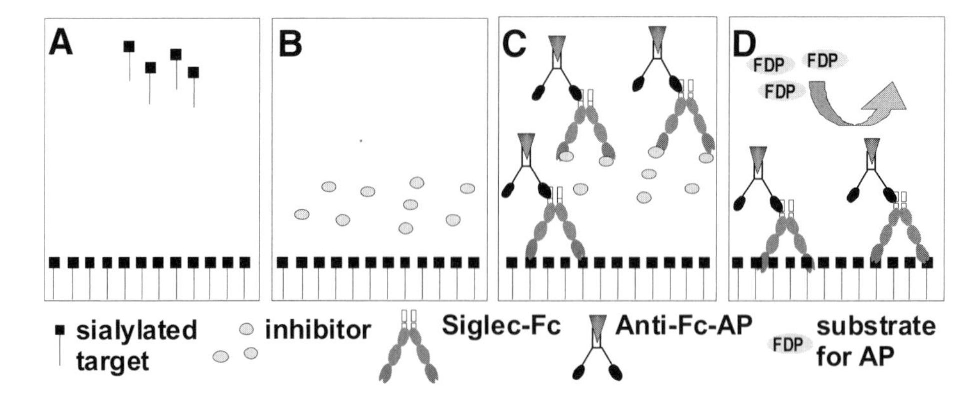

Fig. 1. Hapten inhibition assay with Fc-chimeras and immobilized target sialosides. The experimental steps of the hapten inhibition assay with Fc-chimeras are described in the following methods sections: **(A)** Immobilization of sialylated glyconjugates (*see* **Subheadings 3.4.2.** and **3.4.3.**); **(B)** addition of inhibitor solution (*see* **Subheading 3.4.4.**); **(C)** addition of Fc-chimeras complexed with anti-Fc-antibodies (*see* **Subheadings 3.4.2.–3.4.4.**); **(D)** quantification of bound Fc-chimeras (*see* **Subheadings 3.4.5.** and **3.4.6.**).

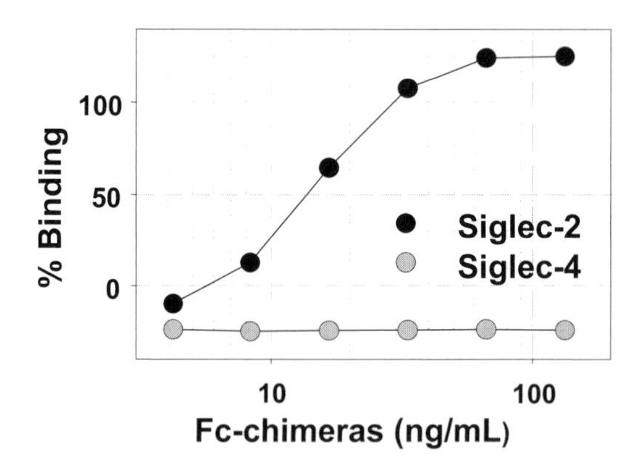

Fig. 2. Specificity of siglecs binding to immobilized bovine IgM. Bovine IgM was immobilized in wells of a 384-well plate and Fc-chimeras of human Siglec-2 (specific for Neu5Ac/Gcα2,6Gal/GalNAc) or murine Siglec-4 (specific for Neu5Acα2,3Gal) at the concentrations indicated were complexed with anti-huIgG-AP (1 : 1000 final concentration) and added to the wells as described in **Subheading 3.4.3.** Binding to control wells without glycoprotein addition was subtracted as nonspecific binding. Whereas human Siglec-2 Fc-chimeras show saturable binding, no binding of murine Siglec-4 Fc-chimeras to IgM is detectable. This is in agreement with the known specificity of these siglecs and the glycan structures found on bovine IgM, i.e., mainly *N*-glycans terminating in Neu5Gcα2,6Gal or Neu5Acα2,6Gal and the absence of significant amounts of α2,3-linked Neu5Ac.

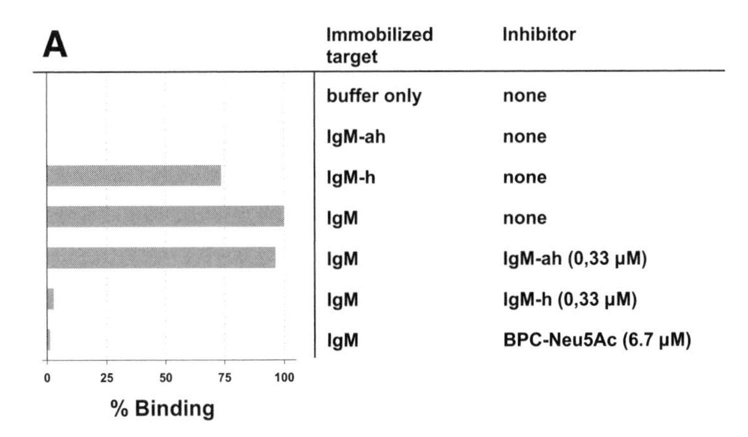

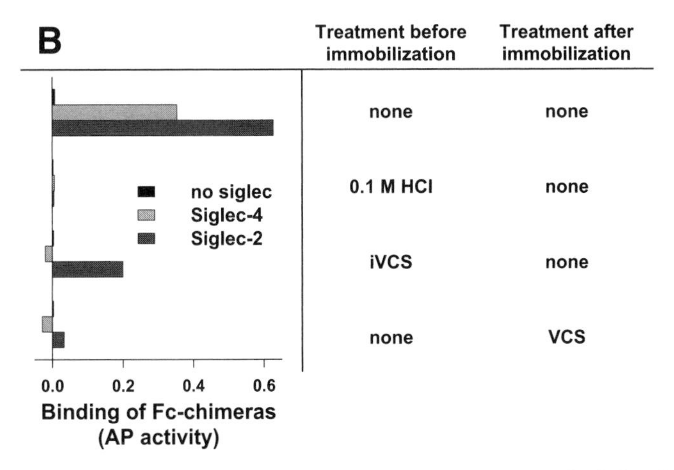

Fig. 3. Testing the sialic acid (Sia)-dependent binding of Siglec-2 and Siglec-4 Fc-chimeras. (**A**) Bovine IgM (IgM), heat-treated IgM (IgM-h; 80°C for 1 h in H_2O) or acid-treated IgM (IgM-ah; 0.1 *M* of HCl at 80°C for 1 h) were immobilized to wells of a 384-well plate as indicated in the left column. Complexed Siglec-2 Fc-chimeras were added in the presence of buffer, IgM-h, IgM-ah, or methyl-α-9-(biphenyl-4-carbonyl)-amino-9-deoxy-Neu5Ac (BPC-Neu5Ac), a potent Sia derivative inhibitor ***(14)***, at the concentrations indicated in the right column. Binding to control wells without glycoprotein (GP) addition was subtracted as NSB, and binding to IgM was taken as 100% binding. (**B**) 100 µg of fetuin was treated for 3 h with iVCS (*see* **Subheading 3.1.2.**) or hydrolyzed with 0.1 *M* of HCl at 80°C for 1 h to remove Sia. Desialylated or untreated fetuin preparations were immobilized in 384-well plates, followed by 3% bovine serum albumin. In some wells immobilized fetuin was treated with 10 mU of VCS at 37°C for 1 h. Anti-huIgG-alkaline phosphatase with or without siglec Fc-chimeras was added and binding determined (*see* **Subheading 3.4.3.**). Binding to control wells without GP addition was subtracted as NSB. The lower cleavage rate for α2,6-linked Sia by VCS is the main reason for the remaining binding of Siglec-2 Fc-chimeras to VCS-treated IgM.

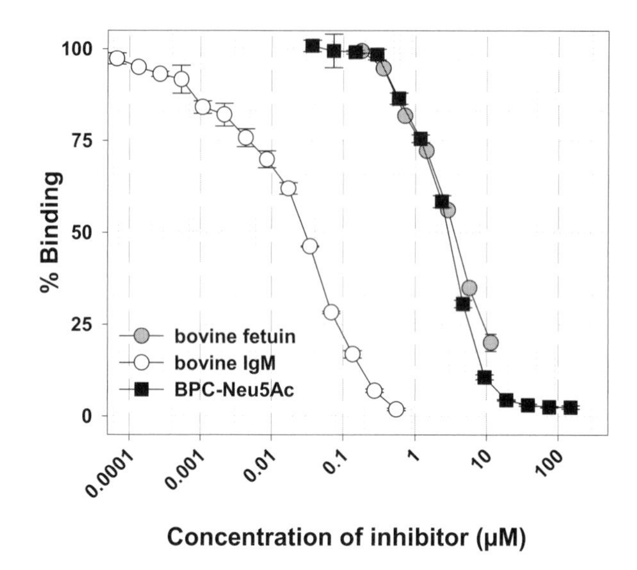

Fig. 4. Competitive inhibition of Siglec-2 binding. IgM was immobilized in a 384-well plate. The binding of complexed Siglec-2 Fc-chimeras was determined in the presence of IgM, fetuin, or methyl-α-9-(biphenyl-4-carbonyl)-amino-9-deoxy-Neu5Ac (BPC-Neu5Ac), a potent sialic acid derivative inhibitor ***(14)***, at the concentrations indicated (*see* **Note 17**). Binding to control wells without glycoprotein addition was subtracted as nonspecific binding, and binding to IgM taken as 100% binding. From these titration experiments the concentrations required for 50% inhibition (IC_{50}) are determined either manually or mathematically.

2. Substrate solution: 20 µ*M* of FDP in AP buffer. This solution should be prepared just before it is needed. Bring AP buffer to the temperature of the assay (usually 37°C), then add the amount of FDP stock solution required just before use.
3. Small-molecular-weight inhibitor stock solutions: Sia derivatives are set up as 10- to 30-m*M* stock solutions in HBS. Because Sia derivatives are acidic compounds, care must be taken to adjust the pH. Therefore, GPs, oligosaccharides, or monosaccharides are dissolved in a small amount of cold H_2O, the pH is checked with 0.5-µL aliquots on pH indicator strips, and adjusted to pH 7.0–7.5 if necessary. Then 10X HBS is added to stabilize the pH, and the volume adjusted with H_2O. Dilutions are prepared from this stock solution as needed for the assays. Stocks and dilutions are kept at –20°C and are stable for years (*see* **Notes 2** and **3**).
4. GPs for immobilization: Fetuin or IgM are commercially available serum proteins, their glycans are well-characterized ***(23–25)***, and they have been used in several lectin-binding studies ***(26–28)***. These solutions should be free of any contaminant proteins, especially other GCs or detergents. Prepare 0.4 mg/mL stock solutions of bovine IgM, bovine fetuin, bovine asialofetuin, or other suitable GP

dissolved in 10 m*M* of HEPES. Store aliquots at –20°C and they will remain stable for years. Immediately prior to use, dilute GP in immobilization buffer to 4 μg/mL final concentration (*see* **Notes 2–4**).

5. Bovine serum albumin (BSA): BSA is commonly used to block NSB sites. However, it is necessary to use a BSA preparation free of GPs. Therefore, check each batch for NSB in your assay system. A stock solution of 10% in H_2O can be kept at –20°C for years (*see* **Note 3**).

2.3. GP Modification

1. 1 U/mL of *Vibrio cholerae* sialidase (VCS; Dade Behring Diagnostics GmbH, Marburg, Germany).
2. 1 U/mL of GlycoCleave® neuraminidase (immobilized *Vibrio cholerae* sialidase [iVCS]; Galab Technologies, Geesthacht, Germany).
3. VCS-buffer: 50 m*M* of acetate (pH 5.5) and 9 m*M* of $CaCl_2$. This buffer is autoclaved and kept at 4°C to reduce microbial growth.
4. Periodate (PI) solution: 100 m*M* of sodium metaperiodate ($NaIO_4$) in PBS. This solution must be kept in the dark and should be prepared fresh.
5. Glycerol stock solutions: A stock solution of 80% glycerol in water is easier to pipet and can be kept at room temperature for a month. Dilute to 8% final concentration with H_2O. This solution should either be prepared fresh or autoclaved and kept under sterile conditions.
6. Sodium borohydrate ($NaBH_4$): Dissolve to a concentration of 20 m*M* in H_2O immediately before use.

2.4. Materials and Reagents for Binding Assays

1. Fc-chimeras of Siglecs: Plasmids coding the Siglec-Fc-chimeras containing the N-terminal three domains of the corresponding Siglec, like Siglec-4_{d1-3}-Fc, and the production of the recombinant proteins are described in **ref. *21***. Fc-chimeras are produced by transient expression of the plasmids in COS cells or by stable transfection in Chinese hamster ovary cells deficient in sialylation. The proteins are purified from tissue culture supernatant by immunoaffinity chromatography on protein A-sepharose. The Fc-chimeras are passed through a filter and stored in 10 m*M* of HEPES (pH 7.4) at 4°C. In most cases the purified proteins are stable for several months under these conditions if kept sterile (*see* **Note 2**).
2. Alkaline phosphatase-conjugated anti-Fc-antibody (anti-huIgG-AP): Available from several sources; mix 1:1 with glycerol and store in aliquots at –20°C. Each lot of antibody conjugate should be checked for suitability in these assays, as some polyclonal anti-Fc antibody preparations interfere with the binding activity of Siglec Fc-chimeras.
3. GlycoWell™ plates are available from Lundonia Biotech, Lund, Sweden. The plates provide a derivatized surface with immobilized carbohydrate structures or with immobilized *N*-acetyl-group (blank) as a control for NSB. For binding of Siglecs, plates containing either Neu5Ac or Neu5Ac2,3Lac can be used. The plates come in the 96-well format as strips with 2 × 8 wells, with a maximum

volume of 200 μL/well. Black plates with U-shaped wells are used for fluorescent assays.

4. High-binding-capacity microtiter plates: Black plates with high binding capacity are used for the immobilization of GPs. Plates with U-shaped wells are preferred. Such plates are available from different suppliers. Each lot should be tested for reliable high binding of the GCs used.

3. Methods

3.1. Desialylation of GPs

Desialylated GPs are usually the best control to determine the Sia-independent binding. Sia of the target GP can be removed enzymatically by VCS or chemically by mild acidic hydrolysis.

3.1.1. Enzymatic Desialylation of Immobilized GPs

1. Block NSB sites with 20 μL of 3% BSA in HBS-T at room temperature for 1 h.
2. Remove the BSA solution.
3. Add 20 μL of VCS diluted to 500 mU/mL in VCS-buffer (*see* **Notes 5** and **6**).
4. Incubate at 37°C for 1 h.
5. Wash the wells five times with HBS-T (*see* **Note 7**).

3.1.2. Desialylation of GP Solutions With Immobilized Sialidase

Immobilized sialidase is a convenient tool for the enzymatic desialylation of GPs, since the enzyme can easily be removed from the incubation mixture by centrifugation or in a small column.

1. Treat 1 mg of GP with 200 μL (packed beads) of iVCS in 2 mL of VCS-buffer in a batch procedure by constant stirring at 37°C for 4 h (*see* **Note 5**).
2. Separate product and iVCS by short centrifugation (30 s at 200*g*).
3. Precipitate the ASGP with acetone if necessary.
4. Dissolve the protein in a suitable buffer.

3.1.3. Mild Acid Hydrolysis of GPs

Removal of Sia by sialidase treatment is not always sufficient to completely prevent Sia-dependent binding. Mild acid hydrolysis (as described in **steps 1–5**) selectively hydrolyzes the glycosidic linkages of Sia. However, it also leads to denaturation of the GP and therefore can also influence Sia-independent interactions.

1. Add HCl (final concentration of 100 m*M*) to the GP solution.
2. Incubate at 80°C for 1 h.
3. Neutralize by adding an equal volume of 100 m*M* of NaOH.
4. Precipitate the ASGP with acetone.
5. Dissolve the protein in a suitable buffer.

3.2. Truncation of Sia Glycerol Side Chains

The binding site of Siglecs interacts with the glycerol side chain of Sia. Therefore, oxidative cleavage of the glycerol side chain possibly provides the most stringent controls for Sia-dependent binding of Siglecs. This process can be done in the microtiter plate well after immobilization or with a GP solution. First, the Sia side chain is oxidized to the C7-aldehyde with PI ($NaIO_4$). The reaction is then stopped with glycerol to quench the PI, and the aldehydes are reduced to the corresponding alcohol with $NaBH_4$.

3.2.1. PI Treatment of Immobilized Sia

1. Incubate the immobilized sialosides with 2 m*M* of $NaIO_4$ in PBS (150 µL/well) at room temperature for 30 min (*see* **Note 8**).
2. Add 20 µL of 8% glycerol.
3. Wash the wells once with 200 µL of PBS.
4. Add 150 µL of 20 m*M* of $NaBH_4$ and incubate at room temperature for 30 min.
5. Wash the wells five times with HBS-T (200 µL each wash) before use.

3.2.2. PI Treatment of GPs in Solution

1. Incubate 100 µg of GP in 200 µL of PBS with 2 m*M* of $NaIO_4$ at room temperature for 30 min in the dark.
2. Add 50 µL of 8% glycerol.
3. Add 100 µL of 20 m*M* of $NaBH_4$ and incubate at room temperature for an additional 30 min in the dark.
4. Adjust the pH to 7.4 with 10% acetic acid.
5. Precipitate the GP with acetone.
6. Dissolve the GP in a suitable buffer.

3.3. Acetone Precipitation of GPs

GPs can be precipitated with acetone to remove cleaved Sia or other reagents and exchange the buffer. However, it is necessary to check whether acetone precipitation has any effect on Sia-independent interactions of the GPs.

1. Add the fourfold volume of ice-cold acetone to the sample and incubate at –20°C for 16 h.
2. Collect the precipitates by centrifugation at $20 \times 10^3 g$ at 0°C for 20 min.
3. Dissolve the sample in an appropriate buffer.
4. Determine the protein content using any standard protein assay.

3.4. Binding and Inhibition Assays

3.4.1. Principles of Binding and Inhibition Assays

The procedures for the binding assays follow the same principle steps of preparing appropriate target surfaces, the binding reaction, and the quantification of binding (*see* **Fig. 1**). The single steps of the protocols depend mainly on the types of target structures for Siglec binding, the presence of potential inhibitors, and the reaction volumes.

The appropriate target surface must be established as a first step in establishing an assay. **Table 1** provides a summary of such targets tested with Siglec Fc-chimeras. Detailed descriptions are provided for two simple binding assays using immobilized Sia in GlycoWell plates (*see* **Subheading 3.4.2.**) or immobilized GPs (*see* **Subheading 3.4.3.**). An example of the relevance of the target structure for binding is shown in **Fig. 2**, demonstrating that human Siglec-2 (CD22) binds well to immobilized bovine IgM whereas murine Siglec-4 does not bind to this GP above background levels. In this context, suitable controls for nonspecific (Sia-independent) binding have to be analyzed. **Figure 3** shows data from such an experiment for Siglec-2 and Siglec-4 using different desialylated GPs as targets or inhibitors to suppress Sia-dependent binding. Further procedures for the detection and quantification of sialoside inhibitor assays are described in separate sections (*see* **Notes 7** and **9–11**).

3.4.2. Binding Assays in GlycoWell Plates

1. Incubate the number of wells needed with 200 μL of HBS-T for 5 min prior use.
2. Use GlycoWell plates with immobilized *N*-acetyl-group as controls for NSB (*see* **Note 12**).
3. Set up complexes of Siglec Fc-chimeras with anti-huIgG-AP (*see* **Note 13**).
4. Wash five times with 200 μL of HBS-T.
5. Add 30 μL of complexes of Fc-chimeras and anti-Fc-antibodies.
6. Centrifuge the plate at 200*g* at 4°C for 1 min.
7. Cover the plate with parafilm and incubate at 4°C for 4 h.
8. Wash five times with 200 μL of HBS-T.
9. Add 50 μL of 20 μ*M* FDP substrate solution equilibrated at 37°C (*see* **Note 14**).
10. Determine the increase in fluorescence at 37°C (*see* **Subheading 3.4.5.**).

3.4.3. Binding Assays With Immobilized Sialylated GCs

1. Use 384-well microtiter plates (black 384-Well, U-shaped well, maximum volume 80 μL; *see* **Notes 10** and **11**).
2. Prepare working solutions of 4 μg/mL of GP in immobilization buffer just before use (*see* **Note 4**).
3. Prepare the corresponding ASGP solution to determine Sia-independent binding (*see* **Notes 6** and **15**).
4. Add 10 μL of GP per well.
5. Centrifuge the plate at 200*g* at 4°C for 1 min.
6. Cover the plate with parafilm and incubate at 4°C for 16 h (*see* **Note 16**).
7. Set up complexes of Siglec Fc-chimeras with anti-huIgG-AP (*see* **Note 13**).
8. Wash five times with 50 μL of HBS-T.

9. Add 10-μL complexes of Siglec Fc-chimeras and anti-huIgG-AP.
10. Centrifuge the plate at 200*g* at 4°C for 1 min.
11. Cover the plate with parafilm and incubate at 4°C for 4 h.
12. Wash five times with 50 μL of HBS-T.
13. Add 20 μL of 20 μ*M* of FDP substrate solution equilibrated at 37°C (*see* **Note 14**).
14. Determine the increase in fluorescence at 37°C (*see* **Subheading 3.4.5.**).

3.4.4. Inhibition Assay

The procedures of the inhibition assay are principally the same as for the binding assays previously described. The steps of an inhibition assay are described for 96-well plates. For 384-well plates, correspondingly smaller volumes must be used (*see* **Notes 7** and **9–11**). Examples of data from such inhibition assays are shown in **Figs. 3** and **4**.

1. Prepare a 96-well plate with suitable target structure as described for binding assays in **Subheadings 3.4.2.** and **3.4.3.**
2. Add 15 μL of the inhibitor solution to each well. Prepare triplicates for each concentration and sample (*see* **Note 17**).
3. Centrifuge the plate at 200*g* at 4°C for 1 min.
4. Add 15 μL of the complexes of Siglec Fc-chimeras and anti-huIgG-AP.
5. Centrifuge the plate at 200*g* at 4°C for 1 min.
6. Cover the plate with parafilm and incubate at 4°C for 4 h.
7. Wash five times with 200 UL of HBS-T.
8. Add 50 μL of 20 μ*M* FDP substrate solution equilibrated at 37°C (*see* **Note 14**).
9. Determine the increase in fluorescence at 37°C (*see* **Subheading 3.4.5.**).

3.4.5. Detection Reaction

1. Equilibrate the empty plate after washing (*see* **Subheadings 3.4.2.–3.4.4.**) at 37°C for 1 min before adding the substrate solution.
2. Add 50 μL of 20 μ*M* of FDP (also equilibrated at 37°C) per well. Shake the plate before starting the reading (*see* **Note 14**).
3. Place the plate in the microtiter plate fluorimeter and start kinetic measurement immediately. Detect the reaction of FDP to fluorescein with excitation at 485 nm and reading emission at 520 nm. Read the plate every 60–120 s in cycles for a total of 15 min (*see* **Note 18**).

3.4.6. Calculation of Binding and Inhibition Rates

1. The amount of bound Siglec is determined from the maximal velocity of the enzymatic reaction as calculated from the maximum change in fluorescence over time (*see* **Note 18**).
2. Calculate the specific binding of the Siglec to its target by subtracting the signal obtained in the corresponding control wells. In most cases the NSB is below 1% and should not be above 5% of the maximal binding. The specific binding in the

absence of inhibitor is taken as 100% binding. Inhibition is calculated for each concentration as a percentage of this value.

3. Estimate the concentration leading to 50% inhibition (IC_{50} values) from inhibition curves by plotting the inhibition against the final inhibitor concentration. The IC_{50} values of Siglec Fc-chimera binding depend to some extent on the amount and quality of the target, the Fc-chimera complexes, and the time of incubation, especially for weak inhibitors.
4. The inhibitory potencies relative to a reference compound (rIPs) can be calculated for each experiment. These rIPs are more reproducible from one experiment to another than the IC_{50} values (*see* **Note 17**).

4. Notes

1. Other buffers, such as HEPES, phosphate, or Tris-HCl can be used as well, if the molecule immobilized is alkaline-sensitive, i.e., if it contains *O*-acetylated Sia.
2. HEPES can be replaced with Tris-HCl or phosphate. However, the buffering capacity at pH 7.5 is low for Tris-HCl, and phosphate must be removed before adding solutions containing Ca^{2+} or Mg^{2+}, such as VCS or AP buffers.
3. Do not use "ice-free" freezers for storage of stock solutions, specifically for small volumes.
4. For the immobilization of precious GPs, 1 μg/mL is usually sufficient to obtain binding.
5. VCS also works in most physiological buffers. However, 9 m*M* of $CaCl_2$ should be added. Therefore, phosphate buffers are not suitable.
6. GPs contain different types and linkages of Sia on their glycans. Some of these are not cleaved well by VCS. Therefore, it is useful to try different sialidases, e.g., from *Arthrobacter ureafaciens*, to optimize the enzymatic desialylation (*see* **Fig. 3**).
7. Wash volumes depend on the capacity of the wells. For 96-well plates use 200 μL, and for 384-well plates use 50 μL per well.
8. The volumes for the PI oxidation given are for 96-well plates; for 384-well plates use 35 μL of 2 m*M* $NaIO_4$, 5 μL of 8% glycerol, and 40 μL of 20 m*M* $NaBH_4$ (50 μL for washes).
9. Keep the microtiter plate chilled on ice in all steps to avoid evaporation. A suitable prechilled metal block on ice is convenient. All washing steps to remove unbound target molecules or antibody–Siglec Fc-complexes are performed at 4°C by washing the wells five times, by adding HBS-T, and by flicking out the buffer afterwards. Washing volumes depend on the maximum well volume. Cold buffer (4°C) is used to minimize dissociation of bound Siglecs during the procedure; the addition of Tween-20 reduces NSB. Avoid additional blocking steps if possible, as blocking reagents could be the source of interfering GCs. Try different concentrations of detergent to reduce NSB to the plate.
10. For assays in 96-well plates, the reaction volume is generally set up in a final volume of 30 μL per well (15 μL of Siglec Fc-chimeras complexed with anti-huIgG-AP and 15 μL of competitive inhibitor or HBS).

11. For assays in 384-well plates, the reaction is generally set up in a final volume of 10 µL per well (5 µL of Siglec Fc-chimeras complexed with anti-huIgG-AP, and 5 µL of inhibitor solution or HBS).
12. Some Siglecs (e.g., Siglec-9) tend to stick to the *N*-acetylated wells. In these cases, alternative control wells prepared by PI oxidation of the Sia in the corresponding wells are recommended.
13. The optimal concentrations of Siglec Fc-chimeras and anti-huIgG-AP have to be determined empirically for each batch of Fc chimeras and antibody conjugates. Usually 1 µg/mL of Fc chimera is fine. The stoichiometric optimal concentration of antibody has to be determined. This is archived by titration of the complexing antibody and the Siglec Fc-chimeras. In most cases it is between 1:250 and 1:5000, depending on the antibody conjugate provided by the supplier.
14. Use a multichannel pipet to add the solution quickly. For 96-well plates use 50 µL, and for 384-well plates add 20 µL of 20 µ*M* FDP per well.
15. Alternative controls are: Blank wells without any immobilized GP, PI-oxidized GPs, or competitive inhibition. If available, a monovalent low-molecular-weight high-affinity inhibitor can be used to determine Sia-independent binding of Siglecs in the presence of a 100-fold excess over the IC_{50}. Under these conditions the Sia-binding site of the Siglec is not available for binding. This approach is particularly useful to determine potential protein–protein interactions. It is recommended to compare at least some of these different controls for NSB when optimizing an assay (*see* **Fig. 3**).
16. Alternatively, immobilization at 37°C for 2 h is sufficient in most cases. However, immobilization at 4°C is more reliable, especially with low protein concentrations.
17. In most cases, seven to eight concentrations in a twofold serial dilution are sufficient to obtain an inhibition titration curve, if the range of concentrations is appropriate (*see* **Fig. 4**). A reference inhibitor compound should be included in each experiment for the calculation of rIPs.
18. The integration times, frequency, and number of cycles used depend on the instrument and the level of binding. Under the conditions described here (substrate and AP concentrations, and so forth), the initial velocity of the AP reaction must be determined accurately within the first 2–5 min for high levels of binding. Therefore, the cycle frequency should not be more than 90 s per cycle. Low levels of binding and NSB are determined more accurately by following the reaction for 15–30 min.

Acknowledgments

We are indebted to Marlies Rusch (Biochemistry Institute, Kiel University) for expert technical help in establishing the first GlycoWell binding assays. We thank Heiko Gäthje, Simone Witt, Sabine Meierhöfer, and Nazila Isakovic for constitutive experimental contributions and suggestions in the solid-phase assays for Fc-chimera production. The financial support of Deutsche

Forschungsgemeinschaft (grant Ke428-3), Volkswagen Foundation and Center for Biotechnology, and Nutrition e. V. is acknowledged.

References

1. Lis, H. and Sharon, N. (1998) Lectins: Carbohydrate-specific proteins that mediate cellular recognition. *Chem. Rev.* **98,** 637–674.
2. Bouckaert, J., Hamelryck, T., Wyns, L., and Loris, R. (1999) Novel structures of plant lectins and their complexes with carbohydrates. *Curr. Opin. Struct. Biol.* **9,** 572–577.
3. Vijayan, M. and Chandra, N. (1999) Lectins. *Curr. Opin. Struct. Biol.* **9,** 707–714.
4. Kelm, S. and Schauer, R. (1997) Sialic acids in molecular and cellular interactions. *Int. Rev. Cytol.* **175,** 137–240.
5. Schauer, R. and Kamerling, J. P. (1997) Chemistry, biochemistry and biology of sialic acids. In *Glycoproteins II* (Montreuil, J., Vliegenthart, J. F. G., and Schachter, H., eds.), Elsevier, Amsterdam, pp. 243–402.
6. Varki, A., Cummings, R., Esko, J., Freeze, H., Hart, G., and Marth, J. (1999) I-type lectins. In *Essentials of Glycobiology* (Varki, A., Cummings, R., Esko, J., Freeze, H., Hart, G., and Marth, J., eds.), Cold Spring Harbor Press, New York, pp. 363–378.
7. Angata, T. and Varki, A. (2002) Chemical diversity in the sialic acids and related alpha-keto acids: an evolutionary perspective. *Chem. Rev.* **102,** 439–469.
8. Kelm, S., Pelz, A., Schauer, R., et al. (1994) Sialoadhesin, myelin-associated glycoprotein and CD22 define a new family of sialic acid-dependent adhesion molecules of the immunoglobulin superfamily. *Curr. Biol.* **4,** 965–972.
9. Crocker, P. R., Clark, E. A., Filbin, M. T., et al. (1998) Siglecs—a family of sialic acid-binding lectins. *Glycobiology* **8,** Glycoforum 2 v–vi.
10. Crocker, P. R. and Kelm, S. (2000) The Siglec family of I-type lectins. In *Carbohydrates in Biology and Chemistry* (Ernst, B., Hart, G. W., and Sinay, P., eds.), Wiley-VCH, Weinheim, pp. 579–595.
11. Kelm, S. (2001) Ligands for Siglecs. In *Mammalian Carbohydrate Recognition Systems* (Crocker, P. R., ed.), Springer, Berlin, pp. 153–176.
12. Crocker, P. R. and Varki, A. (2001) Siglecs in the immune system. *Immunology* **103,** 137–145.
13. Crocker, P. R. (2002) Siglecs: sialic-acid-binding immunoglobulin-like lectins in cell–cell interactions and signalling. *Curr. Opin. Struct. Biol.* **12,** 609–615.
14. Kelm, S., Gerlach, J., Brossmer, R., Danzer, C. P., and Nitschke, L. (2002) The ligand-binding domain of CD22 is needed for inhibition of the B cell receptor signal, as demonstrated by a novel human CD22-specific inhibitor compound. *J. Exp. Med.* **195,** 1207–1213.
15. Ikehara, Y., Ikehara, S. K., and Paulson, J. C. (2004) Negative regulation of T cell receptor signaling by Siglec-7 (p70/AIRM) and Siglec-9. *J. Biol. Chem.* **279,** 43,117–43,125.
16. Kelm, S., Schauer, R., Manuguerra, J. C., Gross, H. J., and Crocker, P. R. (1994) Modifications of cell surface sialic acids modulate cell adhesion mediated by sialoadhesin and CD22. *Glycoconjugate J.* **11,** 576–585.

17. Collins, B. E., Kiso, M., Hasegawa, A., et al. (1997) Binding specificities of the sialoadhesin family of I-type lectins–sialic acid linkage and substructure requirements for binding of myelin-associated glycoprotein, Schwann cell myelin protein, and sialoadhesin. *J. Biol. Chem.* **272,** 16,889–16,895.
18. Blixt, O., Head, S., Mondala, T., et al. (2004) Printed covalent glycan array for ligand profiling of diverse glycan binding proteins. *Proc. Natl. Acad. Sci. USA* **101,** 17,033–17,038.
19. Kelm, S., Brossmer, R., Isecke, R., Gross, H. J., Strenge, K., and Schauer, R. (1998) Functional groups of sialic acids involved in binding to Siglecs (sialoadhesins) deduced from interactions with synthetic analogues. *Eur. J. Biochem.* **255,** 663–672.
20. Strenge, K., Schauer, R., Bovin, N., et al. (1998) Glycan specificity of myelin-associated glycoprotein and sialoadhesin deduced from interactions with synthetic oligosaccharides. *Eur. J. Biochem.* **258,** 677–685.
21. Crocker, P. R. and Kelm, S. (1996) Methods for studying the cellular binding properties of lectin-like receptors. In *Weir's Handbook of Experimental Immunology* (Herzenberg, L. A., Weir, D. M., and Blackwell, C., eds.), Blackwell Science, Cambridge, pp. 166.1–166.11.
22. Lee, Y. C. (1989) Binding modes of mammalian hepatic Gal/GalNAc receptors. *Ciba Found. Symp.* **145,** 80–93, discussion.
23. Brock, J. H., Pineiro, A., and Lampreave, F. (1978) The effect of trypsin and chymotrypsin on the antibacterial activity of complement, antibodies, and lactoferrin and transferrin in bovine colostrum. *Ann. Rech. Vet.* **9,** 287–294.
24. Edge, A. S. and Spiro, R. G. (1987) Presence of an *O*-glycosidically linked hexasaccharide in fetuin. *J. Biol. Chem.* **262,** 16,135–16,141.
25. Green, E. D., Adelt, G., Baenziger, J. U., Wilson, S., and van Halbeek, H. (1988) The asparagine-linked oligosaccharides on bovine fetuin. Structural analysis of *N*-glycanase-released oligosaccharides by 500-megahertz ^{1}H NMR spectroscopy. *J. Biol. Chem.* **263,** 18,253–18,268.
26. Shibuya, N., Goldstein, I. J., Broekaert, W. F., Nsimba-Lubaki, M., Peeters, B., and Peumans, W. J. (1987) The elderberry (Sambucus nigra L.) bark lectin recognizes the Neu5Ac(alpha 2-6)Gal/GalNAc sequence. *J. Biol. Chem.* **262,** 1596–1601.
27. Powell, L. D. and Varki, A. (1994) The oligosaccharide binding specificities of CD22 β, a sialic acid-specific lectin of B cells. *J. Biol. Chem.* **269,** 10,628–10,636.
28. Hanasaki, K., Powell, L. D., and Varki, A. (1995) Binding of human plasma sialoglycoproteins by the B cell-specific lectin CD22. Selective recognition of immunoglobulin M and haptoglobin. *J. Biol. Chem.* **270,** 7543–7550.

Index